Lecture Notes in Computer Scie

T0237978

Commenced Publication in 1973
Founding and Former Series Editors:
Gerhard Goos, Juris Hartmanis, and Jan van Leeuwen

Editorial Board

David Hutchison
  Lancaster University, UK
Takeo Kanade
  Carnegie Mellon University, Pittsburgh, PA, USA
Josef Kittler
  University of Surrey, Guildford, UK
Jon M. Kleinberg
  Cornell University, Ithaca, NY, USA
Alfred Kobsa
  University of California, Irvine, CA, USA
Friedemann Mattern
  ETH Zurich, Switzerland
John C. Mitchell
  Stanford University, CA, USA
Moni Naor
  Weizmann Institute of Science, Rehovot, Israel
Oscar Nierstrasz
  University of Bern, Switzerland
C. Pandu Rangan
  Indian Institute of Technology, Madras, India
Bernhard Steffen
  University of Dortmund, Germany
Madhu Sudan
  Massachusetts Institute of Technology, MA, USA
Demetri Terzopoulos
  University of California, Los Angeles, CA, USA
Doug Tygar
  University of California, Berkeley, CA, USA
Gerhard Weikum
  Max-Planck Institute of Computer Science, Saarbruecken, Germany

John Domingue   Chutiporn Anutariya (Eds.)

The Semantic Web

3rd Asian Semantic Web Conference, ASWC 2008
Bankok, Thailand, December 8-11, 2008
Proceedings.

 Springer

Volume Editors

John Domingue
The Open University Knowledge Media Institute
Walton Hall, Milton Keynes, MK6 7AA, United Kingdom
E-mail: j.b.domingue@open.ac.uk

Chutiporn Anutariya
Shinawatra University
99 Moo 10 Bangtoey, Samkok
Pathum Thani, 12160, Thailand
E-mail: chutiporn@shinawatra.ac.th

Library of Congress Control Number: Applied for

CR Subject Classification (1998): H.4, H.3, C.2, H.5, F.3, I.2, K.4

LNCS Sublibrary: SL 3 – Information Systems and Application, incl. Internet/Web and HCI

ISSN   0302-9743
ISBN-10   3-540-89703-8 Springer Berlin Heidelberg New York
ISBN-13   978-3-540-89703-3 Springer Berlin Heidelberg New York

This work is subject to copyright. All rights are reserved, whether the whole or part of the material is concerned, specifically the rights of translation, reprinting, re-use of illustrations, recitation, broadcasting, reproduction on microfilms or in any other way, and storage in data banks. Duplication of this publication or parts thereof is permitted only under the provisions of the German Copyright Law of September 9, 1965, in its current version, and permission for use must always be obtained from Springer. Violations are liable to prosecution under the German Copyright Law.

Springer is a part of Springer Science+Business Media

springer.com

© Springer-Verlag Berlin Heidelberg 2008

Typesetting: Camera-ready by author, data conversion by Scientific Publishing Services, Chennai, India
Printed on acid-free paper   SPIN: 12571390   06/3180   5 4 3 2 1 0

Preface

This volume contains the main proceedings of the 3rd Annual Asian Semantic Web Conference (ASWC 2008) held in Bangkok, Thailand, during December 8–11, 2008. As such, ASWC 2008 showcased the latest results in the research and application of Semantic Web technologies—applying semantics at a planetary scale.

Over the last few years we have been witnessing a trend in which the Semantic Web has been transforming from a niche research area to the mainstream in academia and industry. The European Semantic Web Conference, held earlier this year in Tenerife, saw a growth in the participation and engagement from semantic start-up companies. This conference, showcasing an Asian perspective and now having established itself, is also a sign that the Semantic Web is moving to the mainstream.

In addition to the emergence in the mainstream, the Semantic Web continues to generate a significant volume of scientifically interesting research articles. Research submissions to ASWC 2008 were scrutinized and filtered via a three-phase reviewing process. First, each submission was evaluated by three members from the Program Committee. Second, papers and the associated reviews were meta-reviewed by members of the Senior Program Committee. In this second phase the meta-reviewers led discussions between reviewers and produced an acceptance recommendation. In the last phase, on the basis of the reviews and associated meta-review recommendations, the final selections were made jointly by the Programme Chairs. Although this process required substantial efforts from the members of the Program Committee, it ensured that only papers of the highest quality were accepted. The final acceptance of 37 papers for publication and presentation at the conference out of the 118 submissions resulted in an acceptance rate of 31%. Of the papers we accepted these are split between Asia (18) and Europe (17), with an additional 2 from the USA. The accepted papers cover topics including: scalable reasoning and logic, ontology mapping, ontology modelling and management, ontologies and tags, human language technologies and machine learning, querying, Semantic Web services and Semantic Web applications.

We would like to thank all members of the Program Committee and Senior Program Committee and the additional reviewers for their considerable and timely efforts in reviewing the submissions, particularly during a period when holidays are traditionally taken in many parts of the globe.

ASWC this year was fortunate to have three very compelling keynotes. Fabio Ciravegna, Professor of Language and Knowledge Technologies at the University of Sheffield, gave a talk entitled "Supporting Knowledge Management in Large Distributed Organizations Using Semantic Web Technologies." In his talk he covered the challenges and requirements for Semantic Web technologies when confronted with the task of knowledge acquisition and sharing in large complex distributed environments giving examples from industrially focused projects.

Amit Sheth, Director of Kno.e.sis Center, at Wright State University, gave the talk "Computing for Human Experience: Sensors, Perception, Semantics, Web N.0, and Beyond." Amit outlined an era of "computing for human experience" incorporating seamless interaction between the physical and virtual worlds facilitated by advances in sensor technology and "edge computing" and linked to emerging areas such as Internet of Things, Intelligence@Interfaces, Humanist Computing, Relationship Web, PeopleWeb, EventWeb, and Experiential Computing. In his presentation Amit highlighted the role that semantics would play in this new era.

In his talk "Common Web Language for Humans and Computers," Hiroshi Uchida, Director of the UNDL Foundation, outlined the Universal Networking Language and discussed its applications in a number of diverse areas of human activity.

Six workshops were accepted at this year's conference covering areas ranging from new forms of reasoning to health care, life sciences, and human factors.

All members of the Organizing Committee were very dedicated to their tasks and deserve our special gratitude. In particular we wish to thank the Conference Chairs, Jerome Euzenat and R.K. Shyamasundar, the Steering Committee Chair, Riichiro Mizoguchi, and all members of the Steering Committee for their valuable advice and constant support to make this conference a success.

We would also like to thank Marco Ronchetti for his organization of the ASWC workshops, Michal Zaremba and Elena Simperl for their work on the Demos and Posters, respectively, and Huajun Chen for his efforts as Metadata Chair at ASWC 2008. Huajun now joins the elite group of "Dogfood Czars" who promote the use of Semantic Web technologies by the Semantic Web community.

On the business side we would like to thank the Industrial Track Chairs, Roberta Cuel, Lyndon Nixon, and Laurentiu Vasiliu, for their innovation in creating a "Software Solutions Track" with commercially inspired criteria and an associated cash prize.

One of the main aspects which made this conference special was its particular location in Bangkok, Thailand. We were three only because of the efforts of our Local Organization Chairs, Asanee Kawtrakul and Rachanee Ungrangsi; our Local Organizing Committee, Photchanan Ratanajaipan and Krissada Maleewong; and our Local Organizing staff.

We thank Alessio Gugliotta for his unswerving support to the Program Chairs in organizing the refereed paper program. We are also grateful to Springer for agreeing to publish the proceedings in its *Lecture Notes in Computer Science* series. Our gratitude also goes to all of our sponsors and STI International for their continued organizational support. Finally, we would like to thank all the colleaques who submitted their papers to the conference, and all the participants who contributed to the interesting presentations and fruitful discussions.

We are happy that ASWC 2008 was a thrilling event and once again showed the high levels of motivation, dedication, creativity, and performance of the Semantic Web community.

December 2008   John Domingue
                                                                        Chutiporn Anutariya

Organization

ASWC 2008 was organized by the Asian Institute of Technology, National Electronics and Computer Technology Center, and Shinawatra University, Thailand.

Steering Committee

Steering Committee Chair

Riichiro Mizoguchi   Osaka Univesity, Japan

Steering Committee

Witold Abramowicz	The Poznan University of Economics, Poland
Dieter Fensel	University of Innsbruck, Austria
Fausto Giunchiglia	University of Trento, Italy
Sung-Kook Han	Wonkwang University, Korea
Hong-Gee Kim	Seoul National University, Korea
Juanzi Li	Tsinghua University, China
Daniel Schwabe	PUC Rio, Brazil
Rudi Studer	Universität Karlsruhe, Germany
R.K. Shyamasundar	Tata Institute of Fundamental Research, India
Vilas Wuwongse	Asian Institute of Technology, Thailand
Ning Zhong	Maebashi Institute of Technology, Japan

Organizing Committee

Conference Chairs

Jerome Euzenat	INRIA Rhône-Alpes, France
R.K. Shyamasundar	Tata Institute of Fundamental Research, India

Program Chairs

John Domingue	Open University, UK
Chutiporn Anutariya	Shinawatra University, Thailand

Workshop Chair

Marco Ronchetti   Trento University, Italy

Industrial Track Chairs

Roberta Cuel	Trento University, Italy
Lyndon J.B. Nixon	Free University of Berlin, Germany
Laurentiu Vasiliu	DERI Galway, Ireland

Demo Chair

Michal Zaremba   STI International, Austria

Poster Chair

Elena Simperl   STI International, Austria

Metadata Chair

Huajun Chen   Zhejiang University, China

Local Organizing Chairs

Asanee Kawtrakul   NECTEC, Thailand
Rachanee Ungrangsi   Shinawatra University, Thailand

Local Organizing Committee

Photchanan Ratanajaipan   Shinawatra University, Thailand
Krissada Maleewong   Shinawatra University, Thailand

Program Committee

Senior Program Committee

Sean Bechhofer   University of Manchester, UK
Dieter Fensel   University of Innsbruck and STI International,
                              Austria
Aldo Gangemi   CNR, Italy
Asun Gomez-Perez   Universidad Politecnica de Madrid, Spain
Guus Schreiber   Vrije Universiteit Amsterdam, The Netherlands
Daniel Schwabe   PUC Rio, Brazil
Amit Sheth   University of Georgia and Semagix, USA
Katia Sycara   Carnegie Mellon University, USA

Program Committee

Harith Alani   University of Southampton, UK
Yuan An   Drexel University, USA
Jurgen Angele   Ontoprise, Germany
Grigoris Antoniou   FORTH, Greece
Lora Aroyo   Free University of Amsterdam, The Netherlands
Budak Arpinar   University of Georgia, USA
Walter Binder   EPFL, Switzerland
Paolo Bouquet   University of Trento, Italy
John Breslin   University of Ireland, Ireland
Francois Bry   University of Munich, Germany
Paul Buitelaar   DFKI Saarbruecken, Germany
Christoph Bussler   Cisco Systems, Inc., USA
Liliana Cabral   Open University, UK

Enhong Chen	University of Science and Technology of China, China
Harry Chen	Image Matters, USA
Oscar Corcho	University of Manchester, UK
Isabel Cruz	University Illinois at Chicago, USA
Bernardo Cuenca Grau	Oxford University, UK
Mike Dean	BBN Technologies, USA
Thierry Declerck	DFKI Kaiserslautern, Germany
Paola Di Maio	MFU, Thailand
Ian Dickinson	Hewlett-Packard Labs, USA
Stefan Dietze	Open University, UK
Ying Ding	Indiana University, USA
Martin Dzbor	Open University, UK
Achille Fokoue	IBM Research, USA
Stefania Galizia	Open University, UK
Fabien Gandon	INRIA Sophia-Antipolis, France
Fausto Giunchiglia	University of Trento, Italy
Marko Grobelnik	J. Stefan Institute, Slovenia
Alessio Gugliotta	Open University, UK
Volker Haarslev	Concordia University, CA
Peter Haase	University of Karlsruhe, Germany
Axel Hahn	University of Oldenburg, Germany
Harry Halpin	University of Edinburgh, UK
Siegfried Handschuh	FZI Karlsruhe, Germany
Maruf Hasan	Shinawatra University, Thailand
Manfred Hauswirth	DERI Galway, Ireland
Martin Hepp	University of Innsbruck and DERI, Austria
Masahiro Hori	Kansai University, Japan
Andreas Hotho	University of Kassel, Germany
Eero Hyvnen	University of Helsinki, Finland
Giovambattista Ianni	University of Calabria, Italy
Ryutaro Ichise	National Institute of Informatics, Japan
Jason Jung	Yeungnam University, Korea
Lalana Kagal	MIT, USA
Bo-Young Kang	Seoul National University, Korea
Vangelis Karkaletsis	NCSR Demokritos, Greece
Vipul Kashyap	Clinical informatics R&D, USA
Takahiro Kawamura	Toshiba, Japan
Yoshinobu Kitamura	Osaka University, Japan
Michel Klein	Vrije Universiteit Amsterdam, The Netherlands
Matthias Klusch	DFKI Saarbruecken, Germany
Manolis Koubarakis	Technical University of Crete, Greece
Georgia Koutrika	Stanford University, USA
Kouji Kozaki	Osaka University, Japan
Ruben Lara	Tecnologia, Informacion y Finanzas, Spain
Alain Leger	France Telecom, France

Juanzi Li	Tsinghua University, China
Alexander Loeser	TU Berlin, Germany
Mihhail Matskin	KTH Stockholm, SE
Yutaka Matsuo	The University of Tokyo, Japan
Diana Maynard	University of Sheffield, UK
Brian McBride	Hewlett Packard, UK
Dennis McLeod	University of Southern California, USA
Riichiro Mizoguchi	Osaka University, Japan
Dunja Mladenic	J. Stefan Institute, Slovenia
Ralf Moeller	Hamburg University of Technology, Germany
Wolfgang Nejdl	University of Hannover and L3S, Germany
Barry Norton	Open University, UK
Ekawit Nantajeewarawat	SIIT, Thailand
Daniel Oberle	SAP AG, Germany
Leo Obrst	MITRE, USA
Daniel Olmedilla	L3S Hannover, Germany
Jeff Z. Pan	University of Aberdeen, UK
Yue Pan	IBM Research Lab, China
Massimo Paolucci	DoCoMo, Germany
Terry Payne	University of Southampton, UK
Carlos Pedrinaci	Open University, UK
Paulo Pinheiro da Silva	Stanford University, USA
Ruzica Piskac	EPFL, Switzerland
Dimitris Plexousakis	University of Crete, Greece
Axel Polleres	DERI, NUI Galway, Ireland
Yuzhong Qu	SouthEast University, China
Sudha Ram	University of Arizona, USA
Ulrich Reimer	University of Konstanz and FHS St. Gallen, Switzerland
Marta Sabou	The Open University, UK
Stefan Schlobach	Vrije Universiteit Amsterdam, The Netherlands
Twittie Senivongse	Chulalongkorn University, Thailand
Elena Simperl	University of Innsbruck, Austria
Michael Stollberg	STI International, Austria
Umberto Straccia	ISTI-CNR, Italy
York Sure	University of Karlsruhe, Germany
Vojtech Svatek	University of Economics, Czech Republic
Valentina Tamma	University of Liverpool, UK
Jie Tang	Tsinghua University, China
Sergio Tessaris	Free University Bozen, Italy
Robert Tolksdorf	Free University Berlin, Germany
Paolo Traverso	ITC/IRST, Italy
Raphael Troncy	CWI Amsterdam, The Netherlands
Victoria Uren	KMi The Open University, UK
Tomas Vitvar	STI Innsbruck, Austria

Holger Wache   Vrije Universiteit Amsterdam, The Netherlands
Krzysztof Wecel   Poznan University of Economics, Poland
Takahira Yamaguchi   Keio University, Japan
Yong Yu   Shanghai Jiao Tong University, China
Hai Zhuge   Institute of Computing Technology, China

Additional Reviewers

Krissada Maleewong   Shinawatra University, Thailand
Photchanan Ratanajaipan Shinawatra University, Thailand
Rachanee Ungrangsi   Shinawatra University, Thailand

Local Organizing Committee

Local Organizing Committee

Photchanan Ratanajaipan Shinawatra University, Thailand
Krissada Maleewong   Shinawatra University, Thailand

Local Organizing Staff

Kornschnok Dittawit   Shinawatra University, Thailand
Nopachat Kalayanapan   Shinawatra University, Thailand
Pawadee Keratichewanun Shinawatra University, Thailand
Panida Kijrattana   Shinawatra University, Thailand

Sponsoring Institutions

Platinum Sponsors

EastWeb-AsiaLink project through Asian Institute of Technology, Thailand
National Electronics and Computer Technology Center, Thailand
Super project and Swing project through DERI-Galway, Ireland

Gold Sponsor

NeOn project through the Open University, UK
Shinawatra University, Thailand

Silver Sponsor

LarKC project through STI-Innsbruck, Austria

Platinum Sponsor

Gold Sponsor

Silver Sponsor

Table of Contents

Ontology Modeling and Management

Ontologies and Tags

Human Language Technologies and Machine Learning

Querying

Semantic Web Services and Semantic Web Applications

A Modularization-Based Approach to Finding All Justifications for OWL DL Entailments

Boontawee Suntisrivaraporn[1], Guilin Qi[2], Qiu Ji[2], and Peter Haase[2]

[1] Theoretical Computer Science, TU Dresden, Germany
meng@tcs.inf.tu-dresden.de
[2] AIFB Institute, University of Karlsruhe, Germany
{gqi,qiji,pha}@aifb.uni-karlsruhe.de

Abstract. Finding the justifications for an entailment (i.e., minimal sets of axioms responsible for it) is a prominent reasoning service in ontology engineering, as justifications facilitate important tasks like debugging inconsistencies or undesired subsumption. Though several algorithms for finding all justifications exist, issues concerning efficiency and scalability remain a challenge due to the sheer size of real-life ontologies. In this paper, we propose a novel method for finding all justifications in OWL DL ontologies by limiting the search space to smaller modules. To this end, we show that so-called locality-based modules cover all axioms in the justifications. We present empirical results that demonstrate an improvement of several orders of magnitude in efficiency and scalability of finding all justifications in OWL DL ontologies.

1 Introduction

Since the Web Ontology Language (OWL) has become a W3C standard, it is widely believed that ontologies play a prominent role in formal representation of knowledge on the Semantic Web. The main advantages of employing OWL in knowledge engineering are twofold. On the one hand, the well-defined semantics of Description Logic (DL), which is the logical underpinning of OWL, helps guarantee that everyone on the Web understands the described knowledge in a consistent way. On the other hand, reasoning services can be exploited to derive implicit knowledge from the one explicitly given. DL systems can, for example, identify unsatisfiable concepts and classify a given ontology, i.e., compute all the subsumption (subconcept–superconcept) relationships between the concepts defined in the ontology. These "standard" reasoning services have proved essential but not sufficient in engineering real-world ontologies. This is because building ontologies is an error-prone endeavor. Although most DL systems can detect an error (an unsatisfiable concept or undesired subsumption) in a given ontology, additional reasoning is needed in order to find its *justifications*, i.e., minimal subsets of the ontology that still have the error.

Several techniques for finding all justifications have been proposed in the literature in the past decade which can be categorized into glass-box approaches and black-box approaches.

J. Domingue and C. Anutariya (Eds.): ASWC 2008, LNCS 5367, pp. 1–15, 2008.
© Springer-Verlag Berlin Heidelberg 2008

Glass-box approaches require the decision (e.g., tableau) procedure to be modified, usually by adding labels to keep track of relevant axioms used during the computation [14,12,11,1,2]. Most of the work in this direction considers specific Description Logics, e.g., $\mathcal{ALC}$, and a specific type of entailment, e.g., concept unsatisfiability. In [14], Schlobach and Cornet proposed an extension to the tableau algorithm for $\mathcal{ALC}$ with unfoldable TBoxes. The extension uses labels to keep track of axioms used during the computation which directly corresponds to justifications. They also coined the name "axiom pinpointing" for the task of finding justifications for an entailment. Since glass-box approaches are based on modifying the internals of a DL reasoning algorithm, an extension has to be developed for each DL. Meyer et al. extended the idea to $\mathcal{ALC}$ with general concept inclusions (GCIs) [12], and Kalyanpur et al. extended it to the more expressive DL $\mathcal{SHIF(D)}$ [11] and $\mathcal{SHOIN(D)}$ [10] which underly the core of OWL. In [1], a general approach for extending a tableau-based algorithm to a pinpointing algorithm is proposed which can be used to find all justifications for a given entailment. Most previous work on glass-box methods considers tableau-based reasoning algorithm. An exception is the work by Baader et al. [2] which extends the polytime classification algorithm in order to compute justifications for a subsumption relation in the lightweight DL $\mathcal{EL}^+$, and also shows that axiom pinpointing is inherently hard, i.e., determining whether there is a justification within a given cardinality bound is NP-complete despite tractability of the underlying DL.

The other class of approaches to axiom pinpointing is known as *black-box*, where a DL reasoner is merely used to test specific entailment queries, and as such its internals need not be modified. With a naïve pruning algorithm, a justification can be computed by invoking the DL reasoner linear number of times [11,2]. The naïve algorithm essentially sweeps through all the axioms in the ontology and tests if the entailment still holds in absence of each axiom. Since this approach is independent from reasoning algorithms, it can be easily implemented on top of any existing DL reasoners. The main disadvantage, however, is that it typically requires several calls to the DL reasoning services that are already computationally expensive. Therefore, several optimization techniques have very recently been proposed that help to reduce the number of calls to the DL reasoner and hence speed up the black-box approach. Examples include the 'sliding window' technique employed in the fast pruning algorithm [10], the 'binary-search' idea adapted to obtain a best-case logarithmic pruning algorithm [3], and the 'relevance-based selection function' that syntactically select relevant axioms from the ontology [9]. Based on a black-box pruning algorithm for computing a single justification, the hitting set tree (HST) algorithm [13,10,9] can be used to recursively compute all justifications.

Recently, ontology modularity and modularization have been studied extensively, with various applications ranging from ontology re-use and optimization of classical reasoning such as subsumption, as well as non-classical reasoning such as incremental classification [5] and axiom pinpointing [3]. Closely related to [9] is the modularization-based approach to axiom pinpointing where relevant

Table 1. Syntax and semantics of $\mathcal{SHOIQ}$ concepts and axioms

Name	Syntax	Semantics
top	$\top$	$\Delta^{\mathcal{I}}$
concept name	A	$A^{\mathcal{I}} \subseteq \Delta^{\mathcal{I}}$
nominal	$\{a\}$	$\{a^{\mathcal{I}}\}$
negation	$\neg C$	$\Delta^{\mathcal{I}} \backslash C^{\mathcal{I}}$
conjunction	$C \sqcap D$	$C^{\mathcal{I}} \cap D^{\mathcal{I}}$
exists restriction	$\exists r.C$	$\{x \in \Delta^{\mathcal{I}} \mid \exists y \in \Delta^{\mathcal{I}} : (x,y) \in r^{\mathcal{I}} \wedge y \in C^{\mathcal{I}}\}$
at-least restriction	$\geq n\, s.C$	$\{x \in \Delta^{\mathcal{I}} \mid \sharp\{y : (x,y) \in s^{\mathcal{I}} \wedge y \in C^{\mathcal{I}}\} \geq n\}$
role name	r	$r^{\mathcal{I}} \subseteq \Delta^{\mathcal{I}} \times \Delta^{\mathcal{I}}$
inverse role	r^{-}	$\{(x,y) \in \Delta^{\mathcal{I}} \times \Delta^{\mathcal{I}} \mid (y,x) \in r^{\mathcal{I}}\}$
role hierarchy	$r \sqsubseteq s$	$r^{\mathcal{I}} \subseteq s^{\mathcal{I}}$
transitivity	$\mathsf{Trans}(r)$	$(x,y), (y,z) \in r^{\mathcal{I}}$ implies $(x,z) \in r^{\mathcal{I}}$
GCI	$C \sqsubseteq D$	$C^{\mathcal{I}} \subseteq D^{\mathcal{I}}$

axioms are precisely those axioms in the module [3]. In order to exploit modularity in black-box axiom pinpointing, Baader and Suntisrivaraporn showed that the reachability-based module [16] covers all justifications for an entailment of interest in $\mathcal{EL}^{+}$ [3].

In the present paper, we combine the relevance-based techniques developed in [9] and the modularization-based techniques in [3] to effectively enhance the HST pinpointing algorithm. Since the results in [3] are w.r.t. reachability-based modules for $\mathcal{EL}^{+}$, we need to adopt the locality-based module [6] for $\mathcal{SHOIQ}$. Our main contributions in the present paper are twofold. In theory, we show that the minimal locality-based module is a *subsumption module* (first defined in [3]), i.e., it covers all justifications. As a consequence, it suffices to focus on axioms in the module when finding *all* justifications and when testing subsumption. In practice, we have implemented the approach using KAON2 as the black-box reasoner and evaluated it on realistic ontologies. Our empirical results demonstrate an improvement of several orders of magnitude in the efficiency and scalability of finding all justifications. The results thus render the black-box approach feasible for application-scale OWL DL ontologies.

2   Preliminaries

In this section, we give formal definitions for $\mathcal{SHOIQ}$ ontologies, justifications and locality-based modules. Then, we introduce selection functions and the HST pinpointing algorithm.

Description logic and justifications

To make the paper self-contained, we first introduce the Description Logic (DL) $\mathcal{SHOIQ}$ [7] which is the underpinning DL formalism of the Web Ontology Language (OWL DL and OWL Lite).

Starting with disjoint sets of concept names CN, role names RN and individuals Ind, a $\mathcal{SHOIQ}$-role is either a role name $r \in$ RN or an inverse role r^- with $r \in$ RN. We denote by Rol the set of all $\mathcal{SHOIQ}$-roles. $\mathcal{SHOIQ}$-concepts can be built using the constructors shown in the upper part of Table 1, where $a \in$ Ind, $r, s \in$ Rol with s a *simple role*[1], n is a positive integer, $A \in$ CN, and C, D are $\mathcal{SHOIQ}$-concepts.[2] We use the standard abbreviations: $\bot$ stands for $\neg\top$; $C \sqcup D$ stands for $\neg(\neg C \sqcap \neg D)$; $\forall r.C$ stands for $\neg(\exists r.\neg C)$; and $\leq ns.C$ stands for $\neg(\geq (n+1)s.C)$. We denote by Con the set of all $\mathcal{SHOIQ}$-concepts.

A $\mathcal{SHOIQ}$ ontology $\mathcal{O}$ is a finite set of *role hierarchy axioms* $r \sqsubseteq s$, *transitivity axioms* Trans(r), and a general concept inclusion axioms (GCIs) $C \sqsubseteq D$ with $r, s \in$ Rol and $C, D \in$ Con.[3] We write CN($\mathcal{O}$), RN($\mathcal{O}$) and Ind($\mathcal{O}$) to denote, respectively, the set of concept names, role names and individuals occurring in the the ontology $\mathcal{O}$, and Sig($\mathcal{O}$) to denote the signature of $\mathcal{O}$, i.e., CN($\mathcal{O}$) $\cup$ RN($\mathcal{O}$) $\cup$ Ind($\mathcal{O}$). Similarly, Sig(r), Sig(C) and Sig(α) are used to denote the signature of a role, a concept and an axiom, respectively.

The DL semantics is defined by means of interpretations $\mathcal{I}$ with a non-empty domain $\Delta^\mathcal{I}$ and a function $\cdot^\mathcal{I}$ that maps each concept $C \in$ Con to a subset of the domain and each role $r \in$ Rol to a binary relation over the domain. An interpretation $\mathcal{I}$ is a *model* of an ontology $\mathcal{O}$ ($\mathcal{I} \models \mathcal{O}$), if the conditions given in the semantics column of Table 1 are satisfied. The main types of entailments are concept satisfiability: C is *satisfiable* w.r.t. $\mathcal{O}$ if there exists a model $\mathcal{I}$ of $\mathcal{O}$ such that $C^\mathcal{I} \neq \emptyset$; and concept subsumption: C *is subsumed by* D w.r.t. $\mathcal{O}$ (written $\mathcal{O} \models C \sqsubseteq D$ or $C \sqsubseteq_\mathcal{O} D$) if, for every model $\mathcal{I}$ of $\mathcal{O}$, $C^\mathcal{I} \subseteq D^\mathcal{I}$. Without loss of generality, we restrict attention to concept subsumption in what follows. Considering an example ontology depicted in Figure 1, all DL reasoners are able to detect that the subsumption $\mathcal{O}_{ex} \models \sigma = $ (Endocarditis $\sqsubseteq$ HeartDisease) holds.

Definition 1 (Justification). *Let $\mathcal{O}$ be a $\mathcal{SHOIQ}$ ontology with an entailment σ (i.e., $\mathcal{O} \models \sigma$). A subset $J \subseteq \mathcal{O}$ is a* justification *for σ in $\mathcal{O}$ if $J \models \sigma$ and, for every $J' \subset J$, $J' \not\models \sigma$.*

Justifications for an entailment need not be unique. Moreover, given an ontology and an entailment, the number of justifications may be exponential in the size of the ontology. For the small example ontology $\mathcal{O}_{ex}$ (see Figure 1), it is not difficult to infer that there are precisely two justifications for σ: one consisting of axioms marked by •, and the other by ⋆.

Modularization

We now introduce the notions of *syntactic locality* and *locality-based module*, which have been first introduced in [6]. Syntactic locality is used to define the notion of module for a signature, i.e., a subset of the ontology that preserves the meaning of names in the signature.

---

[1] A simple role is neither transitive nor a superrole of a transitive role.
[2] Concepts and roles in DL correspond to classes and properties in OWL, respectively.
[3] A concept definition $A \equiv C$ is an abbreviation of two GCIs $A \sqsubseteq C$ and $C \sqsubseteq A$, while ABox assertions $C(a)$ and $r(a, b)$ can be expressed as the GCIs $\{a\} \sqsubseteq C$ and $\{a\} \sqsubseteq \exists r.\{b\}$, respectively.

α_1	Pericardium $\sqsubseteq$ Tissue $\sqcap$ $\exists$part-of.Heart		
α_2	Endocardium $\sqsubseteq$ Tissue $\sqcap$ $\exists$part-of.HeartValve	•	⋆
	$\sqcap$ $\exists$part-of.HeartWall		
α_3	HeartValve $\sqsubseteq$ BodyValve $\sqcap$ $\exists$part-of.Heart	•	
α_4	HeartWall $\sqsubseteq$ BodyWall $\sqcap$ $\exists$part-of.Heart		⋆
α_5	Pericarditis $\sqsubseteq$ Inflammation $\sqcap$ $\exists$has-loc.Pericardium		
α_6	Endocarditis $\sqsubseteq$ Inflammation $\sqcap$ $\exists$has-loc.Endocardium	•	⋆
α_7	Inflammation $\sqsubseteq$ Disease $\sqcap$ $\exists$acts-on.Tissue	•	⋆
α_8	Disease $\sqcap$ $\exists$has-loc.Heart $\sqsubseteq$ HeartDisease	•	⋆
α_9	part-of $\sqsubseteq$ has-loc	•	⋆
α_{10}	Trans(has-loc)	•	⋆

Fig. 1. An example ontology $\mathcal{O}_{ex}$; the minimal locality-based module $\mathcal{O}^{loc}_{Endocarditis}$; and the justifications for Endocarditis $\sqsubseteq_{\mathcal{O}}$ HeartDisease

Definition 2 (Syntactic locality for $\mathcal{SHOIQ}$). *Let* **S** *be a signature. The following grammar recursively defines two sets of concepts* $\mathsf{Con}^{\perp}(\mathbf{S})$ *and* $\mathsf{Con}^{\top}(\mathbf{S})$ *for a signature* **S***:*

$$\mathsf{Con}^{\perp}(\mathbf{S}) ::= A^{\perp} \mid (\neg C^{\top}) \mid (C \sqcap C^{\perp}) \mid (\exists r^{\perp}.C) \mid (\exists r.C^{\perp})$$
$$\mid (\geq n\, r^{\perp}.C) \mid (\geq n\, r.C^{\perp})$$
$$\mathsf{Con}^{\top}(\mathbf{S}) ::= (\neg C^{\perp}) \mid (C_1^{\top} \sqcap C_2^{\top})$$

where $A^{\perp} \notin \mathbf{S}$ *is a concept name,* C *is a* $\mathcal{SHOIQ}$*-concept,* $C^{\perp} \in \mathsf{Con}^{\perp}(\mathbf{S})$, $C_i^{\top} \in \mathsf{Con}^{\top}(\mathbf{S})$ *(for* $i = 1, 2$*), and* $\mathsf{Sig}(r^{\perp}) \nsubseteq \mathbf{S}$*.*

An axiom α *is* syntactically local w.r.t. **S** *if it is of one of the following forms: (i)* $r^{\perp} \sqsubseteq r$*, (ii)* $\mathsf{Trans}(r^{\perp})$*, (iii)* $C^{\perp} \sqsubseteq C$ *or (iv)* $C \sqsubseteq C^{\top}$*. The set of all* $\mathcal{SHOIQ}$*-axioms that are syntactically local w.r.t.* **S** *is denoted by* s_local(**S**)*. A* $\mathcal{SHOIQ}$*-ontology* $\mathcal{O}$ *is* syntactically local w.r.t. **S** *if* $\mathcal{O} \subseteq$ s_local(**S**)*.*

Intuitively, if an axiom α is syntactically local w.r.t. **S**, its interpretation is *directly affected* by that of symbols in **S**, in the sense that α is true in every interpretation $\mathcal{I}$ in which concept and role names from **S** are interpreted with the empty set. Based on this notion, locality-based modules can be defined as follows: Let $\mathcal{O}$ be a $\mathcal{SHOIQ}$ ontology, $\mathcal{O}' \subseteq \mathcal{O}$ a subset of it, and **S** a signature. Then, $\mathcal{O}'$ is a *locality-based module for* **S** *in* $\mathcal{O}$ if every axiom $\alpha \in \mathcal{O} \backslash \mathcal{O}'$ is syntactically local w.r.t. $\mathbf{S} \cup \mathsf{Sig}(\mathcal{O}')$. Given an ontology $\mathcal{O}$ and a signature **S**, there always exists a unique, minimal locality-based module [4], denoted by $\mathcal{O}^{loc}_{\mathbf{S}}$. In the example ontology, it can be easily verified that the underlined axioms are precisely those in $\mathcal{O}^{loc}_{\{Endocarditis\}}$.

The notion of strong subsumption module (first introduced in [3]) is essential for our modularization-based approach.

Definition 3 (Strong subsumption module). *Let* $\mathcal{S} \subseteq \mathcal{O}$ *be* $\mathcal{SHOIQ}$ *ontologies, and* A *a concept name. Then,* $\mathcal{S}$ *is a subsumption module for* A *in* $\mathcal{O}$ *if, for all* $B \in \mathsf{CN}(\mathcal{O})$: $A \sqsubseteq_{\mathcal{O}} B$ *iff* $A \sqsubseteq_{\mathcal{S}} B$.

A subsumption module $\mathcal{S}$ *for* A *in* $\mathcal{O}$ *is called* strong *if, for all* $B \in \mathsf{CN}(\mathcal{O})$: $A \sqsubseteq_{\mathcal{O}} B$ *implies that* $J \subseteq \mathcal{S}$, *for every justification* J *for* $A \sqsubseteq B$ *in* $\mathcal{O}$.

Observe that the *largest* such strong subsumption module is the whole ontology itself, and the *smallest* such module is precisely the union of all justifications J for $A \sqsubseteq B$ in $\mathcal{O}$, for all superconcept B of A. For our purpose, the minimal locality-based module is of interest since it is relative small (though not smallest) and cheap to compute (i.e., quadratic time).

Selection functions
We introduce the notion of selection function in a single ontology given in [8], which will be used in our algorithm to extract a subset of an ontology relevant to a subsumption to some degree. Though applied to arbitrary DL languages, we here restrict attention to $\mathcal{SHOIQ}$:

Definition 4 (Selection function). *Let* $\mathcal{L}$ *be the set of all* $\mathcal{SHOIQ}$ *axioms over a set of signature. Then, a selection function for* $\mathcal{L}$ *is a mapping* $s_{\mathcal{L}}$: $\mathcal{P}(\mathcal{L}) \times \mathcal{L} \times \mathbb{N} \to \mathcal{P}(\mathcal{L})$ *s.t.* $s_{\mathcal{L}}(\mathcal{O}, \alpha, k) \subseteq \mathcal{O}$, *where* $\mathcal{P}(\mathcal{L})$ *is the power set of* $\mathcal{L}$.

Intuitively, a selection function selects a subset of an ontology w.r.t. an axiom at step k. A specific selection function based on *syntactic relevance* is employed in our algorithm. We begin with defining *direct relevance* between two axioms.

Definition 5 (Direct relevance). *Two axioms* α *and* β *are directly relevant iff* $\mathsf{Sig}(\alpha) \cap \mathsf{Sig}(\beta) \neq \emptyset$.

The intuition is that two axioms are directly relevant if they share a common (concept or role) name. Another relevance relation is given in [15]. However, that relevance relation is tailored for *unfoldable* DL $\mathcal{ALC}$, and as such the selection function defined by it cannot be used to find all justifications in our setting, so we do not consider it here.

Based on the notion of direct relevance, we can define the notion of relevance between an axiom and an ontology.

Definition 6. *An axiom* α *is relevant to an ontology* $\mathcal{O}$ *iff there exists an axiom* β *in* $\mathcal{O}$ *such that* α *and* β *are directly relevant.*

We introduce the relevance-based selection function which can be used to find all the axioms in an ontology that are relevant to an axiom to some degree.

Definition 7 (Relevance-based selection function). *Let* $\mathcal{O}$ *be an ontology,* α *be an axiom and* k *be an integer. The* relevance-based selection function, *written* s_{rel}, *is defined inductively as follows:*
$s_{rel}(\mathcal{O}, \alpha, 0) = \emptyset$
$s_{rel}(\mathcal{O}, \alpha, 1) = \{\beta \in \mathcal{O} : \alpha \text{ and } \beta \text{ are directly relevant}\}$
$s_{rel}(\mathcal{O}, \alpha, k) = \{\beta \in \mathcal{O} : \beta \text{ is directly relevant to } s_{rel}(\mathcal{O}, \alpha, k-1)\}$, *where* $k > 1$.

We call $s_{rel}(\mathcal{O}, \alpha, k)$ *the* k-relevant subset *of* $\mathcal{O}$ *w.r.t.* α. *For convenience, we define* $s_k(\mathcal{O}, \alpha) = s_{rel}(\mathcal{O}, \alpha, k) \setminus s_{rel}(\mathcal{O}, \alpha, k-1)$ *for* $k \geq 1$.

Hitting set tree (HST) algorithm

We briefly introduce some notions regarding Reiter's Hitting Set Tree algorithm given in [13] which will be used in our algorithm to find all justifications. We follow the reformulated notions in Reiter's theory in [10]. Given a *universal set* U, and a set $S = \{s_1, ..., s_n\}$ of subsets of U which are *conflict sets*, i.e. subsets of the system components responsible for the error. A *hitting set* T for S is a subset of U such that $s_i \cap T \neq \emptyset$ for all $1 \leq i \leq n$. A *minimal hitting set* T for S is a hitting set such that no $T' \subset T$ is a hitting set for S. A hitting set T is cardinality-minimal if there is no other hitting set T' such that $|T'| < |T|$. Reiter's algorithm is used to calculate minimal hitting sets for a collection $S = \{s_1, ..., s_n\}$ of sets by constructing a labeled tree, called a Hitting Set Tree (HST). In a HST, each node is labeled with a set $s_i \in S$, and each edge is labeled with an element in $\cup_{s_i \in S} s_i$. For each node n in a HST, let $H(n)$ be the set of edge labels on the path from the root of the HST to n. Then the label for n is any set $s \in S$ such that $s \cap H(n) = \emptyset$, if such a set exists. Suppose s is the label of a node n, then for each $\sigma \in s$, n has a successor n_σ connected to n by an edge with σ in its label. If the label of n is the empty set, then we have that $H(n)$ is a hitting set of S. In the case of finding justifications, the universal set corresponds to the ontology and a conflict set corresponds to a justification [10].

3   Justification Coverage in Locality-Based Modules

This section presents the main technical contribution of the paper that lays the foundation of our modularization-based algorithm. We show that a locality-based module for $\mathbf{S} = \{A\}$ in $\mathcal{O}$ is a strong subsumption module for A in $\mathcal{O}$.

Proposition 1. *Let* $\mathbf{S}$ *be a signature, and* $\mathcal{I} = (\Delta^{\mathcal{I}}, \cdot^{\mathcal{I}})$ *an interpretation such that* $x^{\mathcal{I}} = \emptyset$ *for all (concept and role) names* $x \notin \mathbf{S}$. *Then,* $(C^{\perp})^{\mathcal{I}} = \emptyset$ *for every concept* $C^{\perp} \in \mathsf{Con}^{\perp}(\mathbf{S})$, *and* $(C^{\top})^{\mathcal{I}} = \Delta^{\mathcal{I}}$ *for every concept* $C^{\top} \in \mathsf{Con}^{\top}(\mathbf{S})$.

The proof is an easy induction on the structure of the concepts $C^{\perp}$ and $C^{\top}$. Intuitively, every concept in $\mathsf{Con}^{\top}(\mathbf{S})$ ($\mathsf{Con}^{\perp}(\mathbf{S})$, resp.) behaves as if it were the top concept (the bottom concept, resp.) in any interpretation $\mathcal{I}$ with $x^{\mathcal{I}} = \emptyset$ for all $x \notin \mathbf{S}$. It follows that syntactically local axioms of the form $C^{\perp} \sqsubseteq C$ and $C \sqsubseteq C^{\top}$ are vacuously satisfied by such an interpretation $\mathcal{I}$. This property of syntactically local axioms is used to prove the following lemma.

Lemma 1. *Let* $\mathcal{O}$ *be a* $\mathcal{SHOIQ}$ *ontology,* A, B *concept names in* $\mathsf{Sig}(\mathcal{O})$ *such that* $A \sqsubseteq_{\mathcal{O}} B$, $\mathcal{O}_A^{\mathsf{loc}}$ *a locality-based module for* $\{A\}$ *in* $\mathcal{O}$. *If* $A \sqsubseteq_{\mathcal{S}} B$ *for an* $\mathcal{S} \subseteq \mathcal{O}$ *such that* $\mathcal{S} \not\subseteq \mathcal{O}_A^{\mathsf{loc}}$, *then* $A \sqsubseteq_{\mathcal{S}'} B$ *with* $\mathcal{S}' = \mathcal{S} \cap \mathcal{O}_A^{\mathsf{loc}}$.

Proof. We show the contraposition by assuming that $A \not\sqsubseteq_{\mathcal{S}'} B$ and then demonstrating that $A \not\sqsubseteq_{\mathcal{S}} B$. Since $A \not\sqsubseteq_{\mathcal{S}'} B$, there must be a model $\mathcal{I}'$ of $\mathcal{S}'$ and an individual $w \in \Delta^{\mathcal{I}'}$ such that $w \in A^{\mathcal{I}'} \backslash B^{\mathcal{I}'}$. Construct a new interpretation $\mathcal{I}$ based on $\mathcal{I}'$ by setting $x^{\mathcal{I}} := \emptyset$ for all symbols (role or concept names) $x \in \mathsf{Sig}(\mathcal{O}) \backslash \mathsf{Sig}(\mathcal{O}_A^{\mathsf{loc}})$. Obviously, $w \in A^{\mathcal{I}}$ since $\mathcal{I}$ does not change the interpretation of $A \in \mathsf{Sig}(\mathcal{O}_A^{\mathsf{loc}})$. There are two possibilities for B: either $B^{\mathcal{I}} = B^{\mathcal{I}'}$ or $B^{\mathcal{I}} = \emptyset$. In either case, we have that $w \notin B^{\mathcal{I}}$.

It remains to show that $\mathcal{I}$ is a model of $\mathcal{S}$, i.e., satisfies every axiom $\alpha = (\alpha_L \sqsubseteq \alpha_R)$ in $\mathcal{S}$. We make a case distinction as follows:

- $\alpha \in \mathcal{O}_A^{\text{loc}}$. It follows that $\alpha \in \mathcal{S}'$, and thus $\mathcal{I}' \models \alpha$. By construction, both $\mathcal{I}$ and $\mathcal{I}'$ agree on the interpretation of symbols in $\text{Sig}(\mathcal{O}_A^{\text{loc}})$ and thus $\text{Sig}(\alpha)$. Hence, $\mathcal{I} \models \alpha$ as required.
- $\alpha \notin \mathcal{O}_A^{\text{loc}}$. By definition of locality-based modules, α is syntactically local w.r.t. $\mathbf{S} = \text{Sig}(\mathcal{O}_A^{\text{loc}}) \cup \{A\}$. Then, there are four possibilities for α:
  - $\alpha = r^\perp \sqsubseteq r$. First, assume that $r^\perp$ is a role name. Then, $r^\perp \notin \mathbf{S}$ and thus $r^\perp \in \text{Sig}(\mathcal{O}) \backslash \text{Sig}(\mathcal{O}_A^{\text{loc}})$. By construction of $\mathcal{I}$, $(r^\perp)^\mathcal{I} = \emptyset$. Otherwise, $r^\perp$ is an inverse role s^-. Then, $s \in \text{Sig}(r^\perp) \not\subseteq \mathbf{S}$. It follows that $s \in \text{Sig}(\mathcal{O}) \backslash \text{Sig}(\mathcal{O}_A^{\text{loc}})$, and thus $(r^\perp)^\mathcal{I} = s^\mathcal{I} = \emptyset$. In both cases, $\mathcal{I} \models \alpha$ as required.
  - $\alpha = \text{Trans}(r^\perp)$. Analogous to the first case.
  - $\alpha = C^\perp \sqsubseteq C$. By Proposition 1, $(C^\perp)^\mathcal{I} = \emptyset$. Hence, $\mathcal{I} \models \alpha$.
  - $\alpha = C \sqsubseteq C^\top$. By Proposition 1, $(C^\top)^\mathcal{I} = \Delta^\mathcal{I}$. Hence, $\mathcal{I} \models \alpha$.

Since $\mathcal{I}$ is a model of $\mathcal{S}$ such that $w \in A^\mathcal{I} \backslash B^\mathcal{I}$, we have $A \not\sqsubseteq_\mathcal{S} B$, contradicting the premise of the lemma. ❏

Now, we are ready to establish the required property of the modules:

Theorem 1 ($\mathcal{O}_A^{\text{loc}}$ is a strong subsumption module). *Let $\mathcal{O}$ be a $\mathcal{SHOIQ}$ ontology and A a concept name. Then $\mathcal{O}_A^{\text{loc}}$ is a strong subsumption module for A in $\mathcal{O}$.*

Proof. The fact that $\mathcal{O}_A^{\text{loc}}$ is a subsumption module has been shown in [4]. It remains to show that it is strong, i.e., every justification $J \subseteq \mathcal{O}$ for $A \sqsubseteq_\mathcal{O} B$ is contained in $\mathcal{O}_A^{\text{loc}}$, for every concept name $B \in \text{CN}(\mathcal{O})$.

Assume to the contrary that there is a concept name B and a justification J for $A \sqsubseteq_\mathcal{O} B$ that is not contained in $\mathcal{O}_A^{\text{loc}}$. By Lemma 1, the strict subset $J' = J \cap \mathcal{O}_A^{\text{loc}}$ of J is such that $A \sqsubseteq_{J'} B$. Obviously, J is not minimal and hence cannot be a justification for $A \sqsubseteq_\mathcal{O} B$, contradicting the initial assumption. ❏

Intuitively, the (minimal) locality-based module for $\mathbf{S} = \{A\}$ in a $\mathcal{SHOIQ}$-ontology $\mathcal{O}$ contains *all* the relevant axioms for any subsumption $\sigma = (A \sqsubseteq_\mathcal{O} B)$, in the sense that all responsible axioms for σ are included. In other words, in order to find all justifications for a certain entailment in an OWL ontology, it is sufficient to consider only axioms in the locality-based module. Since the *minimal* locality-based modules are relatively very small (see, e.g., [6,16]), our modularization-based approach proves promising. The empirical results on real-life ontologies are described in Section 5.

4   Our Modularization-Based Algorithm

In this section, we propose a new algorithm for finding all justifications based on the relevance-based algorithm and the modularization extraction algorithm. Before we describe our algorithm, we need to recap the relevance-based algorithm given in [9].

Algorithm 1. REL_ALL_JUSTS($A \sqsubseteq B, \mathcal{O}, s$)

Data: An ontology $\mathcal{O}$, a subsumption $A \sqsubseteq B$ and a selection function s.
Result: All justifications $\mathcal{J}$

```
 1  begin
 2  │  Globals : 𝒥 ← ∅;
 3  │  𝒪′ ← HS ← HS_local ← ∅; k ← 1;
 4  │  𝒮 ← s(𝒪, A ⊑ B, k);
 5  │  while 𝒮 ≠ ∅ do
 6  │  │  𝒪′ ← 𝒪′ ∪ 𝒮;
 7  │  │  if HS_local ≠ ∅ then
 8  │  │  │  for P ∈ HS_local do            /* Get global hitting sets */
 9  │  │  │  │  if 𝒪 \ P ⊭ A ⊑ B then
10  │  │  │  │  └  HS ← HS ∪ {P};
11  │  │  │  HS_local ← HS_local \ HS;
12  │  │  │  if (HS_local = ∅) then
13  │  │  │  └  return 𝒥                    /* Early termination */;
14  │  │  │  HS_temp ← HS_local;
15  │  │  │  for P ∈ HS_temp do             /* Expand hitting set tree */
16  │  │  │  │  (𝒥′, HS′_local) ← EXPAND_HST(A ⊑ B, 𝒪′ \ P);
17  │  │  │  │  𝒥 ← 𝒥 ∪ 𝒥′;
18  │  │  │  └  HS_local ← HS_local ∪ {P ∪ P′|P′ ∈ HS′_local} \ {P};
19  │  │  else if 𝒪′ ⊨ A ⊑ B then
20  │  │  └  (𝒥, HS_local) ← EXPAND_HST(A ⊑ B, 𝒪′);
21  │  │  k ← k + 1;
22  │  └  𝒮 ← s_k(𝒪, A ⊑ B);
23  │  return 𝒥
24  end
```

The relevance-based algorithm (Algorithm 1) receives an ontology $\mathcal{O}$, a subsumption $A \sqsubseteq B$ of $\mathcal{O}$ and a selection function s, and outputs the set of all justifications $\mathcal{J}$. We sketch the basic idea of the algorithm and refer to [9] for details of the algorithm. First of all, we find the first k such that $A \sqsubseteq B$ is inferred by the k-relevant subset $\mathcal{O}'$ of $\mathcal{O}$, i.e., the "if" condition in line 19 is satisfied. We then call Algorithm 2 to find a set of justifications for $A \sqsubseteq B$ in $\mathcal{O}'$ and a set of *local hitting sets*, where a local hitting set is a hitting set for all justifications in the selected sub-ontology, i.e., O' in line 20. We then add to the sub-ontology obtained in the previous iteration those axioms that are directly relevant this sub-ontology. For those local hitting sets that are not hitting sets of all justifications in the entire ontology $\mathcal{O}$, we call Algorithm 2 to further expand them, and so on.

To compute a single justification in Algorithm 2, we invoke a sub-procedure SINGLE_JUST($\sigma, \mathcal{O}$) which is a black-box pinpointing algorithm optimized either by the sliding window technique in [10] or by binary search technique in [3].

The correctness of Algorithm 1 follows from Theorem 1 in [9].

Algorithm 2. EXPAND_HST$(A \sqsubseteq B, \mathcal{O})$

Data: An ontology $\mathcal{O}$ and a subsumption $A \sqsubseteq B$ of $\mathcal{O}$
Result: A set of justifications $\mathcal{J}$ for $A \sqsubseteq B$ in $\mathcal{O}$ and a set of hitting sets

1   **begin**
2      $HS \leftarrow HS_1 \leftarrow \emptyset$
3      $J \leftarrow$ SINGLE_JUST$(A \sqsubseteq B, \mathcal{O})$
4      $\mathcal{J} \leftarrow \mathcal{J} \cup \{J\}$
5   **for** $\alpha \in J$ **do**   /* Create all possible branches. */
6        $HS_1 \leftarrow HS_1 \cup \{\{\alpha\}\}$
7   **while** *true* **do**
8        $HS_2 \leftarrow \emptyset$
9   **for** $(P \in HS_1)$ **do**
10   **if** $\mathcal{O} \setminus P \not\models A \sqsubseteq B$ **then**
11            $HS \leftarrow HS \cup \{P\}$
12   **else**
13            $HS_2 \leftarrow HS_2 \cup \{P\}$ /* Branches need to be expanded */
14   **if** $(HS_1 = \emptyset)$ *or* $(HS_2 = \emptyset)$ **then**
15   **return** $(\mathcal{J}, HS)$
16        $HS_1 \leftarrow \emptyset$
17   **for** $P \in HS_2$ **do**
18          $J \leftarrow$ SINGLE_JUST$(A \sqsubseteq B, \mathcal{O} \setminus P)$
19          $\mathcal{J} \leftarrow \mathcal{J} \cup \{J\}$
20   **for** $\alpha \in J$ **do**
21            $HS_1 \leftarrow HS_1 \cup \{P \cup \{\alpha\}\}$

22 **end**

Theorem 2. *Given an ontology $\mathcal{O}$, a subsumption $A \sqsubseteq B$ of $\mathcal{O}$ and a relevance-based selection function s_{rel}, $\mathcal{J}$ returned by Algorithm 1 is the set of all justifications for $A \sqsubseteq B$.*

Based on the algorithms introduced above, we propose our novel algorithm for computing all the justification. The idea of our algorithm is straightforward: to find all justifications for a subsumption $A \sqsubseteq B$ in $\mathcal{O}$, we first extract the locality-based module $\mathcal{O}_A^{loc}$ for $\mathbf{S} = \{A\}$ in $\mathcal{O}$ and then apply Algorithm 1. The method is outlined in Algorithm 3, where EXTRACT_MODULE implements the locality-based extraction algorithm in [4], and s_{rel} is the relevance-based selection function. The correctness of the algorithm can be seen by Theorem 1 and Theorem 2. We illustrate the effectiveness of our algorithm by means of an example:

Example 1. Consider an ontology $\mathcal{O}$ that contains the following axioms:

$$\alpha_{1i} : A_{1i} \sqsubseteq P_{1i} \sqcap Q_{1i} \sqcap Z, \quad \alpha_{2i} : P_{1i} \sqsubseteq A_{2i} \sqcap Z, \quad \alpha_{3i} : Q_{1i} \sqsubseteq A_{2i} \sqcap Z$$
$$\alpha_{4i} : A_{2i} \sqsubseteq P_{2i} \sqcap Q_{2i} \sqcap Z, \quad \alpha_{5i} : P_{2i} \sqsubseteq A_{3i} \sqcap Z, \quad \alpha_{6i} : Q_{2i} \sqsubseteq A_{3i} \sqcap Z,$$

Algorithm 3. MODULE_ALL_JUSTS($A \sqsubseteq B, \mathcal{O}$)

Data: An ontology $\mathcal{O}$ and a subsumption $A \sqsubseteq B$
Result: All justifications $\mathcal{J}$

1 **begin**
2     $\mathcal{O}_A^{loc} \leftarrow$ EXTRACT_MODULE($\mathcal{O}, A$)
3   **return** REL_ALL_JUSTS($A \sqsubseteq B, \mathcal{O}_A^{loc}, s_{rel}$)
4 **end**

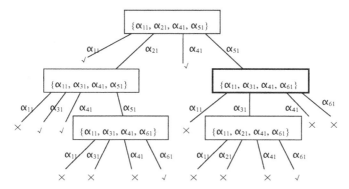

Fig. 2. Finding all justifications by HST algorithm on the locality-based module. Each rectangle represents a justification, and the bold rectangle indicates a justification reuse. '×' means early path termination, while '√' means a hitting set is found.

for $1 \leq i \leq 10000$. Obviously, $\mathcal{O}$ comprises $60\,000$ axioms and entails the subsumption $\sigma = (A_{11} \sqsubseteq A_{31})$. While such an ontology clearly is not a realistic ontology, it well demonstrates the need and potential of search space reduction. If algorithm REL_ALL_JUSTS is applied directly to this ontology, one cannot expect an acceptable performance when finding all justifications. This is because: (i) SINGLE_JUST($\sigma, \mathcal{O}$) has to prune a very large set, and (ii) each subsumption test is w.r.t. the entire ontology $\mathcal{O}$ since all the axioms $\mathcal{O}$ share a common concept Z. In our modularization-based approach, however, we first extract the locality-based module $\mathcal{O}_{A_{11}}^{loc}$ for $\mathbf{S} = \{A_{11}\}$ in $\mathcal{O}$, and then apply REL_ALL_JUSTS to $\mathcal{O}_{A_{11}}^{loc}$ instead of $\mathcal{O}$. Since the module contains only 6 axioms, i.e., $\mathcal{O}_{A_{11}}^{loc} = \{\alpha_{11}, \alpha_{21}, \alpha_{31}, \alpha_{41}, \alpha_{51}, \alpha_{61}\}$, both points above can be achieved in much less time.

Figure 2 illustrates the process of finding all justifications by means of expanding a hitting set tree (HST). To begin with, a justification $\{\alpha_{11}, \alpha_{21}, \alpha_{41}, \alpha_{51}\}$ is computed by SINGLE_JUST($\sigma, \mathcal{O}_{A_{11}}^{loc}$), which is taken as the root of the tree. Since $\mathcal{O}_{A_{11}}^{loc}$ dispensed with α_{11} does not entail σ, $\{\alpha_{11}\}$ is a hitting set. On the other hand, $\mathcal{O}' = \mathcal{O}_{A_{11}}^{loc} \setminus \{\alpha_{21}\}$ still entails σ, and thus another justification can be computed by calling SINGLE_JUST($\sigma, \mathcal{O}'$). The process continues to expand HST until it finds all other justifications for σ: $\{\alpha_{11}, \alpha_{31}, \alpha_{41}, \alpha_{51}\}$, $\{\alpha_{11}, \alpha_{31}, \alpha_{41}, \alpha_{61}\}$, $\{\alpha_{11}, \alpha_{21}, \alpha_{41}, \alpha_{61}\}$. Observe that the node following the branch $\{\alpha_{51}\}$ is a result of the optimization 'justification reuse.'

Table 2. Benchmark ontologies and their characteristics

Ontologies	♯Axioms	♯Concepts	♯Roles	Module size		Extraction time
				Average	Maximum	(sec)
GALEN	4 529	2 748	413	75	530	6
Go	28 897	20 465	1	16	125	40
NCI	46 940	27 652	70	29	436	65

5   Empirical Results

Our algorithm has been realized by using KAON2[4] as the black-box reasoner. Of course, the method (like other black-box approaches) can be applied to any other reasoner, e.g., RacerPro[5] and FaCT++[6]. To fairly compare with the pinpointing algorithm in [10], we re-implemented it with KAON2 API (henceforth referred to as ALL_JUSTS algorithm). The experiments have been performed on a Linux server with an Intel(R) CPU Xeon(TM) 3.2GHz running Sun's Java 1.5.0 with allotted 2GB heap space.

Benchmark ontologies used in our experiments are the GALEN Medical Knowledge Base[7], the Gene Ontology (Go)[8] and the US National Cancer Institute thesaurus (NCI)[9]. The three biomedical ontologies are well-known to both the life science and Semantic Web communities since they are employed in real-world applications and often used as benchmarks for testing DL reasoners. Both Go and NCI are formulated in the lightweight DL $\mathcal{EL}$, while GALEN uses expressivity of the more complex DL $\mathcal{SHF}$. Some information concerning the size and characteristics of the benchmark ontologies are given in the left part of Table 2.

Modularization reveals structures and dependencies of concepts in the ontologies as argued in [4,16]. We extract the (minimal) locality-based module for $\mathbf{S} = \{A\}$ in $\mathcal{O}$, for every benchmark ontology $\mathcal{O}$ and each concept name $A \in \mathsf{CN}(\mathcal{O})$. The size of the modules and the time required to extract them are shown in the last three columns of Table 2. Observe that the modules in GALEN are larger than those in the other two ontologies although the ontology itself is smaller. This suggests that GALEN is more complex in the sense that more axioms in it are non-local (thus relevant) according to Definition 2.

In the experiments, we consider three concept names in $\mathsf{CN}(\mathcal{O})$ for each benchmark ontology $\mathcal{O}$ such that one of them has the largest locality-based module[10]. For the sake of brevity, we denote by $\mathsf{subs}(\mathcal{O})$ the set of all tested subsumptions $A \sqsubseteq B$ in $\mathcal{O}$, with A one of the three concept names mentioned above and B an inferred

---

[4] http://kaon2.semanticweb.org/

[5] http://www.racer-systems.com/

[6] http://owl.man.ac.uk/factplusplus/

[7] http://www.openclinical.org/prj_galen.html

[8] http://www.geneontology.org

[9] http://www.mindswap.org/2003/CancerOntology/nciOntology.owl

[10] The concept name with largest module is hand-picked in order to cover hard cases in our experiments, while the other two are randomly selected.

subsumer of A. For each $\mathcal{O}$ of our benchmark ontologies, we compute *all* justifications for σ in $\mathcal{O}$, where $\sigma \in \mathsf{subs}(\mathcal{O})$. In order to compare with the other existing approaches, we perform the following for each σ and $\mathcal{O}$ to compute all justifications:

1. $\mathsf{ALL\_JUSTS}(\sigma, \mathcal{O})$ (i.e., the algorithm in [10]).
2. $\mathsf{REL\_ALL\_JUSTS}(\sigma, \mathcal{O}, s_{rel})$;
3. $\mathsf{MODULE\_ALL\_JUSTS}(\sigma, \mathcal{O})$;

The justification results by $\mathsf{MODULE\_ALL\_JUSTS}$ are shown in Table 3, where the ontology marked with $\star$ means that some run does not terminate within the two hour time-out. Precisely, there are three subsumptions in GO and one in NCI, for which the computation took more than two hours. The statistics given on the right hand side of the table does not take into account these subsumptions.

Table 3. Justification results using the modularization-based approach

Ontologies	♯Subsumptions	♯Justifications		Justification size	
	$\|\mathsf{subs}(\mathcal{O})\|$	Average	Maximum	Average	Maximum
GALEN	69	1.5	4	9.7	24
GO*	53	3.2	11	5.3	9
NCI*	23	1.6	8	5.4	9

To visualize the time performances of the three algorithms, we randomly selected two subsumptions σ_1 and σ_2 from $\mathsf{subs}(\mathcal{O})$ for each ontology $\mathcal{O}$ and compared their computation time required by the three algorithms. These subsumptions are shown as follows:

$$\begin{aligned}
\text{GALEN:}\sigma_1 \quad & \text{AcuteErosionOfStomach} \sqsubseteq \text{GastricPathology} \\
\text{GALEN:}\sigma_2 \quad & \text{AppendicularArtery} \sqsubseteq \text{PhysicalStructure} \\
\text{GO:}\sigma_1 \quad & \text{GO\_0000024} \sqsubseteq \text{GO\_0007582} \\
\text{GO:}\sigma_2 \quad & \text{GO\_0000027} \sqsubseteq \text{GO\_0044238} \\
\text{NCI:}\sigma_1 \quad & \text{CD97\_Antigen} \sqsubseteq \text{Protein} \\
\text{NCI:}\sigma_2 \quad & \text{APC\_8024} \sqsubseteq \text{Drugs\_and\_Chemicals}
\end{aligned}$$

The chart in Figure 3 depicts the overall computation time required for each algorithm to find all justifications for each tested subsumption. Unlike the time results reported in [10], which excluded the time for satisfiability checking, we report here the overall computation time, i.e. the total time of the algorithm including the time needed by the black-box reasoner for the standard reasoning tasks. Observe that both $\mathsf{ALL\_JUSTS}$ and $\mathsf{REL\_ALL\_JUSTS}$ did not yield results within the time-out of two hours on three out of six tested subsumptions (marked by "TO" on the chart). Comparing these two algorithms (without modularization), $\mathsf{REL\_ALL\_JUSTS}$ performs noticeably better than $\mathsf{ALL\_JUSTS}$ in most cases. For instance, on the subsumptions GALEN:σ_2 and NCI:σ_2, $\mathsf{REL\_ALL\_JUSTS}$ outperforms $\mathsf{ALL\_JUSTS}$ by about 10 and 20 minutes, respectively. On the subsumption GO:σ_2, both algorithms show a similar performance, i.e., time difference is less than a minute. More explanations on the comparison between these two algorithms can be found in [9].

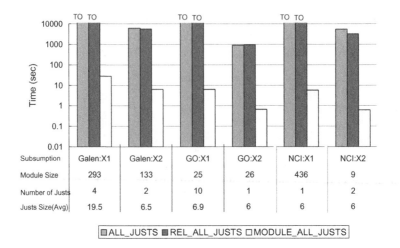

Fig. 3. The time performance of three algorithms for finding all justifications

Interestingly, MODULE_ALL_JUSTS outperforms all the other algorithms on all subsumptions, and the improvement is tremendous as can be seen in all cases in the chart. This empirically confirms our initial conjecture that, given the strongness property (in the sense of Definition 3) and the small size (see Table 2 and [6,16]) of locality-based modules, our optimization should be highly effective. As an example, MODULE_ALL_JUSTS took *only* 0.6 seconds to find all the justifications for $\text{NCI}:\sigma_2$, while REL_ALL_JUSTS needed 3 242 seconds. In this case, the locality-based module for APC_8024 in NCI consists of 9 axioms, whereas the whole ontology has some tens of thousands of axioms. Although the selection function used in REL_ALL_JUSTS also prunes the search space by considering only "k-directly relevant" axioms (see Definition 7) when HST algorithm is executed, several irrelevant axioms (in the sense of syntactic locality) are still considered.

6   Conclusion

In this paper, we proposed a novel approach for finding all justifications for an entailment in OWL DL. The approach is based on the computation of minimal locality-based modules. We first showed that locality-based modules always cover all axioms in all justifications and exploited this property to limit the search space when finding all justifications. Then, we presented a modularization-based pinpointing algorithm that is based on relevance-based techniques and a hitting set tree algorithm. Finally, we reported on several promising empirical results that demonstrate an improvement of several orders of magnitude in efficiency and scalability of finding all justifications in OWL DL ontologies. Our work is based on locality-based modules. As future work, we shall investigate different kinds of modules and selection functions that hopefully produce even more relevant axioms for pinpointing.

Acknowledgements. This work was partially supported by the DFG project under grant BA1122/11-1 and the EU under the IST project NeOn (IST-2006-027595) http://www.neon-project.org.

References

1. Baader, F., Peñaloza, R.: Axiom pinpointing in general tableaux. In: Olivetti, N. (ed.) TABLEAUX 2007. LNCS, vol. 4548, pp. 11–27. Springer, Heidelberg (2007)
2. Baader, F., Peñaloza, R., Suntisrivaraporn, B.: Pinpointing in the description logic $\mathcal{EL}^+$. In: Hertzberg, J., Beetz, M., Englert, R. (eds.) KI 2007. LNCS, vol. 4667, pp. 52–67. Springer, Heidelberg (2007)
3. Baader, F., Suntisrivaraporn, B.: Debugging SNOMED CT using axiom pinpointing in the description logic $\mathcal{EL}^+$. In: Proceedings of KR-MED 2008: Representing and Sharing Knowledge Using SNOMED (2008)
4. Grau, B.C., Horrocks, I., Kazakov, Y., Sattler, U.: Modular reuse of ontologies: Theory and practice. J. of Artificial Intelligence Research (JAIR) 31, 273–318 (2008)
5. Cuenca Grau, B., Halaschek-Wiener, C., Kazakov, Y.: History matters: Incremental ontology reasoning using modules. In: Aberer, K., Choi, K.-S., Noy, N., Allemang, D., Lee, K.-I., Nixon, L., Golbeck, J., Mika, P., Maynard, D., Mizoguchi, R., Schreiber, G., Cudré-Mauroux, P. (eds.) ASWC 2007 and ISWC 2007. LNCS, vol. 4825, pp. 183–196. Springer, Heidelberg (2007)
6. Cuenca Grau, B., Horrocks, I., Kazakov, Y., Sattler, U.: Just the right amount: Extracting modules from ontologies. In: Proc. of WWW 2007, Banff, Canada, pp. 717–726. ACM, New York (2007)
7. Horrocks, I., Sattler, U.: A tableaux decision procedure for $\mathcal{SHOIQ}$. In: Proc. of IJCAI 2005, pp. 448–453 (2005)
8. Huang, Z., van Harmelen, F., ten Teije, A.: Reasoning with inconsistent ontologies. In: Proc. of IJCAI 2005, pp. 254–259 (2005)
9. Ji, Q., Qi, G., Haase, P.: A relevance-based algorithm for finding justifications of DL entailments. In: Technical report, University of Karlsruhe (2008), http://www.aifb.uni-karlsruhe.de/WBS/gqi/papers/RelAlg.pdf
10. Kalyanpur, A., Parsia, B., Horridge, M., Sirin, E.: Finding all justifications of OWL DL entailments. In: Aberer, K., Choi, K.-S., Noy, N., Allemang, D., Lee, K.-I., Nixon, L., Golbeck, J., Mika, P., Maynard, D., Mizoguchi, R., Schreiber, G., Cudré-Mauroux, P. (eds.) ASWC 2007 and ISWC 2007. LNCS, vol. 4825, pp. 267–280. Springer, Heidelberg (2007)
11. Kalyanpur, A., Parsia, B., Sirin, E., Hendler, J.: Debugging unsatisfiable classes in OWL ontologies. Journal of Web Semantics 3(4), 268–293 (2005)
12. Meyer, T., Lee, K., Booth, R.: Knowledge integration for description logics. In: Proc. of AAAI 2005, pp. 645–650. AAAI Press, Menlo Park (2005)
13. Reiter, R.: A theory of diagnosis from first principles. Artificial Intelligence 32(1), 57–95 (1987)
14. Schlobach, S., Cornet, R.: Non-standard reasoning services for the debugging of description logic terminologies. In: Proc. of IJCAI 2003, pp. 355–362 (2003)
15. Schlobach, S., Huang, Z., Cornet, R., van Harmelen, F.: Debugging incoherent terminologies. J. Autom. Reasoning 39(3), 317–349 (2007)
16. Suntisrivaraporn, B.: Module extraction and incremental classification: A pragmatic approach for $\mathcal{EL}^+$ ontologies. In: Bechhofer, S., Hauswirth, M., Hoffmann, J., Koubarakis, M. (eds.) ESWC 2008. LNCS, vol. 5021, pp. 230–244. Springer, Heidelberg (2008)

DL-Lite and Role Inclusions

R. Kontchakov and M. Zakharyaschev

School of Computer Science and Information Systems
Birkbeck College London, UK
http://www.dcs.bbk.ac.uk/∼{roman,michael}

Abstract. We give a classification of the complexity of *DL-Lite* logics extended with role inclusion axioms. We show that the data complexity of instance checking becomes P-hard in the presence of functionality constraints, and CoNP-hard if arbitrary number restrictions are allowed, even with primitive concept inclusions. The combined complexity of satisfiability in this case jumps to ExpTime. On the other side, the combined complexity for the logics without number restrictions depends only on the form of concept inclusions and can range from NLogSpace and P to NP; the data complexity for such logics stays in LogSpace.

1   Introduction

Description logic (DL), a discipline conceived in the 1990s as a family of knowledge representation formalisms, which stemmed from semantic networks and frames, has now been recognised as a 'cornerstone of the Semantic Web' for providing a formal basis for the Web Ontology Language (OWL). *DL-Lite* is part of both OWL 1.1 (currently a W3C member submission) and OWL 2; it belongs to the group of OWL fragments 'that trade expressive power for efficiency of reasoning.' Notably, '*DL-Lite* admits sound and complete reasoning in LogSpace with respect to the size of the data (facts). *DL-Lite* includes most of the main features of conceptual models such as UML class diagrams and ER diagrams.'

Although in many practical cases DL reasoners can cope quite well with tasks of much higher complexity, new challenges are arising that require really tractable reasoning. Typical examples are ontologies with a huge terminology (TBox) or large number of facts (ABox). *DL-Lite* was tailored to provide efficient query answering, which becomes increasingly important for data integration [16], the Semantic Web [13], P2P data management [3] and ontology-based data access [5,7]. E.g., in a standard data integration scenario, the information about objects and relationships between them (ABox) is stored in relational databases, while a special ontology (TBox) defines a new 'logical' view of the stored data so that the user can query the integrated resources in terms of this ontology. In this case, one may be interested in the *combined complexity* of reasoning, when both TBox and ABox are regarded as inputs (e.g., to check consistency), as well as the *data complexity*, i.e., the complexity of solving a problem (say, instance checking or query answering) when the TBox and the query are fixed and only the ABox may vary. *DL-Lite* boasts polynomial-time combined complexity and

J. Domingue and C. Anutariya (Eds.): ASWC 2008, LNCS 5367, pp. 16–30, 2008.
© Springer-Verlag Berlin Heidelberg 2008

LOGSPACE data complexity. Moreover, conjunctive queries to *DL-Lite* ontologies can be rewritten into first-order (or SQL) queries to the underlying databases, so that the existing relational DB engines can be used to evaluate them.

The idea of *DL-Lite* has actually given rise not to a single language but rather a family of related formalisms [6,7,8,10]. Some of them are expressive enough to capture EER and UML class diagrams [1], others enjoy particularly simple procedures for rewriting queries into SQL [8]. Unfortunately, a mechanical union of two languages of the family can easily ruin their nice computation properties. This situation poses a general research problem of investigating the impact of various DL constructs on the computational complexity of reasoning in *DL-Lite* logics. The impact of Boolean operators in concept inclusions as well as arbitrary number restrictions on roles was comprehensively analysed in [2].

Table 1. Complexity of *DL-Lite* logics with role inclusions ($\leq$ and $\geq$ mean upper and lower bounds, respectively; the most important new results are typeset in bold)

language	concept inc.	number restric.	combined comp. of satisfiability	data complexity	
				inst. checking	query answering
$DL\text{-}Lite_{core}^{\mathcal{R}}$	core	–	NLOGSPACE $\leq$ [Th.1]	in LOGSPACE	in LOGSPACE
$DL\text{-}Lite_{core}^{\mathcal{R},\mathcal{F}}$	core	f	**ExpTime** $\geq$ [Th.3]	**P** $\geq$ [Th.6]	P
$DL\text{-}Lite_{core}^{\mathcal{R},\mathcal{N}}$	core	+	EXPTIME	**coNP** $\geq$ [Th.5]	coNP
$DL\text{-}Lite_{krom}^{\mathcal{R}}$	Krom	–	**NLogSpace** $\leq$ [Th.1]	in LOGSPACE	coNP $\geq$[7]
$DL\text{-}Lite_{krom}^{\mathcal{R},\mathcal{F}}$	Krom	f	EXPTIME	**coNP** $\geq$ [Th.4]	coNP
$DL\text{-}Lite_{krom}^{\mathcal{R},\mathcal{N}}$	Krom	+	EXPTIME	coNP	coNP
$DL\text{-}Lite_{horn}^{\mathcal{R}}$	Horn	–	P $\leq$ [Th.1]	in LOGSPACE	in LOGSPACE [8]
$DL\text{-}Lite_{horn}^{\mathcal{R},\mathcal{F}}$	Horn	f	EXPTIME	P $\geq$[8]	P $\leq$[11]
$DL\text{-}Lite_{horn}^{\mathcal{R},\mathcal{N}}$	Horn	+	EXPTIME	coNP	coNP
$DL\text{-}Lite_{bool}^{\mathcal{R}}$	Bool	–	**NP** $\leq$ [Th.1]	in **LogSpace** [Th.2]	coNP
$DL\text{-}Lite_{bool}^{\mathcal{R},\mathcal{F}}$	Bool	f	EXPTIME	coNP	coNP
$DL\text{-}Lite_{bool}^{\mathcal{R},\mathcal{N}}$	Bool	+	EXPTIME $\leq$[12]	coNP	coNP $\leq$[12]

Here we investigate *DL-Lite* languages with *role inclusion axioms*, which are indispensable in data modelling (and are present in RDFS), and give a complete picture of the trade-off between their expressiveness and computational complexity. The obtained new results are as follows (cf. Table 1): (i) One cannot keep the data complexity of instance checking in LOGSPACE and have functionality constraints (or any kind of number restrictions) together with role inclusions in the language: even logics with extremely primitive concept inclusions become P-hard (coNP-hard if arbitrary number restrictions are allowed).[1] (ii) The combined complexity of satisfiability in this case jumps to EXPTIME. On the other hand, for the Horn fragment with functionality constraints, instance checking is P-complete for data complexity. Although this problem is not first-order reducible, it can be reformulated in Datalog [14]. (iii) On the positive side, it turns out that the combined complexity for the logics with role inclusions but without number restrictions depends only on the form of concept inclusions and can range from

---

[1] The proof of [8, Theorem 6, item 2] is incorrect (although the result holds).

NLogSpace and P to NP; the data complexity of instance checking for such logics stays in LogSpace. (iv) Another positive observation is that the addition of conjunction to the left-hands side of concept inclusions does not affect complexity too much: although the combined complexity of satisfiability may rise, the data complexity stays the same: as the latest flavours of *DL-Lite$_A$* [9] contain only 'core' concept inclusions, Table 1 suggests that they can be extended to the Horn languages without damaging their computational properties.

2   The *DL-Lite* Family and Its Neighbours

We begin by defining a description logic that can be regarded as the *supremum* of the original *DL-Lite* family [6,7,8,10] in the lattice of description logics. This *supremum* will be called *DL-Lite$_{bool}^{\mathcal{R},\mathcal{N}}$*. The language of *DL-Lite$_{bool}^{\mathcal{R},\mathcal{N}}$* contains *object names* $a_0, a_1, \ldots$, *atomic concept names* $A_0, A_1, \ldots$, and *atomic role names* $P_0, P_1, \ldots$; its complex *roles* R and *concepts* C are defined as follows:

$$B \quad ::= \quad \bot \quad | \quad A_i \quad | \quad \geq q\, R, \qquad R \quad ::= \quad P_i \quad | \quad P_i^-,$$
$$C \quad ::= \quad B \quad | \quad \neg C \quad | \quad C_1 \sqcap C_2,$$

where $q \geq 1$. The concepts of the form B are called *basic*. A *DL-Lite$_{bool}^{\mathcal{R},\mathcal{N}}$* *TBox*, $\mathcal{T}$, is a finite set of *concept inclusion* and *role inclusion axioms* of the form:

$$C_1 \sqsubseteq C_2 \quad \text{ and } \quad R_1 \sqsubseteq R_2,$$

and an *ABox*, $\mathcal{A}$, is a finite set of assertions of the form $A_k(a_i)$ and $P_k(a_i, a_j)$. Taken together, $\mathcal{T}$ and $\mathcal{A}$ constitute the *DL-Lite$_{bool}^{\mathcal{R},\mathcal{N}}$* *knowledge base* $\mathcal{K} = (\mathcal{T}, \mathcal{A})$.

As usual in description logic, an *interpretation* $\mathcal{I}$ consists of a nonempty domain $\Delta^{\mathcal{I}}$ and an interpretation function $\cdot^{\mathcal{I}}$ such that $A_i^{\mathcal{I}} \subseteq \Delta^{\mathcal{I}}$, $P_i^{\mathcal{I}} \subseteq \Delta^{\mathcal{I}} \times \Delta^{\mathcal{I}}$, and $a_i^{\mathcal{I}} \neq a_j^{\mathcal{I}}$, for all $i \neq j$ (the *unique name assumption*, not adopted in OWL, can be safely removed; it is standard in the DL community and can only make proofs a bit harder). The role and concept constructors are interpreted in $\mathcal{I}$ in the standard way:

$$(P_i^-)^{\mathcal{I}} = \{(y, x) \in \Delta^{\mathcal{I}} \times \Delta^{\mathcal{I}} \mid (x, y) \in P_i^{\mathcal{I}}\}, \qquad \text{(inverse role)}$$
$$\bot^{\mathcal{I}} = \emptyset, \qquad \text{(the empty set)}$$
$$(\geq q\, R)^{\mathcal{I}} = \{x \in \Delta^{\mathcal{I}} \mid \sharp\{y \in \Delta^{\mathcal{I}} \mid (x, y) \in R^{\mathcal{I}}\} \geq q\}, \quad \text{('at least } q \text{ } R\text{-successors')}$$
$$(\neg C)^{\mathcal{I}} = \Delta^{\mathcal{I}} \setminus C^{\mathcal{I}}, \qquad \text{('not in } C\text{')}$$
$$(C_1 \sqcap C_2)^{\mathcal{I}} = C_1^{\mathcal{I}} \cap C_2^{\mathcal{I}}, \qquad \text{('both in } C_1 \text{ and } C_2\text{')}$$

where $\sharp X$ denotes the cardinality of X. We also use the standard abbreviations: $C_1 \sqcup C_2 := \neg(\neg C_1 \sqcap \neg C_2)$, $\top := \neg \bot$, $\exists R := (\geq 1\, R)$ and $\leq q\, R := \neg(\geq q+1\, R)$.

The *satisfaction relation* $\models$ is also standard: $\mathcal{I} \models C_1 \sqsubseteq C_2$ iff $C_1^{\mathcal{I}} \subseteq C_2^{\mathcal{I}}$, $\mathcal{I} \models R_1 \sqsubseteq R_2$ iff $R_1^{\mathcal{I}} \subseteq R_2^{\mathcal{I}}$, $\mathcal{I} \models A_k(a_i)$ iff $a_i^{\mathcal{I}} \in A_k^{\mathcal{I}}$, and $\mathcal{I} \models P_k(a_i, a_j)$ iff $(a_i^{\mathcal{I}}, a_j^{\mathcal{I}}) \in P_k^{\mathcal{I}}$. A knowledge base (KB) $\mathcal{K} = (\mathcal{T}, \mathcal{A})$ is said to be *satisfiable* if there is an interpretation satisfying all the members of $\mathcal{T}$ and $\mathcal{A}$; such an interpretation is called a *model of $\mathcal{K}$*.

We will consider restrictions of $DL\text{-}Lite_{bool}^{\mathcal{R},\mathcal{N}}$ along three axes: Boolean operators $(_{bool})$ on concepts, number restrictions $(\mathcal{N})$ and role inclusions $(\mathcal{R})$. A $DL\text{-}Lite_{bool}^{\mathcal{R},\mathcal{N}}$ TBox $\mathcal{T}$ is a *Krom* TBox if its concept inclusions are of the form:

$$B_1 \sqsubseteq B_2 \quad \text{or} \quad B_1 \sqsubseteq \neg B_2 \quad \text{or} \quad \neg B_1 \sqsubseteq B_2, \qquad \text{(Krom)}$$

where the B_i are basic concepts. $\mathcal{T}$ is called a *Horn* TBox if its concept inclusions are of the form:

$$B_1 \sqcap \cdots \sqcap B_n \sqsubseteq B. \qquad \text{(Horn)}$$

We use $\sqcap_k B_k \sqsubseteq \sqcap_i B_i'$ as an abbreviation for the set of inclusions $\sqcap_k B_k \sqsubseteq B_i'$. Finally, we call $\mathcal{T}$ a *core* TBox if its concept inclusions are of the form:

$$B_1 \sqsubseteq B_2 \quad \text{or} \quad B_1 \sqsubseteq \neg B_2. \qquad \text{(core)}$$

As $B_1 \sqsubseteq \neg B_2$ is equivalent to $B_1 \sqcap B_2 \sqsubseteq \bot$, core TBoxes can be regarded as sitting precisely in the intersection of Krom and Horn TBoxes.

The fragments of $DL\text{-}Lite_{bool}^{\mathcal{R},\mathcal{N}}$ with Krom, Horn and core TBoxes will be denoted by $DL\text{-}Lite_{krom}^{\mathcal{R},\mathcal{N}}$, $DL\text{-}Lite_{horn}^{\mathcal{R},\mathcal{N}}$ and $DL\text{-}Lite_{core}^{\mathcal{R},\mathcal{N}}$, respectively.

Let $\alpha \in \{core, krom, horn, bool\}$. Denote by $DL\text{-}Lite_{\alpha}^{\mathcal{R},\mathcal{F}}$ the fragment of $DL\text{-}Lite_{\alpha}^{\mathcal{R},\mathcal{N}}$ in which number restrictions can occur only in *functionality constraints* of the form $\geq 2\,R \sqsubseteq \bot$ ($R^{\mathcal{I}}$ is functional if $(x,y),(x,z) \in R^{\mathcal{I}}$ implies $y = z$). The fragment of $DL\text{-}Lite_{\alpha}^{\mathcal{R},\mathcal{N}}$ without number restrictions, i.e., concepts of the form $\geq q\,R$, for $q > 1$, is denoted by $DL\text{-}Lite_{\alpha}^{\mathcal{R}}$. The fragments obtained by omitting role inclusions—that is, $DL\text{-}Lite_{\alpha}^{\mathcal{N}}$ (with arbitrary number restrictions), $DL\text{-}Lite_{\alpha}^{\mathcal{F}}$ (with functionality constraints only), and $DL\text{-}Lite_{\alpha}$ (without number restrictions)—have been analysed in [2]. Note that our notation is somewhat different from the original one; cf. [6,7,10,8,2].

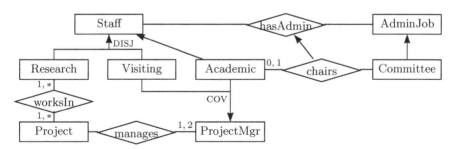

Fig. 1. EER diagram

We illustrate the expressiveness of the *DL-Lite* logics introduced above by considering the EER diagram in Fig. 1. It can be represented in the *DL-Lite* syntax using TBox axioms of the following types:

– $\exists\mathsf{manages} \sqsubseteq \mathsf{ProjectMgr}$ and $\exists\mathsf{manages}^- \sqsubseteq \mathsf{Project}$ to define the domain and range of the relationship 'manages';

- Project $\sqsubseteq\,\geq 1\,\mathsf{manages}^-$ and $\top \sqsubseteq\,\leq 2\,\mathsf{manages}^-$ to impose the cardinality constraints on 'manages';
- Committee $\sqsubseteq$ AdminJob to define the class hierarchy;
- Research $\sqcap$ Visiting $\sqsubseteq \bot$ and ProjectMgr $\sqsubseteq$ Academic $\sqcup$ Visiting to impose the disjointness and covering constraints;
- chairs $\sqsubseteq$ hasAdmin to define the relationship hierarchy.

Note that relation hierarchies can only be expressed in the languages $DL\text{-}Lite_\alpha^\mathcal{R}$ (and their extensions), covering constraints in $DL\text{-}Lite_{bool}$ (and its extensions), cardinality constraints '1,*' in all languages, constraints of the form '0,1' and '1,1' in $DL\text{-}Lite_\alpha^\mathcal{F}$, and arbitrary cardinality constraints in $DL\text{-}Lite_\alpha^\mathcal{N}$.

We will concentrate on three standard reasoning tasks for our logics $\mathcal{L}$:

- *satisfiability*: given an $\mathcal{L}$-KB $\mathcal{K}$, decide whether $\mathcal{K}$ is satisfiable;
- *instance checking*: given an object name a, a basic concept B and an $\mathcal{L}$-KB $\mathcal{K}$, decide whether $a^\mathcal{I} \in B^\mathcal{I}$ whenever $\mathcal{I} \models \mathcal{K}$;
- *query answering*: given a positive existential query $q(\boldsymbol{x})$, an $\mathcal{L}$-KB $\mathcal{K}$ and a tuple $\boldsymbol{a}$ of object names from its ABox, decide whether $\mathcal{K} \models q(\boldsymbol{a})$.

As is well known, many other reasoning tasks for description logics are LOGSPACE reducible to the satisfiability problem; for details see [2]. In particular, this is true of instance checking: an object a is an instance of concept B in every model of $\mathcal{K} = (\mathcal{T}, \mathcal{A})$ iff the KB $(\mathcal{T} \cup \{A_{\neg B} \sqsubseteq \neg B\},\ \mathcal{A} \cup \{A_{\neg B}(a)\})$ is not satisfiable, where $A_{\neg B}$ is a fresh concept name.

Our aim is to investigate (i) the *combined complexity* of the satisfiability problem for the logics of our family, where the whole KB $\mathcal{K}$ is regarded as an input, and (ii) the *data complexity* (or *ABox complexity*) of the instance checking and query answering problems, where the given TBox is assumed to be fixed, while the input ABox can vary.

3 $DL\text{-}Lite_{bool}^\mathcal{R}$ and First-Order Logic with One Variable

First we consider the logic $DL\text{-}Lite_{bool}^\mathcal{R}$ and its fragments. The key observation which clearly explains their computational behaviour is that the satisfiability problem for $DL\text{-}Lite_{bool}^\mathcal{R}$ knowledge bases is LOGSPACE reducible to the satisfiability problem for the *one-variable fragment* $\mathcal{QL}^1$ of first-order logic (without equality and function symbols) and that this reduction preserves the properties of core, Krom, or Horn formulas.

Let $\mathcal{K} = (\mathcal{T}, \mathcal{A})$ be a $DL\text{-}Lite_{bool}^\mathcal{R}$ KB. Denote by $role(\mathcal{K})$ the set of role names occurring in $\mathcal{T}$ and $\mathcal{A}$, by $role^\pm(\mathcal{K})$ the set $\{P_k, P_k^- \mid P_k \in role(\mathcal{K})\}$, and by $ob(\mathcal{A})$ the set of object names in $\mathcal{A}$.

With every $a_i \in ob(\mathcal{A})$ we associate the individual constant a_i of $\mathcal{QL}^1$ and with every concept name A_i the unary predicate $A_i(x)$ from the signature of $\mathcal{QL}^1$. For each pair of roles $P_k, P_k^- \in role^\pm(\mathcal{K})$, we introduce a pair of fresh unary predicates $EP_k(x)$ and $EP_k^-(x)$, which will represent the domain and range of P_k, respectively (in other words, $EP_k(x)$ and $EP_k^-(x)$ are the sets of points with *at least one* P_k-successor and *at least one* P_k-predecessor, respectively). Additionally, for each pair of roles $P_k, P_k^- \in role^{\pm}(\mathcal{K})$, we take a pair of fresh individual constants dp_k and dp_k^- of $\mathcal{QL}^1$, which will serve as 'representatives' of the points from the domains of P_k and P_k^- (provided that they are not empty). Furthermore, for each pair $a_i, a_j \in ob(\mathcal{A})$ and each $R \in role^{\pm}(\mathcal{K})$, we take a fresh *propositional variable* Ra_ia_j of $\mathcal{QL}^1$ to encode $R(a_i, a_j)$. By induction on the construction of a *DL-Lite$_{bool}^{\mathcal{R}}$* concept C we define the $\mathcal{QL}^1$-formula C^*:

$$\bot^* = \bot, \qquad (A_i)^* = A_i(x), \qquad (\exists R)^* = ER(x),$$
$$(\neg C)^* = \neg C^*(x), \qquad (C_1 \sqcap C_2)^* = C_1^*(x) \wedge C_2^*(x).$$

A *DL-Lite$_{bool}^{\mathcal{R}}$* TBox $\mathcal{T}$ corresponds then to the $\mathcal{QL}^1$-sentence:

$$\mathcal{T}^* = \bigwedge_{C_1 \sqsubseteq C_2 \in \mathcal{T}} \forall x \left(C_1^*(x) \rightarrow C_2^*(x) \right) \quad \wedge$$
$$\bigwedge_{R_1 \sqsubseteq R_2 \in \mathcal{T}} \left[\forall x \left(ER_1(x) \rightarrow ER_2(x) \right) \wedge \forall x \left(inv(ER_1)(x) \rightarrow inv(ER_2)(x) \right) \right],$$

where $inv(ER) = EP_k^-$ if $R = P_k$ and $inv(ER) = EP_k$ if $R = P_k^-$. For every role $R \in role^{\pm}(\mathcal{K})$, we also need the following $\mathcal{QL}^1$-sentence:

$$\varepsilon(R) = \forall x \left(ER(x) \rightarrow inv(ER)(inv(dr)) \right),$$

where $inv(dr) = dp_k^-$ if $R = P_k$ and $inv(dr) = dp_k$ if $R = P_k^-$. This sentence says that if the domain of R is not empty then its range is not empty either: it contains the representative $inv(dr)$.

It should be clear how to translate a *DL-Lite$_{bool}^{\mathcal{R}}$* ABox $\mathcal{A}$ into $\mathcal{QL}^1$:

$$\mathcal{A}^{\dagger} = \bigwedge_{A(a_i) \in \mathcal{A}} A(a_i) \quad \wedge \bigwedge_{P(a_i, a_j) \in \mathcal{A}} Pa_ia_j \quad \wedge \bigwedge_{\substack{R \in role^{\pm}(\mathcal{K}) \\ a_i, a_j \in ob(\mathcal{A})}} (Ra_ia_j \rightarrow inv(R)a_ja_i),$$

where $inv(R)a_ja_i$ is the propositional variable $P_k^- a_ja_i$ if $R = P_k$ and $P_ka_ja_i$ if $R = P_k^-$. Finally, for the *DL-Lite$_{bool}^{\mathcal{R}}$* knowledge base $\mathcal{K} = (\mathcal{T}, \mathcal{A})$, we set

$$\mathcal{K}^{\dagger} = \left[\mathcal{T}^* \quad \wedge \bigwedge_{R \in role^{\pm}(\mathcal{K})} \varepsilon(R) \right] \quad \wedge \left[\mathcal{A}^{\dagger} \quad \wedge \bigwedge_{\substack{R \in role^{\pm}(\mathcal{K}) \\ a_i, a_j \in ob(\mathcal{A})}} (Ra_ia_j \rightarrow ER(a_i)) \right].$$

Lemma 1. *A DL-Lite$_{bool}^{\mathcal{R}}$ knowledge base $\mathcal{K}$ is satisfiable iff the $\mathcal{QL}^1$-sentence $\mathcal{K}^{\dagger}$ is satisfiable.*

Proof. Every model for $\mathcal{K}$ gives rise to a model for $\mathcal{K}^{\dagger}$ in the obvious way. The converse can be proved by 'unravelling' a first-order model for $\mathcal{K}^{\dagger}$ similarly to the unravelling construction in [2]. We only note here the main difference from that construction for *DL-Lite$_{bool}^{\mathcal{N}}$*: as *DL-Lite$_{bool}^{\mathcal{R}}$* has no number restrictions, one can create as many R-successors to a point as required without violating the TBox axioms (it was not the case for $DL\text{-}Lite_{bool}^{\mathcal{N}}$; on the other hand, the latter does not have role inclusions, which may force additional R-successors).   □

As $\cdot^{\dagger}$ is computable in LogSpace and $\mathcal{K}^{\dagger}$ is a universal sentence, we can use the known complexity results for the relevant fragments of $\mathcal{QL}^1$ (see, e.g., [4]):

Theorem 1. *The satisfiability problem is* NLogSpace-*complete for* $DL\text{-}Lite_{core}^{\mathcal{R}}$ *and* $DL\text{-}Lite_{krom}^{\mathcal{R}}$, P-*complete for* $DL\text{-}Lite_{horn}^{\mathcal{R}}$ *and* NP-*complete for* $DL\text{-}Lite_{bool}^{\mathcal{R}}$.

Now we show that as far as *data complexity* is concerned, satisfiability of $DL\text{-}Lite_{bool}^{\mathcal{R}}$ KBs can be solved using only *logarithmic space* in the size of the ABox. In what follows, without loss of generality, we assume that all role names of a given KB $\mathcal{K} = (\mathcal{T}, \mathcal{A})$ occur in its TBox and write $role^{\pm}(\mathcal{T})$ instead of $role^{\pm}(\mathcal{K})$. Let $\Sigma(\mathcal{T}) = \{ER(dr) \mid R \in \dot{role}^{\pm}(\mathcal{T})\}$ and, for $\Sigma_0 \subseteq \Sigma(\mathcal{T})$, let

$$core_{\Sigma_0}(\mathcal{T}) = \bigwedge_{ER(dr)\in\Sigma_0} ER(dr) \quad \wedge \bigwedge_{R\in role^{\pm}(\mathcal{T})} \left(\mathcal{T}^*[dr] \wedge \bigwedge_{R'\in role^{\pm}(\mathcal{T})} \varepsilon(R')[dr] \right),$$

$$proj_{\Sigma_0}(\mathcal{K}, a) = \bigwedge_{inv(ER)(inv(dr))\in\Sigma(\mathcal{T})\setminus\Sigma_0} \neg ER(a) \quad \wedge \quad \mathcal{T}^*[a] \quad \wedge \quad \mathcal{A}^\flat(a),$$

where $\mathcal{T}^*[c]$ and $\varepsilon(R')[c]$ are instantiations of the universal quantifier in the respective formulas with the constant c, and

$$\mathcal{A}^\flat(a) = \bigwedge_{A(a)\in\mathcal{A}} A(a) \quad \wedge \bigwedge_{P_k\in role(\mathcal{K})} \left[\bigwedge_{P_k(a,a_1)\in\mathcal{A}} EP_k(a) \wedge \bigwedge_{P_k(a_1,a)\in\mathcal{A}} EP_k^-(a) \right].$$

Lemma 2. $\mathcal{K}$ *is satisfiable iff there is a subset* Σ_0 *of* $\Sigma(\mathcal{T})$ *such that* $core_{\Sigma_0}(\mathcal{T})$ *is satisfiable and* $proj_{\Sigma_0}(\mathcal{K}, a)$ *is satisfiable for every* $a \in ob(\mathcal{A})$.

Proof. If $\mathcal{I} \models \mathcal{K}$, then we take $\Sigma_0 = \{ER(dr) \mid R \in role^{\pm}(\mathcal{T}), (\exists R)^{\mathcal{I}} \neq \emptyset\}$ and the first-order model $\mathfrak{M}$ induced by $\mathcal{I}$. It should be clear that we have $\mathfrak{M} \models core_{\Sigma_0}(\mathcal{T})$ and $\mathfrak{M} \models proj_{\Sigma_0}(\mathcal{K}, a)$, for all $a \in ob(\mathcal{A})$.

Conversely, let $\mathfrak{M}_{\Sigma_0}$ be an Herbrand model of $core_{\Sigma_0}(\mathcal{T})$ and $\mathfrak{M}_a$ an Herbrand model of $proj_{\Sigma_0}(\mathcal{K}, a)$, for $a \in ob(\mathcal{A})$. By definition, the domain of $\mathfrak{M}_{\Sigma_0}$ consists of $|role^{\pm}(\mathcal{T})|$ elements and the domains of the $\mathfrak{M}_a$ are singletons. Clearly, $\mathfrak{M}_{\Sigma_0} \models \mathcal{T}^*$ and $\mathfrak{M}_{\Sigma_0} \models \varepsilon(R)$, for every $R \in role^{\pm}(\mathcal{T})$, and $\mathfrak{M}_a \models \mathcal{T}^*$, for every $a \in ob(\mathcal{A})$. We construct a model $\mathfrak{M}$ by taking the disjoint union of $\mathfrak{M}_{\Sigma_0}$ with all of the $\mathfrak{M}_a$, where we set $P_k a_i a_j^{\mathfrak{M}}$ to be true iff $P_{k'}(a_i, a_j) \in \mathcal{A}$ or $P_{k'}^-(a_j, a_i) \in \mathcal{A}$ for a sub-role $P_{k'}$ of P_k. Let us show that $\mathfrak{M} \models \mathcal{K}^{\dagger}$. We have $\mathfrak{M} \models \mathcal{T}^*$ because $\mathcal{T}^*$ is universal, does not contain constants and is true in every component model. Consider now $\varepsilon(R) = \forall x\, \psi(x)$, where $\psi(x) = (ER(x) \to inv(ER)(inv(dr)))$. We show that, for every d in the domain of $\mathfrak{M}$, we have $\mathfrak{M} \models \psi[d]$. If d is of the form $dr'^{\mathfrak{M}}$, for some $R' \in role^{\pm}(\mathcal{T})$, then clearly $\mathfrak{M} \models \psi[d]$, since $\mathfrak{M}_{\Sigma_0} \models \varepsilon(R)$. If d is of the form $a^{\mathfrak{M}}$, for $a \in ob(\mathcal{A})$, then it trivially holds if $\mathfrak{M}_a \not\models ER(a)$. Otherwise, $\mathfrak{M}_a \models ER(a)$, and so $inv(ER)(inv(dr)) \notin \Sigma(\mathcal{T}) \setminus \Sigma_0$. Therefore, $\mathfrak{M} \models inv(ER)(inv(dr))$ and $\mathfrak{M} \models \psi[d]$. And we clearly have $\mathfrak{M} \models R^{\dagger} \wedge \mathcal{A}^{\dagger}$.   □

This lemma states that satisfiability of a $DL\text{-}Lite^{\mathcal{R}}_{bool}$ KB can be checked locally: one guesses which roles are empty and which are not (i.e., the set Σ_0) and then checks whether each object in the ABox (independently of the others) satisfies the TBox and the role emptiness constraints Σ_0. This observation suggests a high degree of parallelism in the satisfiability check:

Theorem 2. *The data complexity of the satisfiability and instance checking problems for DL-Lite$^{\mathcal{R}}_{bool}$ knowledge bases is in* LOGSPACE.

Proof. Follows from two observations: (i) the size of $core_{\Sigma_0}(\mathcal{T})$ and $proj_{\Sigma_0}(\mathcal{K}, a)$ does not depend on $|\mathcal{A}|$, and (ii) $proj_{\Sigma_0}(\mathcal{K}, a)$ can be computed using extra $\log |\mathcal{A}|$ memory cells whereas $core_{\Sigma_0}(\mathcal{T})$ does not depend on $\mathcal{A}$ at all. $\qquad\square$

In fact, one can improve the above result by showing that the data complexity of satisfiability for $DL\text{-}Lite^{\mathcal{R}}_{bool}$ KBs belongs to the parallel complexity class AC_0 (see, e.g., [17]): $core_{\Sigma_0}(\mathcal{T})$ and $proj_{\Sigma_0}(\mathcal{K}, a)$ can be realised by unbounded fan-in circuits with AND, OR and NOT gates, whose size is polynomial in $|\mathcal{A}|$ and depth does not depend on $|\mathcal{A}|$.

Let us now extend $DL\text{-}Lite^{\mathcal{R}}_{bool}$ and its fragments with number restrictions.

4   Satisfiability: $DL\text{-}Lite^{\mathcal{R},\mathcal{F}}_{core}$ Is ExpTime-Hard

As follows from [12, Theorem 12], satisfiability of $DL\text{-}Lite^{\mathcal{R},\mathcal{N}}_{bool}$ knowledge bases can be decided in EXPTIME. Our aim is to show that this upper bound cannot be improved even for the seemingly rather weak language $DL\text{-}Lite^{\mathcal{R},\mathcal{F}}_{core}$. We need the following observation showing that in certain cases in the core and Krom languages we can actually use intersections in the left-hand side of concept inclusions, which is not allowed by the syntax of $DL\text{-}Lite^{\mathcal{R},\mathcal{F}}_{core}$.

Suppose that a knowledge base $\mathcal{K}$ contains a concept inclusion of the form $A_1 \sqcap A_2 \sqsubseteq C$. Define a new KB $\mathcal{K}'$ by replacing this axiom in $\mathcal{K}$ with the following set of new axioms, where $R_1, R_2, R_3, R_{12}, R_{23}$ are fresh role names:

$$
\begin{array}{llr}
A_1 \sqsubseteq \exists R_1 & A_2 \sqsubseteq \exists R_2, & (1)\\
R_1 \sqsubseteq R_{12}, & R_2 \sqsubseteq R_{12}, & \geq 2\,R_{12} \sqsubseteq \bot, \quad (2)\\
\exists R_1^- \sqsubseteq \exists R_3^-, & \exists R_3 \sqsubseteq C, & (3)\\
R_3 \sqsubseteq R_{23}, & R_2 \sqsubseteq R_{23}, & \geq 2\,R_{23}^- \sqsubseteq \bot. \quad (4)
\end{array}
$$

Lemma 3. (i) *If $\mathcal{I} \models \mathcal{K}'$ then $\mathcal{I} \models \mathcal{K}$, for every interpretation $\mathcal{I}$.*

(ii) *If $\mathcal{I} \models \mathcal{K}$ and $C^{\mathcal{I}} \neq \emptyset$ then there is an extension $\mathcal{I}'$ of $\mathcal{I}$ such that it agrees with $\mathcal{I}$ on every symbol of $\mathcal{K}$ and $\mathcal{I}' \models \mathcal{K}'$.*

Proof. (i) Let $\mathcal{I} \models \mathcal{K}'$ and $x \in A_1^{\mathcal{I}} \cap A_2^{\mathcal{I}}$. By (1), there is y with $(x, y) \in R_1^{\mathcal{I}}$, and so $y \in (\exists R_1^-)^{\mathcal{I}}$, and there is z with $(x, z) \in R_2^{\mathcal{I}}$. By (2), $(x, y), (x, z) \in R_{12}^{\mathcal{I}}$ and thus $y = z$. By (3), $y \in (\exists R_3^-)^{\mathcal{I}}$ and then there is u with $(u, y) \in R_3^{\mathcal{I}}$ and $u \in (\exists R_3)^{\mathcal{I}}$, whence $u \in C^{\mathcal{I}}$ and, by (4), $(u, y) \in R_{23}^{\mathcal{I}}$, and we also have $(x, y) \in R_{23}^{\mathcal{I}}$. It follows from (4) that $u = x$; so $x \in C^{\mathcal{I}}$. Thus, $\mathcal{I} \models \mathcal{K}$.

(ii) Take some point $c \in C^{\mathcal{I}}$ and define an extension $\mathcal{I}'$ of $\mathcal{I}$ to the new role names by setting $R_1^{\mathcal{I}'} = \{(x,x) \mid x \in A_1^{\mathcal{I}}\}$, $R_2^{\mathcal{I}'} = \{(x,x) \mid x \in A_2^{\mathcal{I}}\}$, $R_3^{\mathcal{I}'} = \{(x,x) \mid x \in (A_1 \sqcap A_2)^{\mathcal{I}}\} \cup \{(c,x) \mid x \in (A_1 \sqcap \neg A_2)^{\mathcal{I}}\}$, $R_{12}^{\mathcal{I}'} = R_1^{\mathcal{I}'} \cup R_2^{\mathcal{I}'}$ and $R_{23}^{\mathcal{I}'} = R_2^{\mathcal{I}'} \cup R_3^{\mathcal{I}'}$. It is readily seen that $\mathcal{I}' \models \mathcal{K}'$. ❑

We are now in a position to prove the following:

Theorem 3. *The satisfiability problem for DL-Lite$_{core}^{\mathcal{R},\mathcal{F}}$ KBs is* ExpTime-*hard.*

Proof. First we show how to encode polynomial-space-bounded *alternating Turing machines* (ATMs) by means of DL-Lite$_{horn}^{\mathcal{R},\mathcal{F}}$ KBs. As APSpace = ExpTime, where APSpace is the class of problems accepted by polynomial-space-bounded ATMs (see, e.g., [15]), this will establish ExpTime-hardness of satisfiability for DL-Lite$_{horn}^{\mathcal{R},\mathcal{F}}$. And then we will use Lemma 3 to get rid of the conjunctions in the left-hand side of the concept inclusions involved in this encoding of ATMs.

Without loss of generality, we can only consider ATMs $\mathcal{M}$ with *binary* computational trees. This means that, for every non-halting state q and every symbol a from the tape alphabet, $\mathcal{M}$ has precisely two instructions of the form $(q,a) \leadsto_{\mathcal{M}}^0 (q',a',d')$ and $(q,a) \leadsto_{\mathcal{M}}^1 (q'',a'',d'')$, where $d',d'' \in \{\rightarrow,\leftarrow\}$ and $\rightarrow$ (respectively, $\leftarrow$) means 'move the head right (left) one cell.' We remind the reader that each non-halting state of $\mathcal{M}$ is either an *and-state* or an *or-state*.

Given such an ATM $\mathcal{M}$, a polynomial function $p(n)$ such that any run of $\mathcal{M}$ on any input of length n uses $\leq p(n)$ tape cells, and an input word $\boldsymbol{a} = a_1,\ldots,a_n$, we construct a DL-Lite$_{horn}^{\mathcal{R},\mathcal{F}}$ knowledge base $\mathcal{K}_{\mathcal{M},\boldsymbol{a}}$ with the following properties: (i) the size of $\mathcal{K}_{\mathcal{M},\boldsymbol{a}}$ is polynomial in the size of $\mathcal{M}$, $\boldsymbol{a}$, and (ii) $\mathcal{M}$ accepts $\boldsymbol{a}$ iff $\mathcal{K}_{\mathcal{M},\boldsymbol{a}}$ is not satisfiable. Denote by Q the set of states and by Σ the tape alphabet of $\mathcal{M}$. To encode the instructions of $\mathcal{M}$, we need the following roles:

- S_q, S_q^0, S_q^1, for each $q \in Q$: informally, $x \in \exists S_q^-$ means that x represents a configuration of $\mathcal{M}$ with the state q, and $x \in \exists S_q^k$ that the next state, according to the transition $\leadsto_{\mathcal{M}}^k$, is q, where $k = 0,1$;
- H_i, H_i^0, H_i^1, for each $i \leq p(n)$: informally, $x \in \exists H_i^-$ means that x represents a configuration of $\mathcal{M}$ where the head scans the ith cell, and $x \in \exists H_i^k$ that, according to the transition $\leadsto_{\mathcal{M}}^k$, $k = 0,1$, in the next configuration the head scans the ith cell;
- $C_{ia}, C_{ia}^0, C_{ia}^1$, for all $i \leq p(n)$ and $a \in \Sigma$: informally, $x \in \exists C_{ia}^-$ means that x represents a configuration of $\mathcal{M}$ where the ith cell contains a, and $x \in \exists C_{ia}^k$ that, according to $\leadsto_{\mathcal{M}}^k$, in the next configuration the ith cell contains a.

This intended meaning can be encoded using the following TBox axioms: for every instruction $(q,a) \leadsto_{\mathcal{M}}^k (q',a',\rightarrow)$ of $\mathcal{M}$ and every $i < p(n)$,

$$\exists S_q^- \sqcap \exists H_i^- \sqcap \exists C_{ia}^- \sqsubseteq \exists H_{i+1}^k \sqcap \exists S_{q'}^k \sqcap \exists C_{ia'}^k, \tag{5}$$

and for every instruction $(q,a) \leadsto_{\mathcal{M}}^k (q',a',\leftarrow)$ of $\mathcal{M}$ and every i, $1 < i \leq p(n)$,

$$\exists S_q^- \sqcap \exists H_i^- \sqcap \exists C_{ia}^- \sqsubseteq \exists H_{i-1}^k \sqcap \exists S_{q'}^k \sqcap \exists C_{ia'}^k, \tag{6}$$

To preserve the symbols on the tape that are not in the active cell, we use the following axioms, for $k = 0, 1$, $i, j \leq p(n)$ with $j \neq i$, and $a \in \Sigma$:

$$\exists H_j^- \sqcap \exists C_{ia}^- \sqsubseteq \exists C_{ia}^k \tag{7}$$

and to uniquely identify the head position, state and content of each cell, for $i, j \leq p(n)$ with $i \neq j$, $q', q'' \in Q$ with $q' \neq q''$ and $a', a'' \in \Sigma$ with $a' \neq a''$:

$$\exists H_i^- \sqcap \exists H_j^- \sqsubseteq \bot, \qquad \exists S_{q'}^- \sqcap \exists S_{q''}^- \sqsubseteq \bot, \qquad \exists C_{ia'}^- \sqcap \exists C_{ia''}^- \sqsubseteq \bot. \tag{8}$$

To 'synchronise' our roles, we need two more (functional) roles T_0 and T_1 to represent the 0- and 1-successors of a configuration, and a number of role inclusions are added to the TBox: for all $k = 0, 1$, $i \leq p(n)$, $q \in Q$ and $a \in \Sigma$,

$$C_{ia}^k \sqsubseteq C_{ia}, \qquad H_i^k \sqsubseteq H_i, \qquad S_q^k \sqsubseteq S_q, \tag{9}$$

$$C_{ia}^k \sqsubseteq T_k, \qquad H_i^k \sqsubseteq T_k, \qquad S_q^k \sqsubseteq T_k, \qquad \geq 2\, T_k \sqsubseteq \bot. \tag{10}$$

It remains to encode the acceptance conditions for $\mathcal{M}$ on a. This can be done with the help of role names Y_0, Y_1 and concept names A, D: for $q \in Q$, $k = 0, 1$,

$$\exists S_q^- \sqsubseteq A, \quad q \text{ an accepting state,} \tag{11}$$

$$Y_k \sqsubseteq T_k, \qquad \geq 2\, T_k^- \sqsubseteq \bot, \qquad \exists T_k^- \sqcap A \sqsubseteq \exists Y_k^-, \tag{12}$$

$$\exists S_q^- \sqcap \exists Y_k \sqsubseteq A, \quad q \text{ an or-state,} \tag{13}$$

$$\exists S_q^- \sqcap \exists Y_0 \sqcap \exists Y_1 \sqsubseteq A, \quad q \text{ an and-state,} \tag{14}$$

$$A \sqcap D \sqsubseteq \bot. \tag{15}$$

The TBox $\mathcal{T}$ of the *DL-Lite*$_{horn}^{\mathcal{R},\mathcal{F}}$ knowledge base $\mathcal{K}_{\mathcal{M},a}$ we are constructing consists of axioms (5)–(15). The ABox $\mathcal{A}$ of $\mathcal{K}_{\mathcal{M},a}$ is comprised of the following assertions, for some object name s:

$$s \colon \exists S_{q_0}^-, \quad s \colon \exists H_1^-, \quad s \colon \exists C_{ia_i}^-, \text{ for } i \leq p(n), \quad \text{and} \quad s \colon D, \tag{16}$$

where q_0 is the initial state and a_i the ith symbol on the input tape, $i \leq p(n)$. Clearly, $\mathcal{K}_{\mathcal{M},a} = (\mathcal{T}, \mathcal{A})$ is a *DL-Lite*$_{horn}^{\mathcal{R},\mathcal{F}}$ KB and its size is polynomial in the size of $\mathcal{M}$, a. The proof of the following lemma is routine and left to the reader.

Lemma 4. *The ATM $\mathcal{M}$ accepts a iff the KB $\mathcal{K}_{\mathcal{M},a}$ is not satisfiable.*

Before applying Lemma 3 to eliminate the conjunctions in the left-hand side of (5)–(7), (12)–(14), one has to show that if $\mathcal{K}_{\mathcal{M},a}$ is satisfiable then it is satisfiable in an interpretation $\mathcal{I}$ with $\mathcal{I} \models \mathcal{K}_{\mathcal{M},a}$ and $C_2^{\mathcal{I}} \neq \emptyset$, for any C_2 occurring in an axiom of the form $C_0 \sqcap C_1 \sqsubseteq C_2$ in $\mathcal{K}_{\mathcal{M},a}$. Details are left to the reader. $\quad\square$

5   Instance Checking with Number Restrictions

Theorem 4. *The instance checking problem (and query answering problem) for DL-Lite*$_{krom}^{\mathcal{R},\mathcal{F}}$ *is data complete for* coNP.

Proof. The CONP upper bound follows from [12, Theorem 12]. We prove the matching lower bound by reduction of the non-satisfiability problem for 2+2CNF, which is known to be CONP-complete [18]. Given a 2+2CNF

$$\varphi \; = \; \bigwedge_{k=1}^{n}(a_{k,1} \vee a_{k,2} \vee \neg a_{k,3} \vee \neg a_{k,4}),$$

where each $a_{k,j}$ is one of the propositional variables $a_1, \ldots, a_m$, we construct a DL-Lite$_{krom}^{\mathcal{R},\mathcal{F}}$ KB $(\mathcal{T}, \mathcal{A}_\varphi)$ whose TBox $\mathcal{T}$ does not depend on φ and ABox $\mathcal{A}_\varphi$ is a linear encoding of φ. We will use the object names $f, c_1, \ldots, c_n$ and $a_1, \ldots, a_m$, role names $S, S_{\mathbf{f}}$ and $P_j, P_{j,\mathbf{t}}, P_{j,\mathbf{f}}$, for $1 \leq j \leq 4$, and concept names A, D.

Define $\mathcal{A}_\varphi$ to be the set of the following assertions, for $1 \leq k \leq n$:

$$S(f, c_k), \quad P_1(c_k, a_{k,1}), \quad P_2(c_k, a_{k,2}), \quad P_3(c_k, a_{k,3}), \quad P_4(c_k, a_{k,4}),$$

and let $\mathcal{T}$ consist of the axioms:

$$
\begin{aligned}
\geq 2\, P_j &\sqsubseteq \bot, & &\text{for } 1 \leq j \leq 4, & (17) \\
P_{j,\mathbf{f}} \sqsubseteq P_j, \qquad P_{j,\mathbf{t}} &\sqsubseteq P_j, & &\text{for } 1 \leq j \leq 4, & (18) \\
\neg \exists P_{j,\mathbf{t}} &\sqsubseteq \exists P_{j,\mathbf{f}}, & &\text{for } 1 \leq j \leq 4, & (19) \\
\exists P_{j,\mathbf{f}}^{-} \sqsubseteq \neg A, \qquad \exists P_{j,\mathbf{t}}^{-} &\sqsubseteq A, & &\text{for } 1 \leq j \leq 4, & (20) \\
\exists P_{1,\mathbf{f}} \sqcap \exists P_{2,\mathbf{f}} \sqcap \exists P_{3,\mathbf{t}} \sqcap \exists P_{4,\mathbf{t}} &\sqsubseteq \exists S_{\mathbf{f}}^{-}, & & & (21) \\
\geq 2\, S^{-} \sqsubseteq \bot, \qquad S_{\mathbf{f}} \sqsubseteq S, \qquad \exists S_{\mathbf{f}} &\sqsubseteq D. & & & (22)
\end{aligned}
$$

It should be clear that $(\mathcal{T}, \mathcal{A}_\varphi)$ is LOGSPACE computable (in $|\varphi|$). Note, however, that axiom (21) does not belong to DL-Lite$_{krom}^{\mathcal{R},\mathcal{F}}$ because of the conjunctions in its left-hand side. However, they can be eliminated with the help of Lemma 3. So let us prove that $(\mathcal{T}, \mathcal{A}_\varphi) \models D(f)$ iff φ is not satisfiable.

Suppose first that φ is satisfiable. Then there is an assignment $\mathfrak{a}$ of the truth-values $\mathbf{t}$ and $\mathbf{f}$ to the propositional variables such that $\mathfrak{a}(a_{k,1}) = \mathbf{t}$ or $\mathfrak{a}(a_{k,2}) = \mathbf{t}$ or $\mathfrak{a}(a_{k,3}) = \mathbf{f}$ or $\mathfrak{a}(a_{k,4}) = \mathbf{f}$, for all $k \in \{1, \ldots, n\}$. Consider the interpretation $\mathcal{I}$ with $\Delta^{\mathcal{I}} = \{x_1, \ldots, x_m, y_1, \ldots, y_n, z\}$ and

- $f^{\mathcal{I}} = z$, $\quad c_k^{\mathcal{I}} = y_k$, for $1 \leq k \leq n$, $\quad a_i^{\mathcal{I}} = x_i$, for $1 \leq i \leq m$,
- $A^{\mathcal{I}} = \{x_i \mid \mathfrak{a}(a_i) = \mathbf{t}\} \cup \{y_k \mid 1 \leq k \leq n\} \cup \{z\}$,
  $P_{j,\mathbf{t}}^{\mathcal{I}} = \{(y_k, a_{k,j}^{\mathcal{I}}) \mid 1 \leq k \leq n, \; \mathfrak{a}(a_{k,j}) = \mathbf{t}\} \cup \{(x_i, x_i) \mid \mathfrak{a}(a_i) = \mathbf{t}\} \cup \{(z, z)\}$,
  $P_{j,\mathbf{f}}^{\mathcal{I}} = \{(y_k, a_{k,j}^{\mathcal{I}}) \mid 1 \leq k \leq n, \; \mathfrak{a}(a_{k,j}) = \mathbf{f}\} \cup \{(x_i, x_i) \mid \mathfrak{a}(a_i) = \mathbf{f}\}$,
  $P_j^{\mathcal{I}} = P_{j,\mathbf{t}}^{\mathcal{I}} \cup P_{j,\mathbf{f}}^{\mathcal{I}}$, $\quad$ for $1 \leq j \leq 4$,
- $S^{\mathcal{I}} = \{(z, y_k) \mid 1 \leq k \leq n\}$,
  $S_{\mathbf{f}}^{\mathcal{I}} = \{(z, y_k) \mid 1 \leq k \leq n, \mathfrak{a}(a_{k,1} \vee a_{k,2} \vee \neg a_{k,3} \vee \neg a_{k,4}) = \mathbf{f}\} = \emptyset$,
- $D^{\mathcal{I}} = \{z \mid \mathfrak{a}(\varphi) = \mathbf{f}\} = \emptyset$.

It is not hard to check that $\mathcal{I} \models (\mathcal{T}, \mathcal{A}_\varphi)$, and clearly $\mathcal{I} \not\models D(f)$.

Assume now that φ is not satisfiable and $\mathcal{I} \models (\mathcal{T}, \mathcal{A}_\varphi)$. Define an assignment $\mathfrak{a}$ by taking $\mathfrak{a}(a_i) = \mathbf{t}$ iff $a_i^{\mathcal{I}} \in A^{\mathcal{I}}$. As φ is not satisfiable, there is k, $1 \leq k \leq n$, such that $\mathfrak{a}(a_{k,1}) = \mathfrak{a}(a_{k,2}) = \mathbf{f}$, $\mathfrak{a}(a_{k,3}) = \mathfrak{a}(a_{k,4}) = \mathbf{t}$. In view of (19), for each j, $1 \leq j \leq 4$, we have $c_k^{\mathcal{I}} \in (\exists P_{j,\mathbf{t}})^{\mathcal{I}} \cup (\exists P_{j,\mathbf{f}})^{\mathcal{I}}$, and by (18), $c_k^{\mathcal{I}} \in (\exists P_j)^{\mathcal{I}}$. Therefore, by (17) and (20), $c_k^{\mathcal{I}} \in (\exists P_{j,\mathbf{t}})^{\mathcal{I}}$ if $\mathfrak{a}(a_{k,j}) = \mathbf{t}$ and $c_k^{\mathcal{I}} \in (\exists P_{j,\mathbf{f}})^{\mathcal{I}}$ if $\mathfrak{a}(a_{k,j}) = \mathbf{f}$, and hence, by (21), $c_k^{\mathcal{I}} \in (\exists S_{\mathbf{f}}^-)^{\mathcal{I}}$. Then by (22), we have $f^{\mathcal{I}} \in (\exists S_{\mathbf{f}})^{\mathcal{I}}$ and $f^{\mathcal{I}} \in D^{\mathcal{I}}$. It follows that $(\mathcal{T}, \mathcal{A}_\varphi) \models D(f)$. $\qquad\qquad\square$

If the functionality constraints are relaxed just a bit to allow for axioms of the form $\geq 2\,R \sqsubseteq A$ then the same complexity result holds for the core fragment:

Theorem 5. *The instance checking problem (and query answering problem) for DL-Lite$_{core}^{\mathcal{R},\mathcal{N}}$ is data complete for* CONP.

Proof. The CONP upper bound again follows from [12, Theorem 12], and the matching lower bound is proved by reduction of the non-satisfiability problem for 2+2CNF. The main difference from the previous proof is that DL-Lite$_{core}^{\mathcal{R},\mathcal{N}}$, unlike DL-Lite$_{krom}^{\mathcal{R},\mathcal{F}}$, cannot express 'covering conditions' like (19). It turns out, however, that we can use number restrictions to represent this kind of constraints. Given a 2+2CNF φ, we take the ABox $\mathcal{A}_\varphi$ constructed in the proof of Theorem 4 (and computable in LOGSPACE in $|\varphi|$). The (φ independent) DL-Lite$_{core}^{\mathcal{R},\mathcal{N}}$ TBox $\mathcal{T}$, describing the meaning of any such representation of 2+2CNF ψ in terms of $\mathcal{A}_\psi$, is also defined in the same way as in that proof except that axiom (19) is now replaced by the following set of axioms:

$$T_{j,1} \sqsubseteq T_j, \qquad\qquad T_{j,2} \sqsubseteq T_j, \qquad\qquad T_{j,3} \sqsubseteq T_j, \qquad (23)$$

$$\geq 2\,T_j^- \sqsubseteq \bot, \qquad (24)$$

$$\exists P_j \sqsubseteq \exists T_{j,1}, \qquad\qquad \exists P_j \sqsubseteq \exists T_{j,2}, \qquad (25)$$

$$\exists T_{j,1}^- \sqcap \exists T_{j,2}^- \sqsubseteq \exists T_{j,3}^-, \qquad (26)$$

$$\geq 2\,T_j \sqsubseteq \exists P_{j,\mathbf{t}} \qquad\qquad \exists T_{j,3} \sqsubseteq \exists P_{j,\mathbf{f}}, \qquad (27)$$

where $T_j, T_{j,1}, T_{j,2}, T_{j,3}$ are fresh role names, for $1 \leq j \leq 4$. It should be clear that $(\mathcal{T}, \mathcal{A}_\varphi)$ is LOGSPACE computable (in $|\varphi|$). The conjunctions in the left-hand side of (21) and (26) can be eliminated by using Lemma 3. So it remains to prove that $(\mathcal{T}, \mathcal{A}_\varphi) \models D(f)$ iff φ is not satisfiable.

Suppose first that φ is satisfiable. Then there is an assignment $\mathfrak{a}$ of the truth-values $\mathbf{t}$ and $\mathbf{f}$ to propositional variables such that $\mathfrak{a}(a_{k,1}) = \mathbf{t}$ or $\mathfrak{a}(a_{k,2}) = \mathbf{t}$ or $\mathfrak{a}(a_{k,3}) = \mathbf{f}$ or $\mathfrak{a}(a_{k,4}) = \mathbf{f}$, for all k, $1 \leq k \leq n$. Consider the interpretation $\mathcal{I}$ with $\Delta^{\mathcal{I}} = \{\mathring{x}_1, \ldots, x_m, z\} \cup \{y_k, u_{k,j,1}, u_{k,j,2} \mid 1 \leq j \leq 4, 1 \leq k \leq n\}$ and

- $f^{\mathcal{I}} = z$, $c_k^{\mathcal{I}} = y_k$, for $1 \leq k \leq n$, $a_i^{\mathcal{I}} = x_i$, for $1 \leq i \leq m$,
- $A^{\mathcal{I}} = \{x_i \mid 1 \leq i \leq m, \mathfrak{a}(a_i) = \mathbf{t}\}$,
- $P_{j,\mathbf{t}}^{\mathcal{I}} = \{(y_k, a_{k,j}^{\mathcal{I}}) \mid 1 \leq k \leq n, \mathfrak{a}(a_{k,j}) = \mathbf{t}\}$, for $1 \leq j \leq 4$,
- $P_{j,\mathbf{f}}^{\mathcal{I}} = \{(y_k, a_{k,j}^{\mathcal{I}}) \mid 1 \leq k \leq n, \mathfrak{a}(a_{k,j}) = \mathbf{f}\}$, for $1 \leq j \leq 4$,
- $P_j^{\mathcal{I}} = P_{j,\mathbf{t}}^{\mathcal{I}} \cup P_{j,\mathbf{f}}^{\mathcal{I}}$, for $1 \leq j \leq 4$,
- $T_{j,1}^{\mathcal{I}} = \{(y_k, u_{k,j,1}) \mid 1 \leq k \leq n\}$, for $1 \leq j \leq 4$,
- $T_{j,2}^{\mathcal{I}} = \{(y_k, u_{k,j,2}) \mid 1 \leq k \leq n, \mathfrak{a}(a_{k,j}) = \mathbf{t}\} \cup$
  $\qquad\qquad \{(y_k, u_{k,j,1}) \mid 1 \leq k \leq n, \mathfrak{a}(a_{k,j}) = \mathbf{f}\}$, for $1 \leq j \leq 4$,
- $T_{j,3}^{\mathcal{I}} = \{(y_i, u_{k,j,1}) \mid 1 \leq k \leq n, \mathfrak{a}(a_{k,j}) = \mathbf{f}\}$, for $1 \leq j \leq 4$,

- $T_j^{\mathcal{I}} = T_{j,1}^{\mathcal{I}} \cup T_{j,2}^{\mathcal{I}}$, for $1 \leq j \leq 4$,
- $S_{\mathbf{f}}^{\mathcal{I}}$, $S^{\mathcal{I}}$ and $D^{\mathcal{I}}$ are defined in the same way as in the proof of Theorem 4.

It is not hard to check that $\mathcal{I} \models (\mathcal{T}, \mathcal{A}_\varphi)$, and clearly $\mathcal{I} \not\models D(f)$.

Assume now that φ is not satisfiable and $\mathcal{I} \models (\mathcal{T}, \mathcal{A}_\varphi)$. Define an assignment $\mathfrak{a}$ by taking $\mathfrak{a}(a_i) = \mathbf{t}$ iff $a_i^{\mathcal{I}} \in A^{\mathcal{I}}$. As φ is not satisfiable, there is k, $1 \leq k \leq n$, such that $\mathfrak{a}(a_{k,1}) = \mathfrak{a}(a_{k,2}) = \mathbf{f}$, $\mathfrak{a}(a_{k,3}) = \mathfrak{a}(a_{k,4}) = \mathbf{t}$.

For each j, $1 \leq j \leq 4$, we have $c_k^{\mathcal{I}} \in (\exists P_j)^{\mathcal{I}}$; by (25), $c_k^{\mathcal{I}} \in (\exists T_{j,1})^{\mathcal{I}}, (\exists T_{j,2})^{\mathcal{I}}$. So there are v_1, v_2 such that $(c_k^{\mathcal{I}}, v_1) \in T_{j,1}^{\mathcal{I}}$ and $(c_k^{\mathcal{I}}, v_2) \in T_{j,2}^{\mathcal{I}}$. If $v_1 \neq v_2$ then $c_k^{\mathcal{I}} \in (\geq 2\, T_j)^{\mathcal{I}}$ and, by (27), $c_k^{\mathcal{I}} \in (P_{j,\mathbf{t}})^{\mathcal{I}}$. Otherwise, if $v_1 = v_2 = v$, we have by (26), $v \in (\exists T_{j,3}^-)^{\mathcal{I}}$, and so by (23) and (24), $c_k^{\mathcal{I}} \in (\exists T_{j,3})^{\mathcal{I}}$, from which, by (27), $c_k^{\mathcal{I}} \in (P_{j,\mathbf{f}})^{\mathcal{I}}$. Therefore, $c_k^{\mathcal{I}} \in (\exists P_{j,\mathbf{t}})^{\mathcal{I}} \cup (\exists P_{j,\mathbf{f}})^{\mathcal{I}}$, and by (18), $c_k^{\mathcal{I}} \in (\exists P_j)^{\mathcal{I}}$. Thus, by (17) and (20), $c_k^{\mathcal{I}} \in (\exists P_{j,\mathbf{t}})^{\mathcal{I}}$ if $\mathfrak{a}(a_{k,j}) = \mathbf{t}$ and $c_k^{\mathcal{I}} \in (\exists P_{j,\mathbf{f}})^{\mathcal{I}}$ if $\mathfrak{a}(a_{k,j}) = \mathbf{f}$, and hence, by (21), $c_k^{\mathcal{I}} \in (\exists S_{\mathbf{f}}^-)^{\mathcal{I}}$. Then by (22), we have $f^{\mathcal{I}} \in (\exists S_{\mathbf{f}})^{\mathcal{I}}$ and $f^{\mathcal{I}} \in D^{\mathcal{I}}$. It follows that $(\mathcal{T}, \mathcal{A}_\varphi) \models D(f)$. $\qquad\square$

However, the core fragment with only functionality constraints is data complete for P (the lower bound would follow from [8, Theorem 6, item 2] but the proof there is fallacious).

Theorem 6. *The instance checking problem (and query answering problem) for DL-Lite$_{core}^{\mathcal{R},\mathcal{F}}$ is data complete for P.*

Proof. The polynomial upper bound follows from [11]. We prove the matching lower bound by reduction of the entailment problem for Horn-CNF, which is known to be P-complete (see, e.g., [4, Exercise 2.2.4]). Given a Horn-CNF

$$\varphi = \bigwedge_{k=1}^{n} (\neg a_{k,1} \vee \neg a_{k,2} \vee a_{k,3}) \quad \wedge \quad \bigwedge_{l=1}^{p} a_{l,0},$$

where each $a_{k,j}$ and each $a_{l,0}$ is one of the propositional variables $a_1, \dots, a_m$, we construct a DL-Lite$_{core}^{\mathcal{R},\mathcal{F}}$ knowledge base $(\mathcal{T}, \mathcal{A}_\varphi)$ whose TBox $\mathcal{T}$ does not depend on φ and ABox $\mathcal{A}_\varphi$ is computed in LogSpace from φ. We will need the object names $c_1, \dots, c_n$ and $v_{k,j,i}$, for $1 \leq k \leq n$, $1 \leq j \leq 3$, $1 \leq i \leq m$ (for each variable, we take one object name for each possible occurrence of this variable in each non-unit clause), role names $S, S_{\mathbf{t}}$ and $P_j, P_{j,\mathbf{t}}$, for $1 \leq j \leq 3$, and a concept name A. Define $\mathcal{A}_\varphi$ to be the set containing the assertions:

$S(v_{1,1,i}, v_{1,2,i}), S(v_{1,2,i}, v_{1,3,i}), S(v_{1,3,i}, v_{2,1,i}), S(v_{2,1,i}, v_{2,2,i}), S(v_{2,2,i}, v_{2,3,i}), \dots$

$\dots, S(v_{n,2,i}, v_{n,3,i}), S(v_{n,3,i}, v_{1,1,i})$, for $1 \leq i \leq m$,

$P_j(v_{k,j,i}, c_k)$ iff $a_{k,j} = a_i$,   for $1 \leq i \leq m$, $1 \leq k \leq n$, $1 \leq j \leq 3$,

$A(v_{1,1,i})$ iff $a_{l,0} = a_i$,   for $1 \leq i \leq m$, $1 \leq l \leq p$

(all objects for each variable are organised in an S-cycle and $P_j(v_{k,j,i}, c_k) \in \mathcal{A}_\varphi$ iff the variable a_i occurs in the kth non-unit clause of φ in the jth position).

And let $\mathcal{T}$ consist of the following concept and role inclusions:

$$S_\mathbf{t} \sqsubseteq S, \qquad \geq 2\,S \sqsubseteq \bot, \qquad A \sqsubseteq \exists S_\mathbf{t}, \qquad \exists S_\mathbf{t}^- \sqsubseteq A, \qquad (28)$$

$$P_{j,\mathbf{t}} \sqsubseteq P_j, \qquad \geq 2\,P_j \sqsubseteq \bot, \qquad A \sqsubseteq \exists P_{j,\mathbf{t}}, \qquad \text{for } 1 \leq j \leq 2, \qquad (29)$$

$$P_{3,\mathbf{t}} \sqsubseteq P_3, \qquad \geq 2\,P_3^- \sqsubseteq \bot, \qquad (30)$$

$$\exists P_{1,\mathbf{t}}^- \sqcap \exists P_{2,\mathbf{t}}^- \sqsubseteq \exists P_{3,\mathbf{t}}^-, \qquad (31)$$

$$\exists P_{3,\mathbf{t}} \sqsubseteq A. \qquad (32)$$

It should be clear that $(\mathcal{T}, \mathcal{A}_\varphi)$ is LOGSPACE computable (in $|\varphi|$). As in the previous proofs, here we have an axiom, namely (31), that does not belong to $\textit{DL-Lite}_{core}^{\mathcal{R},\mathcal{F}}$ because of the conjunctions in its left-hand side. As before, this conjunction is eliminated with the help of Lemma 3. Our aim is to show that $(\mathcal{T}, \mathcal{A}_\varphi) \models A(v_{1,1,i_0})$ iff $\varphi \models a_{i_0}$.

Suppose $\varphi \not\models a_{i_0}$. Then there is an assignment $\mathfrak{a}$ with $\mathfrak{a}(\varphi) = \mathbf{t}$ and $\mathfrak{a}(a_{i_0}) = \mathbf{f}$. We construct a model $\mathcal{I}$ for $(\mathcal{T}, \mathcal{A}_\varphi)$ such that $\mathcal{I} \not\models A(v_{1,1,i_0})$. Define $\mathcal{I}$ by taking $\Delta^{\mathcal{I}} = \{x_{k,j,i}, z_{k,j,i} \mid 1 \leq k \leq n, 1 \leq j \leq 3, 1 \leq i \leq m\} \cup \{y_k \mid 1 \leq k \leq n\}$, $c_k^{\mathcal{I}} = y_k$, for $1 \leq k \leq n$, $v_{k,j,i}^{\mathcal{I}} = x_{k,j,i}$, for $1 \leq k \leq n, \leq j \leq 3, 1 \leq i \leq m$. The extensions of the concept and role names are defined as in Fig. 2. It is routine to check that we indeed have $\mathcal{I} \models (\mathcal{T}, \mathcal{A}_\varphi)$ and $\mathcal{I} \not\models A(v_{1,1,i_0})$.

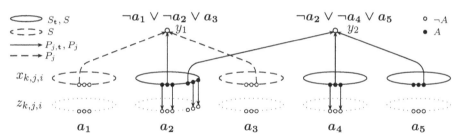

Fig. 2. The model $\mathcal{I}$ satisfying $(\mathcal{T}, \mathcal{A}_\varphi)$, for $\varphi = (\neg a_1 \vee \neg a_2 \vee a_3) \wedge (\neg a_2 \vee \neg a_4 \vee a_5)$

Conversely, assume now that $\varphi \models a_{i_0}$. Consider some $\mathcal{I} \models (\mathcal{T}, \mathcal{A}_\varphi)$ and define $\mathfrak{a}$ to be the assignment such that $\mathfrak{a}(a_i) = \mathbf{t}$ iff $v_{1,1,i}^{\mathcal{I}} \in A^{\mathcal{I}}$, for $1 \leq i \leq m$. By (28), for each i, $1 \leq i \leq m$, we have either $v_{k,j,i}^{\mathcal{I}} \in A^{\mathcal{I}}$, for *all* k, j with $1 \leq k \leq n$, $1 \leq j \leq 3$, or $v_{k,j,i}^{\mathcal{I}} \notin A^{\mathcal{I}}$, for *all* k, j with $1 \leq k \leq n, 1 \leq j \leq 3$.

Now, if we have $\mathfrak{a}(a_{k,1}) = \mathbf{t}$ and $\mathfrak{a}(a_{k,2}) = \mathbf{t}$, for $1 \leq k \leq n$ then, by (29), $c_k^{\mathcal{I}} \in (\exists P_{1,\mathbf{t}}^-)^{\mathcal{I}}, (\exists P_{2,\mathbf{t}}^-)^{\mathcal{I}}$. By (31), $c_k^{\mathcal{I}} \in (\exists P_{3,\mathbf{t}}^-)^{\mathcal{I}}$ and hence, by (30), $v_{k,3,i}^{\mathcal{I}} \in (\exists P_{3,\mathbf{t}})^{\mathcal{I}}$, where $a_{k,3} = a_i$, which means, by (32), that $v_{k,3,i}^{\mathcal{I}} \in A^{\mathcal{I}}$, and so $v_{1,1,i}^{\mathcal{I}} \in A^{\mathcal{I}}$ and $\mathfrak{a}(a_i) = \mathbf{t}$. It follows that $\mathfrak{a}(\varphi) = \mathbf{t}$, and hence $\mathfrak{a}(a_{i_0}) = \mathbf{t}$, which, by definition, means that $v_{1,1,i_0}^{\mathcal{I}} \in A^{\mathcal{I}}$. As $\mathcal{I}$ was an arbitrary model of $(\mathcal{T}, \mathcal{A}_\varphi)$, we can conclude that $(\mathcal{T}, \mathcal{A}_\varphi) \models A(v_{1,1,i_0})$. $\qquad \square$

6   Conclusion

The results obtained in this paper and [2] show the following: (1) One can add either number restrictions or role inclusions to the basic (*core*, *horn*, *krom* and *bool*) *DL-Lite* logics without changing their complexity. (2) However, taken together, these constructs spoil the nice computational properties of the basic *DL-Lite* logics. (3) If both of them are really needed for an application, one should try and restrict their interaction (e.g., by avoiding axioms of the form $R \sqsubseteq P$ with functional role P, as suggested in [9]). Exploring in depth this interaction, as well as the impact of other constructs (transitive roles, Booleans on roles, etc.) on the computational properties of *DL-Lite* logics is an interesting and practically important area for further research.

References

1. Artale, A., Calvanese, D., Kontchakov, R., Ryzhikov, V., Zakharyaschev, M.: Complexity of reasoning over Entity-Relationship models. In: Proc. of DL, pp. 163–170 (2007)
2. Artale, A., Calvanese, D., Kontchakov, R., Zakharyaschev, M.: DL-Lite in the light of first-order logic. In: Proc. of AAAI, pp. 361–366 (2007)
3. Bernstein, P., Giunchiglia, F., Kementsietsidis, A., Mylopoulos, J., Serafini, L., Zaihrayeu, I.: Data management for peer-to-peer computing: A vision. In: Proc. of WebDB (2002)
4. Börger, E., Grädel, E., Gurevich, Y.: The Classical Decision Problem. Perspectives in Mathematical Logic. Springer, Heidelberg (1997)
5. Borgida, A., Brachman, R., McGuinness, D., Alperin Resnick, L.: CLASSIC: A structural data model for objects. In: Proc. of ACM SIGMOD, pp. 59–67 (1989)
6. Calvanese, D., De Giacomo, G., Lembo, D., Lenzerini, M., Rosati, R.: DL-Lite: Tractable description logics for ontologies. In: Proc. of AAAI, pp. 602–607. AAAI Press, Menlo Park (2005)
7. Calvanese, D., De Giacomo, G., Lembo, D., Lenzerini, M., Rosati, R.: Tailoring OWL for data intensive ontologies. In: Proc. of OWLED (2005)
8. Calvanese, D., De Giacomo, G., Lembo, D., Lenzerini, M., Rosati, R.: Data complexity of query answering in description logics. In: Proc. of KR, pp. 260–270 (2006)
9. Calvanese, D., De Giacomo, G., Lembo, D., Lenzerini, M., Poggi, A., Rosati, R.: Linking data to ontologies: the description logic $DL\text{-}Lite_A$. In: Proc. OWLED (2006)
10. Calvanese, D., De Giacomo, G., Lembo, D., Lenzerini, M., Rosati, R.: Tractable reasoning and efficient query answering in description logics: The DL-Lite family. J. of Automated Reasoning 39, 385–429 (2007)
11. Eiter, T., Gottlob, G., Ortiz, M., Šimkus, M.: Query answering in the description logic Horn-$\mathcal{SHIQ}$. In: Proc. of JELIA (2008)
12. Glimm, B., Horrocks, I., Lutz, C., Sattler, U.: Conjunctive query answering for the description logic $\mathcal{SHIQ}$. In: Proc. of IJCAI, pp. 399–404 (2007)
13. Heflin, J., Hendler, J.: A portrait of the Semantic Web in action. IEEE Intelligent Systems 16, 54–59 (2001)
14. Hustadt, U., Motik, B., Sattler, U.: Reasoning in description logics by a reduction to disjunctive Datalog. J. of Automated Reasoning 39, 351–384 (2007)
15. Kozen, D.: Theory of Computation. Springer, Heidelberg (2006)
16. Lenzerini, M.: Data integration: A theoretical perspective. In: Proc. of PODS, pp. 233–246 (2002)
17. Papadimitriou, C.: Computational Complexity. Addison-Wesley, Reading (1994)
18. Schaerf, A.: On the complexity of the instance checking problem in concept languages with existential quantification. J. Intell. Infor. Systems 2, 265–278 (1993)

Temporal Ontology Language for Representing and Reasoning Interval-Based Temporal Knowledge

Sang-Kyun Kim[1], Mi-Young Song[1], Chul Kim[1], Sang-Jun Yea[1],
Hyun Chul Jang[1], and Kyu-Chul Lee[2]

[1] Korea Institute of Oriental Medicine, South Korea
{skkim,smyoung,chulnice,tomita,hcjang}@kiom.re.kr
[2] Dept. of Computer Engineering, Chungnam National University, South Korea
kclee@cnu.ac.kr

Abstract. W3C Web Ontology working group has recently developed OWL as an ontology language for the Semantic Web. However, because OWL does not have the full-fledged semantics for temporal information, it cannot perform reasoning about temporal knowledge. Entities in the real world are changing according to the passage of time and new facts are occurring due to events. If knowledge in the KBs does not have the temporal information, it becomes incomplete and incorrect. Therefore, we in this paper propose an ontology language TL-OWL, which extends OWL to have the temporal semantics in order to represent and reason the temporal information in the Semantic Web.

1 Introduction

Semantic Web has a vision that a machine understands and processes information in web automatically by describing semantics to web. For a machine to process information, knowledge that a machine and humans can share must be described. Semantic Web provides knowledge on web resources by using ontology. OWL is an ontology language for the Semantic Web that has recently been developed by the W3C Web Ontology Working Group. However, since OWL does not have time information, questions depending on time cannot be accurately processed.

For example, let's assume as follows: When four cases happen, each case has time interval x, y, u, v according to when it happens and the relation between time interval is x before y, y overlaps u, and u before v. OWL describes each four cases and time as individuals, and the time relationship can be connected by property of the individuals. However, since OWL cannot perform the transitive reasoning among time relations, the relation of x before v cannot be reasoned. Instead, if the rule-based reasoning such as before(x,v) :- before(x,y) & overlaps(y,u) & before(u,v), questions can be answered. However, generally, the problem of the rule-based reasoning has known to be the semi-decidable. But if semantics on time to ontology can be provided, the temporal reasoning based on ontology such as x before y ∨ y overlaps u ∨ u before v → x before v is possible without using the rule-based reasoning.

J. Domingue and C. Anutariya (Eds.): ASWC 2008, LNCS 5367, pp. 31–45, 2008.
© Springer-Verlag Berlin Heidelberg 2008

In artificial intelligent field, the researches [3] that represent and reason the temporal concepts by using Temporal Description Logics that deals with time based on Description Logics have been suggested. Researches on Temporal Description Logics are classified into Point-based Description Logics [9,12] and Interval-based Description Logics [2,4,5,10] according to how time information can be formalized. However, it is difficult to determine which way has better expression and reasoning. In order to provide the temporal reasoning in Semantics Web, the capability of OWL DL is needed, but both methods do not have that.

Therefore, in order to solve problems of OWL and Temporal Description Logics, we propose an interval-based temporal web ontology language TL-OWL which is extended language of OWL to have semantic on time interval.

The remainder of this paper is organized as following. In section 2, we briefly introduce $\mathcal{TL}$-$\mathcal{ALCF}$. In section 3, we propose an interval-based temporal web ontology language, TL-OWL. In section 4, we compare our work with prior efforts for the related subjects. Finally, in section 5, we summarize this paper.

2   A Temporal Description Logic

In this section, we briefly introduce a class of interval-based temporal Description Logic, $\mathcal{TL}$-$\mathcal{ALCF}$ proposed by Artale and Franconi. They show that the subsumption problem is decidable and supply sound and complete procedures for computing subsumption. $\mathcal{TL}$-$\mathcal{ALCF}$ is composed by the temporal logic $\mathcal{TL}$ which is able to express temporally quantified terms and the non-temporal Description Logic $\mathcal{ALCF}$ [6] extending $\mathcal{ALC}$ with features (i.e., functional roles). In this formalism an action is represented through temporal constraints on world states where each state is a collection of properties of the world holding at a certain time. The intended meaning of $\mathcal{TL}$-$\mathcal{ALCF}$ is explained as the following example.

Reserve-Flight $\doteq \Diamond(x\ y)\ (\sharp\ \text{f}\ x)(\sharp\ \text{m}\ y).\ ((\star\text{TICKET}:$
    Unreserved$)@x \sqcap (\star\text{TICKET}: \text{Reserved})@y)$

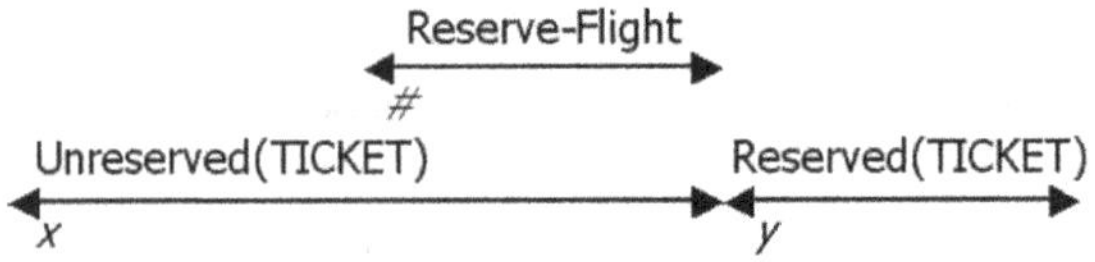

Fig. 1. Temporal dependencies of the intervals in which Reserve-Flight holds

Fig. 1 shows the temporal dependencies of the intervals in which the concept Reserve-Flight holds. Reserve-Flight denotes any action occurring at some interval involving a $\star$TICKET that was once unreserved and then reserved, where $\star$TICKET is a parametric feature and Reserved and Unreserved are non-temporal concepts. The parametric feature $\star$TICKET plays the role of formal parameter of the action, mapping any individual action of type Reserve-Flight to the ticket to be reserved, independently from time. Temporal variables are introduced by the temporal existential quantifier "$\Diamond$" – excluding the special temporal variable

♯, usually called now, and intended as the occurring time of the action type being defined. The temporal constraints $(♯ \text{ f } x)(♯ \text{ m } y)$ state that the interval denoted by x should finish with the interval denoted by ♯ and that ♯ should meet y, where f and m are Allen's temporal relations [1] of Fig. 2.

Relation	Abbr.	Inverse	i	j
before(i, j)	b	a		
meets(i, j)	m	mi		
overlaps(i, j)	o	oi		
starts(i, j)	s	si		
during(i, j)	d	di		
finishes(i, j)	f	fi		

Fig. 2. The Allen's interval relationships

As the evaluation of concept at the interval, $(\star\text{TICKET} : \text{Unreserved})@x$ and $(\star\text{TICKET} : \text{Reserved})@y$ state that $\star\text{TICKET} : \text{Unreserved}$ is qualified at x and $\star\text{TICKET} : \text{Reserved}$ is qualified at y. In the concept description, the operator : is the selection of feature, which is the role quantification that is interpreted as a partial function. The following table shows the syntax of $\mathcal{TL\text{-}ALCF}$. See [2] for the detailed semantic descriptions of $\mathcal{TL\text{-}ALCF}$.

Table 1. The syntax of $\mathcal{TL\text{-}ALCF}$

$\mathcal{TL}$	$E, F \rightarrow C \mid$	(non-temporal concept)
	$\quad E \sqcap F \mid$	(conjunction)
	$\quad E@X \mid$	(qualifier)
	$\quad E[Y]@X \mid$	(substitutive qualifier)
	$\quad \Diamond(\overline{X})\overline{Tc}.E$	(existential quantifier)
	$Tc \rightarrow (♯ \ (V) \ Y) \mid (X \ (V) \ ♯) \mid (X \ (V) \ Y)$	(temporal constraint)
	$\overline{Tc} \rightarrow Tc \mid Tc \ \overline{Tc}$	
	$V, W \rightarrow V , W \mid$	(disjunction)
	$\quad \text{b} \mid \text{a} \mid \text{m} \mid \text{mi} \mid \text{o} \mid \text{oi} \mid$	(Allen's relations)
	$\quad \text{s} \mid \text{si} \mid \text{d} \mid \text{di} \mid \text{f} \mid \text{fi} \mid =$	
	$X, Y \rightarrow \text{x} \mid \text{y} \mid \text{z} \mid \ldots$	(temporal variables)
	$\overline{X} \rightarrow X \mid X \ \overline{X}$	
$\mathcal{ALCF}$	$C, B \rightarrow A \mid$	(atomic concept)
	$\quad \top \mid \bot \mid \neg C \mid C \sqcap B \mid C \sqcup B \mid \forall R.C \mid \exists R.C \mid$	
	$\quad p \downarrow q \mid$	(agreement)
	$\quad p \uparrow q \mid$	(disagreement)
	$\quad p \uparrow \mid$	(undefinedness)
	$\quad p : C \mid$	(selection)
	$p, q \rightarrow f \mid$	(atomic feature)
	$\quad \star g \mid$	(atomic parametric feature)
	$\quad p \circ q \mid$	(path)

3   A Temporal Web Ontology Language

We propose an interval-based temporal ontology language TL-OWL, which adds a temporal language $\mathcal{TL}$ to OWL DL, where $\mathcal{TL}$ is the temporal part$(\mathcal{TL})$ of $\mathcal{TL}$-$\mathcal{ALCF}$ as introduced in Sect. 2.

3.1   Requirements of TL-OWL

W3C Web Ontology Working Group has recently developed OWL which is an ontology language for the Semantic Web. In OWL specification, OWL is defined in two forms of syntax; First, OWL has a frame-like abstract syntax which can be easily understood and created. Second, OWL has a RDF/XML exchange syntax interpreted as RDF graphs since OWL is defined as an extension to RDF. The direct model-theoretic semantic and the RDF-compatible model-theoretic semantic are also defined to provide a formal meaning for the abstract syntax and the exchange syntax, respectively. A mapping from the abstract syntax to RDF graphs is defined and the two model semantics are shown to have the same consequences on OWL ontologies that can be written in the abstract syntax.

OWL however cannot perform the temporal reasoning since it does not have the temporal semantics as introduced in Sect. 1. The reasoning capability is usually provided in the temporal description logics [3], but all of them cannot reach the expressivity of OWL DL. Therefore, in this paper we propose *an interval-based temporal ontology language TL-OWL(TemporaL Web Ontology Language)*, which adds the temporal semantics to OWL DL.

In order to describe TL-OWL, we follow steps of the OWL specification: First, we define a high-level abstract syntax for TL-OWL. Second, we define two formal semantics for TL-OWL. One of these semantics is the direct model-theoretic semantics for TL-OWL ontologies written in the abstract syntax. The other is the RDF-compatible model-theoretic semantics as an extension of the RDF semantics, which provides semantics for TL-OWL ontologies written in the exchange syntax. And third, a mapping from the abstract syntax to RDF graphs is defined and the two model theories are shown to have the same consequences on TL-OWL ontologies. Finally, we show the reasoning in TL-OWL.

3.2   Abstract Syntax

An abstract syntax for OWL is needed since OWL is not very readable when written as RDF triples. This abstract syntax is closer to that of a frame language like OIL. As for a similar way with OWL, in this section, we define an abstract syntax for TL-OWL in the form of EBNF(Extended BNF) in Table 2. OWL syntax is not given in this table, but only OWL constructors required to understand TL-OWL are written as an italic font.

Temporal concepts in $\mathcal{TL}$-$\mathcal{ALCF}$ can be represented of the form: $\diamond(\overline{X})(\overline{Tc})$. $(Q_0 \sqcap Q_1@X_1 \sqcap ... \sqcap Q_n@X_n)$, where $\overline{X}$ is a set of temporal variables, $\overline{Tc}$ is a set of temporal constraints, and $(Q_0 \sqcap Q_1@X_1 \sqcap ... \sqcap Q_n@X_n)$ is a conjunction of qualifiers. However, there is a problem that this normal form cannot be

Table 2. The EBNF version of an abstract syntax

Abstract Syntax
vID (variableID) ::= **URIreference**
featureID ::= **URIreference**
axiom ::= 'TemporalClass(' **classID** ['Deprecated'] { **annotation** } **temporalDescription** ')'
axiom ::= 'TemporalVariable(' **vID** ['Deprecated'] { **annotation** } { **temporalRelation** } ')'
temporalDescription ::= 'intersectionOf(' **Qualification** { **Qualification** } ')'
Qualification ::= 'Qualification(onVariable(' **vID** ') bindVariable(' **description** '))' |
'Qualification(onVariable(' **vID** ') onSubstitutiveVariable(' **vID** ')
bindSubstitutiveVariable(' **description** '))'
temporalRelation ::= 'before(' **vID** ')' | 'after(' **vID** ')' | 'meets(' **vID** ')' | 'metBy(' **vID** ')' |
'overlaps(' **vID** ')' | 'overlappedBy(' **vID** ')' | 'starts(' **vID** ')' | 'startedBy(' **vID** ')' |
'during(' **vID** ')' | 'contains(' **vID** ')' | 'finish(' **vID** ')' | 'finishedBy(' **vID** ')' | 'equal(' **vID** ')'
axiom ::= '*DatatypeProperty(*' *datavaluedPropertyID* ... ['*Functional*' | '*ParametricFunctional*'] ... ')' |
'*ObjectProperty(*' *individualvaluedPropertyID* ... ['*Functional*' | 'ParametricFunctional' |
'*InverseFunctional*' | 'InverseParametricFunctional'] ... ['pathOf(' **featureID featureID** ')'] ')'
description ::= *classID* | *restriction* | *feature* | '*intersectionOf(*' *description* ')' |
'*unionOf(*' { *description* } ')' | '*complementOf(*' { *description* } ')' | '*oneOf(*' { *individualID* } ')'
restriction ::= '*restriction(*' *datavaluedPropertyID dataRestrictionComponent*
{ *dataRestrictionComponent* } ')' |
'*restriction(*' *individualvaluedPropertyID individualRestrictionComponent*
{ *individualRestrictionComponent* } ')'
dataRestrictionComponent ::= '*allValuesFrom(*' *dataRange* ')' | '*someValuesFrom(*' *dataRange* ')' |
'selectValuesFrom(' *dataRange* ')' | '*value(*' *dataLiteral* ')' | *cardinality*
individualRestrictionComponent ::= '*allValuesFrom('description')* ' | '*someValuesFrom('description')*' |
'selectValuesFrom(' *description* ')' | '*value(*' *individualID* ')' | *cardinality*
feature ::= 'agreementOf(' **featureID featureID** ')' | 'disagreementOf(' **featureID featureID** ')' |
'undefinednessOf('**featureID** ')'

represented as RDF triples. Therefore, in this paper we propose *an abstract syntax and a RDF/XML exchange syntax for TL-OWL*, which can be represented as RDF graphs.

TL-OWL has four axioms for TL-OWL classes and properties of TemporalClass, TemporalVariable, DatatypeProperty, and ObjectProperty. A TemporalClass can represent a temporal concept in TL-OWL. A TemporalClass contains one or more temporalDescriptions as properties and a temporalDescription contains a conjunction of Qualifications which bind a temporal variable and a non-temporal concept. A bindSubstitutiveVariable denotes a temporal substitutive qualifier which renames the variable Y to X and supplies a way of making coreference between two temporal variables. A TemporalVariable can represent the constraints among temporal variables. The TemporalVariable is identified with a variableID and has one or more of Allen's temporal relations as properties, where each temporal relation can refer another TemporalVariable. A DatatypeProperty and a ObjectProperty are the axioms defined in OWL, but in TL-OWL the ObjectProperty can have the additional types of ParametricFunctional and InverseParametricFunctional, and pathOf – a construct of the feature logic [6] – to represent a path between two featureIDs. A DatatypeProperty can have only an additional type of ParametricFunctional.

The description of OWL can contain constructs for feature logics which consist of agreementOf, disagreementOf, and undefinednessOf. A selectValuesFrom is declared within the restriction constructor for a selection operator (:).

By using above axioms and constructors, we can see that all the interval-based temporal concepts can be represented as the abstract syntax along with preserving temporal semantics. It is easy to proof by checking the syntax of temporal concepts inductively. For an example, the `Reserve-Flight` concept of Fig. 1 can be represented to an abstract syntax as follows:

```
TemporalClass ( ex:Reserve-Flight
  intersectionOf (
    Qualification (
      onVariable ( ex:x )
      bindVariable (
        restriction (
          onProperty ( ex:TICKET )
          selectValuesFrom ( ex:Unreserved )
  )))
    Qualification (
      onVariable ( ex:y )
      bindVariable (
        restriction (
          onProperty ( ex:TICKET )
          selectValuesFrom ( ex:Reserved )
  )))))
TemporalVariable(ex:x finishedBy NOW)
TemporalVariable(ex:y metBy NOW)
```

Fig. 3. The abstract syntax of the `Reserve-Flight` concept

3.3   Direct Model-Theoretic Semantics

The direct model-theoretic semantics for TL-OWL goes directly from ontologies in the abstract syntax to a standard model theory.

Vocabularies and Interpretations. When considering a TL-OWL ontology, the vocabulary must include all the URI references and literals in that ontology. The following is a definition for a TL-OWL vocabulary.

Definition 1. *A TL-OWL vocabulary V consists of a set of literals V_L and nine sets of URI references, V_C, V_{TC}, V_{TV}, V_D, V_I, V_{DP}, V_{IP}, V_{AP}, and V_O. In any vocabulary V_C, V_{TC}, V_{TV}, and V_D are disjoint and V_{DP}, V_{IP}, V_{AP}, and V_{OP} are pairwise disjoint. V_C, the class names of a vocabulary, contains owl:Thing and owl:Nothing. V_{TC}, the temporal class names of a vocabulary, V_{TV}, the temporal variable names of a vocabulary, V_D, the datatype names of a vocabulary, contains the URI references for the built-in OWL datatypes and rdfs:Literal. V_{AP}, the annotation property names of a vocabulary, contains owl:versionInfo, rdfs:label, rdfs:comment, rdfs:seeAlso, and rdfs:isDefinedBy. V_{IP}, the individual-valued property names of a vocabulary, V_{DP}, the data-valued property names of a vocabulary, V_{OP}, the URI references for the built-in TL-OWL ontology properties, and V_I, the individual names of a vocabulary, V_O, the ontology names of a vocabulary, do not have any required members.*

Definition 2. *A datatype map D is a partial mapping from URI references to datatypes that maps xsd:string and xsd:integer to the appropriate XML Schema datatypes.*

Definition 3. *Let D be a datatype map. An abstract TL-OWL interpretation with respect to D with vocabulary V_L, V_C, V_{TC}, V_{TV}, V_D, V_I, V_{DP}, V_{IP}, V_{AP}, V_O is a tuple of the form: $I = \langle R, T, EC, ER, L, S, LV \rangle$ where (with P being the power set operator)*

Definition. 2 is same as that of the OWL specification. Definition. 1 and Definition. 3 are similar to the definition for the OWL vocabulary and the abstract OWL interpretation except that temporal semantics are added. Thus, in this section we do not explain those of the OWL interpretations in details.

EC provides meaning for URI references that are used as TL-OWL classes and datatypes. ER provides meaning for URI references that are used as TL-OWL properties. As for the formal semantics given in $\mathcal{TL}\text{-}\mathcal{ALCF}$, TL-OWL classes and properties have semantics for temporal structure T such as EC: $V_{TC} \rightarrow P(T \times O)$ and ER: $V_{DP} \rightarrow P(T \times O \times LV)$, where O is URI references and LV is the literal values. L provides meaning for typed literals. S provides meaning for URI references that are used to denote TL-OWL individuals.

Interpretations for Constructs. EC is extended to the syntactic constructs of qualifications, descriptions, temporal relations, features as follows:

NOW, a built-in TL-OWL temporal variable, denotes the current interval of evaluation. The thirteen temporal relations defined as built-in TL-OWL properties can represent the temporal network between two temporal variables. The formal semantics for EC and ER in $\mathcal{TL}\text{-}\mathcal{ALCF}$ are defined as $EC_{V,t,H}$ and $ER_{V,t,H}$, where V is a variable assignment function associating an interval value to a temporal variable, t is an interval, and H is a set of constraints over the assignments. We in this paper omit the subscripts to simplify notations if there are not any misunderstandings. We denote the domain of partial functions by dom, which can be interpreted as a Functional type of properties.

Interpretations for Axioms. An abstract TL-OWL interpretation, I, satisfies TL-OWL axioms in the following table. Optional parts of axioms are given in square brackets ([...]). tr_i $(1 \leq i \leq n)$ can be one out of the thirteen temporal relations in TemporalVariable.

Interpretations of Ontology. The definitions for the satisfiability, consistency, and entailment of TL-OWL ontology are given in this section. These definitions will be used in Definition. 8 and Theorem. 1.

Definition 4. *Let D be a datatype map. An Abstract TL-OWL interpretation, I, with respect to D with vocabulary consisting of V_L, V_C, V_{TC}, V_{TV}, V_D, V_I, V_{DP}, V_{IP}, V_{AP}, V_O,* **satisfies** *a TL-OWL ontology, O, iff*

1. *each URI reference in O used as a class ID (temporal class ID, temporal variable ID, datatype ID, individual ID, data-valued property ID, individual-valued property ID, annotation property ID, annotation ID, ontology ID) belongs to V_C (V_{TC}, V_{TV}, V_D, V_I, V_{DP}, V_{IP}, V_{AP}, V_O, respectively);*

2. *each literal in O belongs to V_L;*
3. *I satisfies each directive in O, except for Ontology Annotations;*
4. *there is some $o \in R$ with $\langle o,S(owl{:}Ontology)\rangle \in ER(rdf{:}type)$ such that for each Ontology Annotation of the form Annotation(p v), $\langle o,S(v)\rangle \in ER(p)$ and that if O has name n, then $S(n) = o$; and*
5. *I satisfies each ontology mentioned in an owl:imports annotation directive of O.*

Definition 5. *A collection of abstract TL-OWL ontologies and axioms and facts is **consistent** with respect to datatype map D iff there is some interpretation I with respect to D such that I satisfies each ontology and axiom and fact in the collection.*

Definition 6. *A collection O of abstract TL-OWL ontologies and axioms and facts **entails** an abstract TL-OWL ontology or axiom or fact O' with respect to a datatype map D if each interpretation with respect to map D that satisfies each ontology and axiom and fact in O also satisfies O'.*

3.4   Mapping to RDF Graphs

We in this section provide a mapping from the abstract syntax to the exchange syntax for TL-OWL. Further, in Sect. 3.5 we show that this mapping preserves the meaning of TL-OWL ontologies.

The exchange syntax for TL-OWL is RDF/XML and the meaning of a TL-OWL ontology in RDF/XML is determined from the RDF graph that results from the RDF parsing of the RDF/XML document. The way of translating a TL-OWL ontology in abstract syntax form into the exchange syntax is by giving a transformation of each directive into a collection of RDF triples. Table reftable:Triples gives the transformation rules that transform the abstract syntax of Table 3 and Table 4 to the TL-OWL exchange syntax. The left column of the table is an abstract syntax, the center column is its transformation into triples, and the right column is an identifier for the main node of the transformation. Repeating components are listed using ellipses, as in Qualification$_1$ ... Qualification$_n$. Optional portions are enclosed in square brackets. The triples in the transformation rules that may or may not be generated are indicated by flagging with [opt]. Some transformations in the table are for directives. Other transformations are for parts of directives. Thus, the transformation rules for the directives call for other rules for components of the directive.

Table 3. Interpretations for Constructs

Abstract Syntax	Interpretations
NOW	EC(NOW) is a current interval
before(x)	$\{[u_1,v_1] \in T \mid EC(x)=[u_2,v_2] \text{ implies } v_1 < u_2\}$
other temporal relations ...	...
qualification(x bindVariable(c))	EC(c), t=V(x)
qualification(x y bindSubstitutiveVariable(c))	EC(c), H=H∪$\{y{\to}V(x)\}$
restriction(p selectValuesFrom(e))	$\{x \in \text{domp} \mid ER(p)(x) \in EC(e)\}$
agreementOf(p q)	$\{x \in \text{domp} \cap \text{domq} \mid ER(p)(x) = ER(q)(x)\}$
disagreementOf(p q)	$\{x \in \text{domp} \cap \text{domq} \mid ER(p)(x) \neq ER(q)(x)\}$
undefinednessOf(p)	O \ domp

Table 4. Interpretations for Axioms

Abstract Syntax	Interpretations
TemporalClass(c q_1 ... q_n)	EC(c) = EC(q_1) $\cap$ ... $\cap$ EC(q_n)
TemporalVariable(c tr_1 ... tr_n)	EC(c) = ER(tr_1) $\cup$ ... $\cup$ ER(tr_n)
DatatypeProperty(p ... [ParametricFunctional])	[ER(p) is parametric functional]
ObjectProperty(p ... [ParametricFunctional])	[ER(p) is parametric functional]
[InverseParametricFunctional])	[ER(p) is inverse parametric functional]
[pathOf(x y])	[$\langle u,v \rangle \in$ ER(x) $\cap$ $\langle v,w \rangle \in$ ER(y) implies $\langle u,w \rangle \in$ ER(p), u$\in$domx, v$\in$domy]

Table 5. Transformation to Triples

Abstract Syntax (and sequences) - S	Transformation - T(S)	Main Node - M(T(S))
vID	vID rdf:type tl:TemporalVariable .	vID
featureID	featureID rdf:type owl:FunctionalProperty . featureID rdf:type tl:ParametricFunctionalProperty [opt] .	featureID featureID
Qualification(vID C)	_:x rdf:type tl:Qualification . _:x rdf:type rdfs:Class [opt] . _:x tl:onVariable T(vID) . _:x tl:bindVariable T(C) .	_:x
Qualification(vID_1 vID_2 C)	*similar*	_:x
restriction(ID selectValuesFrom(selection))	_:x rdf:type owl:Restriction . _:x rdf:type rdfs:Class . [opt] _:x owl:onProperty T(ID) . _:x tl:selectValuesFrom T(selection) .	_:x
TemporalClass(classID [Deprecated] annotation_1 ... annotation_m Qualification_1 ... Qualification_n	classID rdf:type tl:TemporalClass . classID rdf:type rdfs:Class [opt] . [classID rdf:type owl:DeprecatedClass .] classID T(annotation_1) ... classID T(annotation_m) . classID owl:intersectionOf T(SEQ Quantification_1 ... Quantification_n) .	
TemporalVariable(classID [Deprecated] annotation_1 ... annotation_m $\text{temporalRelation}_1$ ... $\text{temporalRelation}_n$	classID rdf:type tl:TemporalVariable . classID rdf:type rdfs:Class [opt] . [classID rdf:type owl:DeprecatedClass .] classID T(annotation_1) ... classID T(annotation_m) . classID owl:unionOf T(SEQ $\text{temporalRelation}_1$ ... $\text{temporalRelation}_n$) .	
ObjectProperty(ID [Deprecated] annotation_1 ... annotation_m [ParametricFunctional| InverseParametricFunctional] pathOf(featureID_1 featureID_2) ...)	ID rdf:type owl:ObjectProperty . ID rdf:type rdf:Property . [opt] [ID rdf:type owl:DeprecatedProperty .] ID T(annotation_1) ... ID T(annotation_m) . [ID rdf:type tl:ParametircFunctional .] [ID rdf:type tl:InverseParametircFunctional .] [ID tl:pathOf T(SEQ featureID_1 featureID_2) .] ...	
DatatypeProperty(ID	*similar*	
agreementOf(featureID_1 featureID_2)	_:x rdf:type rdfs:Class . _:x tl:agreementOf T(SEQ featureID_1 featureID_2) .	_:x
disagreementOf(featureID_1	*similar*	_:x
undefinednessOf(featureID)	*similar*	_:x
before(vID_1 vID_2)	vID_1 rdf:type tl:temporalVariable . vID_2 rdf:type tl:temporalVariable . vID_1 before vID_2 .	
other temporal relations ...	...	

The `Reserve-Flight` concept in an abstract syntax form can be transformed into RDF graphs by using the above transformation rules. The following is the exchange syntax of the `Reserve-Flight` concept in the form of RDF/XML.

```
<tl:TemporalClass rdf:about="#Reserve-Flight">
  <owl:intersectionOf rdf:parseType="Collection">
```

```
<tl:Qualification>
<tl:onVariable rdf:resource="#x"/>
  <tl:bindVariable>
    <owl:Restriction>
      <owl:onProperty rdf:resource="#TICKET"/>
      <tl:selectValuesFrom rdf:resource="#Unreserved"/>
    </owl:Restriction>
  </tl:bindVariable>
</tl:Qualification>
<tl:Qualification>
  <tl:onVariable rdf:resource="#y"/>
  <tl:bindVariable>
    <owl:Restriction>
      <owl:onProperty rdf:resource="#TICKET"/>
      <tl:selectValuesFrom rdf:resource="#Reserved"/>
    </owl:Restriction>
  </tl:bindVariable>
</tl:Qualification>
</owl:intersectionOf>
</tl:TemporalClass>
<tl:TemporalVariable rdf:about="#x">
  <tl:finishedBy rdf:resource="#tl:NOW"/>
</tl:TemporalVariable>
<tl:TemporalVariable rdf:about="#y">
  <tl:metBy rdf:resource="#tl:NOW"/>
</tl:TemporalVariable>
```

Fig. 4. The RDF/XML representation of the `Reserve-Flight` concept

3.5   RDF-Compatible Model-Theoretic Semantics

The model-theoretic semantics for TL-OWL is defined as an extension of the RDF semantics. There is a correspondence between the direct model-theoretic semantics for the abstract syntax and the semantics defined in this section. As a way noted in the OWL specification, if any conflict should ever arise between these two forms, then the direct model-theoretic semantics takes precedence.

From the RDF semantics and the OWL semantics, for V a set of URI references and literals containing the RDF and RDFS vocabulary and D a datatype map, a D-interpretation of V is a tuple $I = \langle R_I, T_I, P_I, EXT_I, S_I, L_I, LV_I \rangle$. R_I is the domain of discourse. T_I is the time intervals of I, P_I is a subset of R_I, the properties of I. EXT_I is used to give meaning to properties, and is a mapping from P_I to $P(R_I \times R_I)$. S_I is a mapping from URI references in V to their denotations in R_I. L_I is a mapping from typed literals in V to their denotations in R_I. LV_I is a subset of R_I that has literal values.

The set of classes C_I is defined as $C_I = \{x \in R_I \mid \langle x, S_I(rdfs:Class) \rangle \in EXT_I$ $(S_I(rdf:type)) \land \langle T_I, x \rangle \in C_I\}$, and the mapping $CEXT_I$ from C_I to $P(R_I)$ is defined as $CEXT_I(c) = \{x \in R_I \mid \langle x, c \rangle \in EXT_I(S_I(rdf:type)) \land \langle T_I, x \rangle \in C_I\}$.

Definition 7. *Let D be a datatype map that includes datatypes for rdf:XML Literal, xsd:integer and xsd:string. A TL-OWL interpretation, $I = \langle R_I, T_I, P_I, EXT_I, S_I, L_I, LV_I \rangle$, of a vocabulary V, where V includes the RDF and RDFS vocabularies, is a D-interpretation of V that satisfies all the constraints in this section.*

The following tables from Table 6 to Table 10 give the constraints of TL-OWL directives and constructs for the RDF-compatible model-theoretic semantics with the D-interpretation.

Table 6. Conditions concerning parts of TL-OWL universe and syntactic categories

If E is	then			Note
	$S_I(E) \in$	$CEXT_I(S_I(E)) =$	and	
tl:TemporalClass	C_I	ITC	$ITC \subseteq C_I$	This defines ITC as the set of TL-OWL classes
tl:TemporalVariable		ITV	$ITV \subseteq C_I$	This defines ITV as the set of TL-OWL temporal variables
tl:Qualification	C_I	ITQ	$ITQ \subseteq C_I$	This defines ITQ as the set of TL-OWL qualifications

Table 7. Characteristics of TL-OWL classes and properties

If E is	then if $e \in CEXT_I(S_I(E))$ then	Note
tl : TemporalClass	$CEXT_I(e) \subseteq IOT$	Instances of TL-OWL classes are TL-OWL individuals.
tl : TemporalVariable	$CEXT_I(e) \subseteq T$	TL-OWL temporal variables are time intervals.
If E is	**then $c \in CEXT_I(S_I(E))$ iff $c \in$ IOOP $\cup$ IODP and**	**Note**
tl : Parametric-FunctionalProperty	$\langle x, y_1 \rangle, \langle x, y_2 \rangle \in EXT_I(c)$ implies $y_1 = y_2$, independently from time	Both individual-valued and datatype properties can be parametric functional properties.
If E is	**then $c \in CEXT_I(S_I(E))$ iff $c \in$ IOOP and**	**Note**
tl : InverseParametric-FunctionalProperty	$\langle x_1, y \rangle, \langle x_2, y \rangle \in EXT_I(c)$ implies $x_1 = x_2$, independently from time	Individual-valued properties can be inverse parametric functional properties.

Table 8. Conditions on TL-OWL restrictions and qualifications

If	then $x \in$ IOR, $y \in$ IOC $\cup$ IDC, $p \in$ IOOP $\cup$ IODP, and $CEXT_I(x) =$
$\langle x, y \rangle \in EXT_I(S_I(\text{tl:selectValuesFrom})) \land \langle x, p \rangle \in EXT_I(S_I(\text{owl:onProperty}))$	$\{ u \in domp \mid EXT_I(p)(u) \in CEXT_I(y) \}$
If	**then $z \in$ ITQ, $d \in$ IOC $\cup$ IDC, $x,y \in$ ITV, and $CEXT_I(z) =$**
$\langle z, d \rangle \in EXT_I(S_I(\text{tl:bindVariable})) \land \langle z, x \rangle \in EXT_I(S_I(\text{tl:onVariable}))$	$\{ u \in IOT \mid u \in CEXT_I(d), t = V(x) \}$
$\langle z, d \rangle \in EXT_I(S_I(\text{tl:bindSubstitutiveVariable})) \land \langle z, x \rangle \in EXT_I(S_I(\text{tl:onVariable})) \land \langle z, y \rangle \in EXT_I(S_I(\text{tl:onSubstitutiveVariable}))$	$\{ u \in IOT \mid u \in CEXT_I(d), H = H \cup \{ y \rightarrow V(x) \} \}$

Table 9. Conditions on TL-OWL features

If E is	then $\langle x, y \rangle \in EXT_I(S_I(E))$ iff
tl : agreementOf	$x \in IOC$, y is a sequence of p,q over IOOP $\cup$ IODP, $CEXT_I(x) = \{ u \in domp \cap domq \mid EXT_I(p)(u) = EXT_I(q)(u) \}$
tl : disagreementOf	$x \in IOC$, y is a sequence of p,q over IOOP $\cup$ IODP, $CEXT_I(x) = \{ u \in domp \cap domq \mid EXT_I(p)(u) \neq EXT_I(q)(u) \}$
tl : undefinednessOf	$x \in IOC$, y is a partial function over IOOP $\cup$ IODP, $CEXT_I(x) = IOT\text{-}domp$
tl : pathOf	$x \in IOOP$, y is a sequence of p,q over IOOP, and $u \in domp$ and $v \in domq$, $\langle u, v \rangle \in EXT_I(p) \cap \langle v, w \rangle \in EXT_I(q)$ implies $\langle u, w \rangle \in EXT_I(x)$

Table 10. Conditions on TL-OWL temporal relations

If E is	then x $\in$ CEXT$_I$(S$_I$(E)) iff
tl : NOW	CEXT$_I$(x) is the current interval
If E is	$\langle$x,y$\rangle$ $\in$ EXT$_I$(S$_I$(E)) iff
tl : before	CEXT$_I$(x)=[u$_1$,v$_1$] $\wedge$ CEXT$_I$(y)=[u$_2$,v$_2$] implies v$_1$ < u$_2$
other temporal relations ...	...

We now show that the two model theories, the direct model-theoretic semantics from Sect. 3.3 and the RDF-compatible model-theoretic semantics from this
section, have the same consequences on TL-OWL ontologies that can be written
in the abstract syntax.

Definition 8. *Let K and Q be collections of RDF graphs and D be a datatype
map. Then K* **TL-OWL entails** *Q with respect to D iff every TL-OWL interpretation with respect to D (of any vocabulary V that includes the RDF and RDFS
vocabularies and the TL-OWL vocabulary) that satisfies all the RDF graphs in
K also satisfies all the RDF graphs in Q. K is* **TL-OWL consistent** *iff there
is some TL-OWL interpretation that satisfies all the RDF graphs in K.*

Theorem 1. *Let O and O' be collections of TL-OWL ontologies and axioms
and facts in abstract syntax form. Given a datatype map D that maps xsd:string
and xsd:integer to the appropriate XML Schema datatypes and that includes the
RDF mapping for rdf:XMLLiteral, then O entails O' with respect to D if and
only if the translation of O TL-OWL entails the translation of O' with respect
to D.*

Proof (sketch): This theorem can be proved by a structural induction for all of
directives and constructs, but the description of its complete proof is too long.
Therefore, in this paper we only introduce the outline of the proof due to the
restriction of pages.

Given a datatype map D, a separated TL-OWL vocabulary is defined into
a set of URI references V' = VO + VC + VTC + VTV + VD + VI + VOP
+ VDP + VAP + VXP. The translation of the separated TL-OWL vocabulary
T(V') consists of all the triples of the form

v rdf:type owl:Ontology. v $\in$ VO, v rdf:type owl:Class. v $\in$ VC, v rdf:type
tl:TemporalClass. v $\in$ VTC, v rdf:type tl:TemporalVariable. v $\in$ VTV, v rdf:type
rdfs:Datatype. v $\in$ VD, v rdf:type owl:Thing. v $\in$ VI, v rdf:type owl:Object
Property. v $\in$ VOP, v rdf:type owl:DatatypeProperty. v $\in$ VDP, v rdf:type
owl:AnnotationProperty. v $\in$ VAP, v rdf:type owl:OntologyProperty. v $\in$ VXP.

Further, a collection of TL-OWL ontologies, axioms, and facts in abstract
syntax form, O, with a separated vocabulary is defined with the new notion of
a separated vocabulary V = VO + VC + VTC + VTV + VD + VI + VOP
+ VDP + VAP + VXP, where all URI references used as ontology names are
taken from VO, class IDs are taken from VC, temporal class IDs are taken from
VTC, temporal variable IDs are taken from VTV, datatype IDs are taken from
VD, individual IDs are taken from VI, individual-valued property IDs are taken

from VOP, data-valued property IDs are taken from VDP, annotation property IDs are taken from VAP, and ontology property IDs are taken from VXP.

Then the above theorem can be paraphrased as the following : Let O and O' be collections of TL-OWL ontologies, axioms, and facts in abstract syntax form. Then O direct entails O' if and only if T(O) TL-OWL entails T(O').

In order to prove this theorem, first, we inductively check whether all the constructs of descriptions and qualifications and all the directives of TemporalClass, TemporalVariable, ObjectProperty and DatatypeProperty satisfy the above theorem.

Suppose O entails O'. Let I be a TL-OWL DL interpretation that satisfies T(O). Then from the above structural induction, there is some direct interpretation I' such that for any abstract TL-OWL ontology or axiom or fact X over V', I satisfies T(X) iff I' satisfies X. Thus I' satisfies each ontology in O. Because O entails O', I' satisfies O', so I satisfies T(O'). Thus T(K),T(V') TL-OWL DL entails T(Q). Conversely, suppose T(O) TL-OWL DL entails T(O'). Let I' be an direct interpretation that satisfies K. Then from the above structural induction, there is some TL-OWL DL interpretation I such that for any abstract TL-OWL ontology X over V', I satisfies T(X) iff I' satisfies X. Thus I satisfies T(O). Because T(O) TL-OWL DL entails T(O'), I satisfies T(O'), so I' satisfies O'. Thus O entails O'. Consequently, by the correspondence of two semantics, we can conclude that O direct entails O' if and only if T(O) TL-OWL entails T(O')

3.6 Reasoning in TL-OWL

Artale and Franconi present a subsumption reasoning for $\mathcal{TL}\text{-}\mathcal{ALCF}$. The calculus is based on the idea of separating the inference on the temporal part($\mathcal{TL}$) from the inference on the DL part($\mathcal{ALCF}$). This is achieved by first looking for a normal form of concepts. The normalization procedure generates a completed existential form of the form: $\Diamond(\overline{X})\overline{Tc}.(Q_0 \sqcap Q_1@X_1 \sqcap ... \sqcap Q_n@X_n)$, where each Q is a non-temporal concept, $\overline{X}$ is a set of temporal variable and $\overline{Tc}$ is a set of temporal constraints. Thanks to the completed existential form, concept subsumption in $\mathcal{TL}\text{-}\mathcal{ALCF}$ can be reduced to concept subsumption between non-temporal concepts and to subsumption between temporal constraint networks, i.e., a labeled directed graph $\langle \overline{X}, \overline{Tc}, \overline{Q@X} \rangle$, where arcs are labeled with a set of arbitrary temporal relationship and nodes are labeled with non-temporal concepts. Moreover, Artale and Franconi show that the subsumption problem in $\mathcal{TL}\text{-}\mathcal{ALCF}$ can be reduced to the subsumption between $\mathcal{ALCF}$ concepts, i.e., the non-temporal part (see Theorem 7.11 in [2]).

The temporal description logic to formalize TL-OWL is $\mathcal{TL}\text{-}\mathcal{SHOIN}(\mathbf{D})$, which extends the non-temporal part($\mathcal{ALCF}$) of $\mathcal{TL}\text{-}\mathcal{ALCF}$ to $\mathcal{SHOIN}(\mathbf{D})$. By this extension this logic can have the expressivity of OWL DL. Moreover, if the completed existential form for $\mathcal{TL}\text{-}\mathcal{SHOIN}(\mathbf{D})$ can be shown by the normalization procedure, the reduction of $\mathcal{TL}\text{-}\mathcal{ALCF}$ can be also used in $\mathcal{TL}\text{-}\mathcal{SHOIN}(\mathbf{D})$.

Most steps of the normalization procedure shown in $\mathcal{TL}\text{-}\mathcal{ALCF}$ is for the temporal part so that it can be sufficient that only simple form[1] of $\mathcal{TL}\text{-}\mathcal{SHOIN}(\mathbf{D})$

---

[1]Artale calls a negation normal form to a simple form.

is defined to obtain the completed existential form for $\mathcal{TL\text{-}SHOIN}(\mathbf{D})$. We define the simple form excluding those of $\mathcal{TL\text{-}ALCF}$ as follows. These are similar to simple forms presented in [7] and the simple form of $\mathcal{TL\text{-}ALCF}$ is also presented in Fig. 14 of [2].

$$\neg \leqslant nR.C \rightarrow \geqslant (n+1)R.C \qquad \neg \exists T.d \rightarrow \forall T.\neg d$$
$$\neg \geqslant (n+1)R.C \rightarrow \leqslant nR.C \qquad \neg \forall T.d \rightarrow \exists T.\neg d$$
$$\neg \geqslant 0R.C \rightarrow C \sqcap \neg C$$

As for $\mathcal{TL\text{-}ALCF}$ case, without loss of generality, the normalization procedure for $\mathcal{TL\text{-}SHOIN}(\mathbf{D})$ reduces the *subsumption problem* in $\mathcal{TL\text{-}SHOIN}(\mathbf{D})$ to the *subsumption* between $\mathcal{SHOIN}(\mathbf{D})$ concepts. It can be also shown that a $\mathcal{TL\text{-}SHOIN}(\mathbf{D})$ concept in completed existential form, $\langle \overline{X}, \overline{Tc}, \overline{Q@X} \rangle$, is *satisfiable* if and only if the non-temporal concepts labeling each node in $\overline{X}$ are *satisfiable*. Moreover, following the above reduction, the subsumption problem of $\mathcal{TL\text{-}SHOIN}(\mathbf{D})$ is decidable because that of $\mathcal{SHOIN}(\mathbf{D})$ is decidable. The proof is similar to the one for $\mathcal{TL\text{-}ALCF}$ (see Proposition 7.8 and Theorem 7,11 in [2]) and we here do not mention it because it is straightforward.

4   Related Work

OWL-Time ontology [13] has recently been proposed to describe the temporal contents of Web pages as well as the temporal properties of Web services. The OWL-Time was formerly the DAML-Time and is currently published as the status of W3C working draft. The ontology provides a vocabulary to represent the topological relations among instants and intervals, along with information about duration and datetime. It is also shown how the ontology can be used within OWL-S by several examples. OWL-Time however is not an ontology language, but a time ontology based on OWL. It therefore have to use the rule-based reasoning to solve the problem as introduced in Sect. 1.

[11] and [8] present a new temporal ontology language based on OWL, which is a similar approach to ours. In order to represent time and temporal aspects such as change in ontologies, they introduce time slices (the temporal parts of an individual) and fluents (properties that hold between timeslices) and define their semantics based on Description Logic. They however do not show the formal reasoning algorithm such as subsumption and entailment. We can not also be convinced that the reasoning for the language is decidable.

5   Conclusions

OWL as an ontology language for Semantic Web can represent and reason the knowledge for information resources on web. It however cannot perform the reasoning for temporal information, since OWL cannot represent the temporal semantics. Therefore, we propose an interval-based temporal ontology language TL-OWL, which can represent and reason time information on Semantic Web. In order to formalize TL-OWL the abstract and the exchange syntax of TL-OWL and their model semantics are defined. These two semantics are also shown to have the same consequences on TL-OWL ontologies that can be written in the abstract syntax. Finally, we show that the resoning in TL-OWL is decidable.

References

1. Allen, J.F.: Maintaining knowledge about temporal intervals. In: Communications of the ACM, vol. 26(11), pp. 832–843. ACM, New York (1983)
2. Artale, A., Franconi, E.: A Temporal Description Logic for Reasoning about Actions and Plans. Journal of Artificial Intelligence Research 9, 463–506 (1998)
3. Artale, A., Franconi, E.: A Survey of Temporal Extensions of Description Logics. Annals of Mathematics and Artificial Intelligence 4(1), 171–210 (2001)
4. Bettini, C.: Time dependent concepts: Representation and reasoning using temporal description logics. Data & Knowledge Engineering 22(1), 1–38 (1997)
5. Halpern, J.Y., Moses, Y.: A Propositional Modal Logic of Time Intervals. Journal of ACM 38(4), 935–962 (1991)
6. Hollunder, B., Nutt, W.: Subsumption Algorithms for Concept Languages, Technical Research Report RR-90-04, DFKI, Germany (1990)
7. Horrocks, I., Sattler, U.: A tableaux decision procedure for SHOIQ. In: Proc. of the 19th International Joint Conference on Artificial Intelligence, pp. 448–453 (2005)
8. Milea, V., Frasincar, F., Kaymak, U., Noia, T.: An OWL-based Approach Towards Representing Time in Web Information Systems. In: Proc. of the 4th International Workshop of Web Information Systems Modeling Workshop, pp. 791–802 (2007)
9. Schild, K.D.: Combining terminological logics with tense logic. In: Proc. of the 6th Portuguese Conference on Artificial Intelligence (1993)
10. Schmiedel, A.: A temporal terminological logic. In: Proc. of the AAAI 1990, pp. 640–645 (1990)
11. Welty, C., Fikes, R., Makarios, S.: A Reusable Ontology for Fluents in OWL. In: Proc. of the International Conference on Formal Ontology in Information Systems, pp. 226–236 (2006)
12. Wolter, F., Zakharyaschev, M.: Temporalizing description logics, Frontiers of Combining Systems. Studies Press-Wiley, Chichester (1999)
13. W3C Working Draft, Time Ontology in OWL (2006), http://www.w3.org/TR/owl-time

A Formal Semantics-Preserving Translation from Fuzzy Relational Database Schema to Fuzzy OWL DL Ontology

Fu Zhang, Z. M. Ma, Hailong Wang, and Xiangfu Meng

College of Information Science & Engineering, Northeastern University,
Shenyang, 110004, China
mazongmin@ise.neu.edu.cn

Abstract. How to construct Web ontologies has become a key technology to enable the Semantic Web, especially how to construct ontologies by extracting domain knowledge from database models such as the relational database model. But in real-world applications, information is often imprecise and uncertain, thus the formal approach to translation from Fuzzy Relational Database Schema (FRDBS) to fuzzy ontology is helpful for extracting domain knowledge from database, which can profitably support fuzzy ontology development and developing data-intensive Semantic Web applications. In this paper, we first give the formal definition of FRDBS. Then, the formal definition and Model-Theoretic semantics of a kind of new fuzzy OWL DL ontology are given in more detail. What's more, we realize the formal translation from FRDBS to fuzzy OWL DL ontology by means of reverse engineering technique. Of course, the correctness of translation is also proved. With an example, it shows that the translation method is semantics-preserving and effective. Finally, the reasoning problem of satisfiability, subsumption, and redundancy of FRDBS may reason automatically through reasoning mechanism of the corresponding fuzzy description logic f-SHOIN(D) of fuzzy OWL DL ontology is also investigated, which can further contribute to constructing fuzzy OWL DL ontologies exactly that meet application's needs well.

Keywords: fuzzy database, Fuzzy Relational Database Schema (FRDBS), fuzzy ontology, fuzzy OWL DL ontology, semantics-preserving translation.

1 Introduction

How to construct Web ontologies that meet applications' needs has become a key technology to enable the Semantic Web [2]. To this end, many kinds of formal methods and ontology tools have been built to constructing Web ontologies. In particularly, the methods by extracting domain knowledge from database models (such as relational database model) have been extensively investigated. Stojanovic [21] presented the mapping rules from relational model to ontology. In [1], the mapping between relational database and ontology was also established. In [24], an approach to creating mappings between relational database schema and OWL ontology is presented. Kasgyap [11] also proposed how to obtain the elementary ontology according to relational database schema. In [23], a formal approach for translating ER schema

J. Domingue and C. Anutariya (Eds.): ASWC 2008, LNCS 5367, pp. 46–60, 2008.
© Springer-Verlag Berlin Heidelberg 2008

into OWL DL ontology is presented. Additionally, Ontology tools such as Protégé-2000 and OntoEdit [9] that can be used for building or reusing ontologies.

However, the above researches were not sufficient for handling vague or ambiguous information. In real-world application, information is often vague or ambiguous. Such as in relational database, the *Inheritance* (*ISA*) relation between two *tables* (Young-Employee and Employee) may be fuzzy with membership degree of [0,1]. Thus, the problems that emerge are how to represent these non-crisp data within the database model and ontology definition. For this purpose, Medina [14] proposed a generalized model of fuzzy relational databases. In [12], Ma presented a conceptual design methodology for fuzzy relational databases.

Also, in order to introduce fuzziness in ontologies, Thomas [22] introduced the need for fuzzy probabilistic formalisms on the Semantic Web. In [17], Parry used fuzzy ontologies, and presented a broad survey of relevant techniques, leading up to the notions of fuzzy search and fuzzy ontologies. Ceravolo [5] presented an approach to build fuzzy ontologies in a bottom-up fashion. In [6], Calegari defined a framework consisting of a fuzzy ontology based on fuzzy Description Logic and fuzzy OWL. In [19], Sanchez introduced a fuzzy ontology structure, Lexicon and Knowledge Base.

Although there have been several kinds of fuzzy extension of relational database model and ontologies, no research on how to translate Fuzzy Relational Database Schema (FRDBS) into fuzzy ontology has been done. If the translation is realized, which will be contribute to constructing Web ontologies by extracting domain knowledge from fuzzy database, it can profitably support fuzzy ontology development and developing data-intensive Semantic Web applications. This is also the motivation of our research.

In this paper, we try to answer the following questions about the relationship between FRDBS and fuzzy OWL DL ontology:

- How to formalize the Fuzzy Relational Database Schema? We solve this problem in section II.
- How to give the formal definition and Model-Theoretic semantics of fuzzy OWL DL ontology? Aiming at the characteristics and requirement of fuzzy ontology, and based on the fuzzy OWL DL, we solve this problem in section III.
- How to establish relationship between FRDBS and fuzzy OWL DL ontology? The answers are (i) In order to realize the translation from FRDBS to fuzzy OWL DL ontology well, firstly, we extract the fuzzy Entity-Relationship model (fuzzy ER model) from FRDBS by reverse engineering technique, and why we take this method is also introduced in section 2.3 in more detail. Then, we realize the translation from FRDBS to fuzzy OWL DL ontology by translating the extracted fuzzy ER model into fuzzy OWL DL ontology with a semantics-preserving translation algorithm. Additionally, to our knowledge, no researches on how to extract fuzzy ER model from FRDBS and how to translate fuzzy ER model into fuzzy ontology have been done respectively. (ii) Proving that the translations are "semantics preserving". (iii) As a fuzzy OWL DL ontology is being equivalent to a description logic f-SHOIN(D) knowledge base [18], the reasoning problem of satisfiability, subsumption, and redundancy of FRDBS may be reasoned automatically through reasoning mechanism of fuzzy description logic f-SHOIN(D) is also investigated, which can further contribute to constructing fuzzy OWL DL ontologies exactly that meet application's needs.

Summarizing, the existence of semantics-preserving translation from FRDBS to fuzzy OWL DL ontology is helpful for constructing Web ontologies by extracting domain knowledge from fuzzy database model, which can profitably support fuzzy ontology development and developing data-intensive Semantic Web applications.

The remainder of this paper is organized as follows. Section II introduces the FRDBS and fuzzy ER model, presents the method that how to extract fuzzy ER model from FRDBS, and why we take this method is also illustrated. Section III introduces the fuzzy OWL DL ontology. In section IV, we realize the formal translation from FRDBS to fuzzy OWL DL ontology. The reasoning problems of FRDBS are also studied. Section V shows the general conclusions and further work.

2 The Fuzzy Relational Database Schema

In this section, we give the formal definition of Fuzzy Relational Database Schema (FRDBS), which is the fuzzy extension of the traditional relational database schema [26]. The fuzzy ER model is introduced, and a kind of new method that how to extract fuzzy ER model from FRDBS is presented. Additionally, why we take the method of extracting fuzzy ER model from FRDBS is also illustrated in detail.

2.1 The Formal Definition of FRDBS

In brief, a Fuzzy Relational Database Schema consists of a set of fuzzy relations FR, each fuzzy relations FR is corresponding to a table, and the columns of a fuzzy relation are called fuzzy attributes denoted as FA. The tuples in fuzzy relations reflect the fuzzy values of fuzzy schema, and they are content of fuzzy database. Here, we assume that all fuzzy relations are in the third normal form.

Definition 1. A Fuzzy Relational Database Schema is a tuple FRDBS = (LT$_{FRDBS}$, unique, pkey, fkey, $\leq_{FRDBS}$), where:

- LT$_{FRDBS}$ = ET$_{FS}$ $\cup$ RT$_{FS}$ $\cup$ D$_{FS}$ is a finite alphabet, where ET$_{FS}$ is a set of fuzzy entity table symbols describing entities in the real word; RT$_{FS}$ is a set of fuzzy relationship table symbols describing the fuzzy relationships between fuzzy entities; D$_{FS}$ is a set of fuzzy domain symbols, each fuzzy domain symbol FD has an associated predefined basic domain, and we assume various basic domains to be pairwise disjoint.
- unique : each fuzzy attribute FA $\in$ attr(FT) has an associated predefined basic domain, where FT $\in$ ET$_{FS}$ $\cup$ RT$_{FS}$, if FA has the unique value, then we have the Boolean function unique(FA)=1, otherwise, unique(FA)=0, where the function attr(FT) acquires the fuzzy attributes contained in a specific FT.
- pkey : for each FT $\in$ ET$_{FS}$ $\cup$ RT$_{FS}$, there is exactly one primary key pkey(FT), if pkey(FT) $\in$ attr(FT), then pkey(FT) is a single-attribute key, if pkey(FT) $\subseteq$ attr(FT), then pkey(FT) is a composite-attribute key.
- fkey : for each FT $\in$ ET$_{FS}$ $\cup$ RT$_{FS}$, there may be 1or n (n > 1) foreign keys fkey(FT, FG), where FG $\in$ ET$_{FS}$, fkey(FT, FG) $\subseteq$ attr(FT), and (fkey(FT,FG)) $\subseteq$ value(pkey(FG) $\cup$ Null). The function fkey(FT, FG) acquires the foreign keys (which point to the FG) in a given fuzzy relation FT, and value(*) denotes the value range of *.

- $\leq_{FRDBS}$ is a fuzzy binary relation over ET_{FS} that models a fuzzy *inheritance* (*ISA*) relation between two fuzzy entity tables ET_1, $ET_2 \in ET_{FS}$. If $pkey(ET_1) \subseteq pkey(ET_2)$, the $pkey(ET_1)$ is also the foreign key of ET_1, and $fkey(ET_1,FG) = fkey(ET_1,ET_2)$, i.e., all the foreign keys of ET_1 point to ET_2, then we have that $ET_1 \leq_{FRDBS} ET_2$, denotes ET_1 is subentity of ET_2, where $FG \in ET_{FS}$, $fkey(ET_1,FG)$ denotes the foreign keys of fuzzy entity table ET_1.

2.2 The Fuzzy ER Model

In order to establish the correspondence between FRDBS and fuzzy OWL DL ontology, in this paper, firstly, we extract the fuzzy Entity-Relationship model (fuzzy ER model) from FRDBS. So, we first give the formal definition and semantics of fuzzy ER model.

The fuzzy ER model [25] is fuzzy extension of traditional ER model [3], which is a graphic tool for concept modeling. The fuzzy ER schema is usually defined using a graphical notation which is particularly useful for an easy visualization of the data dependencies. In the following, for sets X and Y we call a function from a subset of X to Y an X-labeled tuple over Y. The labeled tuple T that maps $x_i \in X$ to $y_i \in Y$, for $i \in \{1,...,k\}$, is denoted $[x_1: y_1,...,x_k: y_k]$. We also write $T[x_i]$ to denote y_i [4].

Definition 2. A fuzzy ER schema is a tuple $FS = (L_{FS}, \leq_{FS}, att_{FS}, rel_{FS}, card_{FS})$, where:

- $L_{FS} = E_{FS} \cup A_{FS} \cup U_{FS} \cup R_{FS} \cup D_{FS}$ is a finite alphabet, where E_{FS} is a set of fuzzy entity symbols, A_{FS} is a set of fuzzy attribute symbols, U_{FS} is a set of fuzzy role symbols, R_{FS} is a set of fuzzy relationship symbols, and D_{FS} is a set of fuzzy domain symbols, each fuzzy domain symbol FD has an associated predefined basic domain D^{FB}, and we assume various basic domains to be pairwise disjoint.
- $\leq_{FS} \subseteq E_{FS} \times E_{FS}$ is a fuzzy binary relation over E_{FS}. $\leq_{FS}$ denotes the *ISA* relationships between two fuzzy entities.
- $att_{FS}: E_{FS} \rightarrow T(A_{FS}, D_{FS})$ is a function that maps each fuzzy entity symbol in E_{FS} to an A_{FS}-labeled tuple over D_{FS}. For simplicity we assume attributes to be single-valued and mandatory.
- rel_{FS} is a function that maps each fuzzy relationship symbol in R_{FS} to an U_{FS}-labeled tuple over E_{FS}. For each fuzzy role $FU \in U_{FS}$, there is a fuzzy relationship FR and a fuzzy entity FE such that $rel_{FS}(FR) = [...,FU:FE,...]$.
- $card_{FS}$ is a function from $E_{FS} \times R_{FS} \times U_{FS}$ to $N_0 \times (N_0 \cup \{\infty\})$ that satisfies the following condition: for a fuzzy relationship $FR \in R_{FS}$ such that $rel_{FS}(FR) = [...,FU:FE,...]$, define the $card_{FS}(FE,FR,FU) = (cmin_{FS}(FE,FR,FU), cmax_{FS}(FE,FR,FU))$. The function $card_{FS}$ specifies cardinality constraints.

Similarly for the semantics of classical ER model [4], the semantics of fuzzy ER model can be given by the fuzzy database state. A fuzzy database state is considered acceptable if it is legal.

Definition 3. Formally, a fuzzy database state FB is constituted by a nonempty finite set Δ^{FB}, assumed to be disjoint from all basic domains, and a function $\bullet^{FB}$ that maps:

- Every domain symbol $FD \in D_{FS}$ to the basic domain D^{FB}, that is $FD^{FB} \in D^{FB}$;
- Every fuzzy entity $FE \in E_{FS}$ to a subset FE^{FB} of Δ^{FB}, that is $FE^{FB} \subseteq \Delta^{FB}$;

- Every fuzzy attribute $FA \in A_{FS}$ to a set $FA^{FB} \subseteq \Delta^{FB} \times \cup FD^{FB}$, where $FD \in D_{FS}$;
- Every fuzzy relationship $FR \in R_{FS}$ to a set FR^{FB} of U_{FS}-labeled tuples over Δ^{FB}.

The elements of FE^{FB}, FA^{FB} and FR^{FB} are called instances of FE, FA, and FR respectively.

Definition 4. A fuzzy database state FB is said to be legal for a fuzzy ER schema FS $= (L_{FS}, \leq_{FS}, att_{FS}, rel_{FS}, card_{FS})$, if it satisfies the following conditions:

- For each pair of fuzzy entities FE_1, $FE_2 \in E_{FS}$ such that $FE_1 \leq_{FS} FE_2$, it holds that $FE_1{}^{FB} \subseteq FE_2{}^{FB}$;
- For each fuzzy entity $FE \in E_{FS}$, if $att_{FS}(FE) = [FA_1:FD_1,..., FA_k:FD_k]$, then for each instance $Fe \in FE^{FB}$ and for each $i \in \{1, ..., k\}$ the following holds: There is exactly one element $Fa_i = <Fe, Fd_i> \in FA_i{}^{FB}$, $Fd_i \in D_i{}^{FB}$, where first component is Fe, and second component of Fa_i is an element of $D_i{}^{FB}$;
- For each fuzzy relationship $FR \in R_{FS}$ such that $rel_{FS}(FR) = [FU_1:FE_1,..., FU_n:FE_n]$, all instances of FR are of the form $[FU_1:Fe_1, ..., FU_n:Fe_n]$, where $Fe_i \in FE_i{}^{FB}$, $i \in \{1, ..., n\}$;
- For each fuzzy relationship $FR \in R_{FS}$ such that $rel_{FS}(FR) = [...,FU:FE,...]$, for each instance $Fe \in FE^{FB}$, it holds that: $cmin_{FS}(FE, FR, FU) \leq \#\{ Fr \in FR^{FB} \mid Fr[FU] = Fe \} \leq cmax_{FS}(FE, FR, FU)$, where $\#\{...\}$ denotes base of the set $\{...\}$.

2.3 Extracting Fuzzy ER Model from FRDBS

To our knowledge, no research on how to create mappings between FRDBS and fuzzy OWL DL ontology has been done. In this paper, firstly, we will take the method of extracting fuzzy ER model from FRDBS for the following reasons.

In [7], it showed that the ER model is richer in semantics than relational database schema. Also, lots of researches focus on the reverse engineering technique, which aims to recover the data model of an existing database in order to apply the data model to a new application setting, thus the database reverse engineering is necessary to semantically enrich and document a database, and to avoid throwing away the huge amounts of data stored in existing legacy databases if the owner of an existing database wants to re-engineer, or maintain and adjust the database design [10]. Additionally, an ER model is the commonly used target of the reverse engineering process [26]. Furthermore, ER model is considered as a graphic tool for database design and the most widespread semantic data model, and the ontology also models the domain by class, attributes, and constraint condition, so the relationship between ER model and ontology is strong [23].

Based on the above analysis, in our method, we first extract the fuzzy ER model from FRDBS by the following Definition 5. Then we realize the formal translation from FRDBS into fuzzy OWL DL ontology by translating the fuzzy ER model into the fuzzy OWL DL ontology with a semantics-preserving translation algorithm in Section IV. The following Definition 5 will show that how to translate the FRDBS into the fuzzy ER model well.

Definition 5. Give a fuzzy relational database schema FRDBS, we can obtain the corresponding fuzzy ER model FS according to the following extracting rules:

Rule 1: Extracting *Entity*

- for each FT $\in$ $ET_{FS} \cup RT_{FS}$, if pkey(FT) only contain one attribute, then FT $\in$ ET_{FS}, that is to say, FT is mapped to a fuzzy entity in fuzzy ER model.
- for each FT $\in$ $ET_{FS} \cup RT_{FS}$, if pkey(FT) contain n attributes ($n > 1$), and there is at least one attribute which is not foreign key of FT, then FT $\in$ ET_{FS}, that is to say, FT is mapped to a fuzzy entity in fuzzy ER model.

Rule 2: Extracting *Associations*

- for each FT $\in$ $ET_{FS} \cup RT_{FS}$, if pkey(FT) contain n attributes ($n > 1$), and all the n attributes are also the foreign keys of FT, then FT $\in$ RT_{FS}, that is to say, FT is mapped to a fuzzy relationship in fuzzy ER model, we notice here that the fuzzy attributes on fuzzy relationships are omitted.
- for each FT $\in$ ET_{FS}, there are fuzzy relationships in fuzzy ER model between FT and fuzzy entities which FT point to them by foreign keys of FT, except for the case of *Rule* 3.

Rule 3: Extracting *Inheritance* relation

- for each ET_1, ET_2 $\in$ ET_{FS}, if pkey(ET_1) $\subseteq$ pkey(ET_2), pkey(ET_1) is also the foreign key of ET_1, and all the foreign keys of ET_1 point to the ET_2, then $ET_1 \leq_{FRDBS}$ ET_2 is mapped to the ET_1 *ISA* ET_2 in fuzzy ER model.

Rule 4: Extracting *cardinality constraints*

- for each ET_1, ET_2 $\in$ ET_{FS} with the associations exists between them, and pkey(ET_1) is the foreign key of ET_2. If foreign key of ET_2 can be empty, that is to say, each fuzzy instance of ET_2 at most corresponding to a fuzzy instance of ET_1, so ET_1 and ET_2 is 1: n; If foreign key of ET_2 can not be empty, that is to say, each fuzzy instance of ET_2 just corresponding to a fuzzy instance of ET_1, so ET_1 and ET_2 is 1:1.
- for each ET_1, ET_2 $\in$ ET_{FS} with the associations exists between them, if there are the foreign keys of ET_2 which point to the ET_1, on the contrary, the foreign keys of ET_2 which point to the ET_1 do not exist, that is to say, each fuzzy instance of ET_2 can correspond to n ($n \geq 1$) fuzzy instance of ET_1, so ET_1 and ET_2 is 1:n.
- for the cardinality constraints of fuzzy relationships extracted according to *Rule* 2, which are $m : n$.

3 Fuzzy OWL DL Ontology

In this section, we first investigate the fuzzy extension of OWL DL, i.e., fuzzy OWL DL. Then, based on fuzzy OWL DL, a kind of fuzzy OWL DL ontology is presented.

3.1 Fuzzy OWL DL

As a suitable ontology language OWL, which consists of three sub-languages of increasing expressive power, namely OWL Lite, OWL DL, and OWL Full, the OWL DL is the language chosen by the major ontology editors because it supports those users who want the maximum expressiveness without losing computational completeness and decidability of reasoning systems [15].

In this paper, our aim is to define a new fuzzy ontology language suitable to implement the fuzzy ontology. So, similarly for [20], we investigate the fuzzy extension of OWL DL, named fuzzy OWL DL in more detail.

As with OWL DL, fuzzy OWL DL has also two types of syntactic form, i.e., the frame-like style abstract syntax [16] and exchange syntax (RDF/XML syntax) [8]. Here, we use abstract syntax which facilitates access to and evaluation of the ontologies [23]. In addition, the fuzzy OWL DL can be approximately viewed as the expressive DL f-SHOIN(D) [18]. So, in Table 1, we give the fuzzy OWL DL abstract syntax, the corresponding Description Logic syntax and Model-Theoretic semantics.

Based on the characteristics of FRDBS and fuzzy ER model, we only take into account of the partial constructors of fuzzy OWL DL abstract syntax, which makes it enough to establish a precise correspondence with FRDBS.

Table 1. Fuzzy OWL DL Abstract Syntax, Description Logics Syntax, and Semantics

fuzzy OWL DL Abstract Syntax	Description Logics Syntax	Model-Theoretic Semantics
fuzzy class description (C)		
A, which is a URIref of a fuzzy class	A	$A^{FI}: \Delta^{FI} \to [0,1]$
owl:Thing	$\top$	$\top^{FI}(d)=1$
owl:Nothing	$\bot$	$\bot^{FI}(d)=0$
restriction (R allValuesFrom(C))	$\forall R.C$	$(\forall R.C)^{FI}(d) = \inf_{d' \in \Delta^{FI}} \{\max\{1 - R^{FI}(d,d'), C^{FI}(d')\}$
restriction (R minCardinality(n))	$\geq nR$	$(\geq nR)^{FI}(d) = \sup_{c_1,\dots,c_n \in \Delta^{FI}} \wedge_{i=1}^{n} R^{FI}(d,c_i)$
restriction (R maxCardinality(n))	$\leq nR$	$(\leq nR)^{FI}(d) = \inf_{c_1,\dots,c_{n+1} \in \Delta^{FI}} \vee_{i=1}^{n+1}(1 - R^{FI}(d,c_i))$
restriction (U allValuesFrom(D))	$\forall U.D$	$(\forall U.D)^{FI}(d) = \inf_{v \in \Delta_D^{FI}} \{\max\{1 - U^{FI}(d,v), D^{FI}(v)\}\}$
fuzzy class axioms		
Class(A partial $C_1 \dots C_n$)	$A \sqsubseteq C_1 \cap \dots \cap C_n$	$A^{FI}(d) \leq \min(C_1^{FI}(d),\dots, C_n^{FI}(d))$
SubClassOf($C_1 C_2$)	$C_1 \sqsubseteq C_2$	$C_1^{FI}(d) \leq C_2^{FI}(d)$
DisjointClasses($C_1 \dots C_n$)	$C_i \neq C_j$	$\min(C_i^{FI}(d), C_j^{FI}(d)) = 0$ $1 \leq i < j \leq n$
fuzzy property axioms		
ObjectProperty (R domain(C_1)…domain(C_m) range(C_1)… range(C_k) [inverseOf(R_0)])	$\geq 1R \sqsubseteq C_i$, $\top \sqsubseteq \forall R.C_i$ $R = (R_0)^{-}$	$R^{FI}(d_1, d_2) \leq C_i^{FI}(d_1)$ i=1,…,m $R^{FI}(d_1, d_2) \leq C_i^{FI}(d_2)$ i=1,…,k $R^{FI}(d_1, d_2) = R_0^{FI}(d_2, d_1)$
DatatypeProperty (U domain(C_1)…domain(C_m) range(D_1)…range(D_k) [Functional])	$\geq 1U \sqsubseteq C_i$, $\top \sqsubseteq \forall U.D_i$ $\top \sqsubseteq \leq 1U$	$U^{FI}(d, v) \leq C_i^{FI}(d)$ i=1,…,m $U^{FI}(d, v) \leq D_i^{FI}(v)$ i=1,…,k $\forall d \in \Delta^{FI}$ #$\{v \in \Delta_D : U^{FI}(d, v) \geq 0\} \leq 1$

The interpretation is given by a pair $< \Delta^{FI}, \bullet^{FI} >$ where Δ^{FI} is a set of objects with empty intersection with the concrete domain Δ_D: $\Delta^{FI} \cap \Delta_D = \varnothing$. A concrete fuzzy domain is a pair $< \Delta_D, \phi_D >$, where Δ_D is an interpretation domain and ϕ_D is the set of concrete fuzzy domain predicates D with a predefined arity n and a fuzzy interpretation D^{FI}: $\Delta_D^n \to [0, 1]$, which is a n-ary fuzzy relation over Δ_D. Individuals d are mapped to objects in Δ^{FI}: $d^{FI} \in \Delta^{FI}$, whereas concrete individuals v are mapped to objects in the concrete domain: $v^{FI} \in \Delta_D$. A concept C is interpreted as a fuzzy set on the domain C^{FI}: $\Delta^{FI} \to [0, 1]$. Abstract roles R and concrete roles U are interpreted as fuzzy binary relations, respectively: R : $\Delta^{FI} \times \Delta^{FI} \to [0, 1]$ and U : $\Delta^{FI} \times \Delta_D \to [0, 1]$. In Table 1, where $\forall$ d^{FI}, $c^{FI} \in \Delta^{FI}$, $v^{FI} \in \Delta_D$, $\bullet^{FI}$ is a fuzzy interpretation function, C denotes fuzzy class description, D denotes fuzzy data range, fuzzy ObjectProperty identifiers and fuzzy DatatypeProperty identifiers are denoted by R and U, respectively, n is a nonnegative integer, #S denotes the cardinality of a set S.

3.2 Fuzzy OWL DL Ontology

The fuzzy OWL DL ontology is fuzzy ontology with ontology language fuzzy OWL DL, which is fuzzy extension of OWL DL ontology [23]. Here, we do not consider the aspects that contain annotation, individual identifiers, individual axioms, RDF Literal, and partial fuzzy OWL constructors, which have no effect on our work. The Model-Theoretic semantics of fuzzy OWL DL ontology can be obtained in Table 1.

Definition 6. A fuzzy OWL DL ontology is a tuple FOO = $(FID_0, FAxiom_0)$, where:

- $FID_0 = FCID_0 \cup FDRID_0 \cup FOPID_0 \cup FDPID_0$ is a finite fuzzy OWL DL identifier set (see Table 1) partitioned into:
  (1) a subset $FCID_0$ of fuzzy class identifiers including user-defined identifiers plus two predefined fuzzy classes owl:Thing and owl:Nothing; fuzzy classes are either fuzzy entity classes describing fuzzy entities or fuzzy relationship classes describing the fuzzy relationships between fuzzy entities.
  (2) a subset $FDRID_0$ of fuzzy data range identifiers; each fuzzy data range identifier is predefined XML Schema fuzzy datatypes, which was discussed simply in [13].
  (3) a subset $FOPID_0$ of fuzzy object property identifiers; fuzzy object properties link individuals (i.e., fuzzy entities) to individuals.
  (4) a subset $FDPID_0$ of fuzzy datatype property identifiers; fuzzy datatype properties link individuals (i.e., fuzzy entities) to fuzzy data values.
- $FAxiom_0$ is a finite fuzzy OWL DL axiom set partitioned into a subset of fuzzy class axioms and a subset of fuzzy property axioms; each fuzzy axiom (see Table 1) is formed by applying fuzzy OWL DL constructs to the identifiers or descriptions that are the basic building blocks of a fuzzy class axiom and describe the fuzzy class either by a fuzzy class identifier or by specifying the extension of an unnamed anonymous fuzzy class via the construct restriction.

From a semantics point of view, a fuzzy OWL DL ontology FOO is a set of fuzzy OWL DL axioms in Table 1. We say that a fuzzy interpretation FI is a model of FOO iff it satisfies all axioms in FOO.

4 Mapping between FRDBS and Fuzzy OWL DL Ontology

This section establishes the relationships between FRDBS and fuzzy OWL DL ontology. Firstly, we realize the translation from FRDBS to fuzzy OWL DL ontology by translating the extracted fuzzy ER model into fuzzy OWL DL ontology with a semantics-preserving translation algorithm. Then, the reasoning problems of FRDBS are also investigated, which can contribute to constructing fuzzy OWL DL ontology.

4.1 Translating FRDBS into Fuzzy OWL DL Ontology

It showed that the semantics was preserved when translating the ER model into Description Logic (such as ALNUI) knowledge bases [4]. Similarly, with a fuzzy OWL DL ontology being equivalent to a fuzzy description logic f-SHOIN(D) knowledge base [18], therefore, there exists a formal and semantics-preserving approach for translating a fuzzy ER schema into a fuzzy OWL DL Ontology.

The following Algorithm 1 gives the formal approach for translating a fuzzy ER schema which is exacted from FRDBS into a fuzzy OWL DL ontology.

Algorithm 1. A semantics-preserving translating algorithm from fuzzy ER Schema which is exacted from FRDBS to fuzzy OWL DL Ontology:

Input: a fuzzy ER schema FS, which is exacted from a FRDBS by Definition 5.
Output: a fuzzy OWL DL Ontology FOO = φ (FS) = (FID$_0$, FAxiom$_0$).
Steps: applying the following transformation rules to fuzzy ER schema FS:
(1) The set FID$_0$ of fuzzy OWL DL identifier of φ(FS) contains following elements:

- For each fuzzy entity FE $\in$ E$_{FS}$, a fuzzy class identifier φ(FE) $\in$ FCID$_0$.
- For each fuzzy relationship FR $\in$ R$_{FS}$, a fuzzy class identifier φ(FR) $\in$ FCID$_0$.
- For each fuzzy attribute symbol FA in FS, a fuzzy datatype property identifier φ(FA) $\in$ FDPID$_0$.
- For each fuzzy domain symbol FD $\in$ D$_{FS}$ is mapped into a fuzzy data range identifier φ(FD) $\in$ FDRID$_0$, where each fuzzy data range identifier is predefined XML Schema fuzzy datatype, which was discussed simply in [13].
- For each fuzzy ER-role FU of fuzzy relationship FR associated with fuzzy entity FE is mapping into a pair of inverse fuzzy object property identifiers φ(FU) $\in$ FOPID$_0$ and FV = invof_φ(FU) $\in$ FOPID$_0$.

(2) The set FAxiom$_0$ of fuzzy OWL DL axiom of φ(FS) as follows:

- For each pair of fuzzy entities FE$_1$, FE$_2 \in$ E$_{FS}$ such that FE$_1 \leq_{FS}$ FE$_2$, a fuzzy class axiom: Class (φ(FE$_1$) partial φ(FE$_2$)).   (i)
- For each fuzzy entity FE $\in$ E$_{FS}$ with att$_{FS}$(FE) = [FA$_1$:FD$_1$,...,FA$_k$:FD$_k$], the following fuzzy axioms:

  A fuzzy class axiom: Class (φ(FE) partial restriction (φ(FA$_1$) allValuesFrom (φ(FD$_1$)) cardinality (1)) ... restriction (φ(FA$_k$) allValuesFrom (φ(FD$_k$)) cardinality (1)));   (ii)

  FOR i=1,2,...,k DO, A fuzzy property axiom: DatatypeProperty (φ(FA$_i$) domain (φ(FE)) range(φ(FD$_i$)) [Functional]).   (iii)

- For each fuzzy relationship $FR \in R_{FS}$ with $rel_{FS}(FR) = [FU_1:FE_1,...,FU_n:FE_n]$, the following fuzzy axioms:

  A fuzzy class axiom: Class ($\varphi(FR)$ partial restriction ($\varphi(FU_1)$ allValuesFrom ($\varphi(FE_1)$) cardinality (1)) ... restriction ($\varphi(FU_n)$ allValuesFrom ($\varphi(FE_n)$) cardinality(1)));   (iv)

  FOR i=1,2,...,n DO

      A fuzzy property axiom: ObjectProperty (FV_i domain ($\varphi(FE_i)$) range ($\varphi(FR)$) inverseOf $\varphi(FU_i)$);   (v)

      A fuzzy class axiom: Class ($\varphi(FE_i)$ partial restriction (FV_i allValuesFrom ($\varphi(FR)$))).   (vi)

- For each fuzzy ER-role FU of fuzzy relationship FR associated with fuzzy entity FE, $rel_{FS}(FR) = [...,FU:FE,...]$, the following fuzzy axioms:

  A fuzzy property axiom: ObjectProperty ($\varphi(FU)$ domain ($\varphi(FR)$) range ($\varphi(FE)$));   (vii)

  If $m = cmin_{FS}(FE, FR, FU) \neq 0$, a fuzzy class axiom: Class ($\varphi(FE)$ partial restriction (FV minCardinality (m)));   (viii)

  If $n = cmax_{FS}(FE, FR, FU) \neq 0$, a fuzzy class axiom: Class ($\varphi(FE)$ partial restriction (FV maxCardinality(n))).   (ix)

- For each pair of symbols X, $Y \in E_{FS} \cup R_{FS} \cup D_{FS}$, such that $X \neq Y$ and $X \in R_{FS} \cup D_{FS}$, a fuzzy class axiom: DisjointClasses ($\varphi(X)$ $\varphi(Y)$).   (x)

The time complexity of Algorithm 1 depends on the data structure of fuzzy ER model, basically, depends on the original fuzzy relational database schema FRDBS. Since all the fuzzy OWL DL identifiers of $\varphi(FS)$ in first step can be obtained by creating the fuzzy axioms in second step, the time complexity of Algorithm 1 can be analyzed as follows: Suppose the scale of fuzzy ER model FS is $N = N_{FE} + N_{FA} + N_{FR} + N_{FU}$, where N_{FE}, N_{FA}, N_{FR}, N_{FU} denotes the cardinality of sets of fuzzy entity symbols, fuzzy attribute symbols, fuzzy role symbols, and fuzzy relationship symbols, respectively. For simplicity we assume attributes to be single-valued and mandatory. The creating times of axiom (i) are $N_{FE}-1$ at most, similarly, axiom (ii) are N_{FE}, axiom (iii) are N_{FA}, axiom (iv) are N_{FR}, axioms (v), (vi), and (vii) are N_{FU}, axioms (viii) and (ix) are N_{FU} at most, and axiom (x) are $N_{FR} \times (N_{FE} + N_{FR}/2 - 1/2)$, so, in worst case, the total running times $T = 2N_{FE} + N_{FA} + N_{FR} + 5N_{FU} - 1 + N_{FR} \times (N_{FE} + N_{FR}/2 - 1/2) < N^2 + 5N$, that is, the time complexity of Algorithm 1 only needs $O(N^2)$ at most.

The following give the correctness of Algorithm 1. Here, based on the idea in [4], the correctness of Algorithm 1 can be proved by the Definition 7 and the Theorem 1.

Definition 7. Let $FS = (L_{FS}, \leq_{FS}, att_{FS}, rel_{FS}, card_{FS})$ be a fuzzy ER schema, $\varphi(FS)$ be a fuzzy OWL DL ontology obtained from FS by Algorithm 1, FI is a relation-fuzzy descriptive model of FOO, if for every fuzzy relationship $FR \in R_{FS}$ with $rel_{FS}(FR) = [FU_1:FE_1,...,FU_n:FE_n]$, for every $r_1, r_2 \in (\varphi(FR))^{FI}$, we have that:

$$(\wedge_{1 \leq i \leq k} \forall \ d \in \Delta^{FI}.(<r_1, d> \in (\varphi(FU_i))^{FI} \leftrightarrow <r_2, d> \in (\varphi(FU_i))^{FI})) \rightarrow (r_1 = r_2)$$

The above Definition 7 is implicit in the semantics of fuzzy ER model (that is, the corresponding FRDBS) that there cannot be two labeled tuples connected through all fuzzy roles of the fuzzy relationship to exactly the same elements of the domain.

Theorem 1. For every fuzzy ER schema FS exacted from FRDBS, φ(FS) be the fuzzy OWL DL ontology according to the Algorithm 1, there exist two mappings α_{FS}, from fuzzy database state FB with respect to FS to relation-fuzzy descriptive model of φ(FS), and β_{FS}, from relation-fuzzy descriptive model of φ(FS) to FB, such that:

- For each legal fuzzy database state FB for FS, there is α_{FS}(FB) is a relation-fuzzy descriptive model of φ(FS).
- For each relation-fuzzy descriptive model FI of φ(FS), there is β_{FS}(FI) is a legal fuzzy database state for FS.

Proof. Note that for each legal database state FB, we only consider the finite set of fuzzy values Δ^{FB}. Then a fuzzy interpretation α_{FS}(FB) of φ(FS) is defined as follows.

- α_{FS} is a mapping from the fuzzy database state corresponding to FS to the relation-fuzzy descriptive interpretation of φ(FS). So we know that the mapping of their corresponding elements also exists. That is, the domain elements $\Delta^{\alpha_{FS}(FB)}$ of fuzzy interpretation of φ(FS) is constituted by fuzzy values of the fuzzy database state FB, i.e., $\Delta^{\alpha_{FS}(FB)} = \dot{\Delta}^{FB} \cup \bigcup_{FR \in R_{FS}} FR^{FB}$. According to Definition 3, we know that each fuzzy relation FR of FS is assigned with a possibly U_{FS}-labeled tuples over Δ^{FB}. Then we have $\bigcup_{FR \in R_{FS}} FR^{FB}$ to explicitly represent the type structure of fuzzy relations in fuzzy OWL DL ontology.
- The fuzzy OWL DL identifier set FID_0 of φ(FS) in Algorithm 1 are defined:
  $(\varphi(X))^{\alpha_{FS}(FB)} = \{X^{FB} \mid X \in D_{FS} \cup E_{FS} \cup A_{FS} \cup R_{FS}\}$,
  and for each fuzzy relationship FR with $rel_{FS}(FR) = [FU_1:FE_1,\dots, FU_n:FE_n]$, we have $(\varphi(FU_i))^{\alpha_{FS}(FB)} = \{<Fr, Fe_i> \in \Delta^{\alpha_{FS}(FB)} \times \Delta^{\alpha_{FS}(FB)} \mid Fr \in FR^{FB} \wedge Fe_i \in FE_i^{FB} \wedge Fr[FU_i] \doteq Fe_i\}$, where $i = 1,\dots, n$.

Based on the above definition of α_{FS}(FB), we prove the first part of Theorem 1 by considering each case of the fuzzy OWL DL axiom set $FAxiom_0$ of φ(FS) in Algorithm 1. Here, we only consider the *Case 2*.

Case 2: Firstly, for a fuzzy instance $Fe \in (\varphi(FE))^{\alpha_{FS}(FB)}$, by definition of of $\alpha_{FS}(FB)$ above, we have $Fe \in FE^{FB}$. Then, according to Definition 4, there is exactly one element $Fa = <Fe, Fd_i> \in FA_i^{FB} = (\varphi(FA_i))^{\alpha_{FS}(FB)}$, $Fd_i \in D_i^{FB} = (\varphi(FD_i))^{\alpha_{FS}(FB)}$, where first component is Fe, and second component of Fa is an element of D_i^{FB}. Furthermore, by Definition 2, all fuzzy relations $\varphi(FA_i)$ corresponding to fuzzy attribute names FA_i is functional. That is, $\alpha_{FS}(FB)$ satisfies the fuzzy class axiom (ii). For the fuzzy class axiom (iii), according to Definition 3, we have $FA_i^{FB} \subseteq \Delta^{FB} \times \cup FD^{FB}$, according to Definition 4, there is exactly one element $<Fe, Fd_i> \in FA_i^{FB}$, $Fd_i \in FD^{FB}$, so we have $FA_i^{FB} \subseteq FE^{FB} \times \cup FD^{FB}$, then by definition of $\alpha_{FS}(FB)$ above, we have $(\varphi(FA_i))^{\alpha_{FS}(FB)} \subseteq (\varphi(FE))^{\alpha_{FS}(FB)} \times (\varphi(FD_i))^{\alpha_{FS}(FB)}$, since FA_i is functional by Definition 2, then $(\varphi(FA_i))^{\alpha_{FS}(FB)}$ is also functional. That is, $\alpha_{FS}(FB)$ satisfies the fuzzy class axiom (iii).

The proof of second part of Theorem 1 is omitted, which can be treated analogously according to the proof of first part, they are mutually inverse process.

Example 1. Firstly, we can obtain the fuzzy ER schema FS_1 in Fig. 1 by applying Definition 5 to the Fuzzy Relational Database Schema $FRDBS_1$ in Table 2. Then we translate the fuzzy ER schema FS_1 into the fuzzy OWL DL ontology $\varphi(FS_1)$ in Fig. 2 by applying Algorithm 1.

Note that: In Table 2, Number-YE $\subseteq$ Number-E denotes pkey(Young-Employee) $\subseteq$ pkey(Employee), Number-CL $\subseteq$ Number-L denotes pkey(Chief-Leader) $\subseteq$ pkey(Leader). In Fig. 1, the attributes of Employee and Leader are omitted. For brevity, the disjoint class axioms in Fig. 2 are omitted, and Fig. 2 is in the next page.

Table 2. The Fuzzy Relational Database Schema $FRDBS_1$

Table name	Attribute	Primary key	Foreign key & Reference
Employee	Number-E	Number-E	No
Leader	Number-L	Number-L	No
Young-Employee	Number-YE, Age	Number-YE	Number-YE (Number-E)
Chief-Leader	Number-CL	Number-CL	Number-CL (Number-L)
Mainly-Manage	No	Number-YE, Number-CL	Number-YE (Number-E), Number-CL (Number-L)

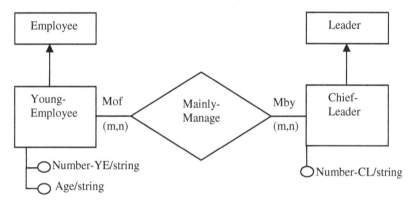

Fig. 1. The fuzzy ER schema FS_1 with respect to $FRDBS_1$

4.2 Reasoning Problems on FRDBS

In general, the reasoning problems of FRDBS include satisfiability, subsumption, and redundancy. At present, in the fuzzy database modeling, the designers usually need to check the above reasoning problems by hand. In [4] D. Calvanese has pointed out that reasoning problems of ER model may reason automatically through reasoning mechanism of Description Logics.

As the aforementioned analyses, a fuzzy OWL DL ontology being equivalent to a DL f-SHOIN(D) knowledge base [18], so a fuzzy OWL DL ontology $\varphi(FS)$ such as in Fig. 2 is corresponding to a DL f-SHOIN(D) knowledge base $\varphi'(FS)$, i.e., fuzzy class identifier $\varphi(FE)$ in $\varphi(FS)$ corresponding to concept $\varphi'(FE)$ in $\varphi'(FS)$, and so on. Here, we omitted the detailed correspondence between $\varphi(FS)$ and $\varphi'(FS)$. Therefore, it is convenient and worth it checking the reasoning problems of FRDBS through reasoning mechanism of f-SHOIN(D), which will be further contribute to constructing fuzzy OWL DL ontologies exactly that meet application's needs.

According to the Algorithm 1, we can translate the fuzzy ER schema FS_1 which is extracted from $FRDBS_1$ into fuzzy OWL DL ontology $\varphi(FS_1)$.

$\varphi(FS_1) = (FID_0, FAxiom_0)$, where:

$FID_0 = \{ FCID_0 \cup FDRID_0 \cup FOPID_0 \cup FDPID_0 \}$.

$FCID_0 = \{$ Employee, Young-Employee, Leader, Chief-Leader, Mainly-Manage $\}$.

$FDRID_0 = \{$ xsd:string $\}$. $FDPID_0 = \{$ Number-YE, Age , Number-CL $\}$.

$FOPID_0 = \{$ Mof, Mby, invof_Mof, invof_ Mby $\}$.

$FAxiom_0 = \{$ Class (Young-Employee partial Employee) ;

Class (Chief - Leader partial Leader) ;

Class (Young-Employee partial restriction (Number-YE allValuesFrom (xsd:string)
        cardinality(1)) restriction (Age allValuesFrom (xsd:string) cardinality(1))) ;

DatatypeProperty (Number-YE domain(Young-Employee) range(xsd:string)
        [Functional]) ;

DatatypeProperty (Age domain(Young-Employee) range(xsd:string) [Functional]) ;

Class (Chief-Leader partial restriction (Number-CL allValuesFrom (xsd:string)
        cardinality(1))) ;

DatatypeProperty (Number-CL domain(Chief-Leader) range(xsd:string) [Functional]) ;

Class (Mainly-Manage partial restriction (Mof allValuesFrom (Young-Employee)
        cardinality(1)) restriction (Mby allValuesFrom (Chief-Leader) cardinality(1))) ;

ObjectProperty (invof_Mof domain(Young-Employee) range(Mainly-Manage)
    inverseOf Mof) ;

ObjectProperty (invof_Mby domain(Chief-Leader) range(Mainly-Manage)
    inverseOf Mby) ;

Class (Young-Employee partial restriction (invof_Mof allValuesFrom
    (Mainly-Manage))) ;

Class (Chief-Leader partial restriction (invof_Mby allValuesFrom (Mainly-Manage))) ;

ObjectProperty (Mof domain(Mainly-Manage) range(Young-Employee)) ;

ObjectProperty (Mby domain(Mainly-Manage) range(Chief-Leader)) ;

Class (Young-Employee partial restriction (invof_Mof minCardinality (m))) ;

Class (Young-Employee partial restriction (invof_ Mof maxCardinality (n))) ;

Class (Chief-Leader partial restriction (invof_Mby minCardinality (m))) ;

Class (Chief-Leader partial restriction (invof_Mby maxCardinality (n))) ; $\}$.

Fig. 2. The fuzzy OWL DL ontology $\varphi(FS_1)$ derived from fuzzy ER schema FS_1 in Fig. 1

Based on the Theorem 4.9 [4], the correctness of Theorem 2 and Theorem 3 can be proved similarly.

Theorem 2 (satisfiability). Let FRDBS = $(LT_{FRDBS}$, unique, pkey, fkey, $\leq_{FRDBS})$ be a Fuzzy Relational Database Schema, $\varphi'(FS)$ be the corresponding fuzzy description logic f-SHOIN(D) knowledge base of fuzzy OWL DL ontology $\varphi(FS)$, and FE (FR) be a fuzzy entity (relation) of FRDBS, then FE (FR) is satisfiable, iff: $\varphi'(FS) \not\models$ $\varphi'(FE) \sqsubseteq \bot$ ($\varphi'(FS) \not\models \varphi'(FR) \sqsubseteq \bot$) .

Theorem 3 (subsumption). Let FRDBS = $(LT_{FRDBS}$, unique, pkey, fkey, $\leq_{FRDBS})$ be a Fuzzy Relational Database Schema, $\varphi'(FS)$ be the corresponding fuzzy description logic f-SHOIN(D) knowledge base of fuzzy OWL DL ontology $\varphi(FS)$, FE_1, FE_2 be two fuzzy entities of FRDBS, then $FE_1 \leq_{FRDBS} FE_2$, iff: $\varphi'(FS) \models \varphi'(FE_1) \sqsubseteq \varphi'(FE_2)$.

Definition 8. Let FRDBS = $(LT_{FRDBS}$, unique, pkey, fkey, $\leq_{FRDBS})$ be a Fuzzy Relational Database Schema, for two fuzzy entities FE_1 and FE_2 of FRDBS, if $FE_1 \leq_{FRDBS} FE_2$ and $FE_2 \leq_{FRDBS} FE_1$ satisfy, then we say that FRDBS is redundant.

Theorem 4 (redundancy). Let FRDBS = $(LT_{FRDBS}$, unique, pkey, fkey, $\leq_{FRDBS})$ be a Fuzzy Relational Database Schema, $\varphi'(FS)$ be the corresponding fuzzy description logic f-SHOIN(D) knowledge base of fuzzy OWL DL ontology $\varphi(FS)$, FE_1 and FE_2 be two fuzzy entities of FRDBS, then FRDBS is redundant, iff: the following conditions satisfy: $\varphi'(FS) \vDash \varphi'(FE_1) \sqsubseteq \varphi'(FE_2)$ and $\varphi'(FS) \vDash \varphi'(FE_2) \sqsubseteq \varphi'(FE_1)$.

The correctness of Theorem 4 can be proved according to Theorem 3 easily.

5 Conclusion

We have presented the formal definition of Fuzzy Relational Database Schema. Additionally, the formal definition and Model-Theoretic semantics of a kind of new fuzzy OWL DL ontology were given. The more significant contributions of the work are as follows: (i) extracting the fuzzy Entity-Relationship model (fuzzy ER model) from FRDBS, based on this, we realized the formal translation from FRDBS to fuzzy OWL DL ontology by translating the extracted fuzzy ER model into fuzzy OWL DL ontology with a semantics-preserving translation algorithm, which will be beneficial to the development of constructing Web ontologies; (ii) investigating the reasoning problems of FRDBS automatically through reasoning mechanism of the corresponding fuzzy description logic f-SHOIN(D) of fuzzy OWL DL ontology, which can contribute to constructing fuzzy OWL DL ontology exactly that meet application's needs well.

In the future, we aim at developing the other methods of constructing ontologies.

Acknowledgements. This work was supported by the *Program for New Century Excellent Talents in University* (NCET-05-0288) and in part by the *National Natural Science Foundation of China* (60873010) and *MOE Funds for Doctoral Programs* (20050145024).

References

1. An, Y., Borgida, A., Mylopoulos, J.: Refining Semantic Mappings from Relational Tables to Ontologies. In: Bussler, C.J., Tannen, V., Fundulaki, I. (eds.) SWDB 2004. LNCS, vol. 3372, pp. 84–90. Springer, Heidelberg (2005)
2. Berners-Lee, T., Hendler, J., Lassila, O.: The Semantic Web. Scientific American 284(5), 34–43 (2001)
3. Chen, P.P.: The entity-relationship model—toward a unified view of data. ACM Trans. Database Systems 1(1), 9–36 (1976)
4. Calvanese, D., Lenzerini, M., Nardi, D.: Unifying class-based representation formalisms. J. Artificial Intelligence Research 11(2), 199–240 (1999)
5. Ceravolo, P., Corallo, A., Damiani, E., Elia, G., Viviani, M., Zilli, A.: Bottom-up Extraction and Maintenance of Ontology-based Metadata. In: Sanchez, E. (ed.) Fuzzy Logic and the Semantic Web, pp. 265–282. Elsevier, Amsterdam (2006)
6. Calegari, S., Ciucci, D.: Fuzzy ontology, fuzzy description logics and fuzzy-OWL. In: Masulli, F., Mitra, S., Pasi, G. (eds.) WILF 2007. LNCS, vol. 4578, pp. 118–126. Springer, Heidelberg (2007)

7. Juric, D., Skočir, Z.: Building OWL ontologies by analyzing relational database schema concepts and WordNet semantic relations. In: Juric, D., Skočir, Z. (eds.) Proceedings of the 9th Int. Conference on Telecommunications, ConTEL, Zagreb, Croatia, pp. 235–242 (2007)

8. Dean, M., Schreiber, G. (eds.): OWL Web Ontology Language Reference. W3C Recommendation, 10 Feb (2004),
   `http://www.w3.org/TR/2004/REC-owl-ref-20040210/`

9. Fensel, D., Pérez, A. (eds.): A Survey on Ontology Tools. IST Project OntoWeb Deliverable 1.3 (May 2002),
   `http://ontoweb.aifb.uni-karlsruhe.de/About/`
   `Deliverables/D13_v1-0.zip`

10. Jesus, L., P-de., S.P.: Selection of Reverse Engineering Methods for Relational Databases. In: Proceedings of the Third European Conference on Software Maintenance and Reengineering, pp. 194–197 (2003)

11. Kashyap, V.: Design and creation of ontologies for environmental information retrieval. In: Proc. of the Workshop on Knowledge Acquisition, Modeling and Management (1999)

12. Ma, Z.M.: A conceptual design methodology for fuzzy relational databases. Journal of Database Management 16(2), 66–83 (2005)

13. Ma, Z.M., Zhang, W.J., Ma, W.Y.: Extending object-oriented databases for fuzzy information modeling. Information Systems 29(5), 421–435 (2004)

14. Medina, J.M., Pons, O., Vila, M.A.: GEFRED: a generalized model of fuzzy relational databases. Information Sciences 76, 87–109 (1994)

15. OWL: Ontology Web Language, `http://www.w3.org/2004/OWL/`

16. Patel-Schneider, P.F., Hayes, P., Horrocks, I. (eds.): OWL Web Ontology Language Semantics and Abstract Syntax. W3C Recommendation (Feburary 10, 2004),
   `http://www.w3.org/TR/2004/REC-owl-semantics-20040210/`

17. Parry, D.: Fuzzy ontologies for information retrieval on the WWW. In: Sanchez, E. (ed.) Fuzzy Logic and the Semantic Web (2004)

18. Straccia, U.: Towards a fuzzy description logic for the semantic Web. In: Proc. of the 2nd European Semantic Web Conf. (2005)

19. Sanchez, E., Yamanoi, T.: Fuzzy ontologies for the semantic web. In: Larsen, H.L., Pasi, G., Ortiz-Arroyo, D., Andreasen, T., Christiansen, H. (eds.) FQAS 2006. LNCS, vol. 4027, pp. 691–699. Springer, Heidelberg (2006)

20. Stoilos, G., Stamou, G., Tzouvaras, V., Pan, J.Z., Horrocks, I.: Fuzzy OWL: Uncertainty and the Semantic Web. In: International Workshop of OWL: Experiences and Directions (OWL-ED2005), Galway, Ireland (2005)

21. Stojanovic, L., Stojanovic, N., Volz, R.: Migrating data-intensive web sites into the semantic Web. In: Proc. of the 17th ACM Symp. on Applied Computing, pp. 1100–1107. ACM Press, New York (2002)

22. Thomas, C., Sheth, A.: On the Expressiveness of the Languages for the Semantic Web- Making a Case for 'A Little More'. In: Sanchez, E. (ed.) Fuzzy Logic and the Semantic Web. Elsevier, Amsterdam (March 2006)

23. Xu, Z., Cao, X., Dong, Y.: Formal approach and automated tool for translating ER schemata into OWL ontologies. In: Dai, H., Srikant, R., Zhang, C. (eds.) PAKDD 2004. LNCS, vol. 3056, pp. 464–475. Springer, Heidelberg (2004)

24. Xu, Z., Zhang, S., Dong, Y.: Mapping between Relational Database Schema and OWL Ontology for Deep Annotation. In: Proceedings of the 2006 IEEE/WIC/ACM International Conference on Web Intelligence, December 18-22, pp. 548–552 (2006)

25. Zvieli, A., Chen, P.P.: Entity-relationship modeling and fuzzy databases. In: Proc. 1986 IEEE Int. Conf. Data Engineering, Los Angeles, California, USA, pp. 320–327 (1986)

26. Zhao, S., Chang, E.: From database to semantic web ontology: An overview. In: Meersman, R., Tari, Z., Herrero, P. (eds.) OTM-WS 2007, Part II. LNCS, vol. 4806, pp. 1205–1214. Springer, Heidelberg (2007)

A Tableau Algorithm for Possibilistic Description Logic $\mathcal{ALC}$

Guilin Qi[1] and Jeff Z. Pan[2]

[1] Institute AIFB, University of Karlsruhe, Germany
gqi@aifb.uni-karlsruhe.de
[2] Department of Computing Science, The University of Aberdeen
jpan@csd.abdn.ac.uk

Abstract. Uncertainty reasoning and inconsistency handling are two important problems that often occur in the applications of the Semantic Web. Possibilistic description logics provide a flexible framework for representing and reasoning with ontologies where uncertain and/or inconsistent information is available. Although possibilistic logic has become a popular logical framework for uncertainty reasoning and inconsistency handling, its role in the Semantic Web is underestimated. One of the challenging problems is to provide a practical algorithm for reasoning in possibilistic description logics. In this paper, we propose a tableau algorithm for possibilistic description logic $\mathcal{ALC}$. We show how inference services in possibilistic $\mathcal{ALC}$ can be reduced to the problem of computing the *inconsistency degree* of the knowledge base. We then give tableau expansion rules for computing the inconsistency degree of a possibilistic $\mathcal{ALC}$ knowledge. We show that our algorithm is sound and complete. The computational complexity of our algorithm is analyzed. Since our tableau algorithm is an extension of a tableau algorithm for $\mathcal{ALC}$, we can reuse many optimization techniques for tableau algorithms of $\mathcal{ALC}$ to improve the performance of our algorithm so that it can be applied in practice.

1 Introduction

Uncertainty reasoning and inconsistency handling are two important problems that often occur in the applications of the Semantic Web, such as the areas like medicine and biology [21,16]. Recently, there is an increasing interest to extend Web Ontology Language OWL to represent uncertain knowledge. Most of the work is based on Description Logics (DL) that provide important formalisms for representing and reasoning with ontologies. A DL knowledge base is then extended by attaching each axiom in it with a degree of belief. The degree of belief can have several meanings depending on the semantics of the logic. For example, in probabilistic description logics, the degree of belief can be explained as degree of overlap between two concepts (see [21]) or probability of a concept given another one (see [9,16]), and in possibilistic description logics, the degree of belief is explained as the necessity degree or certainty degree (see [11,18]). Inconsistency handling in DL is another problem that has attracted a lot of attention. Inconsistency can occur due to several reasons, such as modeling errors,

J. Domingue and C. Anutariya (Eds.): ASWC 2008, LNCS 5367, pp. 61–75, 2008.
© Springer-Verlag Berlin Heidelberg 2008

migration or merging ontologies, and ontology evolution. When an ontology is inconsistent, an ontology language which has first-order features, such as a description logic, cannot be applied to infer non-trivial conclusions.

Let us consider an uncertain medical ontology which is modified from the medical example given in [16].

Example 1. Given an ontology consisting of the following terminological axioms attached with weights:

$ax_1 :$ $(Heartpatient \sqsubseteq HighBloodPressure, 1)$

$ax_2 :$ $(PacemakerPatient \sqsubseteq \neg HighBloodPressure, 1)$

$ax_3 :$ $(HeartPatient \sqsubseteq MalePacemakerPatient, 0.4)$

$ax_4 :$ $(HeartPatient \sqsubseteq \exists HasHealthInsurance.PrivateHealth, 0.9)$

$ax_5 :$ $(PacemakerPatient(Tom), 0.8).$

Suppose we use possibilistic logic, then ax_1 means that "it is absolute certain that heart patients suffers from high blood pressure", ax_2 means that "it is absolute certain that pacemaker patients do not suffer from high blood pressure", ax_3 says that "it is a little certain that heart patient are male pacemaker patient", ax_4 says "it is highly certain that heart patients have a private insurance", and finally ax_5 states that "it is quite certain that Tom is a pacemaker patient". Suppose we learn that Tom is a heart patient with degree 0.5 (ax_6: $(HeartPatient(Tom),0.5))$, i.e., it is somewhat certain that Tom is a heart patient, and we add this axiom to the ontology, then the ontology will become inconsistent. From this updated ontology, we may want to query if Tom suffers from high blood pressure and to ask to what degree we can infer this conclusion?

Possibilistic description logics, first proposed by Hollunder in [11], are extension of description logics with possibilistic semantics. It is well-known that possibilistic logic is a powerful logical framework for dealing with uncertainty and handling inconsistency. Possibilistic description logics inherit these two nice properties and have very promising applications in the Semantic Web. A possibilistic DL knowledge base consists of a set of weighted axioms of the form (ϕ, α), where ϕ is a DL axiom such as an assertional axiom of the form $C(a)$ and α is an element of the semi-open real interval (0,1] or of a finite total ordered scale. A weighted axiom (ϕ, α) encodes the constraint $N(\phi) \geq \alpha$, where N is a necessity measure [7], with the intended meaning that the necessity degree of ϕ is at least α. One critical difference between possibilistic description logics and probabilistic description logics is that the weight attached to an axiom is not absolute and can be be replaced by another number as long as the ordering between two weights is not changed. Therefore, possibilistic description logics provide a more flexible way to represent uncertain information than probabilistic description logics. In Example 1, if the weight of ax_4 is changed to 0.85, the changed possibilistic DL knowledge base is query-equivalent to the original one. That is, the answer to an arbitrary query over the original knowledge bases is the same that over the changed knowledge base. In contrast, if we apply the probabilistic DLs in [16], axioms in Example 1 are considered as conditional constraints which are interpreted as conditional probabilities of a concept w.r.t. another concept. So the weights of axioms should be precisely given.

Although possibilistic logic has become a popular logic framework for uncertainty reasoning, its role in the Semantic Web is underestimated. Until now, there has been very few work on extending ontology languages with possibilistic semantics. One of the challenging problems is to provide an algorithm for reasoning in possibilistic description logics that can be applied in practice. In [18], an algorithm is provided for the inference services in possibilistic description logics. We show the complexity of the algorithm for checking the *inconsistency degree* of a possibilistic DL knowledge base, i.e., it needs to call polynomial times of a satisfiability check of a DL reasoner w.r.t. the number of distinct confidence values. Therefore, the algorithm may become very inefficient if there is a large number of distinct confidence values (some preliminary results have been shown in Table 2 of [18]). This will severely restrict the applicability of the system based on the algorithm.

In this paper, to alleviate the above problem, we propose a tableau algorithm for possibilistic DL $\mathcal{ALC}$. First, we show how inference services in possibilistic $\mathcal{ALC}$ can be reduced to the problem of computing the *inconsistency degree* of a possibilistic DL knowledge base. We then give tableau expansion rules for computing the inconsistency degree of a possibilistic $\mathcal{ALC}$-ABox. The idea is that we attach a weight to each concept in the completion forest and propagate the weight in the tableau expansion rule. A weighted concept with higher necessity degree should be expanded first by a tableau rule if applicable. We show that our algorithm is sound and complete. We also show that checking the inconsistency degree of a possibilistic $\mathcal{ALC}$-ABox has the same complexity as that of a tableau algorithm in $\mathcal{ALC}$. After that, we propose tableau expansion rules for a possibilistic $\mathcal{ALC}$ knowledge base with acyclic terminologies and general TBox and show that the tableau algorithm is still sound and complete. Although $\mathcal{ALC}$ has a high worst-case complexity, there exist optimization techniques for the tableau algorithm so that the system based on it performs well with problems that occur in realistic applications. Since our tableau algorithm is an extension of tableau algorithm for $\mathcal{ALC}$, we can reuse many optimization techniques to improve the performance of our algorithm so that it can be applied in practice.

2   Possibilistic Description Logic $\mathcal{ALC}$

The syntax and semantics of possibilistic Description Logics have been given in [18]. However, in this work, we restrict the underlying Description Logic (DL) language to DL $\mathcal{ALC}$ because tableau algorithms for $\mathcal{ALC}$ are the basis of tableau algorithms for many important and more expressive DLs. We first briefly introduce DL $\mathcal{ALC}$, then extend it to possibilistic DL $\mathcal{ALC}$ and define inference tasks in it.

2.1   DL $\mathcal{ALC}$

A DL knowledge base $\Sigma = (\mathcal{T}, \mathcal{A})$ consists of a set $\mathcal{T}$ (TBox) of concepts axioms and a set $\mathcal{A}$ (ABox) of individual axioms. Concept axioms have the form $C \sqsubseteq D$ where C and D are (possibly complex) concept descriptions. The ABox contains *concept assertions* of the form $a : C$ where C is a concept and a is an individual name, and *role assertions* of the form $\langle a, b \rangle : R$, where R is a role, and a and b are individual names. A *concept description* (or simply *concept*) of the smallest propositionally closed DL

$\mathcal{ALC}$ is defined by the following syntactic rules, where C is a concept name, R is a role, C, C_1 and C_2 are concept descriptions:

$$\top \mid \bot \mid C \mid \neg C_1 \mid C_1 \sqcap C_2 \mid C_1 \sqcup C_2 \mid \exists R.C \mid \forall R.C.$$

An interpretation $\mathcal{I} = (\Delta^{\mathcal{I}}, \cdot^{\mathcal{I}})$ consists of the domain of the interpretation $\Delta^{\mathcal{I}}$ (a non-empty set) and the interpretation function $\cdot^{\mathcal{I}}$, which maps each concept name C to a set $C^{\mathcal{I}} \subseteq \Delta^{\mathcal{I}}$, each role name R to a binary relation $R^{\mathcal{I}} \subseteq \Delta^{\mathcal{I}} \times \Delta^{\mathcal{I}}$ and each individual a to an object in the domain $a^{\mathcal{I}}$. The interpretation function can be extended to give semantics to concept descriptions. An interpretation $\mathcal{I}$ *satisfies* a concept axiom $C \sqsubseteq D$ (a concept assertion $a : C$ and a role assertion $\langle a, b \rangle : R$, resp.) if $C^{\mathcal{I}} \subseteq D^{\mathcal{I}}$ ($a^{\mathcal{I}} \in C^{\mathcal{I}}$ and $\langle a^{\mathcal{I}}, b^{\mathcal{I}} \rangle \in R^{\mathcal{I}}$ resp.). An interpretation $\mathcal{I}$ *satisfies* a knowledge base Σ if it satisfies all axioms in Σ; in this case, we say $\mathcal{I}$ is an interpretation of Σ. A knowledge base is *consistent* if it has an interpretation. A concept is unsatisfiable in Σ iff it is interpreted as an empty set by all the interpretation of Σ.

2.2 Syntax and Semantics of Possibilistic DL $\mathcal{ALC}$

Syntax. A *possibilistic axiom* is a pair (ϕ, α) consisting of a DL axiom ϕ and a weight $\alpha \in (0, 1]$. A *possibilistic TBox* (resp., *ABox*) is a finite set of possibilistic axioms (ϕ, α), where ϕ is a DL $\mathcal{ALC}$ concept (resp., assertional) axiom. A possibilistic DL knowledge base $\mathcal{B} = (\mathcal{T}, \mathcal{A})$ consists of a possibilistic TBox $\mathcal{T}$ and a possibilistic ABox $\mathcal{A}$. We use $\mathcal{T}^*$ to denote the classical DL axioms associated with $\mathcal{T}$, i.e., $\mathcal{T}^* = \{\phi_i : (\phi_i, \alpha_i) \in \mathcal{T}\}$ ($\mathcal{A}^*$ can be defined similarly). The classical base $\mathcal{B}^*$ of a possibilistic DL knowledge base is $\mathcal{B}^* = (\mathcal{T}^*, \mathcal{A}^*)$. A possibilistic DL knowledge base $\mathcal{B}$ is inconsistent if and only if $\mathcal{B}^*$ is inconsistent.

Given a possibilistic DL knowledge base $\mathcal{B} = (\mathcal{T}, \mathcal{A})$ and $\alpha \in (0, 1]$, the α-cut of $\mathcal{T}$ is $\mathcal{T}_{\geq \alpha} = \{\phi \in \mathcal{B}^* \mid (\phi, \beta) \in \mathcal{T} \text{ and } \beta \geq \alpha\}$ (the α-cut of $\mathcal{A}$, denoted as $\mathcal{A}_{\geq \alpha}$, can be defined similarly). The strict α-cut of $\mathcal{T}$ (resp., $\mathcal{A}$) can be defined similarly as the strict cut in possibilistic logic. The α-cut (resp., strict α-cut) of $\mathcal{B}$ is $\mathcal{B}_{\geq \alpha} = (\mathcal{T}_{\geq \alpha}, \mathcal{A}_{\geq \alpha})$ (resp., $\mathcal{B}_{>\alpha} = (\mathcal{T}_{>\alpha}, \mathcal{A}_{>\alpha})$). The *inconsistency degree* of $\mathcal{B}$, denoted $Inc(\mathcal{B})$, is defined as $Inc(\mathcal{B}) = max\{\alpha_i : \mathcal{B}_{\geq \alpha_i} \text{ is inconsistent}\}$ with $Inc(\mathcal{B}) = 0$ if $\mathcal{B}$ is consistent.

Semantics. The semantics of possibilistic DL is defined by a *possibility distribution* π over the set $\mathbf{I}$ of all interpretations of a DL language, i.e., $\pi : \mathbf{I} \to [0, 1]$. $\pi(\mathcal{I})$ represents the degree of compatibility of interpretation $\mathcal{I}$ with available information. For two interpretations $\mathcal{I}_1$ and $\mathcal{I}_2$, $\pi(\mathcal{I}_1) > \pi(\mathcal{I}_2)$ means that $\mathcal{I}_1$ is preferred to $\mathcal{I}_2$ according to the available information. Given a possibility distribution π, we can define the possibility measure Π and necessity measure N as follows: $\Pi(\phi) = sup\{\pi(\mathcal{I}) : \mathcal{I} \in \mathbf{I}, \mathcal{I} \models \phi\}$ and $N(\phi) = 1 - max\{\pi(\mathcal{I}) : \mathcal{I} \not\models \phi\}$. Given two possibility distributions π and π', we say that π is more specific (or more informative) than π' iff $\pi(\mathcal{I}) \leq \pi'(\mathcal{I})$ for all $\mathcal{I} \in \mathbf{I}$. A possibility distribution π satisfies a possibilistic axiom (ϕ, α), denoted $\pi \models (\phi, \alpha)$, iff $N(\phi) \geq \alpha$. It satisfies a possibilistic DL knowledge base $\mathcal{B}$, denoted $\pi \models \mathcal{B}$, iff it satisfies all the possibilistic axioms in $\mathcal{B}$. According to [18], a least specific possibility distribution $\pi_\mathcal{B}$ an be defined from $\mathcal{B}$ such that $\mathcal{B}$ is consistent iff there exists an interpretation $\mathcal{I}$ such that $\pi_\mathcal{B}(\mathcal{I}) = 1$.

2.3 Possibilistic Inference in Possibilistic DLs

The following inference services have been proposed in possibilistic DLs (see [18] and [11]). To save space, we do not provide semantical definition of these inference services but define them by using standard DL inference services.

- **Instance checking:** an individual a (resp. a pair of individuals (a, b)) is a *plausible* instance of a concept C (resp. a role R) with respect to a possibilistic DL knowledge base $\mathcal{B}$, written $\mathcal{B} \models_P C(a)$ (resp. $\mathcal{B} \models_P R(a, b)$), if $\mathcal{B}_{>Inc(\mathcal{B})} \models C(a)$ (resp. $\mathcal{B}_{>Inc(\mathcal{B})} \models R(a, b)$).
- **Subsumption:** a concept C is subsumed by a concept D with respect to a possibilistic DL knowledge base $\mathcal{B}$, written $\mathcal{B} \models_P C \sqsubseteq D$, if $\mathcal{B}_{>Inc(\mathcal{B})} \models C \sqsubseteq D$.
- **Instance checking with necessity degree:** an individual a (resp. a pair of individuals (a, b)) is an instance of a concept C (resp. a role R) to degree α with respect to $\mathcal{B}$, written $\mathcal{B} \models_\pi (C(a), \alpha)$ (resp. $\mathcal{B} \models_\pi (R(a, b), \alpha)$), if the following conditions hold: (1) $\mathcal{B}_{\geq\alpha}$ is consistent, (2) $\mathcal{B}_{\geq\alpha} \models C(a)$ (resp. $\mathcal{B}_{\geq\alpha} \models R(a, b)$), (3) for all $\beta > \alpha$, $\mathcal{B}_{\geq\beta} \not\models C(a)$ (resp. $\mathcal{B}_{\geq\beta} \not\models R(a, b)$).
- **Subsumption with necessity degree:** a concept C is subsumed by a concept D to a degree α with respect to a possibilistic DL knowledge base $\mathcal{B}$, written $\mathcal{B} \models_\pi (C \sqsubseteq D, \alpha)$, if the following conditions hold: (1) $\mathcal{B}_{\geq\alpha}$ is consistent, (2) $\mathcal{B}_{\geq\alpha} \models C \sqsubseteq D$, (3) for all $\beta > \alpha$, $\mathcal{B}_{\geq\beta} \not\models C \sqsubseteq D$.
- **Possibilistic instance checking:** given a possibilistic assertion $(C(a), \alpha)$ (resp. $(R(a, b), \alpha)$), it can be inferred from $\mathcal{B}$, written $\mathcal{B} \models (C(a), \alpha)$ (resp. $\mathcal{B} \models (R(a, b), \alpha)$), if $\alpha > Inc(\mathcal{B})$ and $\mathcal{B}_{\geq\alpha} \models C(a)$ (resp. $\mathcal{B}_{\geq\alpha} \models R(a, b)$).
- **Possibilistic subsumption:** given a possibilistic concept axiom $(C \sqsubseteq D, \alpha)$, it can be inferred from $\mathcal{B}$, written $\mathcal{B} \models (C \sqsubseteq D, \alpha)$, if $\alpha > Inc(\mathcal{B})$ and $\mathcal{B}_{\geq\alpha} \models C \sqsubseteq D$.

Note that the task of instance checking with necessity is different from that of possibilistic instance checking because the former is to check to what degree an assertion holds whilst the latter is to check if a possibilistic assertion holds. The first and the second inference services are similar to standard DL inference services, but they are inconsistency-tolerant. The other inference services are more powerful than the first and the second ones as they allow us to deal with uncertainty. For example, instance checking with necessity degree allows us to infer to what degree an individual can be non-trivially inferred from a possibilistic DL knowledge base. We define every inference service by reducing it to classical inference. For Example 1 in Introduction, to query if Tom suffers from high blood pressure is an instance checking problem, i.e., check if $B \models HighBloodPressure(Tom)$ holds or not, where B is the updated ontology, whilst to ask to what degree we can infer $HighBloodPressure(Tom)$ is a problem of instance checking with necessity degree, i.e., check if $B \models_\pi (HighBloodPressure(Tom), \alpha)$ holds.

3 Related Work

Hollunder in [11] proposes a possibilistic extension of DL $\mathcal{ALCN}$. A proof method for first-order possibilistic logic is given and is then applied to check possibilistic subsumption or possibilistic assertion. However, they have not provide a proof method for othermore interesting inference services, such as instance checking. In [8], a discussion between possibilistic description logics and fuzzy description logics is given. In [18], we provide two algorithms for the inference services in possibilistic description logics and also report preliminary evaluation results on the algorithms. These algorithms take a DL reasoner as a black box, so they are independent of DL reasoners. The advantage of the black-box algorithms is that they can be easily implemented based on any DL reasoner. However, according to [18], the algorithm for checking inconsistency degree of a possibilistic DL knowledge base needs to call polynomial times of satisfiability check of a DL reasoner w.r.t. the number of distinct confidence values. In contrast, our tableau algorithm is a generalization of the tableau algorithm for ALC therefore checking the inconsistency degree of a possibilistic DL knowledge base has the same complexity as one for ALC. In parallel to our work, Couchariere et.al. in [5] proposed some tableau expansion rules for possibilistic DL $\mathcal{ALC}$. Their tableau expansion rules are in essence the same as ours. However, their work is preliminary and does not discuss the following important issues: (1) the reduction from inference services in a possibilistic DL knowledge base to computing its inconsistency degree; (2) a tableau algorithm that computes the inconsistency degree of a possibilistic $\mathcal{ALC}$-ABox in PSpace; (3) the proof of soundness and completeness of the algorithm; (4) tableau expansion rules for general TBoxes. All these missed issues are discussed in our paper.

Another family of important approaches that extend description logics with uncertainty reasoning are probabilistic description logics, such as those given in [12,10,13,9,16]. We list some major differences between possibilistic extension and probabilistic extension. First, possibilistic logic is based on possibility measures, whilst probabilistic extensions are based on probabilistic measures. Second, unlike probabilistic logic, the weight attached to an axiom in possibilistic logic is not absolute and can be be replaced by another number as long as the ordering between two weights is not changed. Probabilistic DLs given in [12] and [10] are based on probabilistic reasoning in probabilistic logic and are much less popular than the probabilistic DLs presented in [16]. In [16], Lukasiewicz proposes expressive probabilistic DLs P-$\mathcal{SHIF}(\mathbf{D})$ and P-$\mathcal{SHOIN}(\mathbf{D})$ which are semantically based on the notion of probabilistic lexicographic entailment from probabilistic default reasoning given in [15]. Possibilistic DL considered in this paper is based on necessity measure. So each axiom in the knowledge base is attached with a confidence value interpreted as certainty degree of the axiom. However, in expressive probabilistic DLs, each *conditional constraint* in the probabilistic knowledge base is attached with an interval. A nice feature of possibilistic description logics is that when the weights of all axioms in the knowledge base are 1, then the possibilistic DL knowledge base is reduced to a standard DL knowledge base. This feature is not captured by probabilistic DLs given in [15]. Another important difference between possibilistic DLs and expressive probabilistic DLs is that all the inference services in expressive possibilistic DLs can be reduced to the problem of computing the inconsistency degree of the knowledge base, whilst inference services in expressive probabilistic DLs are dependent on the reasoning tasks in probabilistic default theory. It has been shown in [15] that complexity of decision problems in some probabilistic DLs, such as probabilistic DL-Lite, is harder than that in corresponding DLs. In contrast, complexity of decision problems in possibilistic DLs remain the same as that in the corresponding DLs. The work on combining DLs with Bayesian Network (such as [13]) is less relevant to our work.

Arguably, fuzzy description logics can be used to deal with uncertainty (e.g., [20,19]). In possibilistic DLs, the truth value of an axiom is still two-valued, whilst in fuzzy DLs, the truth value of an axiom is multi-valued. So the semantics of possibilistic DLs is different from that of fuzzy DLs. Apart from this major difference, the tableau expansion rules proposed in this paper has some similarity with the tableau expansion rules for fuzzy DL $\mathcal{ALC}$ given in [20].

4   Tableau Algorithms for Inference in Possibilistic DL $\mathcal{ALC}$

In this section, we extend tableau expansion rules for DL $\mathcal{ALC}$ to possibilistic DL $\mathcal{ALC}$. The introduction of the tableau rules for DL $\mathcal{ALC}$ can be found in [4,3]. Our tableau rules are inspired from the resolution rules in possibilistic logic [7]. However, our tableau algorithm is a trivial extension of tableau algorithm in $\mathcal{ALC}$ because we need to take into account of the weights of the axioms in the possibilistic DL knowledge base when we define our tableau rules and when we choose a concept to expand. Although our tableau rules have some similarity with resolution rules in possibilistic logic [7], our algorithm is specific to DL $\mathcal{ALC}$ and introduces novel techniques to ensure its soundness and completeness. Due to the space limitation, we do not give proofs of all theorems in this paper but refer the reader to our technical report that can be found at http://www.aifb.uni-karlsruhe.de/WBS/gqi/papers/poss-DL.pdf.

4.1   The Reduction

We consider only the reduction of instance checking and instance checking with necessity degree to compute the inconsistency degree of a knowledge base. The reduction of other inference tasks can be done similarly. Given a possibilistic $\mathcal{ALC}$ knowledge base $\mathcal{B}$, to check if an assertion $C(a)$ can be inferred from $\mathcal{B}$, we first need to compute the inconsistency degree of $\mathcal{B}$, then compute the inconsistency degree of $\mathcal{B}' = \{(\phi_i, 1) : (\phi_i, \alpha_i) \in \mathcal{B}, \alpha_i > Inc(\mathcal{B})\} \cup \{(\neg C(a), 1)\}$. It is easy to see that $Inc(\mathcal{B}') = 1$ if and only if $\mathcal{B}'$ is inconsistent if and only if $\mathcal{B} \models_P C(a)$.

To infer to what degree an individual a is an instance of a concept C with respect to $\mathcal{B}$, we first need to compute the inconsistency degree of $\mathcal{B}$. We then compute the inconsistency degree of possibilistic $\mathcal{ALC}$ knowledge base $\mathcal{B}' = \mathcal{B} \cup \{(\neg C(a), 1)\}$. Next, we show that if $Inc(\mathcal{B}') > Inc(\mathcal{B})$, then we can infer $C(a)$ with degree $Inc(\mathcal{B}')$ with respect to $\mathcal{B}$.

Proposition 1. *Let $\mathcal{B}$ be a possibilistic $\mathcal{ALC}$ knowledge base and $C(a)$ an ABox assertion. Let $\mathcal{B}' = \mathcal{B} \cup \{(\neg C(a), 1)\}$. Then $\mathcal{B} \models_\pi (C(a), \alpha)$ if and only if $Inc(\mathcal{B}') > Inc(\mathcal{B})$ and $Inc(\mathcal{B}') = \alpha$.*

4.2   Tableau Expansion Rules for Computing the Inconsistency Degree of a Possibilistic $\mathcal{ALC}$-ABox

Let $\mathcal{A}$ be a possibilistic $\mathcal{ALC}$-ABox. Without loss of generality, we assume that all the concepts appearing in $\mathcal{A}$ are in *negation normal form* (NNF), i.e., that negation occurs only directly in front of concept names. For an arbitrary concept, we can transform it to an equivalent one in NNF by pushing negation inwards using a combination of de Morgan's laws and the duality between existential and universal restrictions, i.e., $\neg \exists R.C \equiv \forall R.\neg C$ and $\neg \forall R.C \equiv \exists R.\neg C$.

To compute the inconsistency degree of $\mathcal{A}$, we construct a completion forest $\mathcal{F}_\mathcal{A}$ from $\mathcal{A}$. Each node x in the completion forest is labelled with a set of weighted concepts $\mathcal{L}(x)$ and each edge $\langle x, y \rangle$ is labelled with a set of weighted role names $\mathcal{L}(\langle x, y \rangle)$. If a weighted concept C^α is in $\mathcal{L}(x)$, it means that x belongs to concept C with necessity degree α. Similar comment is applied to weighted role names. We say that a weighted concept C^α is subsumed by another one C^β if $\beta \geq \alpha$. The completion forest $\mathcal{F}_\mathcal{A}$ is initialized such that it contains a root node x_a, with $\mathcal{L}(x_a) = \{C^\alpha | ((C(a), \alpha) \in \mathcal{A}\}$, for each individual a occurring in $\mathcal{A}$, and an edge $\langle x_a, x_b \rangle$, with $\mathcal{L}(\langle x_a, x_b \rangle) = \{r^\alpha : (r(a, b), \alpha) \in \mathcal{A}\}$, for each pair (a, b) of individual names for which $\mathcal{L}(\langle x_a, x_b \rangle)$ is non-empty. We then apply the following expansion rules:

- $\sqcap$-rule: if
  - [1] $(C_1 \sqcap C_2)^\alpha \in \mathcal{L}(x)$, x is not blocked, and
  - [2] there are no $\beta \geq \alpha$ and $\gamma \geq \alpha$ such that $\{(C_1)^\beta, (C_2)^\gamma\} \subseteq \mathcal{L}(x)$
  - then set $\mathcal{L}(x) = \mathcal{L}(x) \cup \{(C_1)^\alpha, (C_2)^\alpha\}$
- $\sqcup$-rule: if
  - [1] $(C_1 \sqcup C_2)^\alpha \in \mathcal{L}(x)$, x is not blocked, and
  - [2] there are no $\beta \geq \alpha$ and $\gamma \geq \alpha$ such that $\{(C_1)^\beta, (C_2)^\gamma\} \subseteq \mathcal{L}(x)$
  - then set $\mathcal{L}(x) = \mathcal{L}(x) \cup \{C^\alpha\}$ for some $C \in \{C_1, C_2\}$
- $\exists$-rule: if
  - [1] $(\exists r.C)^\alpha \in \mathcal{L}(x)$, x is not blocked, and
  - [2] there is no y such that $r^\beta \in \mathcal{L}(\langle x, y \rangle)$ where $\beta \geq \alpha$ and $C^\gamma \in \mathcal{L}(y)$ where $\gamma \geq \alpha$,
  - then create a new node y with $\mathcal{L}(\langle x, y \rangle) = \{r^\alpha\}$ and $\mathcal{L}(y) = \{C^\alpha\}$
- $\forall$-rule: if
  - [1] $(\forall r.C)^\alpha \in \mathcal{L}(x)$, x is not blocked, and
  - [2] there is a y such that $r^\beta \in \mathcal{L}(\langle x, y \rangle)$ with $C^\gamma \notin \mathcal{L}(y)$ for $\gamma \geq min(\alpha, \beta)$,
  - then set $\mathcal{L}(y) = \mathcal{L}(y) \cup \{C^{min(\alpha, \beta)}\}$

The meaning of "blocked" in our tableau expansion rules is similar to that in tableau expansion rules for $\mathcal{ALC}$. A node x is *directly blocked* if there is an ancestor y of x such that for each $C^\alpha \in \mathcal{L}(x)$, there is a $C^\beta \in \mathcal{L}(y)$ such that $\alpha \leq \beta$. x is *blocked* if x is directly blocked or there is an ancestor z of x such that z is directly blocked. Note that blocking is not required if the possibilistic $\mathcal{ALC}$-TBox is empty because there is no cycle in any completion forest.

We briefly explain the $\sqcap$-rule and the $\forall$-rule. Other rules can be explained similarly. The $\sqcap$-rule says that if a weighted concept $(C_1 \sqcap C_2)^\alpha$ is in $\mathcal{L}(x)$ and x is not blocked, and $(C_1)^\alpha$ (resp. $(C_2)^\alpha$) is not subsumed by any weighted concept $(C_1)^\beta$ (resp. $(C_2)^\gamma$) in $\mathcal{L}(x)$, then we add both $(C_1)^\alpha$ and $(C_2)^\alpha$ to $\mathcal{L}(x)$. The $\forall$-rule says that if $(\forall r.C)^\alpha$ is in $\mathcal{L}(x)$ and x is not blocked, and there is another individual y which is related to x with weighted role r^β and $C^{min(\alpha, \beta)}$ is not subsumed by any weighted concept in $\mathcal{L}(y)$, then we add $C^{min(\alpha, \beta)}$ to $\mathcal{L}(y)$. The idea of using *min* to aggregate α and β is taken from automated reasoning in possibilistic logic in [6]. Note that the $\sqcup$-rule leads to *non-determinism* of the algorithm as it adds *either* $(C_1)^\alpha$ or $(C_2)^\alpha$ to $\mathcal{L}(x)$. If the choice of one disjunct, such as $(C_1)^\alpha$, leads to a clash, then the second one must be explored. This is equivalent to say that after application of the $\sqcup$-rule, we get two completion forests from the original one. In our tableau algorithm, after we add a new weighted concept C^α to $\mathcal{L}(x)$ by applying a tableau rule, we then delete every weighted concept C^β which is subsumed by C^α because it is redundant. After a tableau rule is applied to expand a node in a completion forest, this node or another node related to it is updated, or a new node is created.

The tableau algorithm stops when it encounters a clash: a completion forest $\mathcal{F}$ in which $\{A^\alpha, (\neg A)^\beta\} \subseteq \mathcal{L}(x)$ for some node x and some concept name A, where $\alpha, \beta > 0$. In this case, the completion forest contains an inconsistency and the inconsistency degree, denoted $d_{Inc}(\mathcal{F})$, is $min(\alpha, \beta)$. If the algorithm stops and all of the forest[1] i ($i = 1, ..., n$) contain an inconsistency with inconsistency degree α_i (in this case $\alpha_i > 0$), then the inconsistency degree of $\mathcal{A}$ is $min(\alpha_1, ..., \alpha_n)$. Otherwise, if the algorithm stops and there is a forest that does not contain an inconsistency, then the possibilistic $\mathcal{ALC}$-ABox is consistent.

In tableau algorithm for $\mathcal{ALC}$, the choice of which concept in a given completion forest to expand is "don't care" non-deterministic. However, in our algorithm, it is very critical to expand the weighted concept in a right order. That is, we apply tableau rules to expand those concepts with the highest weight first, then those concepts with the second highest weight, and so on. For example, if $(C_1 \sqcap C_2)^{0.7}$ and $(C_3 \sqcap C_4)^{0.6}$ belong to $\mathcal{L}(x)$ for a node x, then the weighted concept $(C_1 \sqcap C_2)^{0.7}$ must be expanded first by the $\sqcap$-rule. For each individual name in $\mathcal{A}$, the forest contains a root node, which will be called as *old nodes*. The nodes that are created by the $\exists$-rule are called *new nodes*. To ensure that our algorithm only requires space polynomial in $|\mathcal{A}|$, we apply tableau expansion rules in the following order:

- apply the $\sqcap$-rule and the $\sqcup$-rule to old nodes as long as possible and check for clash;
- treat each old node in turn, generate all the necessary direct successors of it in a depth first manner by applying the $\exists$-rule and the $\forall$-rule, and check for clash;
- successively handle the successors in the same way.

4.3 Soundness and Completeness

We show the termination of the algorithm.

Proposition 2. *Let $\mathcal{A}$ be a finite possibilistic $\mathcal{ALC}$-ABox and $\mathcal{F}_\mathcal{A}$ the completion forest constructed from $\mathcal{A}$ without application of any tableau rule. Then there is no infinite sequence of rule applications*

$$\mathcal{F}_\mathcal{A} \rightarrow \mathcal{S}_1 \rightarrow \mathcal{S}_2 \rightarrow \cdots .$$

According to Proposition 2, each branch of the completion forest has finite nodes. Therefore, the tableau algorithm must terminate.

Soundness. To show the soundness and completeness of the algorithm, we need to define the notion of inconsistency degree of a set of completion forests. Given a complete forest $\mathcal{F}_\mathcal{A}$, we can transform it to a possibilistic $\mathcal{ALC}$-ABox $\mathcal{A}_\mathcal{F}$ as follows $\mathcal{A}_\mathcal{F} =$

---

[1] For each application of the $\sqcup$-rule, if we explore both disjuncts, then we get two different completion forests. So there may have more than one completion forest.

$\cup_{\mathcal{L}(x_a)\in\mathcal{F}_\mathcal{A}}\{(C(a),\alpha):C^\alpha\in\mathcal{L}(x_a)\}\cup_{\mathcal{L}(\langle x_a,x_b\rangle)\in\mathcal{F}_\mathcal{A}}\{(r(a,b),\alpha):r^\alpha\in\mathcal{L}(\langle x_a,x_b\rangle)\}$.
The inconsistency degree of $\mathcal{F}_\mathcal{A}$, denoted as $Inc(\mathcal{F}_\mathcal{A})$, is defined as $Inc(\mathcal{A}_\mathcal{F})$.

Definition 1. *Let $\mathcal{M} = \{\mathcal{F}^1_\mathcal{A}, ..., \mathcal{F}^n_\mathcal{A}\}$ be a set of complete forests constructed from $\mathcal{A}$ by application of tableau expansion rules. Then the inconsistency degree of $\mathcal{M}$, denoted as $Inc(\mathcal{M})$, is defined as $min_{\mathcal{F}^i_\mathcal{A}} Inc(\mathcal{F}^i_\mathcal{A})$.*

Theorem 1. *(Soundness) Let $\mathcal{M} = \{\mathcal{F}^1_\mathcal{A}, ..., \mathcal{F}^n_\mathcal{A}\}$ be a set of complete forests constructed from $\mathcal{A}$ by application of tableau expansion rules. Suppose $\mathcal{M}'$ is a set of complete forests obtained from $\mathcal{M}$ by application of a tableau expansion rule. If $Inc(\mathcal{M}) = \alpha$ then $Inc(\mathcal{M}') = \alpha$.*

Proof. The following lemma shows that application of any tableau rule to a complete forest will not change its inconsistency degree.

Lemma 1. *Let $\mathcal{F}_\mathcal{A}$ be a complete forest constructed from $\mathcal{A}$ and by application of tableau expansion rules. The following conclusions hold. (1) Assume that $\mathcal{F}'_\mathcal{A}$ is obtained from $\mathcal{F}_\mathcal{A}$ by applying the $\sqcap$-rule, the $\exists$-rule or the $\forall$-rule. If $Inc(\mathcal{F}_\mathcal{A}) = \alpha$, then $Inc(\mathcal{F}'_\mathcal{A}) = \alpha$. (2) Assume that the $\sqcup$-rule is applied to $\mathcal{F}_\mathcal{A}$, then it can be applied in such a way that it yields a completion forest $\mathcal{F}'_\mathcal{A}$ such that if $Inc(\mathcal{F}_\mathcal{A}) = \alpha$, then $Inc(\mathcal{F}'_\mathcal{A}) = \alpha$.*

Proof. (sketch) Due to page limit, we consider only the $\sqcup$-rule. Other rules can be discussed similarly.

Assume that the $\sqcup$-rule is applied to $\mathcal{F}_\mathcal{A}$. Let $\mathcal{F}'_\mathcal{A}$ and $\mathcal{F}''_\mathcal{A}$ be two complete forests obtained from $\mathcal{F}_\mathcal{A}$ by applying the $\sqcup$-rule. Suppose on the contrary that $Inc(\mathcal{F}'_\mathcal{A}) > \alpha$ and $Inc(\mathcal{F}''_\mathcal{A}) > \alpha$. Similar to the analysis in (1), we have that $\mathcal{A}_{\mathcal{F}'} = \mathcal{A}_\mathcal{F} \cup \{(C_1(a),\beta)\}$ and $\mathcal{A}_{\mathcal{F}''} = \mathcal{A}_\mathcal{F} \cup \{(C_2(a),\beta)\}$. Suppose $\beta \leq \alpha$, then $Inc(\mathcal{A}_{\mathcal{F}'}) = \alpha$ (resp. $Inc(\mathcal{A}_{\mathcal{F}''}) = \alpha$), which is a contradiction. Assume that $\beta > \alpha$. Since $Inc(\mathcal{F}'_\mathcal{A}) > \alpha$ and $Inc(\mathcal{F}''_\mathcal{A}) > \alpha$, both $(\mathcal{A}_{\mathcal{F}'})_{>\alpha}$ and $(\mathcal{A}_{\mathcal{F}''})_{>\alpha}$ are inconsistent. Since $Inc(\mathcal{F}_\mathcal{A}) = \alpha$, $(\mathcal{A}_\mathcal{F})_{>\alpha}$ is consistent. Therefore, $(\mathcal{A}_\mathcal{F})_{>\alpha}$ is in conflict with both $C_1(a)$ and $C_2(a)$ and also with $(C_1 \sqcup C_2)(a)$. However, $(C_1 \sqcup C_2)(a) \in (\mathcal{A}_\mathcal{F})_{>\alpha}$, this is a contradiction.

The proof of Theorem 1 follows from Lemma 1 and definition of the inconsistency degree of a set of complete forests.

According to Theorem 1, after we apply the tableau expansion rules, the inconsistency degree of the newly constructed set of complete forests does not change.

Completeness. Assume that $\mathcal{M} = \{\mathcal{F}^1_\mathcal{A}, ..., \mathcal{F}^n_\mathcal{A}\}$ be a set of complete forests constructed from $\mathcal{A}$ by application of tableau expansion rules until no more rules apply. Let $d_{Inc}(\mathcal{F}^i_\mathcal{A}) = \alpha_i$ and $d_{Inc}(\mathcal{M}) = min\{\alpha_1, ..., \alpha_n\}$. We need to show that $d_{Inc}(\mathcal{M}) = Inc(\mathcal{M})$. We first need a lemma.

Lemma 2. $d_{Inc}(\mathcal{F}^i_\mathcal{A}) = Inc(\mathcal{F}^i_\mathcal{A})$ *for $i = 1, n$.*

Proof. (sketch) Suppose $\mathcal{A}_{\mathcal{F}_i}$ is the corresponding ABox of $\mathcal{F}^i_\mathcal{A}$. It is clear that $Inc(\mathcal{A}_{\mathcal{F}_i}) \geq \alpha_i$ because $(C(a),\alpha_i), (\neg C(a),\alpha_i) \in \mathcal{A}_{\mathcal{F}_i}$ for some $C(a)$. So we only need to show that $(\mathcal{A}_{\mathcal{F}_i})_{>\alpha_i}$ is consistent. The proof is similar to the proof of completeness of tableau algorithm for DL $\mathcal{ALC}$ by considering the following facts: (1) $(\mathcal{A}_{\mathcal{F}_i})_{>\alpha_i}$ contains all the ABox assertions that cannot be expanded by the tableau rules, (2) there is no clash $\{A^\beta, A^\gamma\}$ in $\mathcal{F}_A^i$ such that $min(\beta, \gamma) > \alpha$ and (3) a weighted concept with higher necessity degree should always be expanded by an expansion rule first if applicable. Condition (3) is very critical to ensure the completeness of the algorithm as it ensures that we find the maximum weight α such that α-cut of the ABox is inconsistent. Suppose we do not have this requirement and we want to apply the tableau algorithm to a possibilistic $\mathcal{ALC}$-ABox $\mathcal{A} = \{(A(a), 0.7), ((C \sqcap \neg A)(a), 0.6), ((B \sqcap \neg A)(a), 0.5)\}$. We first construct the completion forest $\mathcal{F}_A = \{\mathcal{L}_{x_a}\}$ where $\mathcal{L}_{x_a} = \{A^{0.7}, (C \sqcap \neg A)^{0.6}, (B \sqcap \neg A)^{0.5}\}$. Now suppose we apply the $\sqcap$-rule to expand $(B \sqcap \neg A)^{0.5}$, then we have $\mathcal{L}_{x_a} = \{A^{0.7}, C \sqcap \neg A^{0.6}, (B \sqcap \neg A)^{0.5}, B^{0.5}, (\neg A)^{0.5}\}$. There is a conflict in $\mathcal{F}_A = \{\mathcal{L}_{x_a}\}$ and the algorithm will terminate and return the inconsistency degree 0.5. Since there is only one completion forest for $\mathcal{A}$, the inconsistency degree of $\mathcal{A}$ is 0.5, which is wrong.

Theorem 2. *(Completeness)* $d_{Inc}(\mathcal{M}) = Inc(\mathcal{M})$

Theorem 1 and Theorem 2 together show that our tableau algorithm output the correct inconsistency degree of $\mathcal{A}$.

4.4 Complexity

We consider computational issues of our tableau algorithm. The following theorem gives the complexity of checking the inconsistency degree of a possibilistic $\mathcal{ALC}$-ABox .

Theorem 3. *Checking if γ is the inconsistency degree of possibilistic $\mathcal{ALC}$-ABox $\mathcal{A}$ is a PSpace-complete problem.*

By Theorem 3, we have that instance checking for a consistent possibilistic $\mathcal{ALC}$-ABox is a PSpace-complete problem.

Corollary 1. *Let $\mathcal{A}$ be a consistent possibilistic $\mathcal{ALC}$-ABox. Then checking if $\mathcal{A} \models_P C(a)$ or if $\mathcal{A} \models_\pi (C(a), \alpha)$ is a PSpace-complete problem.*

4.5 Terminological Axioms

In this subsection, we extend the tableau algorithm for computing the inconsistency degree of a possibilistic $\mathcal{ALC}$-ABox by considering a non-empty TBox. We first consider acyclic terminologies and then general terminologies.

Acyclic Terminologies. A possibilistic TBox $\mathcal{T}$ is an *acyclic terminology* iff $\mathcal{T}^*$ is an acyclic terminology, i.e., $\mathcal{T}^*$ is a set of concept definitions that neither contains *multiple definitions* nor *cyclic definitions*. Therefore, we do not allow an atomic concept appearing on the left side twice. We call a concept axiom of the form $(A \equiv C, \alpha)$ in $\mathcal{T}$ a *weighted definition*.

Given a possibilistic DL $\mathcal{ALC}$ knowledge base $\mathcal{B} = \langle \mathcal{T}, \mathcal{A} \rangle$, when $\mathcal{T}$ is acyclic, then the problem of computing the inconsistency degree of $\mathcal{B}$ can be reduced to the problem of computing the inconsistency degree of a possibilistic $\mathcal{ALC}$-ABox by *unfolding* the weighted definitions. For any concept C, we use $\neg C$ to denote the NNF of $\neg C$. We extend the *lazy unfolding* rules given in [2] by proposing the following two rules:

- $\equiv_1$-rule: if
  - [1] $(A \equiv C, \alpha) \in \mathcal{T}$, $A^\beta \in \mathcal{L}(x)$, and
  - [2] there does not exist $C^\gamma \in \mathcal{L}(x)$ such that $\gamma \geq min(\alpha, \beta)$
  - then set $\mathcal{L}(x) = \mathcal{L}(x) \cup \{C^{min(\alpha,\beta)}\}$;
- $\equiv_2$-rule: if
  - [1] $(A \equiv C, \alpha) \in \mathcal{T}$, $\neg A^\beta \in \mathcal{L}(x)$, and
  - [2] there does not exist $\neg C^\gamma \in \mathcal{L}(x)$ such that $\gamma \geq min(\alpha, \beta)$
  - then set $\mathcal{L}(x) = \mathcal{L}(x) \cup \{\neg C^{min(\alpha,\beta)}\}$.

The $\equiv_1$-rule (resp. $\equiv_2$-rule) says that if the tableau algorithm encounters a weighted concept A^β (resp. $\neg A^\beta$) and there exists a weighted definition $(A \equiv C, \alpha) \in \mathcal{T}$, and $C^{min(\alpha,\beta)}$ (resp. $\neg C^{min(\alpha,\beta)}$) is not subsumed by any C^γ (resp. $\neg C^\gamma \in \mathcal{L}(x)$) in $\mathcal{L}(x)$, then it adds the weighted concept $C^{min(\alpha,\beta)}$ (resp. $\neg C^{min(\alpha,\beta)}$) to $\mathcal{L}(x)$. The order of application of the lazy unfolding rules is very important to ensure correctness of our algorithm. Like the expansion of weighed concepts, a weighted definition with higher weight should be unfolded first. Furthermore, if the weight of a weighed definition is greater than that of a weighted concept, it should be unfolded before the expansion of the weighted concept, vice versa.

Similar to the proof of Theorem 1 in [17], we can show that the extended lazy unfolding yields in a PSpace-algorithm for computing the inconsistency degree of a a possibilistic $\mathcal{ALC}$ knowledge base with possibilistic acyclic terminology.

Theorem 4. *Given a possibilistic DL $\mathcal{ALC}$ knowledge base $\mathcal{B} = \langle \mathcal{T}, \mathcal{A} \rangle$, when $\mathcal{T}$ is acyclic, checking if γ is the inconsistency degree of $\mathcal{B}$ is a PSpace-complete problem.*

General TBoxes. When general possibilistic TBoxes are considered, we cannot apply the technique of unfolding any more. We extend the $\sqsubseteq$-rule given in [3].

- $\sqsubseteq$-rule: if
  - [1] $(C_1 \sqsubseteq C_2, \alpha) \in \mathcal{T}$, x is not blocked, and
  - [2] there is no $\beta \geq \alpha$ such that $(C_2 \sqcup \neg C_1)^\beta \in \mathcal{L}(x)$
  - then set $\mathcal{L}(x) = \mathcal{L}(x) \cup \{(C_2 \sqcup \neg C_1)^\alpha\}$

The $\sqsubseteq$-rule says that if $(C_1 \sqsubseteq C_2, \alpha)$ is in $\mathcal{T}$ and x is not blocked, and $(C_2 \sqcup \neg C_1)^\alpha$ is not subsumed by a weighted concept in $\mathcal{L}(x)$, then we add $(C_2 \sqcup \neg C_1)^\alpha$ to $\mathcal{L}(x)$. Blocking is required to ensure termination of the tableau algorithm. A possibilistic concept axiom with higher weight should be expanded first. If the weight of a possibilistic concept axiom is greater than that of a weighted concept, it should be expanded before the expansion of the weighted concept, vice versa.

Let us consider Example 1 given in Introduction again. We construct a completion forest $\mathcal{F}_A = \{\mathcal{L}(x_{Tom})\}$, where
$$\mathcal{L}(x_{Tom}) = \{PacemakerPatient^{0.8}, HeartPatient^{0.5}\}.$$
We first expand ax_1 and ax_2 using the $\sqsubseteq$-rule and we get
$$\mathcal{L}(x_{Tom}) = \{(HighBloodPressure \sqcup \neg Heartpatient)^1,$$
$$(\neg HighBloodPressure \sqcup \neg PacemakerPatient)^1,$$
$$PacemakerPatient^{0.8}, HeartPatient^{0.5}\}.$$

We then apply the $\sqcup$-rule to expand $(HighBloodPressure \sqcup \neg Heartpatient)^1$ and $(\neg HighBloodPressure \sqcup \neg PacemakerPatient)^1$. Assume that we get $\mathcal{L}(x_{Tom})$ $= \{HighBloodPressure^1, \neg HighBloodPressure^1, PacemakerPatient^{0.8},$ $HeartPatient^{0.5}\}$, which contains a clash. So the algorithm stop expanding this forest and the inconsistency degree is 1. The algorithm continues another alternative $\mathcal{L}(x_{Tom})$ $= \{HighBloodPressure^1, \neg PacemakerPatient^1, PacemakerPatient^{0.8},$ $HeartPatient^{0.5}\}$, which contains a clash and the inconsistency degree is 0.5. Similarly we can get another completion forest with inconsistency degree 0.8. Therefore, the inconsistency degree of the possibilistic DL knowledge base is 0.5.

We show the termination of our tableau algorithm with a general possibilistic TBox.

Proposition 3. *Let $\mathcal{B} = \langle \mathcal{T}, \mathcal{A} \rangle$ be a finite possibilistic $\mathcal{ALC}$ knowledge base and $\mathcal{F}_A$ the completion forest constructed from $\mathcal{A}$ without application of any tableau rule. Then there is no infinite sequence of rule applications*

$$\mathcal{F}_A \rightarrow \mathcal{S}_1 \rightarrow \mathcal{S}_2 \rightarrow \cdots .$$

We show that the tableau algorithm is sound and complete by incorporating the $\sqsubseteq$-rule when a general TBox is available.

Given a completion forest $\mathcal{F}_A$ and a general possibilistic TBox $\mathcal{T}$, the inconsistency degree of the pair $\langle \mathcal{F}_A, \mathcal{T} \rangle$, denoted $Inc(\langle \mathcal{F}_A, \mathcal{T} \rangle)$, is $Inc(\mathcal{A}_\mathcal{F} \cup \mathcal{T})$. Similar to Definition 1, we have the following definition.

Definition 2. *Let $\mathcal{M} = \{\mathcal{F}_A^1, ..., \mathcal{F}_A^n\}$ be a set of completion forests constructed from $\mathcal{A}$ and $\mathcal{T}$ by application of tableau expansion rules (including the $\sqsubseteq$-rule). Then the inconsistency degree of the pair $\langle \mathcal{M}, \mathcal{T} \rangle$, denoted as $Inc(\langle \mathcal{M}, \mathcal{T} \rangle)$, is defined as $min_{\mathcal{F}_A^i} Inc(\langle \mathcal{F}_A^i, \mathcal{T} \rangle)$.*

We have the following theorem that shows the soundness of our algorithm.

Theorem 5. *(Soundness) Let $\mathcal{M} = \{\mathcal{F}_A^1, ..., \mathcal{F}_A^n\}$ be a set of completion forests constructed from $\mathcal{A}$ and $\mathcal{T}$ by application of tableau expansion rules. Suppose $\mathcal{M}'$ is a set of completion forests obtained from $\mathcal{M}$ by application of a tableau expansion rule. If $Inc(\langle \mathcal{M}, \mathcal{T} \rangle) = \alpha$ then $Inc(\langle \mathcal{M}', \mathcal{T} \rangle) = \alpha$.*

Assume that $\mathcal{M} = \{\mathcal{F}_A^1, ..., \mathcal{F}_A^n\}$ is a set of completion forests constructed from $\mathcal{A}$ and $\mathcal{T}$ by application of tableau expansion rules until no more rules apply. Let $d_{Inc}(\langle \mathcal{F}_A^i, \mathcal{T} \rangle)$ $= \alpha_i$ $(i = 1, ..., n)$ and $d_{Inc}(\langle \mathcal{M}, \mathcal{T} \rangle) = min\{\alpha_1, ..., \alpha_n\}$. Similar to Theorem 2, we can show the following theorem.

Theorem 6. *(Completeness) $d_{Inc}(\langle \mathcal{M}, \mathcal{T} \rangle) = Inc(\langle \mathcal{M}, \mathcal{T} \rangle)$.*

4.6   Optimization Techniques

We discuss how some important optimization techniques applied to the tableau algorithm for $\mathcal{ALC}$ given in Chapter 9 of [1] can be reused for our tableau algorithm.

Normalisatoin and *Absorption* in the preprocessing step can be easily reused as they are applied to concepts. For example, a possibilistic axiom of the form $(A \sqcap D \sqsubseteq D', \alpha)$ can be rewritten as $(A \sqsubseteq D' \sqcup \neg D, \alpha)$. The *semantic branching* technique can be also reused. That is, instead of choosing an unexpanded weighted disjunction such as $(A \sqcup B, \alpha)$ in $\mathcal{L}(x)$, a single disjunct is chosen from one of the unexpanded weighted disjunction in $\mathcal{L}(x)$. Suppose A from $(A \sqcup B, \alpha)$ is chosen, then two possible sub-trees obtained by adding either (A, α) or $(\neg A, \alpha)$ to $\mathcal{L}(x)$ are searched. *Local simplification* is to simplify disjunctions if possible before any non-deterministic expansion of a node. An important simplification is called Boolean constraint propagation (BCP) which is based on inference rules such as $\frac{\neg C_1, ..., \neg C_n, C_1 \sqcup ... \sqcup C_n \sqcup D}{D}$. The BCD simplification can be adapted by extending the inference rules based on the idea of possibilistic resolution given in [14], for example, $\frac{(\neg C_1, \alpha_1), ..., (\neg C_n, \alpha_n), (C_1 \sqcup ... \sqcup C_n \sqcup D, \alpha)}{(D, min(\alpha_1, ..., \alpha_n, \alpha))}$. The *backjumping* technique can be also adapted. Similar to backjumping, we can label each weighted concept in a node and each weighted role in an edge with a dependence set that indicates the non-deterministic expansion choices on which they depend. Using this dependence set, our algorithm can avoid fruitless exploration.

5   Conclusion and Future Work

In this paper, we presented a tableau algorithm for possibilistic $\mathcal{ALC}$. The contributions of this paper can be summarized as follows:

- First, we showed how inference services, such as instance checking with necessity degree, in possibilistic $\mathcal{ALC}$ can be reduced to the problem of computing the inconsistency degree of a posssibilistic DL knowledge base.
- We gave tableau expansion rules for computing the inconsistency degree of a possibilistic $\mathcal{ALC}$-ABox by extending the tableau rules for $\mathcal{ALC}$. We showed that the tableau algorithm terminates in finite steps and it is complete and sound.
- We showed that checking the inconsistency degree of a possibilistic $\mathcal{ALC}$-ABox is a PSpace-complete problem. As a corollary, we can infer that instance checking for a consistent possibilistic $\mathcal{ALC}$-ABox is PSpace-complete.
- We proposed tableau rules for a possibilistic $\mathcal{ALC}$ knowledge base with acyclic terminologies and general TBox and showed that the tableau algorithm is still sound and complete, and it is decidable to check the inconsistency degree of a possibilistic $\mathcal{ALC}$ knowledge base.

Several interesting problems are left for future work: (1) We will implement the proposed tableau algorithm and provide evaluation results. Our theoretical analysis shows that our tableau algorithm is more promising than the black-box algorithm given in [18]. This needs to be further justified by the empirical evaluation. (2) We will consider conjunctive query answering for expressive possibilistic DLs by extending the results of query answering on expressive DLs.

Acknowledgement

Guilin Qi is partially supported by the EU under the IST project NeOn and the X-Media project, and Jeff Z. Pan is partially supported by the EU MOST project.

References

1. Baader, F., Calvanese, D., McGuinness, D.L., Nardi, D., Patel-Schneider, P.F. (eds.): The description logic handbook: theory, implementation, and applications. Cambridge University Press, New York (2003)
2. Baader, F., Hollunder, B., Nebel, B., Profitlich, H.-J., Franconi, E.: An empirical analysis of optimization techniques for terminological representation systems, or making kris get a move on. In: Proc. of KR 1992, pp. 270–281 (1992)
3. Baader, F., Horrocks, I., Sattler, U.: Description Logics. In: van Harmelen, F., Lifschitz, V., Porter, B. (eds.) Handbook of Knowledge Representation. Elsevier, Amsterdam (to appear, 2007)
4. Baaderand, F., Sattler, U.: An overview of tableau algorithms for description logics. Studia Logica 69(1), 5–40 (2001)
5. Couchariere, O., Lesot, M.-J., Bouchon-Meunier, B.: Consistency checking for extended description logics. In: Description Logics (2008)
6. Dubois, D., Lang, J., Prade, H.: Automated reasoning using possibilistic logic: Semantics, belief revision, and variable certainty weights. IEEE Trans. Knowl. Data Eng. 6(1), 64–71 (1994)
7. Dubois, D., Lang, J., Prade, H.: Possibilistic logic. In: Handbook of logic in Aritificial Intelligence and Logic Programming, 3rd edn., pp. 439–513. Oxford University Press, Oxford (1994)
8. Dubois, D., Mengin, J., Prade, H.: Possibilistic uncertainty and fuzzy features in description logic: A preliminary discussion. In: Capturing Intelligence: Fuzzy Logic and the Semantic WEb, pp. 101–113. Elsevier, Amsterdam (2006)
9. Giugno, R., Lukasiewicz, T.: P-$\mathcal{SHOQ}(d)$: A probabilistic extension of $\mathcal{SHOQ}(d)$ for probabilistic ontologies in the semantic web. In: Flesca, S., Greco, S., Leone, N., Ianni, G. (eds.) JELIA 2002. LNCS, vol. 2424, pp. 86–97. Springer, Heidelberg (2002)
10. Heinsohn, J.: Probabilistic description logics. In: Proc. of UAI 1994, pp. 311–318 (1994)
11. Hollunder, B.: An alternative proof method for possibilistic logic and its application to terminological logics. In: Proc. of UAI 1994, pp. 327–335 (1994)
12. Jaeger, M.: Probabilistic reasoning in terminological logics. In: Proc. of KR 1994, pp. 305–316 (1994)
13. Koller, D., Levy, A.Y., Pfeffer, A.: P-classic: A tractable probablistic description logic. In: Proc. of AAAI/IAAI 1997, pp. 390–397 (1997)
14. Lang, J.: Possibilistic logic: complexity and algorithms. In: Handbook of Defeasible Reasoning and Uncertainty Management Systems, pp. 179–220. Kluwer, Dordrecht (1998)
15. Lukasiewicz, T.: Probabilistic default reasoning with conditional constraints. Ann. Math. Artif. Intell. 34(1-3), 35–88 (2002)
16. Lukasiewicz, T.: Expressive probabilistic description logics. Artif. Intell. 172(6-7), 852–883 (2008)
17. Lutz, C.: Complexity of terminological reasoning revisited. In: LPAR 1999, pp. 181–200 (1999)
18. Qi, G., Pan, J.Z., Ji, Q.: Extending description logics with uncertainty reasoning in possibilistic logic. In: Mellouli, K. (ed.) ECSQARU 2007. LNCS, vol. 4724, pp. 828–839. Springer, Heidelberg (2007)
19. Stoilos, G., Stamou, G., Pan, J.Z., Tzouvaras, V., Horrocks, I.: Reasoning with very expressive fuzzy description logics. J. Artif. Intell. Res. 30, 273–320 (2007)
20. Straccia, U.: Reasoning within fuzzy description logics. J. Artif. Intell. Res. 14, 137–166 (2001)
21. Udrea, O., Deng, Y., Hung, E., Subrahmanian, V.S.: Probabilistic ontologies and relational databases. In: Proc. of CoopIS/DOA/ODBASE 2005, pp. 1–17 (2005)

SAOR: Authoritative Reasoning for the Web*

Aidan Hogan, Andreas Harth, and Axel Polleres

Digital Enterprise Research Institute, National University of Ireland, Galway

Abstract. In this paper we discuss the challenges of performing reasoning on large scale RDF datasets from the Web. We discuss issues and practical solutions relating to reasoning over web data using a rule-based approach to forward-chaining; in particular, we identify the problem of ontology hijacking: new ontologies published on the Web re-defining the semantics of existing concepts resident in other ontologies. Our solution introduces consideration of authoritative sources. Our system is designed to scale, comprising of file-scans and selected lightweight on-disk indices. We evaluate our methods on a dataset in the order of a hundred million statements collected from real-world Web sources.

1 Introduction

Data attainable through the Web is unique in terms of scale and diversity. Millions of data sources contribute billions of statements to a giant data graph. The Semantic Web technology stack includes means to supplement instance data being published using the Resource Description Framework (RDF) with ontologies described in RDF Schema (RDFS) [1] and the Web Ontology Language (OWL) [16], allowing people to formally specify a domain of discourse, and providing machines a more sapient understanding of the data. While there exists a large body of work in the area of reasoning algorithms and systems that work and scale well in confined environments, the distributed and loosely coordinated creation of a world-wide knowledge base creates new challenges.

Reasoning over aggregated Web data is useful, for example: to infer new assertions using terminological knowledge from ontologies and therefore provide a more complete dataset; to unite fractured knowledge about individuals collected from disparate sources; and to execute mappings between domain descriptions and therefore provide translations from one conceptual model to another. Our work on reasoning is motivated by the requirements of the Semantic Web Search Engine (SWSE) project[1], within which we strive to offer search, querying and browsing over the Semantic Web.

Reasoning on Web data poses a number of requirements:

- the system has to perform on web-scale, with implications on the completeness of the reasoning procedure, algorithms and optimisations

---

* This work has been supported by Science Foundation Ireland (SFI/02/CE1/I131), European FP6 project inContext (IST-034718) and COST Action "Agreement Technologies" (IC0801).
[1] http://swse.deri.org/

J. Domingue and C. Anutariya (Eds.): ASWC 2008, LNCS 5367, pp. 76–90, 2008.
© Springer-Verlag Berlin Heidelberg 2008

- the method has to perform on collaboratively created knowledge bases, which has implications on trust and the privileges of data publishers
- the web search scenario requires sub-second response times, which has implications on the reasoning and query processing strategy

We present SAOR – Scalable Authoritative OWL Reasoner – which focuses on performing best-effort RDFS and OWL reasoning on Web data. SAOR is designed to accept as input a web knowledge-base in the form of a body of statements as produced by a web-crawl and to output by forward-chaining a knowledge-base enhanced by a given fragment of OWL reasoning. Discussion of the end consumption of such a reasoned knowledge-base is outside of the scope of this paper.

Specifically, we make the following contributions in this paper:

- We apply only a selected subset of OWL reasoning, to i) avoid an explosion of inferred statements and ii) to protect existing specifications from undesirable contributions made in independent locations. Our system implements a positive fragment of OWL Full which has roots in ter Horst's pD^* [17] entailment. We describe an analysis of the authority of sources to counter-act the highlighted problem of *ontology hijacking* in web data (Section 2).
- We describe a scalable, optimised method for performing rule-based forward chaining reasoning through means of file-scan and lightweight dynamic data structures. In particular, our algorithm capitalises on the low volume of structural T-Box data relative to A-Box instance data (Section 3).
- We show experimentally that a forward-chaining materialisation approach is feasible on Web data from 315k sources and in the order of 100m triples: with our confined reasoning strategy we found that the knowledge base only roughly doubles in size (Section 4) by cautious materialisation.

We discuss related work in Section 5 and conclude with Section 6.

2   Pragmatic Inferencing for the Web

We begin by stating a couple of observations regarding the feasibility of reasoning on the Web. Firstly, most OWL data crawlable on the Web is OWL Full. Idealised assumptions made in OWL DL, such as disallowing subclassing or defining subproperties of the OWL and RDF(S) vocabularies, are violated by even very commonly used ontologies[2]. Secondly, consistency cannot be expected on the Web. For instance, a past web-crawl of ours revealed the following:[3]

```
<timbl> a foaf:Person; foaf:homepage <http://w3.org/> .
<w3c> a foaf:Organization; foaf:homepage <http://w3.org/> .
foaf:homepage a :InverseFunctionalProperty.
foaf:Organization :disjointWith foaf:Person .
```

---

[2] E.g., one of the reasons why the commonly used FOAF vocabulary falls into OWL Full is that `foaf:name` is defined as a subproperty of `rdfs:label` [19].

[3] Throughout this paper, we assume that `http://www.w3.org/2002/07/owl#` is the default namespace and use well understood prefixes for other namespaces.

These triples together infer that Tim Berners-Lee is the same as the W3C and cause an inconsistency. However, despite such examples which arise from misunderstanding of the FOAF vocabulary, there might be cases where different parties deliberately make contradictive statements.

These two points already suggest that complete inference at the instance level is neither feasible nor desirable: firstly, for the computational infeasibility of complete OWL Full reasoning, and secondly, since we do not deem the explosive nature of contradiction in classical logics desirable in a Web reasoning scenario.

Thus, rather than striving for complete inference, we adopt a "best effort" reasoning strategy, optimising inference based on the following principles:

1. We assume a separation of T-Box from A-Box.
2. We trade completeness for implementational feasibility following a rule-based, finite, forward-chaining approach to OWL inference.
3. We trade completeness for producing a much smaller subset of inferred statements; i.e, we deliberately ignore (i) the explosive behaviour of classical inconsistency, (ii) arguably "void" statements in terms of non-standard use of the RDF(S) and OWL vocabularies, (iii) *non-authoritative* T-Box statements.

2.1   Separating A-Box from T-Box

In SAOR, we separate terminological knowledge from assertional data according to their use of the RDF(S) and OWL vocabulary; we call these the "A-Box" and the "T-Box" respectively (loosely borrowing Description Logics terminology).

Table 1 provides a list of graph patterns in RDF graphs we consider to be part of the T-Box. Note that when retrieving graphs from the Web, the instances of these patterns are all of the T-Box statements we consider in our reasoning process: triples that do not match one of these patterns are not considered being part of the T-Box, but are treated purely as assertional "data" triples.

The materialisation of axiomatic statements and completing the entire T-Box may create a bulk of statements with little practical utility. In fact, we deliberately accept the omission of T-Box inference rather as an optimisation: we focus on answering queries over A-Box data rather than, e.g., inferring all members of `:Class`.

SAOR does not support metamodelling, except by conceptually separating the instance- class- or property-meanings of a resource: by separating the T-Box and A-Box segment of the knowledge base, we do not support all possible entailments from the simultaneous description of both a class and an instance. Particularly, we treat URIs in the context they appear, in the spirit of "punning"[4]. We do subject the T-Box data to reasoning analogously to the A-Box, but only store results in the A-Box. For example, we do not carry over `:sameAs` inferences to the T-Box – this is in-line with first-order-logic point of view, where equalities do not affect predicates.

We filter out further triples when extracting the T-Box; namely, we ignore nonstandard use of RDF in our reasoning efforts. Non-standard use of RDF

---

[4] http://www.w3.org/2007/OWL/wiki/Punning

Table 1. Allowed T-Box constructs for each rule. Bold type indicates that the element must be authoritatively spoken for. Where they appear, at least one of the italic type elements must be authoritatively spoken for (see 2.3).

No	DL Syntax	Corresponding RDF graph pattern
01	$C \sqsubseteq D$	**?C** rdfs:subClassOf ?D .
02_a	$C \equiv D$	**?C** :equivalentClass ?D .
02_b		?C :equivalentClass **?D** .
03	$P \sqsubseteq Q$	**?P** rdfs:subPropertyOf ?Q .
04_a	$P \equiv Q$	**?P** :equivalentProperty ?Q .
04_b		?P :equivalentProperty **?Q** .
05_a	$P \equiv Q^{-}$	**?P** :inverseOf ?Q .
05_b		?P :inverseOf **?Q** .
06	$\top \sqsubseteq \forall P^{-}.C$	**?P** rdfs:domain ?C .
07	$\top \sqsubseteq \forall P.C$	**?P** rdfs:range ?C .
08	$P \equiv P^{-}$	**?P** a :SymmetricProperty .
09_a	$\exists P.x$	?C :hasValue ?x; :onProperty **?P** .
09_b		**?C** :hasValue ?x; :onProperty ?P .
10	$C_1 \sqcup ... \sqcup C_n$	?C :unionOf (**?C**$_1$ ... **?C**$_i$ ... ?C$_n$) .
11_a	$C_1 \sqcap ... \sqcap C_n$	**?C** :intersectionOf (?C$_1$ ... ?C$_n$) .
11_b		?C :intersectionOf (*?C$_1$* ... *?C$_n$*) .
12	$\top \sqsubseteq \forall \leq 1P$	**?P** a :FunctionalProperty .
13	$\top \sqsubseteq \forall \leq 1P^{-}$	**?P** a :InverseFunctionalProperty .
14	$P^{+} \sqsubseteq P$	**?P** a :TransitiveProperty .
15	$\exists P.D$	?C :someValuesFrom *?D*; :onProperty *?P* .
16	$\forall P.D$	*?C* :allValuesFrom *?D*; :onProperty *?P* .
17_a	$(\leq 1P)$	*?C* :maxCardinality 1; :onProperty *?P* .
17_b	$(= 1P)$	*?C* :cardinality 1; :onProperty *?P* .
18	$\{x_i....x_n\}$	?C :oneOf (?x1 ... ?xN) .

briefly equates to the use of properties and classes which make up the RDF(S) vocabulary in locations where they have not been intended, cf. [2,13]. We adapt the definition of non-standard use for our purposes and only restrict the usage of the vocabulary for the T-Box: let $\mathcal{P}$ = { rdf:type, rdf:domain, rdf:range, rdfs:subClassOf, rdfs:subPropertyOf, :equivalentClass, :equivalentProperty, :inverseOf, :onProperty, :hasValue, :someValuesFrom, :allValuesFrom, :intersectionOf, :unionOf, :maxCardinality, :cardinality, :oneOf }, and $\mathcal{C}$ = { :FunctionalProperty, :InverseFunctionalProperty, :TransitiveProperty, :SymmetricProperty }. We omit from the T-Box any triples with non-standard use of the properties and classes in $\mathcal{P} \cup \mathcal{C}$, that is, triples where properties in $\mathcal{P}$ appear in a position other than the predicate position or where classes in $\mathcal{C}$ appear in a position other than the object of an rdf:type triple.

In summary, our view of a web knowledge base $\mathcal{KB}$ consists of the RDF merge of a set of source graphs. $\mathcal{KB}$ is separated into a T-Box $\mathcal{T}$ mentioning classes and properties $\mathcal{C}_{\mathcal{KB}}$ and $\mathcal{P}_{\mathcal{KB}}$ and an A-Box $\mathcal{A}$ consisting of class and property membership assertions possibly using identifiers in $\mathcal{C}_{\mathcal{KB}} \cup \mathcal{P}_{\mathcal{KB}}$.

2.2 Rule-Based OWL Reasoning

Reasoning in SAOR is inspired by previous approaches, particularly the pD^{*} fragment defined by ter Horst [17], to cover large parts of OWL by positive inference rules which can be implemented in a forward-chaining engine.

Table 2 lists all of the currently supported rules. Although certain triples matched in the antecedents come from the T-Box and others come from the

Table 2. Supported rules with N3-style syntax used for triple patterns with T-Box statements italicised and A-Box in plain font

	DL Syntax	Rule
01	$C \sqsubseteq D$	*?C rdfs:subClassOf ?D* . ?s a ?C . $\Rightarrow$?s a ?D .
02$_a$	$C \equiv D$	*?C :equivalentClass ?D* . ?s a ?C .$\Rightarrow$?s a ?D .
02$_b$		*?C :equivalentClass ?D* . ?s a ?D .$\Rightarrow$?s a ?C .
03	$P \sqsubseteq Q$	*?P rdfs:subPropertyOf ?Q* . ?s ?P ?o . $\Rightarrow$?s ?Q ?o .
04$_a$	$P \equiv Q$	*?P :equivalentProperty ?Q* . ?s ?P ?o . $\Rightarrow$?s ?Q ?o .
04$_b$		*?P :equivalentProperty ?Q* . ?s ?Q ?o . $\Rightarrow$?s ?P ?o .
05$_a$	$P \equiv Q^-$	*?P :inverseOf ?Q* . ?s ?P ?o . $\Rightarrow$?o ?Q ?s .
05$_b$		*?P :inverseOf ?Q* . ?s ?Q ?o . $\Rightarrow$?o ?P ?s .
06	$\top \sqsubseteq \forall P^-.C$	*?P rdfs:domain ?C* . ?s ?P ?o . $\Rightarrow$?s a ?C .
07	$\top \sqsubseteq \forall P.C$	*?P rdfs:range ?C* . ?s ?P ?o . $\Rightarrow$?o a ?C .
08	$P \equiv P^-$	*?P a :SymmetricProperty* . ?s ?P ?o . $\Rightarrow$?o ?P ?s .
09$_a$	$\exists P.x$	*?C :hasValue ?x; :onProperty ?P* . ?y ?P ?x . $\Rightarrow$?y a ?C .
09$_b$		*?C :hasValue ?x; :onProperty ?P* . ?y a ?C . $\Rightarrow$?y ?P ?x .
10	$C_1 \sqcup ... \sqcup C_n$	*?C :unionOf (?C$_1$ ... ?C$_i$ ... ?C$_n$)* . ?x a ?C$_i$^{5} . $\Rightarrow$?x a ?C .
11$_a$	$C_1 \sqcap ... \sqcap C_n$	*?C :intersectionOf (?C$_1$ ... ?C$_n$)* . ?y a ?C . $\Rightarrow$?y a ?C$_1$, ..., ?C$_n$.
11$_b$		*?C :intersectionOf (?C$_1$ ... ?C$_n$)* . ?y a ?C$_1$, ..., ?C$_n$. $\Rightarrow$?y a ?C .
12	$\top \sqsubseteq \forall \leq 1P$	*?P a :FunctionalProperty* . ?s ?P ?x , ?y . $\Rightarrow$?x :sameAs ?y .
13	$\top \sqsubseteq \forall \leq 1P^-$	*?P a :InverseFunctionalProperty* . ?x ?P ?o . ?y ?P ?o . $\Rightarrow$?x :sameAs ?y .
14	$P^+ \sqsubseteq P$	*?P a :TransitiveProperty* . ?x ?P ?y . ?y ?P ?z . $\Rightarrow$?x ?P ?z .
15	$\exists P.D$	*?C :someValuesFrom ?D; :onProperty ?P* . ?x ?P ?y . ?y a ?D . $\Rightarrow$?x a ?C .
16	$\forall P.D$	*?C :allValuesFrom ?D; :onProperty ?P* . ?x ?P ?y; a ?C . $\Rightarrow$?y a ?D .
17$_a$	$(\leq 1P)$	*?C :maxCardinality 1; :onProperty ?P* . ?x a ?C; ?P ?y, ?z . $\Rightarrow$?y :sameAs ?z .
17$_b$	$(= 1P)$	*?C :cardinality 1; :onProperty ?P* . ?x a ?C; ?P ?y, ?z . $\Rightarrow$?y :sameAs ?z .
18	$\{o_i....o_n\}$	*?C :oneOf (?o$_1$ ... ?o$_n$)* . $\Rightarrow$?o$_1$ ... ?o$_n$ a ?C .
19$_a$	$x = y$	?x :sameAs ?y . ?x ?p ?o .$\Rightarrow$?y ?p ?o .
19$_b$		?x :sameAs ?y . ?s ?p ?x .$\Rightarrow$?s ?p ?y .
19$_c$		?x :sameAs ?y . $\Rightarrow$?y :sameAs ?x .
19$_d$		?x :sameAs ?y . ?y :sameAs ?z . $\Rightarrow$?x :sameAs ?z .

A-Box, in contrast to ter Horst's original rules, inferences are stored in the A-Box only. Thus, on exhaustive application of the rules, the T-Box remains unchanged.

Next, we only support inferences in one direction for :someValuesFrom and :allValuesFrom, as we do not apply any inference rules that involve invention of new blank nodes. Like ter Horst, we do not support inequalities or disjointness; i.e., SAOR operates monotonically without "explosive" reaction in inconsistency. In addition to *pD** we support functional cardinality constraints, as well as limited support for enumerated classes (:oneOf).

Some of the rules in SAOR differ from their versions in *pD**, e.g. (05$_{a,b}$), (19$_{a,b}$). The alert reader may also miss rules to infer transitivity of rdfs:sub-ClassOf, rdfs:subPropertyOf, :equivalentClass, :equivalentProperty as well as symmetry of the equivalence and inverse-of properties. Whereas symmetry is covered by symmetric rules (02$_{a,b}$), (04$_{a,b}$), (05$_{a,b}$), transitivity is handled by SAOR via path traversals over internal data structures representing the subclass and subproperty hierarchies, following the RDF ground entailment algorithm outlined in [13].

Note that SAOR does not materialise any axiomatic triples [8]. Axiomatic A-Box statements are not produced by SAOR so as to avoid a bulk of syntactic statements (for example, rdf:type rdf:Resource statements and reflexive :sameAs statements). Axiomatic T-Box statements are not considered as we

---

^{5} $?C_i \in \{?C_1, ..., ?C_n\}$

have a concretely defined T-Box which is extracted by means of a single scan and does not support updates.

Finally, let us point out that there is good reason for excluding non-standard usage of the ontology vocabulary: non-standard RDF could have unpredictable results even under our simple rule-based entailment. One may consider a finite combination of only four non-standard triples that, upon naive reasoning, would explode all web resources R by inferring $|R|^3$ triples, namely:

```
rdfs:subClassOf rdfs:subPropertyOf rdfs:Resource.
rdfs:subClassOf rdfs:subPropertyOf rdfs:subPropertyOf.
rdf:type rdfs:subPropertyOf rdfs:subClassOf.
rdfs:subClassOf rdf:type :SymmetricProperty.
```

The exhaustive application of standard RDFS inference rules plus standard inference rule for property symmetry together with the typical inference for class membership in `rdfs:Resource` for all collected resources in typical rulesets lead to inference of any possible triple $(r_1\ r_2\ r_3)$ for arbitrary $r_1, r_2, r_3 \in R$.

Having introduced our rule set, we are able to define our notion of closure.

Definition 1 (Closure). *We denote by $Cl_{\mathcal{T}}(\mathcal{A})$ the closure of $\mathcal{A}$, i.e., the union of $\mathcal{A}$ with the set of statements inferred by exhaustive application of rules (01)-(19) from Table 2 with respect to T-Box $\mathcal{T}$.*

2.3   Authoritative Reasoning against Ontology Hijacking

SAOR is designed to counter-act a behaviour we discovered from initial evaluation which we term *ontology hijacking*. We counter such non-authoritative extensions of ontologies by ignoring problematic statements during T-Box generation. Before defining ontology hijacking, let us give some preliminary definitions:

Definition 2 (Authoritative Source)
   A graph $s \in \mathcal{KB}$ speaks authoritatively about a concept $c \in \mathcal{C}_{\mathcal{KB}} \cup \mathcal{P}_{\mathcal{KB}}$ if c appears in a triple of s and one of the following holds true:

1. *c is not identified by a URI (i.e., identified by a blank node)*
2. *s is retrievable from a URI which coincides with (or redirects to) the namespace[6] of the URI identifying c.*

Firstly, all sources are authoritative for anonymous classes or properties defined in that source. The second condition is designed to support best practices as currently adopted by web ontology publishers[7].

Let $s \in \mathcal{KB} = (\mathcal{T}, \mathcal{A})$ and $\mathcal{KB}' = (\mathcal{T}', \mathcal{A}') = \mathcal{KB} \setminus \{s\}$ be the knowledge base constructed from all graphs in $\mathcal{KB}$ except s. By *Ontology Hijacking* we now mean that a source s speaks non-authoritatively about a concept $c \in \mathcal{C}_{\mathcal{KB}} \cup \mathcal{P}_{\mathcal{KB}}$ (i.e., where c appears in $\mathcal{T}'$), in such a way that $Cl_{\mathcal{T}}(\mathcal{A}') \neq Cl_{\mathcal{T}'}(\mathcal{A}')$.

Ontology hijacking is the re-definition or extension of a definition of a legacy concept (class or property) in a non-authoritative source such that performing

---

[6] Here, slightly abusing XML terminology by "namespace" of a URI we mean the prefix of the URI obtained from stripping off the final NCname.
[7] See Appendix A&B of http://www.w3.org/TR/swbp-vocab-pub/

reasoning on legacy A-Box data results in a change in inferencing. One particular method of ontology hijacking is defining new super-concepts of legacy concepts. As a concrete example, if one were to publish today a property in an ontology (in a non-authoritative location for FOAF), `my:name`, within which the following was stated: `foaf:name rdfs:subClassOf my:name .`, that person would be hijacking the `foaf:name` property and effecting the translation of all `foaf:name` statements in the web knowledge base into `my:name` statements as well.

Ontology hijacking is problematic in that it vastly increases the amount of statements that are materialised and can potentially harm inferencing on data contributed by other parties. With respect to materialisation, the former issue becomes prominent: instance data published using concepts from popular/core ontologies get translated into a plethora of conceptual models described in obscure ontologies; we quantify the problem in Section 4. However, taking precautions against harmful ontology hijacking is growing more and more important as the Semantic Web features more and more attention; motivation for spamming and other malicious activity propagates amongst certain parties with ontology hijacking being a prospective avenue. With this in mind, we assign sole responsibility for the concepts and thus the semantics of their instances of the concepts to those who maintain the authoritative specification.

Related to the idea of ontology highjacking is the idea of "non-conservative extension" described in the Description Logics literature: cf. [11]. However, the notion of a conservative extension was defined with a slightly different objective in mind: according to the notion of deductively conservative extensions, an ontology O_B is only considered malicious towards O_A if it causes additional inferences with respect to the intersection of the signature of the original O_A with the newly inferred statements. Returning to the `ex:myName` example from above, the super-classing of `foaf:Name` alone would still constitute a conservative extension. However, further stating that `ex:myName a :InversefunctionalProperty.` would indeed violate the conservative extension property, since instances of `ex:myName` might then cause equalities in other remote ontologies as side-effects, independent from the newly defined signature. Summarising, we can state that every non-conservative extension (with respect to our notion of deductive closure) constitutes a case of ontology highjacking, but not vice versa.

In SAOR, we avoid the effects of ontology hijacking and non-conservative extensions by disregarding possibly harmful nonauthoritative use of concepts directly during T-Box construction. Table 1 shows how non-authoritative statements are disregarded upon T-Box construction. The source containing the concept description must be authoritative for the elements highlighted in boldface. Where multiple elements are italicised, at least one such element must be authoritatively spoken for. One can verify from Table 1 and Table 2 that, in the antecedent of each rule, the T-Box axioms must be authoritative for at least one of the class/property names appearing in the A-Box. Thereby, we protect A-Box reasoning from the influence of non-authoritative T-Box axioms. For Rules 02, 04, 05, 09 & 11 if the given source of data is only authoritative for one element, inferencing will only be executed in the direction "away" from that element. For example, for the statements `foaf:Person :equivalentClass ex:New-Class . ex:NewClass :equivalentClass foaf:Person .` described in a source only authoritative for the `ex:` namespace, inferencing will only translate `ex:NewClass` instances into `foaf:Person` and not in the other direction.

When publishing OWL or RDFS descriptions on the Web, we recommend that people avoid ontology-hijacking as defined in this section. We encourage extension of existing concepts where possible so that instance data in the newly published domain get translated into existing domains. Indeed, from brief analysis of some prominent specifications (specifically FOAF, DC, SIOC, SKOS), we found that they were entirely compliant with our restrictive reasoning.

3   Reasoning Algorithm

In the following we firstly present observations on web data that influenced the design of the algorithm, then give an overview of the algorithm, and next discuss details of how we handle T-Box information, perform statement-wise reasoning, and deal with ground equality.

3.1   Characteristics of Web Data

The design of our algorithm is motivated by observations on our Web dataset:

1. Reasoning accesses a large slice of data in the index: around 41% of statements produced uniquely inferred statements.
2. Relative to A-Box data, the volume of T-Box data on the Web is small: only around 2.5% of statements were classifiable as T-Box statements[8].
3. The T-Box is the most frequently accessed segment of data for reasoning: all but Rules 19_{a-d} (`:sameAs`) require access to T-Box information.

Following from the first observation, we employ a file-scan approach which is more efficient in this scenario than query processing lookups. Thus, we avoid the overhead of indexing the data and running full query processing; also we avoid probing the same statements repeatedly for different rules at the low cost of scanning a given percentage of statements not useful for reasoning.

Following from the second and third observations, we optimise by placing T-Box data in a separate data structure accessible by the reasoning engine. Currently, we hold the T-Box data in-memory, but the algorithm can be generalised to provide for an on-disk structure or a distributed in-memory structure as needs require.

3.2   Algorithm Overview

The algorithm involves three scans over the data as illustrated in Figure 1:

1. SCAN 1: separate T-Box information and build in-memory representation
2. PRE SCAN 2: execute rules with only T-Box patterns in the antecedent (Rule 18)
3. SCAN 2: perform reasoning in a statement-wise manner:

---

[8] Includes RDF collection fragments which may not be part of a class description.

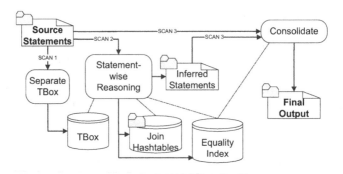

Fig. 1. High-level architecture

- Execute rules with only a single A-Box pattern in the antecedent (Rules 1-11_a); join A-Box statement with in-memory T-Box; write inferred statements immediately.
- Execute rules with two or more A-Box patterns in the antecedent (Rules 11_b-17); join indices are maintained for such A-Box patterns. When a statement matches one of the A-Box patterns for these rules and the necessary T-Box join exists, the statement is written to the join index and the join index is checked to see if all other A-Box patterns have been previously satisfied; if they have, the rule is fired.
- Execute rules which involve equality reasoning (Rules 19_{a-d}); lists of equivalent individuals are maintained in a hashtable structure (equality index). Newly identified equivalences are immediately reflected in the join indices for Rules 11_b-17 whereby new A-Box joins may form and fire rules.
4. SCAN 3: consolidate source data along with inferred statements according to the equality index and write to final output.

3.3   Handling Structural Data

In the following, we describe how to separate the T-Box data and how to create the data structures for representing the T-Box.

T-Box data from RDFS and OWL specifications can be acquired either from conventional crawling techniques, or by accessing the locations indicated by the dereferenced URIs of classes and properties in the instance data. We assume for brevity that all T-Box data are already present in the input data. If T-Box data are sourced via different means we can build an in-memory representation directly, without requiring the first scan of the input data.

During the scan, all statements relating to the supported T-Box constructs are identified and stored in an in-memory representation of classes and properties. The data structure holds the necessary information to infer new statements given a class or role membership assertion from the A-Box. We employ separate hashtables with URIs as keys and values containing a Java representation of the classes or properties, as follows:

- Property objects contain the property URI and references to objects representing equivalent properties, super properties, inverse properties, domain classes and range classes. Pointers are also kept to restrictions where the property in question is the object of an :onProperty statement.
- Class objects contain the class URI and references to objects representing equivalent classes, super classes and classes for which this class is a component of a union or intersection. On top of these core elements, different types of objects are created for different types of class description:
  - union and intersection classes store references to their constituent class objects
  - enumerated classes store references to constituent individuals
  - restriction classes store a reference to the property the restriction applies to and also, as applicable to the type of restriction:
    * the :cardinality or :maxCardinality value
    * the class identified by :allValuesFrom or :someValuesFrom
    * the value of a :hasValue restriction

Some class descriptions rely on the rdf:Collection construct, namely: unions, intersections and enumerations. To construct in-memory representations of these descriptions, the algorithm performs in-memory joining of rdf:Collection segments as the data are scanned according to rdf:first and rdf:rest properties. Any collections not relevant to the T-Box segment of the knowledge base are discarded at the end of loading the input data.

For each statement, the authority of the source for the given subject and object are inspected. If the statement is allowed as enumerated in Table 1, the statement is added to the T-Box.

3.4   Reasoning by Statement-Wise Scan

Having loaded the structural data, the SAOR engine is now prepared for reasoning by statement-wise scan of the data.

We firstly analyse rules which do not require A-Box joins to compute. There are two distinct types of statements which require different handling, namely rdf:type statements and general non-rdf:type statements. The rdf:type statements are subject to class-based entailment reasoning (Rules 1-2 & 9_b-11_a : rules with a single rdf:type A-Box pattern), and require joins with class descriptions in the T-Box. The non-rdf:type statements are subject to property-based entailments (Rules 3-9_a : rules with a single non-rdf:type A-Box pattern) and thus require joins with T-Box property descriptions.

We assume disjointness between the statement categories: we know that the defined semantics of rdf:type do not require any property-based entailment and we further do not allow any external extension of the core rdf:type semantics (non-standard use). Thus, we do not subject rdf:type statements to entailment of Rules 3-9_a.

The reasoning scan process can be described as recursive depth-first reasoning whereby each unique statement produced is input immediately for reasoning. Statements produced thus far for the original input statement are kept in a set to provide uniqueness testing and avoid cycles; a uniquing function is maintained on a resource level ensuring that statements are only produced once for a given common subject group. Once all of the statements produced by a rule have been themselves recursively analysed, the reasoner moves on to analysing the proceeding rule.

Rules 11_b-17 cannot be computed solely on a statement-wise basis. Instead, for each rule, we assign an on-disk persistent data structure with an in-memory MRU cache. Each index stores a representation of statements which may contribute to satisfying the antecedent of its pertinent rule. During the scan, if a statement satisfies the necessary T-Box join for a rule, it is written to the index for that rule. When a statement is added which completes the pattern of an antecedent for that rule, the rule is fired.

3.5   Equality Reasoning

In the following we discussed :sameAs entailment as encoded in Rules 19_{a-d} of Table 2. For Rules $19_{a,b}$, we employ an in-memory index for storing equivalence of individuals. We store the identifiers for equivalent individuals in lists and also store the lists which individuals are in using a hashtable. Thus, we can perform a lookup for an individual in the hashtable to get a reference to a list of equivalent individuals. The list structure maintains the transitivity and symmetric properties of equivalence. Usually, :sameAs entailment on individuals results in multiple individuals with the same data attached; however, we select a "pivot element" to reduce the number of inferred statements. The pivot element of each list is used to keep a consistent identifier for the set of equivalent individuals: the first one encountered is chosen. For alternative choices of pivot identifiers on web data see [9]. We use the pivot identifier to consolidate data by rewriting all occurrences of equivalent identifiers to the pivot identifier (effectively merging the equivalent set into one individual).

The in-memory equivalence index is filled from raw input :sameAs statements and also from inferencing performed on Rules 12, 13 and 17. For the purposes of the A-Box scan, we need not be immediately concerned about equality reasoning for closure: no joins are present on the individual level. However, for the join index reasoning, equality reasoning is paramount for closure. The join indices are immediately updated to reflect new equality knowledge; identifiers for equivalent individuals are rewritten to their pivot identifiers are soon as equivalence is determined. Rewriting of indices can lead to new inferences whereby the rewritten identifiers align under the pivot identifier to form a new join, thus firing a rule.

Based on the equivalence knowledge attained during the second scan, the inferred output and input data are finally scanned once more to ensure proper consolidation. All statements are rewritten so that they only contain pivot identifiers, producing the final output.

4   Evaluation and Discussion

We now provide evaluation of the SAOR methodology, firstly with quantitative analysis of the importance of authoritative reasoning, and secondly we provide some performance measurements, discussion and some insights into the fecundity of each rule wrt. reasoning over web data. Throughout, we use a 106M statement web-crawl dataset from mid-April 2008, taken from 315k sources.

To show the effects of ontology hijacking we constructed two T-Boxes with and without authoritative analysis. We then ran reasoning on single membership assertions for the top five classes and properties found natively in our dataset. Table 3 summarises the results. Taking `foaf:Person` as an example, with an authoritative T-Box, six statements are output for every input `rdf:type foaf:Person` statement. With the non-authoritative T-Box, 362 statements are output for every such input statement. Considering that there are 2.4M such statements in the input dataset, overall output for `rdf:type foaf:Person` input statements alone approach 1 billion statements for non-authoritative reasoning. With authoritative reasoning, we only produce 14M output statements: a 64.8x savings on materialised statements.

Table 3. Comparison of authoritative and non-authoritative reasoning for the number of inferred statements produced w.r.t. the five most frequently occurring classes and properties in the input data

Class URI	$\mathcal{A}$	$\mathcal{NA}$	n	$n * \mathcal{A}$	$n * \mathcal{NA}$
http://purl.org/rss/1.0/item	0	356	2,550,664	0	908,036,384
http://xmlns.com/foaf/0.1/Person	6	389	2,410,331	14,461,986	937,618,759
http://xmlns.com/foaf/0.1/Document	1	355	1,497,132	1,497,132	531,481,860
http://xmlns.com/wordnet/1.6/Person	0	236	1,097,415	0	258,989,940
http://xmlns.com/foaf/0.1/chatEvent	0	0	1,097,265	0	0
TOTAL	7	1,336	8,652,807	15,959,118	2,636,126,943
Property URI	$\mathcal{A}$	$\mathcal{NA}$	n	$n * \mathcal{A}$	$n * \mathcal{NA}$
http://purl.org/dc/elements/1.1/title	0	250	4,222,957	0	1,055,739,250
http://xmlns.com/foaf/0.1/name	5	664	3,753,791	18,768,955	2,492,517,224
http://purl.org/dc/elements/1.1/date	0	625	3,677,251	0	2,298,281,875
http://xmlns.com/foaf/0.1/nick	0	637	3,100,733	0	1,975,166,921
http://purl.org/dc/elements/1.1/description	0	631	30,138,087	0	19,017,132,897
TOTAL	5	2,807	44,892,819	18,768,955	26,838,838,167

We measured the performance of applying only the rules which do not require A-Box joins $(1\text{-}11_a)$ and for applying all rules. The results of the evaluation on a 2.2 GhZ AMD Opteron machine with 3G of Java heap-space is shown in Figure 2. Please note that the trend with respect to statements read/input statements processed is very similar to that presented for written statements/statements output. Also, please observe that applying rules without A-Box joins exhibits perfectly linear scaling behaviour, while using all rules slows down the algorithm after inferring about 120m output statements. For implementing the on-disk A-Box join indices we employ BerkeleyDB[9], which slows down considerably if the index size exceeds a certain limit (depending on caching policy and main memory available to the JVM). We are currently investigating alternatives to dynamic data structures for Rules $11_b\text{-}17$.

Table 4 lists the number of times the rules for a given primitive were fired during reasoning over all rules. Interestingly, from Figure 2 and Table 4 we can deduce that the bulk of current web reasoning is covered by those rules $(1\text{-}12_a)$ which exhibit linear scale.

---

[9] http://www.oracle.com/database/berkeley-db/je/

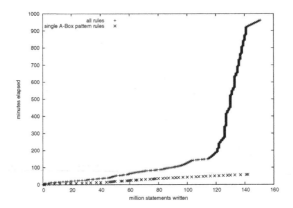

Fig. 2. Performance of the inferencing algorithm

Table 4. Count of number of statements inferred for each primitive

No	Primitive	Inferred Count
01	rdfs:subClassOf	66,283,568
02	:equivalentClass	7,325,048
03	rdfs:subPropertyOf	7,314,815
04	:equivalentProperty	6,943,803
05	:inverseOf	8,485,414
06	rdfs:domain	29,850,430
07	rdfs:range	19,466,468
08	:SymmetricProperty	458,467
09	:hasValue	9,938
10	:unionOf	5,676,861
11	:intersectionOf	13,239
12	:FunctionalProperty	15,218
13	:InverseFunctionalProperty	1,379,003
14	:TransitiveProperty	2,862,631
15	:someValuesFrom	51,403
16	:allValuesFrom	460,031
17	:cardinality	265
18	:oneOf	5,898

5   Related Work

OWL reasoning, specifically Query Answering over OWL Full, is not tackled by typical DL Reasoners; such as FaCT++ [18], RACER [7] or Pellet [15]; which focus on complex reasoning tasks such as subsumption checking and provable completeness of reasoning. Likewise, KAON2 [12], which reports better results on query answering, is limited to OWL-DL expressivity due to completeness requirements. Despite being able to deal with complex ontologies in a complete manner, these systems are not tailored for the particular challenges of processing large amounts of RDF data.

Conversely, incomplete (wrt. OWL Full) rule-based inference, as we advocate it in this paper, may be considered to have greater potential for scale. Several rule expressible non-standard OWL fragments; namely OWL-DLP [6], OWL⁻ [5] (which is a slight extension of OWL DLP), OWLPrime [20], pD^* [17], or *intentional OWL* [4, Section 9.3]; have been defined in the literature and enable incomplete but sound RDFS and OWL inferences. Amongst those fragments, the fragment we support here is most closest to [17].

Systems such as Triple [14], JESS[10], or Jena[11] support rule representable RDFS or OWL fragments as we do, but only work in-memory whereas our framework is focused on conducting scalable reasoning using persistent storage.

Analogous to our approach, [3] introduce certain restrictions for axioms accepted by a reasoning engine, however, lacking a rigorous treatment of acceptable axioms.

The OWLIM [10] family of systems allows reasoning over a number of rule-representable OWL fragments using the TRREE: Triple Reasoning and Rule Entailment Engine. Besides the in-memory version SwiftOWLIM, which uses TRREE, there is also a version offering query-processing over a persistent image of the repository, BigOWLIM, which comes closest technically to our approach despite focusing on different fragments of OWL, including those inferring inconsistencies. Whereas similarly to BigOWLIM, we employ persistent materialisation of inferred triples, our reasoning approach strictly focuses on sensible reasoning for web data; we only consider a positive fragment of OWL-Horst and analyse the authority of T-Box statements. We deliberately sacrifice logical completeness for what we believe to be a more cautious, but still sound approach for the web data use-case.

6   Conclusion and Future Work

We have presented SAOR: a reasoning methodology for performing reasoning over Web data based on primitives known to scale. To keep the resulting knowledge base manageable, both in size and quality, we made the following modifications to traditional reasoning procedures:

- allow only standard use of RDF and disallow metamodelling
- allow extension of classes and properties only from authoritative sources (no ontology hijacking)
- use pivot identifiers instead of full materialisation of equality

We envision extensions to our system along two lines: scalability enhancements by replacing the dynamic on-disk data structure with a more scalable scans/sort approach and distributing the system using a hash-based placement or T-Box replication strategy.

References

1. Brickley, D., Guha, R.: Rdf vocabulary description language 1.0: Rdf schema. W3C Recommendation (February 2004), http://www.w3.org/TR/rdf-schema/
2. de Bruijn, J., Heymans, S.: Logical foundations of (e)RDF(S): Complexity and reasoning. In: Aberer, K., Choi, K.-S., Noy, N., Allemang, D., Lee, K.-I., Nixon, L., Golbeck, J., Mika, P., Maynard, D., Mizoguchi, R., Schreiber, G., Cudré-Mauroux, P. (eds.) ASWC 2007 and ISWC 2007. LNCS, vol. 4825, pp. 86–99. Springer, Heidelberg (2007)

---

[10] http://herzberg.ca.sandia.gov/
[11] http://jena.sourceforge.net/

3. Cheng, G., Ge, W., Wu, H., Qu, Y.: Searching semantic web objects based on class hierarchies. In: Proceedings of Linked Data on the Web Workshop (2008)
4. de Bruijn, J.: Semantic Web Language Layering with Ontologies, Rules, and Meta-Modeling. PhD thesis, University of Innsbruck (2008)
5. de Bruijn, J., Polleres, A., Lara, R., Fensel, D.: OWL$^-$. Final draft d20.1v0.2, WSML (2005)
6. Grosof, B., Horrocks, I., Volz, R., Decker, S.: Description logic programs: Combining logic programs with description logic. In: 13th International Conference on World Wide Web (2004)
7. Haarslev, V., Möller, R.: Racer: A core inference engine for the semantic web. In: International Workshop on Evaluation of Ontology-based Tools (2003)
8. Hayes, P.: RDF Semantics. W3C Recommendation (February 2004), http://www.w3.org/TR/rdf-mt/
9. Hogan, A., Harth, A., Decker, S.: Performing object consolidation on the semantic web data graph. In: 1st I3 Workshop: Identity, Identifiers, Identification Workshop (2007)
10. Kiryakov, A., Ognyanov, D., Manov, D.: Owlim - a pragmatic semantic repository for owl. In: Web Information Systems Engineering Workshops, New York, USA, November 2005. LNCS, pp. 182–192 (2005)
11. Lutz, C., Walther, D., Wolter, F.: Conservative extensions in expressive description logics. In: Proc. of IJCAI-2007, pp. 453–459 (2007)
12. Motik, B.: Reasoning in Description Logics using Resolution and Deductive Databases. PhD thesis, Forschungszentrum Informatik, Karlsruhe, Germany (2006)
13. Muñoz, S., Pérez, J., Gutiérrez, C.: Minimal deductive systems for rdf. In: Franconi, E., Kifer, M., May, W. (eds.) ESWC 2007. LNCS, vol. 4519, pp. 53–67. Springer, Heidelberg (2007)
14. Sintek, M., Decker, S.: Triple - a query, inference, and transformation language for the semantic web. In: 1st International Semantic Web Conference, pp. 364–378 (2002)
15. Sirin, E., Parsia, B., Grau, B.C., Kalyanpur, A., Katz, Y.: Pellet: A practical owl-dl reasoner. Journal of Web Semantics 5(2), 51–53 (2007)
16. Smith, M.K., Welty, C., McGuinness, D.L.: Owl web ontology language guide. In: W3C Recommendation (February 2004), http://www.w3.org/TR/owl-guide/
17. ter Horst, H.J.: Combining RDF and part of owl with rules: Semantics, decidability, complexity. In: 4th International Semantic Web Conference, pp. 668–684 (2005)
18. Tsarkov, D., Horrocks, I.: Fact++ description logic reasoner: System description. In: International Joint Conf. on Automated Reasoning, pp. 292–297 (2006)
19. Wang, T.D., Parsia, B., Hendler, J.: A survey of the web ontology landscape. In: Cruz, I., Decker, S., Allemang, D., Preist, C., Schwabe, D., Mika, P., Uschold, M., Aroyo, L.M. (eds.) ISWC 2006. LNCS, vol. 4273, pp. 682–694. Springer, Heidelberg (2006)
20. Wu, Z., Eadon, G., Das, S., Chong, E.I., Kolovski, V., Annamalai, M., Srinivasan, J.: Implementing an Inference Engine for RDFS/OWL Constructs and User-Defined Rules in Oracle. In: 24th International Conference on Data Engineering. IEEE, Los Alamitos (to appear, 2008)

Scalable Distributed Ontology Reasoning
Using DHT-Based Partitioning

Qiming Fang, Ying Zhao*, Guangwen Yang, and Weimin Zheng

Tsinghua National Laboratory for Information Science and Technology
Department of Computer Science and Technology,
Tsinghua University, Beijing 100084, China
fangqiming@gmail.com, {yingz,ygw,zwm-dcs}@tsinghua.edu.cn

Abstract. Ontology reasoning is an indispensable step to fully exploit the implicit semantics of Semantic Web data. The inherent distribution characteristic of the Semantic Web and huge amount of ontology instance data necessitates efficient and scalable distributed ontology reasoning. Current researches on distributed ontology reasoning mainly focus on dealing with the heterogeneity of different ontologies but pay little attention to the performance of distributed reasoning and have not presented practical approaches and systems. Our goal is to propose an efficient and scalable distributed ontology reasoning approach, making it practical in real semantic applications. We propose an approach in this paper, in which Description Logic reasoners for TBox reasoning are combined with rule engines for ABox reasoning to support both expressive ontologies and large amount of instance data. The published data from each node is distributed using a DHT-based partitioning and stored in well-designed relational databases to support convenient and efficient reasoning through cooperation of the distributed nodes. A practical distributed ontology reasoning and querying system called DORS is developed based on our proposed approach. Our experiments both in LANs and on PlanetLab using University Ontology Benchmark show high efficiency of DORS compared with the centralized OWL ontology reasoning system Minerva as well as good scalability with respect to the number of nodes and volume of data in the system.

1 Introduction

The Semantic Web [1] carries out the vision of a Web of data usable for both humans and machines. This web consists of inter-connected instance data annotated with possibly expressive ontologies. Ontologies provide formal and precise conceptualization of specific domains that can be used to describe resources on the Web and thus enable the reuse, sharing and portability of information and knowledge, coupled with a better conceptual understanding and analysis of a certain knowledge domain. Ontologies can enhance the current Web with the possibility of automated reasoning about knowledge, which makes it possible to derive new and only implicitly available knowledge. So ontologies are fundamental to the realization of the Semantic Web and play a central

---

* Corresponding author.

J. Domingue and C. Anutariya (Eds.): ASWC 2008, LNCS 5367, pp. 91–105, 2008.
© Springer-Verlag Berlin Heidelberg 2008

role for the success of the Semantic Web. W3C has recommended two standards for publishing and sharing ontologies on the Web: RDF/RDFS [2] and OWL [3], whose logical foundation is Description Logic (DL) [4].

Ontology reasoning is an indispensable step to fully exploit the implicit semantics of Semantic Web data. The DL-based ontology reasoning is comprised of TBox reasoning (i.e., reasoning with concepts) and ABox reasoning (i.e., reasoning with instances). Many ontology reasoners have been developed, e.g., Pellet [5], Racer [6], FaCT++ [7], KAON2 [8], OWLIM [9], Minerva [10] and Instance Store [12]. Unfortunately, all these reasoners cannot well perform distributed ontology reasoning which we believe is needed to promote knowledge discovering and sharing, data integration and interoperation on the Semantic Web, because the Semantic Web is inherently distributed like the current Web and the ontologies and data are distributed among many nodes which makes centralized reasoning difficult or even impossible in many situations. For example, we are currently developing a semantic retrieval system, which crawls semantic data of a certain domain, e.g., travel information, from the Web and provide semantic retrieval service on the crawled data. An important observation is that resources on the Web are likely to be annotated with relatively lightweight ontologies (low number of concepts), but the number of resources annotated with these ontologies is likely to be very large (large instance sets) [13]. Limited by the capability of single machines, we use many crawlers located on various machines to crawl the large amount of semantic data and store them on many machines. In this case, to discover and make use of the semantic relationships between the data located on different machines, an efficient distributed ontology reasoning algorithm is needed.

Some research efforts have been taken to solve the problem of distributed ontology reasoning [15,16,17]. These researches focus on distributed reasoning on different ontologies and their main idea is to overcome the heterogeneity of ontologies through ontology mapping. Their researches provide a mechanism to do reasoning on different ontologies but have not paid much attention to the performance of reasoning, e.g., time efficiency, communication overhead and system scalability. Performance is a key factor for a reasoning system to be practical in real world semantic applications, especially when the applications need to process huge amount of semantic data, which is very normal in the Semantic Web. To the best of our knowledge, there is no practical distributed ontology reasoning system currently available. These contexts motivate us to propose an efficient and scalable distributed ontology reasoning algorithm and develop a practical reasoning system. We believe that the exhaustive solution of distributed ontology reasoning will enable more distributed semantic applications and facilitate the success of the Semantic Web.

Unlike previous works that was putting emphasis on applying ontology mapping to deal with the ontology heterogeneity to provide a mechanism for distributed reasoning on different ontologies, we focus on the distributed ontology reasoning algorithm itself, expecting to present an efficient and scalable distributed reasoning algorithm to deal with huge amount of distributed semantic data. Therefore, we do not pay much attention to the ontology heterogeneity which can be addressed by ontology mapping, instead, we assume the distributed nodes follow the same ontology, i.e., the nodes have the same TBox while the ABoxes can be different.

In this paper, we propose a practical distributed ontology reasoning approach. We combine DL reasoners for TBox reasoning with rule engines for ABox reasoning. The

DL reasoners infer complete subsumption relationships between classes and properties of the ontology and based on these relationships the rule engines conduct ABox reasoning following the predefined logic rules. This combination exploits the particular advantage of DL reasoners, i.e., supporting of expressive ontologies in terms of logic, as well as the advantage of rule engines, i.e., capability of handling huge amount of instance data with the help of mature relational database storage. To provide convenient and efficient support for distributed reasoning, we employ the efficient data organizing method DHT (Distributed Hash Table) to well organize the instance data. On top of the DHT-based partitioning and organizing of the data, the ABox reasoning on the huge amount of instance data can be carried out through the cooperation of all nodes in a well-organized manner. We implement a distributed ontology reasoning and querying system called DORS based on FreePastry, an open source implementation of the well-known DHT-based peer-to-peer network Pastry [24]. DORS implements the Description Logic Programs (DLP) [23] rules and thus its reasoning is sound and complete on Description Horn Logic (DHL, a subset of OWL-DL) ontologies. To investigate the performance of our approach, experimental evaluation is conducted on DORS deployed both in LANs environment and on PlanetLab (http://www.planet-lab.org/) using UOBM (University Ontology Benchmark) [25]. Experimental results show that DORS has better time efficiency than the centralized OWL ontology reasoning system Minerva [10] and good scalability w.r.t. the number of nodes and volume of data.

The rest of the paper is organized as follows. In next section we propose our approach for distributed ontology reasoning in details. Section 3 describes the developed DORS system employing the proposed approach. In Section 4 a detailed experimental evaluation is conducted. Then the related work is presented in Section 5 and finally we conclude the paper and introduce the possible future work in the last section.

2 Distributed Ontology Reasoning

As mentioned before, we focus on the problem of distributed ontology reasoning in the situation that all nodes follow the same ontology, i.e., all nodes have the same TBox while the ABoxes can be different. In this environment, we propose a practical approach for distributed ontology reasoning. In this section, we will describe the approach in details.

2.1 The Approach Overview

The main idea of our approach resides in three aspects: the combination of DL reasoners for TBox reasoning with rule engines for ABox reasoning, the relational database storage and the DHT-based data organization.

Reasoning Method. The mainstream ontology language, OWL, is based on Description Logic (DL). The DL-based ontology reasoning is comprised of TBox reasoning and ABox reasoning, so our approach consists of these two parts. We use DL reasoners to accomplish TBox reasoning and rule engines to ABox reasoning. The DL reasoners infer complete subsumption relationships between classes and properties of the ontology and based on these relationships the rule engines conduct ABox reasoning following the predefined logic rules.

The reasons for this combination of DL reasoners with rule engines are twofold. On one hand, existing DL reasoners can not provide efficient mechanisms for distributed ontology reasoning and often provide limited support in dealing with large number of instances. Previous work [14] has demonstrated that DL reasoners are able to cope with TBox reasoning of real world ontologies but the extremely large number of instances of real ontologies makes it difficult for DL reasoners to deal with ABox reasoning. We think rule engines can possibly solve these problems: they can handle large amount of assertion facts with the help of mature database storage and provide possible solution of efficient distributed ontology reasoning as the experiments indicate in latter section. On the other hand, rule engines are in principle limited in terms of the logic they are able to support, while many mature DL reasoners can support expressive ontologies. So we combine DL reasoners with rule engines to support expressive ontologies as well as large number of instances, achieving an effective solution for scalable distributed ontology reasoning.

Data Storage. Storage and reasoning are considered as inseparable in a complete ontology reasoning system. We employ relational database to store the ontology data and carefully design the database schema to effectively support convenient and efficient ontology reasoning as well as to obtain good scalability regarding large amount of instance data.

Data Organization. To effectively and efficiently support distributed reasoning on the ontology data published by many nodes, it is needed to well organize the data in the distributed system. We employ a DHT-based partitioning approach to organize the data. The nodes form a DHT-based peer-to-peer network, and the data is distributed in the DHT network. The reasoning is conducted with the cooperation of all nodes on the well-organized distributed data. Finally, the reasoning results are distributed and materialized in the databases to support efficient query processing.

In next subsections, we will introduce in details the processes of TBox reasoning and ABox reasoning in our approach respectively.

2.2 TBox Reasoning

For TBox reasoning, we employ DL reasoners to obtain all class and property subsumption relationships. There are some well-known and widely used DL reasoners such as Pellet [5], Racer [6], FaCT++ [7], KAON2 [8], etc., each has its own pros and cons. Some benchmarking and comparison work has been done to analyze and evaluate the characteristics, applicability and performance of different reasoners [21,22]. In our approach, since all nodes follow the same TBox and each node can finish the entire TBox reasoning task independently, they can choose their own DL reasoners appropriately. This brings more flexibility to the approach. Certainly, the TBox reasoning can also be taken in only one node and then distribute the reasoning results to other nodes. This approach reduces the reasoning consumption while increasing the distributing consumption, in other words, it is to trade communication for computation.

2.3 ABox Reasoning

In our approach, ABox reasoning is conducted by rule engines deployed in each node. The rule engine can produce new assertions using the predefined logic rules from TBox axioms and existing assertions, including the original ones and the previously inferred ones.

Rule Set Definition. To use the rule engine, we must firstly define the rule set. Currently we use the same logic rule set with Minerva [10]. Minerva uses some DLP rules that cover all DHL axioms. DHL is an expressive fragment of Description Logic (DL) and includes RDFS. DLP is the Logic Programs (LP)-correspondent of DHL axioms. The definition of DHL and DLP makes it practicable to do efficient reasoning of large-scale ontology using rule engines. The defined DLP rules are listed in Table 1.

Table 1. The DLP rules used by rule engines for ABox reasoning (Rel stands for Relationship)

DLP rules	DHL Axioms
Group 1:	
$Rel(y,P,x) \leftarrow Rel(x,P,y) \cdot Symmetric(P)$	$P \equiv P^-$
$Rel(y,Q,x) \leftarrow Rel(x,P,y) \cdot InversePropertyOf(P,Q)$	$P \equiv Q^-$
$Rel(x,P,z) \leftarrow Rel(x,P,y) \cdot Rel(y,P,z) \cdot Transitive(P)$	$P^+ \equiv P$
$Rel(x,Q,y) \leftarrow Rel(x,P,y) \cdot SubPropertyOf(P,Q)$	$P \sqsubseteq Q$
Group 2:	
$TypeOf(x,C) \leftarrow Rel(x,P,y) \cdot Domain(P,C)$	$T \sqsubseteq \forall P^-.C$
$TypeOf(y,C) \leftarrow Rel(x,P,y) \cdot Range(P,C)$	$T \sqsubseteq \forall P.C$
Group 3:	
$TypeOf(x,D) \leftarrow TypeOf(x,C) \cdot SubClassOf(C,D)$	$C \sqsubseteq D$
$TypeOf(x,C) \leftarrow Rel(x,P,y) \cdot TypeOf(y,D) \cdot SomeValuesFrom(C,P,D)$	$\exists P.D \sqsubseteq C$
$TypeOf(y,D) \leftarrow Rel(x,P,y) \cdot TypeOf(x,C) \cdot AllValuesFrom(C,P,D)$	$C \sqsubseteq \forall P.D$
$TypeOf(x,C) \leftarrow TypeOf(x,D_1) \cdot IntersectionMemberOf(D_1,C)$	
$\cdot TypeOf(x,D_2) \cdot IntersectionMemberOf(D_2,C) \cdots$	$D_1 \sqcap D_2 \cdots \sqcap D_n \sqsubseteq C$
$\cdot TypeOf(x,D_n) \cdot IntersectionMemberOf(D_n,C)$	

In the DLP rules, TBox axioms are expressed by facts of a fixed number of predicates, e.g., *SubPropertyOf*, *SubClassOf* and *Domain*. All class and property instances are expressed by facts of two predicates: *TypeOf* and *Relationship*. Of all the predicates, facts of *Symmetric*, *InversePropertyOf*, *Transitive*, *Domain*, *Range*, *SomeValuesFrom*, *AllValuesFrom* and *IntersectionMemberOf* totally come from the original ontology data; facts of *SubPropertyOf* and *SubClassOf* are from both original data and inferred data produced by the DL reasoners; facts of *TypeOf* and *Relationship* are obtained from original data as well as reasoning results of the rule engines.

Having defined these rules, the ABox reasoning can be accomplished by iteratively execute the rules until no new assertions can be derived. To reduce the cost and speed up the reasoning, the rules are categorized into three groups based on their dependency so that rules in lower group will not be influenced by rules in higher group, and then the rule engines can process each group of rules sequentially until no new results can be generated for each group of rules.

Data Storage and Distribution. We employ relational databases to store the ontology data, including TBox data and ABox data, for databases have been demonstrated to be capable to deal with large-scale ontology data [10]. To support convenient and efficient ABox reasoning, we carefully design our database schema, in which each predicate in the rules has a corresponding table in the database. Therefore, the rules can be easily translated into sequences of relational algebra operations. In our approach, all the DLP rules can be executed via table joins in databases. For example, the first rule of group 1 in Table 1 can be executed via a simple join between tables *Relationship* and *Symmetric*.

The instance data, i.e., tables *Relationship* and *TypeOf*, is distributed using DHT-based partitioning. For table *Relationship* we use the *Property URI* (in practical system it may be encoding id) as the hash key, that is to say, the triples with the same property will be stored in the same node. Similarly, the *Class URI* is used as the hash key for table *TypeOf* and the records with the same class will be put in the same node. In this way, the tables *Relationship* and *TypeOf* are partitioned by the hash keys using the DHT and the ABox reasoning tasks are also partitioned by the nodes. The DHT-based distribution well organizes the instance data and guarantees efficient lookup of target data which is necessary in ABox reasoning process. The selection of hash keys can support convenient join operations on tables.

ABox Reasoning Process. Since the data is distributed using DHT-based partitioning, the reasoning must be done in a distributed manner through collaboration of the nodes, for the reasoning on one node many influence the reasoning on another node. The influence is twofold. On one hand, one node may need some data stored in another node to execute some rules. For example, to execute the rule 2 of group 3 in Table 1, assuming that the *Relationship* triples with property P is stored in node A and the *TypeOf* records with class D is stored in node B, it is necessary to deliver data between the two nodes to execute the rule on these data fragments. To solve this data dependency, we add a prefetch procedure to retrieve the needed data before the rule engine starts a reasoning procedure. On the other hand, reasoning results of one node may trigger the rule engine of another node to generate new assertions, which is an obvious influence. To cover this influence, the ABox reasoning on each node must be designed to be a multi-step iterative algorithm. After finishing reasoning on current data it stores, the node will distribute the results to other nodes using the DHT-based partitioning. When a node has received some amount of data, it should start a new reasoning process to possibly produce new assertions.

In summary, the ABox reasoning in each node is a multi-step iterative process, and each complete reasoning step is comprised of four stages: preparation, prefetch, rule inference and distributing, here the preparation stage represents the possible procedure of waiting and receiving reasoning results distributed by other nodes. The complete ABox reasoning process can be depicted by Figure 1. In the figure, we use node A as a representative node to depict the reasoning process of each node. All nodes have the same reasoning process as node A and they collaborate to finish the reasoning tasks together. Our experiments show that the iterative process usually finishes in 3 to 7 steps in practice if all nodes start reasoning roughly at the same time, very few nodes may take more than 10 steps to complete the reasoning, mainly depending on the data distribution.

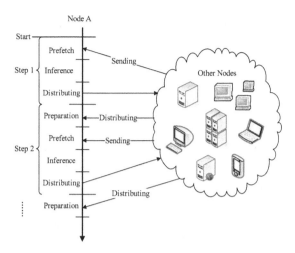

Fig. 1. The ABox reasoning process

3   The DORS System

Based on the proposed approach, we developed a distributed ontology reasoning and querying system called DORS, aiming to meet scalability requirements of real applications and provide practical reasoning capability as well as high query answering performance. The DORS system is comprised of many ontology data providers (nodes) that follow the same ontology and can publish different instance data. It can perform distributed reasoning on the distributed data and answer user queries based on the reasoning results. Figure 2 shows the architecture of DORS. The system is deployed on top of a DHT-based P2P network, e.g., Chord, CAN or Pastry. Besides the P2P infrastructure, the components of each node consist of five modules: Importing Module, Distributing Module, Storage Module, Reasoning Module and Querying Module. The design principles and implementation details of these modules are described below.

Importing Module. In DORS, the nodes publish ontology data through OWL documents and the importing module is responsible for importing the data into the reasoning system from original OWL documents. DORS uses relational databases to store ontology data, including its concepts and instances, so the importing module needs to convert the ontology data described in OWL language into relational database records according to the database schema. The importing module consists of an OWL parser, a TBox translator and an ABox translator. The OWL parser parses OWL documents and extracts the ontology concepts and instances. The TBox translator populates all TBox axioms into the TBox reasoner, obtain reasoning results and insert them into the database after taking schema translation. The ABox translator translates all ABox assertions into database records which then will be distributed to some nodes and stored into their databases by the data distributor.

Distributing Module. This module deals with the matter of data distributing, including sending and receiving data. The data distributed is mainly ABox data, involving not only the original data imported from OWL documents but also the reasoning results produced by the ABox reasoner, and TBox data also can be distributed if necessary. The data is distributed using DHT-based partitioning which uses *Property URI* as the hash key for table *Relationship* and *Class URI* for table *TypeOf*.

Storage Module. This module takes charge of ontology data storage in the system, including original TBox and ABox data as well as reasoning results from TBox and ABox reasoners. Currently, we use MySQL as the storage database.

Reasoning Module. This is the dedicated module to complete the reasoning tasks. The module is comprised of a TBox Reasoner and an ABox Reasoner, which are a DL reasoner and a rule engine respectively. Currently, we use Pellet as the DL reasoner and the rule engine implements all of the DLP rules in Table 1, so DORS can provide sound and complete reasoning w.r.t. the semantics of DHL which covers RDFS semantics and most practical OWL semantics.

Querying Module. This module gets user queries, executes queries and presents the query results to users. The user queries are answered by directly executing SQL statements in a distributed manner on top of the relational databases located on many nodes. The query response time is expected to be reduced by the materialization of reasoning results because there is no need to do time consuming reasoning tasks during the query processing stage. We only completed a prototype implementation in which the queries are written in SQL statements directly but we are planning to support some widely used ontology query languages, e.g., SPARQL [27].

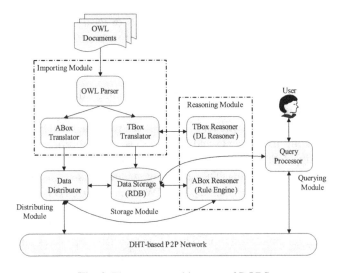

Fig. 2. The system architecture of DORS

4 Experimental Evaluation

4.1 Experimental Settings

We implemented a DORS system based on FreePastry using MySQL database for storage. FreePastry is an open source implementation of Pastry [24], a well-know DHT-based P2P infrastructure. We use FreePastry to construct a P2P substrate which can provide automatic maintenance to the network topology and efficient routing and location which must be utilized in distributed reasoning and query processing. The FreePastry version used in our experiments is 2.0_03 which can be downloaded from the FreePastry website (http://freepastry.rice.edu/FreePastry/). The DORS system is deployed in some computers located in two LANs connected by IP tunnel as well as on PlanetLab, a famous global research network residing in real Internet environment. We use different numbers of nodes in different settings, e.g., DORS-4 denotes a 4-nodes DORS. Moreover, in all experiments we employ the tableaux-based DL reasoner Pellet to do the TBox reasoning.

The evaluation is conducted on University Ontology Benchmark (UOBM) [25], a direct extension of the well-known Lehigh University Benchmark (LUBM) [26] in terms of expressiveness. Ma et al [25] build the UOBM based on their conclusion that LUBM is insufficient to evaluate the inference capability and less effective to reflect the scalability of an ontology system. The UOBM extends the LUBM by adding extra TBox axioms to support both OWL Lite and OWL DL ontologies covering a complete set of OWL Lite and DL constructs respectively. It consists of university domain ontologies, customizable and repeatable synthetic data, a set of test queries and corresponding answers. In our experiments, we use 3 datasets: OWL Lite-1, OWL Lite-5 and OWL Lite-10 (the parameter number denotes the number of universities in the dataset), which can be downloaded from the IBM Integrated Ontology Development Toolkit (IODT) website (http://www.alphaworks.ibm.com/tech/semanticstk). The statistics of the 3 datasets is listed in Table 2. We can see that the 3 datasets have the same number of classes and properties and different numbers of instances and triples; this is because they follow the same OWL Lite TBox but different ABoxes.

Table 2. The statistics of the UOBM datasets used in our experiments

	OWL Lite-1	OWL Lite-5	OWL Lite-10
Classes	52	52	52
Properties	54	54	54
Instances	25272	114054	223947
Triples	245864	1075060	2096973

In our experiments, we firstly add n nodes to construct a DORS system. Then the UOBM ontology dataset (in form of OWL documents) is divided into n parts of roughly equal size, each part includes data of one or more departments of the universities in the dataset. The divided data is put into the n nodes equally, so each node will be responsible for data of one or some departments, trying to model the real situation of semantic applications. Then the nodes import and distribute the data using the DHT-based partitioning. Finally the reasoning process is started which is an iterative procedure that runs until no new results can be generated, indicating that the reasoning on the dataset has been finished.

4.2 Experimental Results

Load Time. We use the load time to represent the time elapsed from the start of the importation of the OWL documents until the end of the reasoning. We run experiments on DORS of various numbers of nodes (4, 8, 16, 32) on the 3 datasets and the results are listed in Table 3 in which the load time of the centralized reasoning system Minerva presented in [10] is also listed as a comparison. Note that the machines used in our experiments (PCs with Intel Xeon CPU 2.66GHz or Intel Pentium D CPU 3.00GHz, 1GB memory and 512MB Java VM memory) are comparable with the machines used by Minerva's experiments (a PC with Intel Pentium IV CPU 2.66GHz, 1GB memory and 512MB Java VM memory). It can be seen that for the same OWL dataset, the load time of DORS in each setting is less than Minerva. Another obvious observation is that the load time of DORS decreases with the increasing of the number of nodes for the same dataset, this is due to the task division and parallel execution of DORS.

In fact, the load time of DORS is comprised of three parts: importing time, distributing time and reasoning time. In a load process, the data is imported from the OWL documents, distributed to the nodes using DHT-based partitioning, and then the reasoning on the distributed data is conducted. To further investigate the time efficiency of DORS, we present the importing time, distributing time and reasoning time on DORS of different numbers of nodes on OWL Lite-10 in Figure 3. We can see that the importing time, distributing time and reasoning time all decrease when the number of nodes in the system grows. This is a benefit from the parallelism of DORS. All nodes of DORS can import and distribute their own parts of the dataset in parallel, and this can reduce the importing and distributing time significantly. For reasoning, our distributed reasoning algorithm essentially employs the idea of task division and the tasks on different nodes can be executed in parallel, reducing the reasoning time when the number of nodes increases. As a result, our distributed reasoning approach can reduce the load time of the system and perform better than the centralized system.

Table 3. The load time of DORS compared with Minerva (the unit is second)

	OWL Lite-1	OWL Lite-5	OWL Lite-10
Minerva	868	5469	9337
DORS-4	760	2850	5064
DORS-8	559	1944	3151
DORS-16	445	1423	2291
DORS-32	358	1226	1707

We also conducted some experiments on PlanetLab and the results are presented in Table 4 and Figure 4 in which DORS-4-P denotes a 4-nodes DORS system deployed on PlanetLab. The load time of DORS deployed on PlanetLab is more than that in LANs, mainly because of the smaller bandwidth of PlanetLab compared to LANs, which increases the distributing and reasoning time illustrated in Figure 4. However, the load time of DORS on PlanetLab is still more than 20% less than Minerva except for the smallest dataset OWL Lite-1 (almost equal). This can be an indication that our approach is to some extent practical in real applications deployed over the Internet.

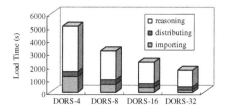

Fig. 3. The load time of DORS on OWL Lite-10

Table 4. The load time of DORS on PlanetLab (the unit is second)

	OWL Lite-1	OWL Lite-5	OWL Lite-10
DORS-4-P	898	4196	7350

Fig. 4. The load time of DORS on OWL Lite-10

Scalability. Scalability is a very important requirement for Semantic Web techniques to be usable in real world applications. For a distributed system like DORS, the scalability concerns mainly include two aspects: how does it perform when the volume of data increases and when the number of nodes increases. We investigate these two issues through experimental analysis.

We run DORS on the 3 datasets and various numbers of nodes. Figure 5 shows the reasoning time and preparation time in different settings. Here we use preparation time to denote the sum of the importing time and distributing time. Both the reasoning time and preparation time increase linearly or sub-linearly when the volume of data increases for a certain number of nodes and decreases with the increasing of the number of nodes for a certain dataset. So the load time, the sum of the preparation time and reasoning time, will certainly has the same features. These features indicate good scalability of DORS w.r.t. the volume of data and number of nodes.

Communication cost is a factor that greatly influences the performance of distributed algorithms and systems. We measure the total and average communication traffic of DORS in different settings and the results are presented in Figure 6. We can see that along with the increase of the number of nodes, the total communication traffic increases slowly for each dataset, but the average communication traffic per node decreases fast. This implies that the system can support larger datasets through adding more nodes to apportion the total communication traffic, making the system scalable.

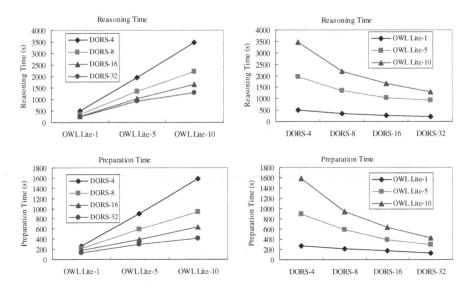

Fig. 5. The reasoning and preparation time of DORS

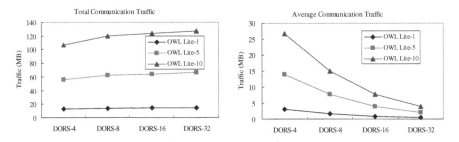

Fig. 6. The total and average communication traffic of DORS

5   Related Work

Some efforts have been made to solve the problem of distributed ontology reasoning. Serafini et al [15,16,17] discuss the problem of reasoning with multiple ontologies interrelated with semantic mappings. They propose a distributed reasoning approach in which reasoning is the result of a combination, via semantic mappings, of local reasoning chunks performed in single ontologies. A tableau-based distributed reasoning procedure is presented which is sound and complete w.r.t. Distributed Description Logics (DDL) [11], the formal framework used to represent multiple semantically connected ontologies. A distributed reasoning system called DRAGO implementing the distributed tableaux procedure is proposed. DRAGO represents a peer-to-peer like architecture in which every peer registers a set of ontologies and mappings, and the reasoning is implemented using local reasoning in the registered ontologies and by coordinating with other peers when local ontologies are semantically connected with the ontologies registered in other peers.

Bao et al [18][19] describe a distributed reasoning algorithm for Package-based Description Logics (P-DL), a modular ontology language that extends DL. This algorithm adopts a federated approach to reasoning with modular ontologies [20] wherein each ontology module is associated with a local reasoner. The local reasoners communicate with each other as needed in an asynchronous fashion.

The approach for distributed ontology reasoning proposed in this paper is basically a distributed extension of Minerva [10], a centralized storage, inference and querying system for large-scale OWL ontologies on top of relational databases. Minerva aims to meet scalability requirements of real applications and provide practical reasoning capability as well as high query performance. It combines DL reasoners for the TBox inference with logic rules for the ABox inference. It customizes the database schema based on inference requirements and user queries are answered by directly retrieving materialized results from the back-end database. In our approach, we borrow many basic ideas from Minerva and extend them into a distributed network, resulting in wider applicability, better efficiency and scalability.

6   Conclusion and Future Work

The inherently distributed Semantic Web necessitates efficient and scalable distributed ontology reasoning to exploit the implicit semantics of the huge amount of distributed semantic data. In this paper, we propose a practical distributed ontology reasoning approach, assuming the distributed semantic data follows the same ontology. This approach combines DL reasoners for TBox reasoning with rule engines for ABox reasoning, exploiting the particular advantages of each one, to support expressive ontologies and large amount of instance data. The data is distributed using a DHT-based partitioning and stored in well-designed relational databases to support convenient and efficient reasoning through cooperation of the distributed nodes. A practical reasoning system DORS is implemented based on the proposed approach. Our experiments both in LANs and on PlanetLab show the high efficiency of DORS compared with the centralized reasoning system Minerva as well as good scalability w.r.t. the number of nodes and volume of data.

Our approach is still primary and some improvements can be made. The reasoning process does not address efficiently the ontology update problem. The current strategy is to rerun the reasoning process if the ontology is updated. This is the simplest method but it brings a heavy overhead. Although ontology update is infrequent in real applications, we hope to find some methods to greatly reduce the update overhead. Currently, we implement DLP rules in DORS, so the reasoning is sound and complete on DHL (a subset of OWL-DL) ontologies. We are investigating the feasibility of adding more rules to enhance the reasoning capability. Load balance is a key factor to influence the performance of P2P systems. Previous work [28] has shown the load-imbalance in P2P based RDF stores and presented some load-balancing strategies. We are planning to examine the load balance of nodes in DORS. Furthermore, we plan to make some optimization to our approach, e.g., compressed transfers which can reduce the communication cost. A more detailed experimental evaluation on larger datasets and larger number of nodes is also part of the plan, by which we expect to further prove the practicality of the proposed approach in the real Semantic Web.

Acknowledgements

This work is supported by ChinaGrid project of Ministry of Education of China, Natural Science Foundation of China (60573110, 60673152, 90612016), National Key Basic Research Project of China (2004CB317007, 2003CB318000), National High Technology Development Program of China (2006AA01A101, 2006AA01A108, 2006AA01A111, 2006AA01A117), EU IST programme and Asia Link programme.

References

1. Berners-Lee, T., Hendler, J., Lassila, O.: The Semantic Web. Scientific American, 28–37 (2001)
2. Brickley, D., Guha, R. (eds.): RDF Vocabulary Description Language 1.0: RDF Schema. W3C Recommendation (2004), http://www.w3.org/TR/rdf-schema/
3. Bechhofer, S., Harmelen, F., Hendler, J., Horrocks, I., McGuinness, D.L., Patel-Schneider, P.F., Stein, L.A. (eds.): OWL Web Ontology Language Reference. W3C Recommendation (2004), http://www.w3.org/TR/owl-ref/
4. Baader, F., Calvanese, D., McGuinness, D.L., Nardi, D., Patel-Schneider, P.F. (eds.): The Description Logic Handbook: Theory, Implementation, and Applications. Cambridge University Press, Cambridge (2003)
5. Sirin, E., Parsia, B., Grau, B., Kalyanpur, A., Katz, Y.: Pellet: A Practical OWL-DL Reasoner. Journal of Web Semantics 5(2), 51–53 (2007)
6. Haarslev, V., Moller, R.: Racer: A Core Inference Engine for the Semantic Web Ontology Language (OWL). In: Proceedings of the 2nd International Workshop on Evaluation of Ontology-based Tools (EON 2003), pp. 27–36 (2003)
7. Tsarkov, D., Horrocks, I.: FaCT++ description logic reasoner: System description. In: Furbach, U., Shankar, N. (eds.) IJCAR 2006. LNCS, vol. 4130, pp. 292–297. Springer, Heidelberg (2006)
8. Motik, B., Studer, R.: KAON2 - A Scalable Reasoning Tool for the Semantic Web. In: ESWC 2005, Heraklion, Greece (2005)
9. Kiryakov, A., Ognyanov, D., Manov, D.: OWLIM – A pragmatic semantic repository for OWL. In: Dean, M., Guo, Y., Jun, W., Kaschek, R., Krishnaswamy, S., Pan, Z., Sheng, Q.Z. (eds.) WISE 2005 Workshops. LNCS, vol. 3807, pp. 182–192. Springer, Heidelberg (2005)
10. Zhou, J., Ma, L., Liu, Q., Zhang, L., Yu, Y., Pan, Y.: Minerva: A scalable OWL ontology storage and inference system. In: Mizoguchi, R., Shi, Z.-Z., Giunchiglia, F. (eds.) ASWC 2006. LNCS, vol. 4185, pp. 429–443. Springer, Heidelberg (2006)
11. Borgida, A., Serafini, L.: Distributed Description Logics: Assimilating Information from Peer Sources. Journal of Data Semantics, 153–184 (2003)
12. Bechhofer, S., Horrocks, I., Turi, D.: The OWL instance store: System description. In: Nieuwenhuis, R. (ed.) CADE 2005. LNCS, vol. 3632, pp. 177–181. Springer, Heidelberg (2005)
13. Weithoner, T., Liebig, T., Luther, M., Bohm, S.: What's Wrong with OWL Benchmarks? In: Proceedings of the Second International Workshop on Scalable Semantic Web Knowledge Base Systems (SSWS 2006), pp. 101–114 (2006)

14. Haarslev, V., Moller, R.: High Performance Reasoning with Very Large Knowledge Bases: A Practical Case Study. In: International Joint Conference on Artificial Intelligence (IJCAI 2001), pp. 161–168. Morgan-Kaufmann, San Francisco (2001)
15. Serafini, L., Tamilin, A.: DRAGO: Distributed reasoning architecture for the semantic web. In: Gómez-Pérez, A., Euzenat, J. (eds.) ESWC 2005. LNCS, vol. 3532, pp. 361–376. Springer, Heidelberg (2005)
16. Serafini, L., Borgida, A., Tamilin, A.: Aspects of Distributed and Modular Ontology Reasoning. In: International Joint Conference on Artificial Intelligence (IJCAI 2005), pp. 570–575 (2005)
17. Serafini, L., Tamilin, A.: Local Tableaux for Reasoning in Distributed Description Logics. In: International Workshop on Description Logics (DL 2004), pp. 100–109 (2004)
18. Bao, J., Caragea, D., Honavar, V.: A Distributed Tableau Algorithm for Package-based Description Logics. In: Proceedings of the 2nd International Workshop on Context Representation and Reasoning, CRR (2006)
19. Bao, J., Caragea, D., Honavar, V.: A Tableau-based Federated Reasoning Algorithm for Modular Ontologies. In: Proceedings of the 2006 IEEE/WIC/ACM International Conference on Web Intelligence (WI 2006), pp. 404–410 (2006)
20. Bao, J., Caragea, D., Honavar, V.G.: Modular ontologies - A formal investigation of semantics and expressivity. In: Mizoguchi, R., Shi, Z.-Z., Giunchiglia, F. (eds.) ASWC 2006. LNCS, vol. 4185, pp. 616–631. Springer, Heidelberg (2006)
21. Weithoner, T., Liebig, T., Luther, M., Bohm, S., Henke, F., Noppens, O.: Real-world reasoning with OWL. In: Franconi, E., Kifer, M., May, W. (eds.) ESWC 2007. LNCS, vol. 4519, pp. 296–310. Springer, Heidelberg (2007)
22. Rock, J., Haase, P., Ji, Q., Volz, R.: Benchmarking OWL Reasoners. In: Bechhofer, S., Hauswirth, M., Hoffmann, J., Koubarakis, M. (eds.) ESWC 2008. LNCS, vol. 5021, pp. 1–15. Springer, Heidelberg (2008)
23. Grosof, B., Horrocks, I., Volz, R., Decker, S.: Description Logic Programs: Combining Logic Programs with Description Logic. In: Proceedings of the 12th International Conference on World Wide Web (WWW 2003), pp. 48–57. ACM, New York (2003)
24. Rowstron, A., Druschel, P.: Pastry: Scalable, decentralized object location, and routing for large-scale peer-to-peer systems. In: Guerraoui, R. (ed.) Middleware 2001. LNCS, vol. 2218, pp. 329–349. Springer, Heidelberg (2001)
25. Ma, L., Yang, Y., Qiu, Z., Xie, G.T., Pan, Y., Liu, S.: Towards a complete OWL ontology benchmark. In: Sure, Y., Domingue, J. (eds.) ESWC 2006. LNCS, vol. 4011, pp. 125–139. Springer, Heidelberg (2006)
26. Guo, Y., Pan, Z., Heflin, J.: An evaluation of knowledge base systems for large OWL datasets. In: McIlraith, S.A., Plexousakis, D., van Harmelen, F. (eds.) ISWC 2004. LNCS, vol. 3298, pp. 274–288. Springer, Heidelberg (2004)
27. Prud'hommeaux, E., Seaborne, A. (eds.): SPARQL Query Language for RDF. W3C Recommendation (2008), http://www.w3.org/TR/rdf-sparql-query/
28. Battre, D., Heine, F., Hoing, A., Kao, O.: Load-balancing in P2P Based RDF Stores. In: Proceedings of the 2nd International Workshop on Scalable Semantic Web Knowledge Base Systems (SSWS 2006), pp. 29–42. Springer, Heidelberg (2006)

Versatile Semantic Modeling of Frame Logic Programs under Answer Set Semantics

Mario Alviano, Giovambattista Ianni, Marco Marano, and Alessandra Martello

Dipartimento di Matematica, Università della Calabria, I-87036 Rende (CS), Italy
lastname@mat.unical.it

Abstract. This work introduces the framework of Frame Answer Set programs (FAS). FAS programs are a frame logic-like language working under answer set semantics augmented with higher order constructs.

The syntax of the language includes the possibility to manipulate nested molecules, class hierarchies, basic method signatures and contexts (called *framespaces*). Semantics is defined in terms of a corresponding stable model semantics, paving the way to model object ontologies and their semantics under this well known paradigm.

The language is purposely designed so that inheritance behavior and other features of the language can be easily customized by the introduction of specialized axiomatic modules, which can be modeled on purpose by advanced developers of ontology languages. Also, contexts allow to model hybrid systems integrating multiple data sources working under different entailment regimes. Properties and relationship with original F-logic semantics of some of the presented axiomatizations are given. A system prototype has been implemented and is available for evaluation.

1 Introduction

Frame Logic (F-logic) [17,33] is a knowledge representation and ontology modeling language which combines the declarative semantics and expressiveness of deductive database languages with the rich data modeling capabilities supported by the object oriented data model.

As such, F-logic constitutes both an important methodology and a tool for modeling ontologies in the context of Semantic Web. This is witnessed by projects which focussed in F-logic as representation language, such as WSMO [9,27]. Also, F-logic features play a crucial role in the ongoing activity of the RIF Working group [3,2][1]. F-logic was originally defined under first-order semantics [17], while a well-founded semantics, satisfactorily dealing with nonmonotonic inheritance can be found in [33].

The stable model semantics (nowadays better known as Answer Set Programming – ASP), has some attractive feature which make interesting to consider the possibility of defining a frame-based language under this setting. ASP is

---

[1] http://www.w3.org/2005/rules/wiki/RIF_Working_Group

J. Domingue and C. Anutariya (Eds.): ASWC 2008, LNCS 5367, pp. 106–121, 2008.
© Springer-Verlag Berlin Heidelberg 2008

nowadays a mature field, offering languages and systems[2], based on a strongly assessed model-theoretic semantics [15]. ASP allows to model declaratively non-determinism and gives the possibility to specify, in a declarative way, search spaces, preferences, strong and soft constraints [6], and more). ASP shares with F-logic under well-founded semantics the possibility to reason about ontologies using nonmonotonic constructs, included nonmonotonic inheritance, as it is done in some ASP extensions conceived for modeling ontologies [26].

This paper aims at closing the gap between F-logic based languages and Answer Set Programming, in both directions: on one hand, Answer Set Programming misses the useful F-logic syntax, its higher order reasoning capabilities, and the possibility to focus knowledge representation on objects, more than on predicates. On the other hand, manipulating F-logic ontologies under stable model semantics opens a variety of modeling possibilities, given the higher expressiveness of the latter with respect to well-founded semantics.

Our approach is set in between a pure model theoretic semantics (proper of F-logic and many of its extensions [17,33]), and a pure "rewriting" semantics, in which inheritance is specified by means of an ad-hoc translation to logic programming [16].

In the former case, semantics is given in a clean and sound manner: however, the way inheritance (and in general, the semantics of the language) is modeled is hardwired within the logic language at hand, and cannot be easy subject of modifications. In the latter case, semantics is enforced by describing a rewriting algorithm from theories to appropriate logic programs. In such a setting the semantics of the overall language can be better tuned by changing the rewriting strategy. It is however necessary to have knowledge of internal details about how the language is mapped to logic programming, making the process of designing semantics cumbersome and virtually reserved to the authors of the language only.

In this work, we define a basic stable model semantics for FAS programs which does not purposely fix a special meaning for the traditional operators of F-logic, such as class membership ": " and subclass containment "::". Indeed, FAS programs are conceived as a test-bed on which an advanced ontology designer is allowed to choose the behavior of available operators from a predefined library, or to design her own semantics from scratch. The ability to customize the semantics of the language is crucial especially in presence of inheritance constructs. In fact, when one has to model a particular problem, a specific semantics for inheritance may be more suitable than another, and it is often necessary to manipulate and/or combine the predefined behaviors of the language.

The contributions of our paper are highlighted next:

1. We present the family of Frame Answer Set Programs (FAS programs), allowing usage of frame-like constructs, and of higher order atoms. Interestingly, positively *nested frames* may appear both in the head and in the body of rules. The language allows to reason in multiple *contexts* which are called *framespaces*.
2. We provide the model-theoretic semantics of FAS programs in terms of their *answer sets*.

---

[2] Among the variety of such systems we recall here DLV [19] and smodels [29].

3. We show how semantics features can be introduced on top of the basic semantics of the language by adding an appropriate axiomatization. Structural, behavioral, and arbitrary semantic for inheritance can be easily designed and coupled with user ontologies. In some cases, we show how these axiomatizations relate with F-logic under first order semantics.

4. We illustrate in which terms contexts can be exploited for manipulating hybrid knowledge bases having many data sources working under different entailment regime.

5. The language has been implemented within the DLT system, a front-end for answer set solvers. Besides the fragment of language herein presented, DLT allows negated nested molecules, in the spirit of [20], and re-usable template programs. If coupled with a proper answer set solver, the same front-end allows usage of complex terms (e.g. functions, lists, sets), and external predicates [12].

The remainder of the paper is structured as follows: Section 2 introduces the syntax of the language FAS (Frame Answer Set). Section 3 contains a formalization of the semantics of FAS programs, while Section 4 describes how to use the language for modeling and axiomatizing knowledge, and proves some properties of the axiomatic modules presented. The system supporting FAS programs is described in Section 6; related works are discussed in Section 7 and conclusions are then drawn.

2   Syntax

We present here the syntax of FAS programs. Informally, the language allows disjunctive rules with negation as failure in the body; with respect to ordinary Ans-Prolog (the basic language of Answer Set Programming), there are three crucial differences. First, besides traditional atoms and predicates, the language supports *frame molecules* in both the body and the head of rules, following the style of F-logic [17]. When representing knowledge, frame molecules allow to focus on objects, more than on predicates. An object can belong to *classes*, and have a number of *property* (attribute) values. As an example, the following is a frame molecule:

$$brown : employee\,[\ surname \rightarrow \text{``Mr. Brown''},$$
$$skill \twoheadrightarrow \{java, asp\},$$
$$salary \rightarrow 800,$$
$$gender \rightarrow male,$$
$$married \rightarrow pink\,]$$

The above molecule defines membership of the *subject* of the molecule (*brown*) to the *employee* class and asserts some values corresponding to the *properties* (which we will call also *attributes*) bound to this object. This frame molecule states that *brown* is *male* (as expressed by the value of the attribute *gender*), and is *married* to another employee identified by the subject *pink*. *brown* knows *java* and *asp* languages, as the values of the *skill* property suggest, while he has a *salary* equal to *800*. Intuitively, one can see a class membership statement in form $x:c$ as similar to a unary predicate $c(x)$. Accordingly, $x[m \rightarrow v]$ can be seen has a binary predicate $m(x,v)$.

As a second important difference, higher order reasoning is a first class citizen in the language: in other words, it is allowed quantification over predicate, class and property names. For instance, $C(brown)$ is meant to have the variable C ranging over the Herbrand universe, thus having $employee(brown)$ as possible ground instance.

Finally, our language allows the use of *framespaces* to place atoms and molecules in different contexts. For example, suppose there are two *Mr. Brown*, one working for *Sun* and the other for *Ibm*. We can use two different assertions, related to two different framespaces to distinguish them, e.g. $brown : employee@sun$ and $brown : employee@ibm$.

We formally define the syntax of the language next.

Let $\mathcal{C}$ be an infinite and countable set of distinguished constant and predicate symbols. Let $\mathcal{X}$ be a set of variables. We conventionally denote variables with uppercase first letter (e.g. X, $Project$), while constants will be denoted with lowercase first letter (e.g. x, $brown$, $nonWantedSkill$). A *term* is either a constant or a variable.

Atoms can be either *standard atoms* or *frame atoms*. A standard atom is in the form $t_0(t_1, \ldots, t_n)@f$, where $t_0, \ldots, t_n, f$ are *terms*, t_0 represents the *predicate name* of the atom and f the *context* (or *framespace*) in which the atom is defined.

A *frame atom*, or *molecule*, can be in one of the following three forms:

- $s[v_1, \ldots, v_n]@f$
- $s \diamond c@f$
- $s \diamond c[v_1, \ldots, v_n]@f$

where s, c and f are terms, and $v_1, \ldots, v_n$ is a list of *attribute expressions*. Here and in the following, the allowed values for the meta-symbol $\diamond$ are ":" (*instance operator*), or "::" (*subclass operator*). Moreover, s is called the *subject* of the frame, while f represents the *context* (or *framespace*).

To simplify the notation, whenever the context term f is omitted, we will assume $f = d$, for $d \in \mathcal{C}$ a special symbol denoting the *default* context.

An attribute expression is in the form p, $p \rightharpoonup v_1$ or $p \relbar\joinrel\rightharpoonup \{v_1, \ldots, v_n\}$, where p (*the property/attribute name*) is a term, and $v_1, \ldots, v_n$ (*the attribute values*) are either terms or frame molecules. Here and in the following, the meta-symbols $\rightharpoonup$ and $\relbar\joinrel\rightharpoonup$ are intended to range respectively over $\{\rightarrow, \bullet\!\!\rightarrow\}$ and $\{\Rightarrow, \twoheadrightarrow, \Rrightarrow, \bullet\!\!\rightarrow\!\!\rightarrow\}$. Note that, according to this definition, when used within attribute expressions, the symbols in the set $\{\Rightarrow, \twoheadrightarrow, \Rrightarrow, \bullet\!\!\rightarrow\!\!\rightarrow\}$ allow sets of attribute values on their right hand side, while $\rightarrow$ and $\bullet\!\!\rightarrow$ allow single values.

A *literal* is either an atom p (positive literal), or an expression of the form $\neg p$ (strongly negated literal or, simply, negated literal), where p is an atom. A *naf-literal* (negation as failure literal) is either of the form b (positive naf-literal), or of the form *not* b (negative naf-literal), where b is a literal.

A *formula* is either a naf-literal, a conjunction of formulas or a disjunction of formulas.

A *simple atom* is either a standard atom, or a frame atom in the forms $s \diamond c @ f$, $s[p \rightharpoonup v]@f$ or $s[p \rightharpoonup\!\!\!\!\rightharpoonup \{v\}]@f$, for s, c, p, v and f terms of the language. The notion of simple literal and of simple naf-literal are defined accordingly on top of the notion of simple atom.

A Frame Answer Set *program* (FAS program) is a set of *rules*, of the form

$$a_1 \vee \cdots \vee a_n \leftarrow b_1, \ldots, b_k, not\, b_{k+1}, \ldots, not\, b_m.$$

where $a_1, \ldots, a_n$ and $b_1, \ldots, b_k$ are literals, $not\, b_{k+1}, \ldots, not\, b_m$ are naf-literals, and $n \geq 0$, $m \geq k \geq 0$. The disjunction $a_1 \vee \cdots \vee a_n$ is the head of r, denoted by $H(r)$, while the conjunction $b_1 \wedge \cdots \wedge b_k \wedge not\, b_{k+1} \wedge \ldots, \wedge not\, b_m$ is the body of r, denoted by $B(r)$. A rule with empty body will be called *fact*, while a rule with empty head is a *constraint*.

A *plain higher order* FAS program contains only standard atoms, while a *plain* FAS program contains only standard atoms with a constant predicate name. A *positive* FAS program do not contain negation as failure and strongly negated atoms. In the following, we will assume to deal with *safe* FAS programs, that is, programs in which each variable appearing in a rule r appears in at least one positive naf-literal in $B(r)$.

Example 1. The following one rule program is a valid FAS program. Intuitively, it represents the fact that each person is *male* or *female*.

$$P[gender \rightarrow \text{``male''}] \vee P[gender \rightarrow \text{``female''}] \leftarrow P : person.$$

3   Semantics

Semantics of FAS programs is defined by adapting the traditional Gelfond-Lifschitz reduct, originally given for a ground disjunctive logic program with strong and default negation [15], to the case of FAS programs.

Given a FAS program P, its ground version $grnd(P)$ is given by grounding rules of P by all the possible substitutions of variables that can be obtained using consistently elements of $\mathcal{C}$.[3] A ground rule thus contains only ground atoms; the set of all possible simple ground literals that can be constructed combining predicates and terms occurring in the program is usually referred to as *Herbrand base* (B_P). We remark that the grounding process substitutes also nonground predicates names with symbols from $\mathcal{C}$ (e.g., a valid ground instance of the atom $H(brown, X)$ is $married(brown, pink)$, while a valid ground instance of $brown[H \rightarrow yellow]$ is $brown[color \rightarrow yellow]$).

An *interpretation* for P is a set of simple ground literals, that is, an interpretation is a subset $I \subseteq B_P$. I is said to be *consistent* if $\forall a \in I$ we have that $\neg a \notin I$.

We define the following entailment notion with respect to an interpretation I. For a a ground atom:

---

[3] As shown next, our semantics implicitly assumes that elements of $\mathcal{C}$ are mapped to themselves in any interpretation, thus embracing the unique name assumption.

$(E1)$ If a is simple, then $I \models a$ iff $a \in I$;

$(E2)$ $I \models not\, a$ iff $I \not\models a$.

For $l_1, \ldots, l_n$ ground literals:

$(E3)$ $I \models l_1 \wedge \cdots \wedge l_n$ iff $I \models l_i$, for each $1 \leq i \leq n$;

$(E4)$ $I \models l_1 \vee \cdots \vee l_n$ iff $I \models l_i$ for some $1 \leq i \leq n$.

For s, p, f ground terms, and $m_1, \ldots, m_n$ ground frame molecules:

$(E5)$ $I \models s[p \twoheadrightarrow \{m_1, \ldots m_n\}]@f$ iff $I \models s[p \twoheadrightarrow \{m_i\}]@f$, for each $1 \leq i \leq n$.

For s, s', c, p, f, f' ground terms, and $\overline{v} = \{v_1, \ldots, v_n\}$ a set of ground attribute value expressions:

$(E6)$ $I \models s[v_1, \ldots, v_n]@f$ iff $I \models s[v_1]@f \wedge \cdots \wedge s[v_n]@f$;

$(E7)$ $I \models s \diamond c[\overline{v}]@f$ iff $I \models s \diamond c@f \wedge s[\overline{v}]@f$;

$(E8)$ $I \models s[p \rightarrow s'[\overline{v}]]@f$ iff $I \models s[p \rightarrow s']@f \wedge s'[\overline{v}]@f$;

$(E9)$ $I \models s[p \twoheadrightarrow \{s'[\overline{v}]\}]@f$ iff $I \models s[p \twoheadrightarrow \{s'\}]@f \wedge s'[\overline{v}]@f$;

$(E10)$ $I \models s[p \rightarrow s'[\overline{v}]@f']@f$ iff $I \models s[p \rightarrow s']@f \wedge s'[\overline{v}]@f'$;

$(E11)$ $I \models s[p \twoheadrightarrow \{s'[\overline{v}]@f'\}]@f$ iff $I \models s[p \twoheadrightarrow \{s'\}]@f \wedge s'[\overline{v}]@f'$.

Note that rules $(E8)$ and $(E9)$ force $s'[\overline{v}]$, which does not have an explicit framespace, to belong to the context f of the molecule containing it. On the contrary, $s'[\overline{v}]@f'$ in $(E10)$ and $(E11)$ has a proper framespace f', and the entailment rules take care of this fact. Then, rules $(E6)$ to $(E11)$ define the context of a frame molecule as the *nearest* framespace explicitly specified.

For a rule r:

$(E12)$ $I \models r$ iff $I \models H(r)$ or $I \not\models B(r)$.

A *model* for P is an interpretation M for P such that $M \models r$ for every rule $r \in grnd(P)$. A model M for P is *minimal* if no model N for P exists such that N is a proper subset of M. The set of all minimal models for P is denoted by $\mathrm{MM}(P)$.

Given a program P and an interpretation I, the *Gelfond-Lifschitz (GL) transformation* of P w.r.t. I, denoted P^I, is the set of positive rules of the form $\{a_1 \vee \cdots \vee a_n \leftarrow b_1, \cdots, b_k\}$ such that $\{a_1 \vee \cdots \vee a_n \leftarrow b_1, \cdots, b_k, not\, b_{k+1}, \cdots, not\, b_m\}$ is in $grnd(P)$ and $I \models not\, b_{k+1} \wedge \cdots \wedge not\, b_m$. An interpretation I for a program P is an *answer set* for P if $I \in \mathrm{MM}(P^I)$ (i.e., I is a minimal model for the positive program P^I) [24,15]. The set of all answer sets for P is denoted by $ans(P)$. We say that $P \models a$ for an atom a, if $M \models a$ for all $M \in ans(P)$. P is *consistent* if $ans(P)$ is non-empty.

For a positive program P allowing only the term d in context position, we define the F-logic first-order semantics in terms of its *F-models*. A *F-model* M_f is a model of P subject to the conditions

$(F1)$ "::" encodes a partial order in M_f;
$(F2)$ if $a : b \in M_f$ and $b :: c \in M_f$ then $a : c \in M_f$;
$(F3)$ if $a[m \rightharpoonup v] \in M_f$ and $a[m \rightharpoonup w] \in M_f$ then $v = w$, for $\rightharpoonup \in \{\rightarrow, \bullet\!\!\rightarrow\}$;
$(F4)$ if $a[m \approx\!\!> v] \in M_f$ and $b :: a$ then $b[m \approx\!\!> v] \in M_f$, for $\approx\!\!> \in \{\Rightarrow, \Rrightarrow\}$;
$(F5)$ if $c[m \Rightarrow v]$, $a : c$ and $a[m \rightarrow w] \in M_f$ then $w : v \in M_f$;
$(F6)$ if $c[m \Rrightarrow v]$, $a : c$ and $a[m \twoheadrightarrow w] \in M_f$ then $w : v \in M_f$.

We say that $P \models_f a$ for an atom a if $M_f \models a$ for all F-models of P.

Example 2. The program in Example 1 together with the fact *brown : person.*
has two answer sets, $M_1 = \{ brown : person, brown[gender \rightarrow \text{"male"}] \}$
and $M_2 = \{ brown : person, brown[gender \rightarrow \text{"female"}] \}$. Both M_1 and M_2 are
F-models. Note that $M_3 = \{ brown : person, brown[gender \rightarrow \text{"female"}],$
$brown[gender \rightarrow \text{"male"}] \}$ is neither an F-model nor an answer set for different reasons: it is not an F-model because of condition $(F3)$ given above,
while it is not an answer set because it is not minimal. Note also that disjunctive rules trigger in general the existence of multiple answer sets, while the
presence of constraints may eliminate some or all constraints: for instance, the
same program enriched with the constraints $\leftarrow brown[gender \rightarrow \text{"male"}]$ and
$\leftarrow brown[gender \rightarrow \text{"female"}]$ has no answer set[4].

4   Modeling Semantics and Inheritance

Given the basic semantics for a FAS program P, it is then possible to enforce a specific behavior for operators of the language by adding to P specific
"axiomatic modules". An *axiomatic module* A is in general a FAS program. Given
a union of axiomatic modules $S = A_1 \cup \cdots \cup A_n$, we will say that P entails a
formula ϕ under the axiomatization S ($P \models_S \phi$) if $P \cup S \models \phi$. The answer sets
of P under axiomatization S are defined as $ans_S(P) = ans(P \cup S)$.
    We illustrate next some basic axiomatic modules.

Basic class taxonomies. The axiomatic module $\mathcal{C}$, shown next, associates to
" : " and " :: " the usual meaning of monotonic class membership and subclass
operator.

$c_1 : A :: B \leftarrow A :: C, C :: B.$
$c_2 : A :: A \leftarrow X : A.$
$c_3 : \leftarrow A :: C, C :: A, A \neq C.$
$c_4 : X : C \leftarrow X : D, D :: C.$

Rules c_1 and c_2 enforce transitivity and reflexivity of the subclass operator,
respectively. Rule c_3 prohibits cycles in the class taxonomy, while c_4 implements
the class inheritance for individuals by connecting the " :: " operator to the " : "
operator. The acyclicity constraint can be relaxed if desired: we define in this
case $\mathcal{C}'$ as $\mathcal{C} \setminus \{c_3\}$[5].

---

[4] A constraint $\leftarrow c$ can be seen as a rule $f \leftarrow c, not\ f$, for which there is no model
    containing c.
[5] Note that the atom $A \neq C$ amounts to *syntactic* inequality between A and C.

Single valued attributes. Under standard F-logic, the operators $\rightarrow$ and $\bullet\!\!\!\rightarrow$ are associated to families of single valued functions: indeed, in a F-model M it can not hold both $a[m \rightarrow v]$ and $a[m \rightarrow w]$, unless $v = w$. Under unique names assumption, we can state the above condition by the set $\mathcal{F}$ of constraints:

$f_5 : \leftarrow A[M \rightarrow V], A[M \rightarrow W], V \neq W.$
$f_6 : \leftarrow A[M \bullet\!\!\!\rightarrow V], A[M \bullet\!\!\!\rightarrow W], V \neq W.$

Structural and behavioral inheritance. We show here how to model some peculiar types of inheritance, such as structural and behavioral inheritance.

Structural inheritance is usually associated to the operator $\Rightarrow$. Let P_1 be the following example program:

$webDesigner::javaProgrammer.\ javaProgrammer::programmer.$
$webDesigner::htmlProgrammer.\ javaProgrammer[salary \Rightarrow medium].$
$htmlProgrammer[salary \Rightarrow low].$

For short, we denote in the following *webDesigner* as *wd*, *javaProgrammer* as *jp* and *htmlProgrammer* as *hp*.

Under structural inheritance, as defined in [17], property values of superclasses are "monotonically" added to subclasses. Thus, since c_1 is subclass of c_2 and c_4, one expects that $P_1 \models_{\mathcal{C}\cup\mathcal{S}} webDesigner[salary \Rightarrow \{low, medium\}]$ for some axiomatic module $\mathcal{S}$.

The axiomatic module $\mathcal{S}$ shown next, associates this behavior to the operators $\Rightarrow$ and $\Rrightarrow$.

$s_7 : D[A \Rightarrow T] \leftarrow D::C,\ C[A \Rightarrow T].$
$s_8 : D[A \Rrightarrow T] \leftarrow D::C,\ C[A \Rrightarrow T].$

Note that s_5 (resp. s_6) do not enforce any relationship between "$\Rightarrow$" and "$\rightarrow$" (resp. "$\Rrightarrow$" and "$\bullet\!\!\!\rightarrow$") as in [17]. We will discuss this issue later in the section.

Behavioral inheritance [33], allows instead nonmonotonic overriding of property values. Overriding is a common feature in object-oriented programming languages like Java and C++: when a more specific definition (value, in our case) is introduced for a method (a property, in our case), the more general one is overridden. In case different information about an attribute value can be derived from several inheritance paths, inheritance is *blocked*. Let us assume to add to P_1 the assertions $jp[income \bullet\!\!\!\rightarrow 1000]$ and $hp[income \bullet\!\!\!\rightarrow 1200]$.

Under behavioral inheritance regime [33][6], the assertions $jp[income \bullet\!\!\!\rightarrow 1000]$ and $hp[income \bullet\!\!\!\rightarrow 1200]$ would be considered in conflict when inherited from *wd*. Indeed, both $wd[income \bullet\!\!\!\rightarrow 1000]$ and $wd[income \bullet\!\!\!\rightarrow 1200]$ under the three-valued semantics of [33] are left *undefined*. Under FAS semantics it is then expected to have some axiomatic module $\mathcal{B}$ where neither $P_1 \models_{\mathcal{B}\cup\mathcal{F}\cup\mathcal{C}} wd[income \bullet\!\!\!\rightarrow 1000]$ nor $P_1 \models_{\cap\mathcal{B}\cup\mathcal{F}\cup\mathcal{C}} wd[income \bullet\!\!\!\rightarrow 1200]$ hold.

The above behavior can be enforced by defining $\mathcal{B}$ as follows

---

[6] Note that in [33] the above semantics is conventionally associated to the $\rightarrow$ operator, while we will use $\bullet\!\!\!\rightarrow$

b_9 : $overridden(D, M, C) \leftarrow E[M \bullet\!\!\rightarrow V], C::E, E::D, C \neq E, E \neq D.$
b_{10} : $inheritable(C, M, D) \leftarrow C::D, D[M \bullet\!\!\rightarrow V], not\, overridden(D, M, C).$
b_{11} : $C[M \bullet\!\!\rightarrow V] \vee C[M \bullet\!\!\rightarrow V]@false \leftarrow inheritable(C, M, D), D[M \bullet\!\!\rightarrow V].$
b_{12} : $exists(C, M) \leftarrow C[M \bullet\!\!\rightarrow V].$
b_{13} : $\leftarrow inheritable(C, M, D), not\, exists(C, M).$
b_{14} : $existsSubclass(A, C) \leftarrow A:C, A:D, D::C, C \neq D.$
b_{15} : $A[M \rightarrow V]@candidate \leftarrow A:C, C[M \bullet\!\!\rightarrow V], not\, existsSubclass(A, C).$
b_{16} : $A[M \rightarrow V] \vee A[M \rightarrow V]@false \leftarrow A[M \rightarrow V]@candidate.$
b_{17} : $exists'(A, M) \leftarrow A[M \rightarrow V].$
b_{18} : $\leftarrow inheritable(C, M, C), A:C, not\, exists'(A, M).$

The above module makes usage of stable model semantics for modeling multiple inheritance conflicts. By means of rule b_{11} and b_{16} it is triggered the existence of multiple answer set in the presence of inheritance conflicts, one for each possible way to solve the conflict itself.

Note that $ans_{\mathcal{BUFUC}}(P_1)$ contains two different answer sets M_1 and M_2 which respectively are such that $M_1 \models wd[income \bullet\!\!\rightarrow 1200]$ and $M_2 \models wd[income \bullet\!\!\rightarrow 1000]$. However, both assertions do not hold in all the possible answer sets. Thus, similarly to "well-founded optimism" semantics, we obtain that $P_1 \not\models_{\mathcal{CUB}} wp[income \bullet\!\!\rightarrow X]$ for any X.

Constructive vs well-typed semantics. The operator $\Rightarrow$ is traditionally associated to $\rightarrow$. For instance if both $jp[keyboard \Rightarrow americanLayout]$ and $jim : jp[keyboard \rightarrow ibm1050]$ hold, one might expect that $ibm1050 : americanLayout$.

However, one might wonder whether to implement the above required behavior under a *constructive* or a *well-typed* semantics.

The two type of semantics differ in the way incomplete information is dealt with. In a "well-typed" flavored semantics, most axioms are seen as hard constraints, which, if not fulfilled, make the theory at hand inconsistent.

In the first case, it may be desirable to use the "$\Rightarrow$" operator for defining strong desiderata about range and domain of properties, while the "$\rightarrow$" could be used to denote actual instance values such as in the following program P_2:

$programmer[salary \Rightarrow integer].$
$g : programmer[salary \rightarrow aSalary].$
$\leftarrow X : programmer[salary \rightarrow Y], not\, Y : integer.$[7]

Note that $ans(P_2)$ is empty, unless it is not *explicitly* asserted (well-typed) the fact $aSalary : integer$.

On the other hand one may want to interpret *constructively* desiderata about domain and range of properties, as it is typical, e.g. of RDFS[31]. Consider the program P_3:

$programmer[salary \Rightarrow integer].$
$g : programmer[salary \rightarrow aSalary].$
$Y : integer \leftarrow X : programmer[salary \rightarrow Y].$

Here P_3 has a single answer set containing the fact $aSalary : integer$.

---

[7] With some liberality we use here "integer" as class name more than a concrete datatype, without losing the sense of our example.

The two types of semantics stem from profound philosophical differences: well-typedness is commonly (but not necessarily) associated to modeling languages inspired from database systems, living under a single model semantics and Closed World Assumption. To a large extent one can instead claim that first order logics (and descendant formalisms, such as descriptions logics and RDFS), is much more prone to deal constructively with incomplete information.

It is however worth noting that despite their conceptual difference, constructive and well-typed semantics are often needed together. As a matter of example, modeling in Java (as well as C++ and F-logic) needs both flavors. Constructiveness comes into play in inheritance within class taxonomies (e.g., if $A::B$ and $B::C$ hold, the information $A::C$ does not need to be well-typed and is inferred automatically), but well-typedness is required in several other contexts, (e.g. strong type-checking prescribes that a function having a given signature can not be invoked using actual parameters which are not *explicitly known* to fulfil the function signature).

Whenever required, FAS programs can be coupled with axiomatic modules encoding both well-typed and constructive axioms.

The following axiomatic module $\mathcal{CO}$ encodes constructively how the operators $\Rightarrow$ and $\rightarrow$ can be related each other:

$$co_{15} : V:T \leftarrow C[A \Rightarrow T], I:C, I[A \rightarrow V].$$

while $\mathcal{W}$, shown next, encodes the same relation under a well-typed semantics.

$$w_{16} : \leftarrow C[A \Rightarrow T], I:C, I[A \rightarrow V], not\ V:T.$$

5   Properties of FAS Programs

FAS programs have some property of interest. First, F-logic entailment can be modeled on top of FAS programs by means of the axiomatic modules $\mathcal{C}, \mathcal{S}, \mathcal{F}$, and $\mathcal{CO}$. Let $\mathcal{A} = \mathcal{C} \cup \mathcal{S} \cup \mathcal{F} \cup \mathcal{CO}$.

Theorem 1. *Given a positive, non-disjunctive, FAS program P with default contexts only, and a formula ϕ, then $P \models_{\mathcal{A}} \phi$ iff $P \models_f \phi$.*

Proof. (Sketch). ($\Rightarrow$) Assume $P \cup \mathcal{A}$ is inconsistent. Given that P is a positive program, then inconsistency amounts to the violation of some instance of constraints c_3, f_5 or f_6. We can show that, accordingly, there is no F-model for P. On the other hand, if $P \cup \mathcal{A}$ is consistent, one can show that the unique answer set of P is the least F-model of P.

($\Leftarrow$) It can be shown that if P has no F-model, then $P \cup \mathcal{A}$ is inconsistent. Viceversa, if P has some F-model its least model corresponds to the unique answer set of $P \cup \mathcal{A}$.   □

One might wonder at the significance of $\models_{\mathcal{A}}$-entailment for disjunctive programs with negation. This entailment regime diverges quickly from the behavior of monotonic logic as soon as negation as failure and disjunction is considered, and is thus incomparable with first order F-logic. It is matter of future research to investigate on the relationship between FAS programs and F-logic under well-founded semantics.

As a second important property, we show that contexts can be exploited for modeling hybrid environments in which more than one semantics has to be taken in account. For instance one might desire a context s in which only $\mathcal{C} \cup \mathcal{S}$ hold as axiomatic modules (this is typical e.g. of RDFS reasoning restricted to ρ-DF [22]), while in a context b we would like to have a different entailment regime, taking in account e.g. $\mathcal{B}$ and $\mathcal{F}$.

We will say that an axiomatic module (resp. a program, a formula) $\mathcal{A}$ is defined at context c if for each rule $r \in \mathcal{A}$, each atom $c \in r$ has context c. If an axiomatic module (resp. a program, or a formula) $\mathcal{A}$ is defined at the default context d, then the axiomatic module $\mathcal{A}@c$, defined at context c, is obtained by replacing each atom a appearing in $\mathcal{A}$ with $a@c$.

Example 3. Consider the program P_4 defined as follows. P_4 has two contexts, rdf and inh. P_4 contains knowledge coming from an RDF triplestore defined in term of the facts $t(gb, rdf{:}type, hp)@rdf$, $t(gb, name,$ "Gibbi"$)@rdf$, etc. Also P_4 contains the rules $X{:}C@rdf \leftarrow t(X, rdf{:}type, C)@rdf$, $X[M \rightarrow V]@rdf \leftarrow t(X, M, V)@rdf$, $C{::}D@rdf \leftarrow t(C, rdfs{:}subClassOf, D)@rdf$. Then, we add to P_4 the program $P_1@inh$ where P_1 is taken from Section 4, plus the rule $X : C@inh \leftarrow X : C@rdf$.

We want that $\mathcal{C}$ and $\mathcal{S}$ hold under the rdf context, while $\mathcal{C}$ and $\mathcal{B}$ hold under the inh context. This can be obtained by defining $\mathcal{A} = (\mathcal{C} \cup \mathcal{S})@rdf \cup (\mathcal{C} \cup \mathcal{B})@inh$ and evaluating P_4 under $\models_{\mathcal{A}}$-entailment.

For instance, $P_4 \models_{\mathcal{A}} gb{:}[income \bullet\!\!\rightarrow 1000]@inh$.

We clarify next how contexts interact each other. First, we consider programs in which contexts are strictly separated: that is, each rule in a program contains only atoms either with context a or only atoms with context b. This way, a program can be seen as composed by two separate modules, one defining a and the other defining b. The following proposition shows that programs defined in separated context behave separately under their axiomatic regime.

Proposition 1. *It is given a program $P = P'@a \cup P''@b$, and axiomatic modules $A@a$ and $B@b$. Then, for formulas $\phi@a$ and $\psi@b$, we have that, if $P \cup A@a \cup B@b$ is consistent,*

$$P \models_{A@a \cup B@b} \phi@a \wedge \psi@b \Leftrightarrow P' \models_A \phi \wedge P'' \models_B \psi$$

Contexts can be seen in some sense as separate knowledge sources, each of which having its own semantics for its data. In such a setting, it is however important to consider cases in which knowledge flows bidirectionally from a context to another and viceversa.

This situation is typical of languages implementing hybrid semantics schemes. For instance, $\mathcal{DL}{+}log$ [28] is a rule language where each knowledge base combines a description logic base D (living under first order semantics), with a rule program P (living under answer set semantics). D and P can mutually exchange knowledge: in the case of $\mathcal{DL}{+}log$, predicates of D can appear in P, allowing flow of information from D to P.

Similarly, we are assuming to have a program P, two contexts a and b, each of which coupled with axiomatic modules $A@a$ and $B@b$. The program P freely combines atoms with context a with atoms with context b, possibly in the same rule.

For simplicity, the following theorem is given for programs containing simple naf-literals only.

Given an interpretation I we define I_a as the subset of I containing only atoms with context a. The *extended reduct* P^{*I_a} of a ground program P is given by modifying each rule $r \in P$ in the following way:

- if $l@a \in H(r)$ and $l@a \notin I_a$ then delete $l@a$ from r;
- if $l@a \in H(r)$ and $l@a \in I_a$ then delete r;
- if $l@a \in B(r)$ and $l@a \in I_a$ then delete $l@a$ from r;
- if $l@a \in B(r)$ and $l@a \notin I_a$ then delete r;
- if $not\ l@a \in B(r)$ and $l@a \notin I_a$ then delete $not\ l@a$ from r;
- if $not\ l@a \in B(r)$ and $l@a \in I_a$ then delete r.

Theorem 2. *Let P be a program containing only atoms with context a and b, and $A@a$ and $B@b$ be two axiomatic modules.*
   Then,

$$M \in ans_{A@a \cup B@b}(P) \Leftrightarrow M_a \in ans_{A@a}(P^{*M_b}) \wedge M_b \in ans_{B@b}(P^{*M_a})$$

Roughly speaking, the above theorem states that from the point of view of context a one can see atoms from context b as external facts, and viceversa. An answer set M of the overall program is found when, assuming M_a as the set of true facts for a, we obtain that M_b is the answer set of $P^{*M_a} \cup B@b$, i.e. an answer set of the program obtained by assuming facts in M_a true. Viceversa, if one assumes M_b as the set of true facts for context b, one should obtain M_a as the answer set of $P^{*M_b} \cup A@a$.

Proof. (Sketch). ($\Rightarrow$) Assume $M \in ans(P \cup A@a \cup B@b)$, it is easy, yet tedious, to construct M_a and M_b and verify that $M_a \in ans(P^{*M_b} \cup A@a)$ and $M_b \in ans(P^{*M_a} \cup B@b)$. Given $P_a = P^{*M_b} \cup A@a$ and $P_b = P^{*M_a} \cup B@b$, the proof is conducted by showing that M_a (resp. M_b) is a minimal model of $P_a^{M_a}$ (resp. $P_b^{M_b}$).
   ($\Leftarrow$) Given M_a and M_b such that $M_a \in ans(P^{*M_b} \cup A@a)$ and $M_b \in ans(P^{*M_a} \cup B@b)$, the proof is carried out by showing that $M = M_a \cup M_b$ is a minimal model of $P \cup A@a \cup A@b^M$. □

6   System Overview

FAS programs have been implemented within the DLT environment [8]. The current version of the system is freely available on the DLT Web page[8], together with examples, a tutorial, and the axiomatic modules herein presented.

---

[8] http://dlt.gibbi.com

DLT works as a front-end for an answer set solver of choice S. Programs are rewritten in the syntax of S and then processed. Resulting answer sets in the format of S are then processed back and output in DLT format. DLT is compatible with most of the languages of the DLV family such as DLV [19], dlvhex [13] and the recent DLV-complex[9]. The native features of the solver of choice are made available to the DLT programmer: this way features such as soft constraints, aggregates (DLV), external predicates (dlvhex), and function, list and set terms (DLV-complex) are accessible. Limited support is given also for other ASP solvers.

DLT allows the syntax presented in this paper and implements the presented semantics. Atoms without context specification are assumed to have the default context d. In order to avoid typing, the default implicit context can be switched by using a directive in the form @*name*., which sets the implicit context to *name* for the rules following the directive.

We overview next some of the other features of DLT, which, for space reasons, can not be focused in the present work.

Complex nested expression. DLT allows the usage of negated attribute expressions. From the operational point of view, if a frame literal in the body of a rule r has subject o and a negative attribute *not m*, our prototype removes *not m* from the attributes of o, adds *not a* to the body of r, where a is a fresh auxiliary atom, and adds a new rule $a \leftarrow o[m]$. to the program. This procedure can be iterated until no negated attribute appears in the program. Then, the answer sets of the original program are the answer sets of the rewritten program without auxiliary atoms. Since negated attributes can appear in negative literals and can be nested, they behave like the nested expressions of [20], allowing in many case to represent information in a more succinct way. The model-theoretical semantics of this aspect of the language is not focused in this paper and is matter of future work.

Example 4. The following rule states that a programmer P is suitable for project p_3 if P knows $c++$ and *perl*, but is not married to another programmer knowing $c++$ and *perl*.

$$P[suitable \twoheadrightarrow p_3] \leftarrow X:programmer,$$
$$P:programmer[skills \twoheadrightarrow \{\text{``c++''}, \text{``perl''}\},$$
$$not \; married \rightarrow X[skills \twoheadrightarrow \{\text{``c++''}, \text{``perl''}\}].$$

Template definitions. A DLT program may contain *template atoms*, that allow to define intensional predicates by means of a subprogram, where the subprogram is generic and reusable. This feature provides a succinct and elegant way for quickly introducing new constructs using the DLT language, such as predefined search spaces, custom aggregates, etc. Differently from higher order constructs, which can be used for the same purpose, templates are based on the notion of generalized quantifier, and allow more versatile usage. Syntax and semantics of template atoms are described in [8].

---

[9] http://www.mat.unical.it/dlv-complex

7 Related Work

Stable vs well-founded semantics. FAS programs have some peculiar differences with respect to the original F-logic. Importantly, while well-founded semantics [14] is at the basis of the nonmonotonic semantics of F-logic, FAS programs live under stable model semantics. The two semantics are complementary in several respects. The well-founded semantics is preferable in terms of computational costs: at the same time, this limits expressiveness with respect to the stable model semantics, which for disjunctive programs can express any query in the computational class Σ_2^p.

On the other hand, the well-founded semantics is three-valued. Having a third truth value as first class citizen of the language is an advantage in several scenarios, such as just in the case of object inheritance. Indeed, the undefined value is exploited in F-Logic when inheritance conflicts can not be solved with a clear truth value. Note, however, that the stable model semantics gives finer grained details in situations in which the well-founded semantics leaves truth values undefined. The reader can find a thorough comparison of the two semantics in [14]. FAS answer sets should not be confused with the notion of *stable object model* given in [33].

Semantic Web languages. Since F-logic features a natural way for manipulating ontologies and web data, it has been investigated for a long as suitable basis for representing and reasoning on data on the web. The two main F-Logic systems Flora and Florid ([32,21]) share with FAS programs the ability to work both on the level of concepts and attributes and on instances.

Several Semantic Web initiatives point to F-logic as rule-based language core, like SWSL ([1]) and WSML ([11]) which in its more powerful variants is based on F-logic layered on top of Description Logic [10].

F-logic has been investigated as a logical way to provide reasoning capability on top of RDF in the system TRIPLE ([30]) that has native support for contexts (called *models*), URIs and namespaces. It is possible also to personalize semantics either via rule axiomatization (e.g. one can simulate RDFS reasoning by means of TRIPLE rules) or by means of interfacing external reasoners. The semantics of the full TRIPLE language has not been clearly formalized: its positive, non-higher order fragment coincides with Horn logic.

The possibility to define custom rule set for specifying the semantics which best fits the concrete application context is also allowed in OWLIM ([18]).

Answer Set Programming. Several works share some point in common with this paper in the field of Answer Set Programming. An inspiring first definition of F-logic under stable model semantics can be found in [10]. The fragment considered focuses on first order F-logic with class hierarchies, and do not explicitly axiomatize structural inheritance with constructive semantics and single valued attributes. Higher order reasoning is present in dlvhex [12]. Contexts were investigated under stable model semantics also in [23]. In this setting, context atoms are exploited to give meaning to a form of scoped negation, useful in Semantic

Web applications where data sources with complete knowledge need to be integrated with sources expected to work under Open World Assumption. Similarly to our work, multi-context systems of [4] are used in order to define hybrid system with a logic of choice. Contexts can transfer knowledge each other by means of *bridge rules*, while in our setting it is not necessary a clear distinction between knowledge bases and bridge rules.

Nested attribute expressions behave like nested expressions as in [20], although we do not allow the use of negation in the head of rules. A different approach to nonmonotonic inheritance in the context of stable model semantics was proposed in [5], in which modules (which can be overridden each other) are associated with each object, and objects are partially sorted by an *isa* relation. The idea of defining an object-oriented modeling language under stable model semantics has been also subject of research in [26] and [25].

References

1. Battle, S., et al.: Semantic Web Services Language,
   `http://www.w3.org/Submission/SWSF-SWSL/`
2. Boley, H., Kifer, M.: Rif core design. W3C Editor's Draft (2007)
3. Boley, H., Kifer, M., Pătrânjan, P.-L., Polleres, A.: Rule interchange on the web. In: Antoniou, G., Aßmann, U., Baroglio, C., Decker, S., Henze, N., Patranjan, P.-L., Tolksdorf, R. (eds.) Reasoning Web 2007. LNCS, vol. 4636, pp. 269–309. Springer, Heidelberg (2007)
4. Brewka, G., Eiter, T.: Equilibria in heterogeneous nonmonotonic multi-context systems. In: AAAI, pp. 385–390 (2007)
5. Buccafurri, F., Faber, W., Leone, N.: Disjunctive Logic Programs with Inheritance. TPLP 2(3) (May 2002)
6. Buccafurri, F., Leone, N., Rullo, P.: Enhancing Disjunctive Datalog by Constraints. IEEE TKDE 12(5), 845–860 (2000)
7. Calimeri, F., Cozza, S., Ianni, G., Leone, N.: Computable functions in ASP: Theory and implementation (unpublished, 2008)
8. Calimeri, F., Ianni, G.: Template programs for disjunctive logic programming: An operational semantics. AI Communications 19(3), 193–206 (2006)
9. de Bruijn, J., et al.: WSMO Final Draft (2005),
   `http://www.wsmo.org/TR/d2/v1.2/`
10. de Bruijn, J., Heymans, S.: Translating ontologies from predicate-based to frame-based languages. RuleML, 7–16 (2006)
11. de Bruijn, J., Lausen, H., Polleres, A., Fensel, D.: The web service modeling language WSML: An overview. In: Sure, Y., Domingue, J. (eds.) ESWC 2006. LNCS, vol. 4011, pp. 590–604. Springer, Heidelberg (2006)
12. Eiter, T., Ianni, G., Tompits, H., Schindlauer, R.: A uniform integration of higher-order reasoning and external evaluations in answer set programming. In: IJCAI, pp. 90–96 (2005)
13. Eiter, T., Ianni, G., Tompits, H., Schindlauer, R.: Effective Integration of Declarative Rules with External Evaluations for Semantic Web Reasoning. In: Sure, Y., Domingue, J. (eds.) ESWC 2006. LNCS, vol. 4011, pp. 273–287. Springer, Heidelberg (2006)

14. van Gelder, A., Ross, K.A., Schlipf, J.S.: The well-founded semantics for general logic programs. J. ACM 38(3), 620–650 (1991)
15. Gelfond, M., Lifschitz, V.: Classical Negation in Logic Programs and Disjunctive Databases. New Generation Computing 9, 365–385 (1991)
16. Jamil, H.M.: Implementing abstract objects with inheritance in datalog¬. In: VLDB, pp. 56–65 (1997)
17. Kifer, M., Lausen, G., Wu, J.: Logical foundations of object-oriented and frame-based languages. Journal of the ACM 42(4), 741–843 (1995)
18. Kiryakov, A., Ognyanov, D., Manov, D.: Owlim - a pragmatic semantic repository for OWL. In: WISE Workshops, pp. 182–192 (2005)
19. Leone, N., Pfeifer, G., Faber, W., Eiter, T., Gottlob, G., Perri, S., Scarcello, F.: The DVL system for knowledge representation and reasoning. ACM TOCL 7(3), 499–562 (2006)
20. Lifschitz, V., Tang, L.R., Turner, H.: Nested Expressions in Logic Programs. AMAI 25(3–4), 369–389 (1999)
21. Ludäscheret, B., et al.: Managing semistructured data with florid: A deductive object-oriented perspective. Inf. Syst. 23(8), 589–613 (1998)
22. Muñoz, S., Pérez, J., Gutierrez, C.: Minimal deductive systems for RDF. In: Franconi, E., Kifer, M., May, W. (eds.) ESWC 2007. LNCS, vol. 4519, pp. 53–67. Springer, Heidelberg (2007)
23. Polleres, A., Feier, C., Harth, A.: Rules with contextually scoped negation. In: Sure, Y., Domingue, J. (eds.) ESWC 2006. LNCS, vol. 4011, pp. 332–347. Springer, Heidelberg (2006)
24. Przymusinski, T.C.: Stable Semantics for Disjunctive Programs. New Generation Computing 9, 401–424 (1991)
25. Ricca, F., et al.: OntoDLV: an ASP-based System for Enterprise Ontologies. Journal of Logic and Computation (Forthcoming, 2008)
26. Ricca, F., Leone, N.: Disjunctive logic programming with types and objects: The dlv⁺ system. J. Applied Logic 5(3), 545–573 (2007)
27. Roman, D., et al.: Web service modeling ontology. Applied Ontology 1(1), 77–106 (2005)
28. Rosati, R.: Dl+log: Tight integration of description logics and disjunctive datalog. In: KR, pp. 68–78 (2006)
29. Simons, P., Niemelä, I., Soininen, T.: Extending and implementing the stable model semantics. Artif. Intell. 138(1-2), 181–234 (2002)
30. Sintek, M., Decker, S.: TRIPLE - an RDF query, inference, and transformation language. In: Horrocks, I., Hendler, J. (eds.) ISWC 2002. LNCS, vol. 2342, pp. 364–378. Springer, Heidelberg (2002)
31. RDF Core Working Group. The Resource Description Framework (2006), http://www.w3.org/RDF/
32. Yang, G., Kifer, M., Zhao, C.: Flora-2: A rule-based knowledge representation and inference infrastructure for the semantic web. In: CoopIS/DOA/ODBASE, pp. 671–688 (2003)
33. Yang, G., Kifer, M.: Inheritance in Rule-Based Frame Systems: Semantics and Inference. Journal on Data Semantics 7, 79–135 (2006)

Deriving Concept Mappings through Instance Mappings

Balthasar A.C. Schopman, Shenghui Wang, and Stefan Schlobach

Vrije Universiteit Amsterdam

Abstract. Ontology matching is a promising step towards the solution to the interoperability problem of the Semantic Web. Instance-based methods have the advantage of focusing on the most active parts of the ontologies and reflect concept semantics as they are actually being used. Previous instance-based mapping techniques were only applicable to cases where a substantial set of instances shared by both ontologies. In this paper, we propose to use a lexical search engine to map instances from different ontologies. By exchanging concept classification information between these mapped instances, an artificial set of common instances is built, on which existing instance-based methods can apply. Our experiment results demonstrate the effectiveness and applicability of this method in broad thesaurus mapping context.

1 Introduction

The problem of semantic heterogeneity and the resulting problems of interoperability and information integration have been an important hurdle to the realisation of the Semantic Web. Different communities use different ontologies and are unable to intercommunicate easily. Solving matching problems is one step to the solution of the interoperability problem. To address it, the Database and Semantic Web communities have invested significant efforts over the past few years [1,2,3].

Instance-based ontology matching techniques determine the similarity between concepts of different ontologies by examining the extensional information of concepts [4,5], that is, the instance data they classify. The idea behind such instance-based matching techniques is that similarity between the extensions of two concepts reflects the semantic similarity of these concepts. A first and straightforward way is to measure the *common extension* of the concepts — the set of objects that are simultaneously classified by both concepts [6,7]. This method has a number of important benefits. Contrary to lexical methods, it does not depend on the concept labels, which is particularly important when the ontologies or thesauri where written in a multi-lingual setting. Moreover, as opposed to structure-based methods, it does not depend on a rich ontology structure; this is important in the case of thesauri, which often have a very weak, and sometimes even almost flat structure.

However, measuring the common extension of concepts requires the existence of sufficient amounts of shared instances, something which is often not the case.

J. Domingue and C. Anutariya (Eds.): ASWC 2008, LNCS 5367, pp. 122–136, 2008.
© Springer-Verlag Berlin Heidelberg 2008

Therefore, in this paper, we aim at enriching one ontology by instances from the other ontology which it should be mapped to and vice versa. Such enrichment is carried out through mappings between instances, that is, similar instances should be classified to the same or similar concepts. In this way, an artificial common ontology extension is built so that many current instance-based methods, such as those in [7], can apply.

Research questions. In this paper, we are experimenting to answer the following research questions:

1. Can an ontology be automatically enriched by instances from another ontology using the similarity between instances?
2. Can the artificially built dually classified instances produce reasonable mappings between two enriched ontologies?

Method and experiments. We use the Lucene search engine[1] to match instances from two different ontologies. For each instance i_t of an ontology T, the most similar instance i_s from the to-be-mapped ontology S is automatically classified to the concepts which i_t also belongs to. After the enrichment, we preserve the instances of each concept from both thesauri, which include their original instances and the ones populated from the ontology to be mapped. Based on such artificially built extensional information of concepts, we calculate a similarity (in our case, the simple Jaccard similarity) to measure the overlap between the extensions of two concepts, which in the end leads to mappings between them, *i.e.*, the higher similarity, the higher probability they should be mapped.

We applied this method on two different cases of thesaurus mapping, a special but frequent mapping problem:

1. mapping GTT and Brinkman whose instances are all books from the National Library of the Netherlands. These are homogeneous instances with the same meta-data fields.
2. mapping GTT/Brinkman and GTAA. The latter thesaurus is used to annotate broadcast materials in the Dutch archive for Sound and Vision. These are heterogeneous instances with different meta-data descriptions.

Evaluation. We first measure the quality of the instance mappings, using the first case and then evaluate the concept mappings in both cases to check the effectiveness and applicability of our method to both homogeneous and heterogeneous collections.

Relation to our previous work. In [7], the similarity between two concepts is measured based on the overlap of their instance sets. This method relies on the existence of a set of common instances and therefore limits itself not applicable if there are no common instances. In [8], all instances of each concept are

---

[1] http://lucene.apache.org/

aggregated to form a unified representation of this concept. A probabilistic classifier is trained to model the correlation between the similarity between such aggregated representations and the mapping between two concepts. In this paper, we directly use the similarity between individual instances and assume similar instances should be classified to similar concepts. An artificial set of common instances will be built, based on which the similarity between concepts is measured by applying the methods used in [7]. This is an extension of the work done in [7][2] and in parallel with the learning method used in [8].

The rest of the paper is structured as following: Section 2 describes the two application problems in more details. Section 3 introduce our method of using instance mappings to derive concept mappings, including how to use the lexical search engine, Lucene, to achieve instance mappings. In Section 4, we present the results of our experiments. Section 5 introduces some related work, and finally, Section 5 concludes this paper and discusses the future work.

2   Application Problems

Our research has been motivated by practical problems in the Cultural Heritage domain, an interoperability problem within National Library of the Netherlands (Koninklijke Bibliotheek, or KB), and the problem of unified access to two heterogeneous collections, one from the KB, one from the Dutch archive for Sound and Vision (Nederlands Instituut voor Beeld en Geluid, or BG).

2.1   Homogeneous Collections with Multiple Thesauri

Our first task is to match the GTT and Brinkman thesauri, which contain 35K and 5K concepts respectively. The average concept depths are 0.689606 and 1.03272 respectively.[3] Both thesauri have similar coverage but differ in granularity. These two thesauri are individually used to annotate two book collections in KB: the *Scientific Collection* annotated mainly by GTT concepts and the *Deposit Collection* annotated mainly by Brinkman concepts.

In order to improve the interoperability between these two collections, for example, using GTT concepts to search books annotated only with Brinkman concepts, we need to find mappings between these two thesauri.[4] Among nearly 1M books whose subjects are annotated by concepts from these two thesauri, 307K books are annotated with GTT concepts only, 490K with Brinkman concepts only and 222K with both. The books in both collections are described using the same metadata structure, more specially, using an extension of the Dublin Core metadata standard.[5]

---

[2] See Section 4.2 for detailed comparison.

[3] Nearly 20K GTT concepts have no parents.

[4] Descriptions of different scenarios of using mappings, the requirements on mappings and various evaluation methods can be found in [9].

[5] http://dublincore.org/documents/dces/

2.2   Heterogeneous Collections with Multiple Thesauri

Our second task is to match the Brinkman thesaurus from the KB to the GTAA thesaurus, which is used to annotate the multimedia collection in the BG. The BG serves as the archive of the Dutch national broadcasting corporations. All radio and television programmes that are broadcast by these corporations are continuously added to the archive. Besides over 700K hours of material, the BG also houses 2M still images and the largest music library of the Netherlands. Each object in the BG collection is annotated by one or several concepts from the GTAA thesaurus. The GTAA thesaurus contains 160K concepts in total, including 3868 from the subject facet which are interesting to map with the KB thesauri. The concept hierachy of the subject facet has an average depth of 1.30817.

Mapping GTAA to one or both of the KB thesauri is very interesting from a Cultural Heritage (CH) perspective, as interoperability across collections has become an urgent practical issue in this domain. For example, one could be interested to search for some broadcasts from the BG about the author of the book he is reading in the KB. Aligning these thesauri with which the collections are annotated provides a promising solution to achieve this interoperability. Different from the KB case, the meta-data structure of instances differs significantly across collections.

In both cases, each of the thesauri to be mapped contains a large amount of concepts, which many current matching tools could not even load. The concepts within the thesaurus are poorly structured or rather in a nearly flat list, which makes the structural matching techniques not really applicable. Luckily, the instances of those concepts are available which allows us to apply instance-based methods, as done in our previous work [7,8]. In this paper, we continue exploring the instance-based method at the meta-data level.

3   Method: From Instance Mappings to Concept Mappings

Our task is to map two thesauri, each of which is used to annotate a collection of objects (books or multimedia materials). Thesaurus concepts are used to annotated the subject of these objects and we consider an object is annotate by a concept as the instance of this concept. Each object may be annotated by multiple concepts, therefore, one object can be the instance of multiple concepts.

On top of their subject feature, instances also have other features, such as title, abstract, creator, *etc.* These features together uniquely represent an instance. All instances are virtually projected into a space where the distance between them can be measured, *e.g.*, using the Euclidean distance in the feature space.

Instances that are close in this space could potentially be classified to similar concepts. Based on this hypothesis, for one concept in one ontology, if instances in the other ontology are similar to its own instances, we can add those instances as its *virtual* instances. Therefore, these instances can be seen as common instances shared by this concept and those they really belong to, *i.e.*, the concept(s) in the other ontology. Once this artificial set of common instances is built, the existing instance-based methods can be applied to generate concept mappings.

Let us formally describe the (rather simple) idea: let S (for source) and T (for target) be two thesauri we want to map, and I_s and I_t be their finite sets of instances. Let $ann_s(i) = \{C \in S \mid i \in ext(C)\}$ be the annotation of an instance i, which contains a set of concepts from S. These concepts have instance i in their extension $ext(C)$.

Suppose we have a similarity function sim between instances (across I_s and I_t). For each instance $i \in I_s$, we look for an instance $j \in I_t$ which is the most similar to i. That is,

$$j = \operatorname*{argmax}_{t} sim(t, i).$$

We can now simply add j to the extension of all concepts $C \in ann_s(i)$. The same process is carried out in both directions. This way, we create a virtual dually annotated corpus. This section remains to explain how we calculate the similarity between instances, and to recall how we calculate concept mappings from dually annotated corpora.

3.1 Matching Instances

Based on the above hypothesis, we use the Lucene search engine to achieve instance mappings. Lucene is a high-performance and scalable information retrieval library through which any piece of textual data can be indexed and made searchable. Indexing with Lucene can be divided into three main phases: (i) converting data to text, (ii) analysing the text and (iii) saving the text to an index. We feed instance data in Lucene, stored in the form of a *Document*. A *Lucene document* (LD) consists of a collection of *fields*. Every field contains the content of the corresponding instance features, such as "title," "abstract," "creator," *etc.* Additionally, each instance has a "subject" field which contains the labels or unique identifiers of the concepts they belong to. Lucene allows keyword-based search and search results (on the form of LDs) are collected within Lucene *Hits*. Each LD contained in the Hits, has an associated score value (between 0 and 1) that indicates its similarity to the search key. Lucene scoring schema is based on the Vector Space Model [10] of information retrieval. The benefits of using Lucene are very fast response time, shown in [11], and complexity almost hidden to the users.

The instance matching process is as follows. Let I_s and I_t be the two instance sets of two ontologies, *e.g.*, two book collections annotated by the GTT and Brinkman thesauri. First we populate the Lucene database (Ldb) with a collection I_s. Each instance is stored as a LD with its fields containing information about this instance. Since Lucene operates on a lexical level, we use the textual representations of fields where possible, such as "title," "subject," "abstract," "descriptions," *etc.*

Then we execute a query for every instance i_t in the other collection I_t. The Lucene search engine allows us to search for information in specific fields of the documents, by specifying one or more keywords and one or more Fields to search within. We can use words in, for example, the "title" field of instance i_t as keywords and search the "title" fields of all LDs in the Ldb. It is also possible to carry out cross-field queries. That is, for example, using words in the "title" field to search the "subject" or "abstract" fields, or vice versa. We can also construct queries by concatenating multiple fields. In this construction we create Lucene documents with a single field containing the concatenation of certain fields, for instance the "title" and "subject" fields. Then we execute single-field queries to match these concatenations with each other.

For every query Lucene returns a list of hits, which is ordered by relevance. We take the most similar instance and observe which concepts it belongs to, *i.e.*, concepts in the "subject" field. We then classify instance i_t as an instance of these concepts, by adding these concepts into its "subject" field.[6] The same process is carried out from collection I_s to I_t, *i.e.*, populating the Ldb with collection I_t and enriching the instances of collection I_s. In the end, each instance in both collections will be classified against concepts from both thesauri, which means an artificial set of common instances is created.

If instances in different collections are homogeneously structured, *i.e.*, the same features are available across different collections, such as the two collections in the KB, we can use Lucene to directly map instances. However, in more cases, different collections have different structures to represent/store their instances, such as the different collections in the KB and the BG. Similarly, we can feed different collections to the Lucene database, using their own features. However, we need to specify corresponding query fields in order to run Lucene queries and map instances afterwards. Different from constructing queries for the homogeneous case, when it is clear that two fields are good for query, such as "title" to "title" or "title" to "subject," in the heterogeneous case, we need to anticipate these potentially good pairs of fields. Readers are referred to [8] for different ways of automatically choosing such pairs.

3.2 Matching Concepts

Once the artificial common instance set is built, we can apply existing instance-based techniques to compute mappings between concepts. Our previous work has shown simple measures of similarity between instances suffice to produce sensible mappings [7]. In this paper, we use the Jaccard similarity measure to determine whether two concepts can be mapped or not. Specifically, each concept corresponds to a set of instances which are annotated by this concept. For all possible pairs of concepts, we measure the Jaccard similarity between their instance sets. This is a measure of similarity between the extensional semantics of those concepts. Pairs of concepts with a high Jaccard similarity are considered as a mapping.

---

[6] By adding concepts into the "subject" field of an instance, this instance will be considered as an instance of each added concepts.

There are two parameters which need to be taken into account:

1. The minimum similarity: a threshold that determines how similar the two concepts must be in order to define a mapping between them.
2. The minimal number of instances shared by two concepts. If a concept has very few instances, using these instances to determine its extensional semantics is not sufficient. Sometimes, it may mislead the similarity judgement. For example, if two concepts each have one instance and by chance this instance is shared. This will result in a Jaccard measure of 1, but it should actually carry less weight than the case that two concepts have 1000 instance in common and a few not.

These two parameters in practice are set in an empirical way. Based on some evaluation criteria, we can set them up to optimise the performance. Note, this is obviously a biased solution, as the evaluation criteria may vary due to different mapping usage scenarios, see [9] for more details.

4   Experiment and Evaluation

We will study the following questions:

1. How good is our method for finding similar instances?
2. How does our proposed method perform on homogeneous data collections?
3. How does our proposed method perform on heterogeneous data collections?

4.1   Evaluation of the Quality of Instance Mappings

In the KB case, we have 222K books which have been previously dually annotated with two thesauri to be matched. This gives us an opportunity to evaluate whether similarity of the descriptions of instances (books) indeed leads to valid instance mappings.

For this purpose, we split the original dually annotated instance set into two parts, noted as I_G and I_B. By hiding the GTT annotation of each book in I_B and the Brinkman annotation of the books in I_G, we created two collections annotated by only one thesaurus.

We first populated the Lucene database (Ldb) with I_G and using the method introduced in Section 3.1, each instance in I_B finds the most similar book in I_G and adopts this book's GTT annotation as its new GTT annotation. Similarly, each book in I_G also borrows the Brinkman annotations from the most similar book in I_B. By comparing its original manually created annotation and this new GTT/Brinkman annotation automatically obtained from the mapped instance, we can evaluate the basic hypothesis of our method.

We calculated the similarity of the original annotations with the artificial ones built from the instance mappings. As the Jaccard similarity is the most common

Table 1. Performance of using different query configurations

Query fields	Sim_a
title	0.244
title, subjects	0.324
title, subjects (cross query)	0.318
title, subjects (concatenated)	0.310

way of comparing sets, we use it again for this purpose,[7] more concretely: the quality of a prediction for each book in I_G and I_B is calculated as following:

$$ sim_a = \frac{|S_m \cap S_n|}{|S_m \cup S_n|} \tag{1} $$

where S_m is the manual (original) annotation and S_n is the annotation built by the instance mapping. Clearly, a higher similarity implies a higher chance for this method to produce a reasonable set of common instances. We then take the average of this Jaccard similarity over all dually annotated books as the final measure. Different ways of query configurations, as discussed in Section 3.1, perform differently, shown in Table 1.

From Table 1, we can see that the new annotations obtained by querying the "title" and "subject" fields separately, on average, are the most similar to the original manual ones, with a Jaccard measure of 0.324. This may seem to be a low value, but the following experiments will show that the predicative power of these (now artificially dually annotated) instances is almost as high as of the original ones. It is also worth noting that the values given in Table 1 only refer to the original annotations which are, in our experience, also not necessarily perfect, and often incomplete. This means that this measure may under-estimate the correctness of the new annotations. A proper manual evaluation of this is impossible due to the size of the corpus, and the specialised nature of the annotation task in a library.

4.2   Mapping Thesauri over Homogeneous Collections

In order to evaluate our proposed mapping method for thesauri over homogeneous collections we repeat the experiments of [7] to map Brinkman and GTT, but now based on the full set of instances (not just the doubly annotated corpus). We used Jaccard similarity measure to generate mappings based on instances. We applied this measure on the real singly annotated datasets. In this case, I_G and I_B contain 307K and 490K book instances, respectively.

---

[7] The reader should not confuse our use of the Jaccard measure to calculate similarity of concepts, and to evaluate the quality of the artificial annotations. Here, the Jaccard similarity measures how similar the artificial annotation is to the original one, while in the former case, the Jaccard similarity measures the overlap of the common extension of two concepts.

Table 2. Comparison with results from the real dually annotated dataset, where the recovered mappings are those found by our method which are also found from the real dually annotated dataset and the percentage in the bracket is the corresponding proportion.

Query fields	Recovered mappings	New mappings
title	426 (31%)	15
title, subjects	549 (40%)	390
title, subjects (cross query)	640 (47%)	564
title, subjects (concatenated)	1140 (84%)	429

The task to be performed is a book-reindexing scenario, which influences the way the experiments need to be evaluated.[8] The minimal number of instances shared by two concepts is 10 and the lowest threshold was set to 0.001. The performance varies with the choice of threshold, as depicted in Fig. 1. In our case, the Jaccard measure is the one we would like to optimise — a high Jaccard similarity means the translated annotation covers most of its manual annotation without introducing many errors.

We can see from this figure that the optimum Jaccard measure is achieved by taking the threshold around 0.1. Fig. 1 (b) is the performance of mappings generated from the real dually annotated dataset, which can be seen as an upper bound performance of these mappings in this scenario. Mappings generated from the real singly annotated dataset performs at a similar level to what the OAEI'2007 participants did on the same dataset.[9] This is very encouraging, because it indicates that our method does not have the constraints on the existence of the explicit dually annotated instances, and still, performs as well as the state-of-art tools do.

In Table 2, we compared mappings generated from the real singly annotated dataset (i.e., 307K books with only GTT annotations and 490K books with only Brinkman annotations) with those generated from the dually annotated dataset (i.e., 222K books with both Brinkman and GTT annotations). In the base case — generate artificial common instances using concatenated "title" and "subject" — we found 84% pairs which are found from the dually annotated dataset, both using the threshold of 0.1. A manual evaluation has shown that 97% of the 429 new mappings are correct. This comparison confirms that our method can to a large extent recover the mappings generated from a dually annotated dataset if it is available; also the high precision of the new mappings indicates that our method makes use of the information which was not usable before. It means that even when dually annotated instances are available, using our method with singly annotated instances can improve current mapping results.

---

[8] Technical details can be found in [9].

[9] See http://oaei.inrialpes.fr/2007/results/library/ for more details of the results of OAEI'2007 participants.

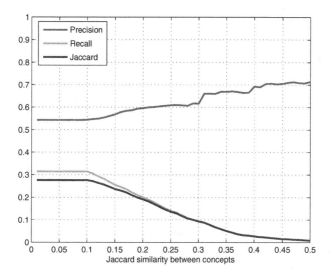

(a) Real singly annotated dataset

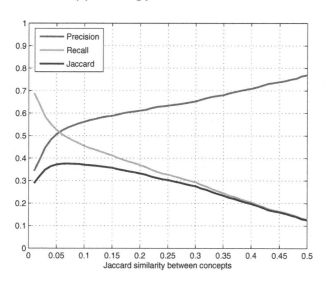

(b) Original dual annotated dataset (upper bound)

Fig. 1. Performance evaluated in book-reindexing scenario

4.3   Mapping Thesauri over Heterogeneous Collections

Now we map GTAA with Brinkman using disjoint and heterogeneous collections.
As we introduced earlier, GTAA is used for annotating multimedia materials in
the Dutch archive for Sound and Vision (BG), Brinkman for books in the KB.
In our dataset from the BG collection, there are nearly 60K instances and their

subjects are annotated against 3593 GTAA concepts. The task here is to map these GTAA concepts with 5207 Brinkman concepts.

As we discussed earlier in Section 3.1, we first map instances in order to build an artificial common extension of these concepts. Each collection was fed into the Lucene database and the query fields were set up manually.

We specify queries on concatenated "kb:title" + "kb:subject" and "bg:title" + "bg:subject" + "bg:description." Each KB instance will be classified to one or several GTAA concepts through the mapping between instances, and similarly, each BG instance will be classified to one or several Brinkman concepts. Then, all instances of one concept were put together as the extensional representation of this concept. The Jaccard similarity was measured between the instance sets of all possible pairs of Brinkman and GTAA concepts. All pairs were then ranked by their Jaccard similarity into an ordered list, with the most promising mappings on the top.

Ideally, the generated mappings should be evaluated against a reference alignment for a global view of the precision and recall. Unfortunately, obtaining a complete list of possible mappings is not practically possible. We therefore compare the obtained mappings with results from a lexical mapper and then manually measure the precision of the top K mappings.

Using a simple lexical mapper,[10] we obtained 1458 lexical equivalent mappings between 5207 Brinkman concepts and 3593 GTAA concepts from the subject facet. One or both concepts of 115 lexical mappings do not have any instances and therefore cannot be measured by our method.

Moving from the top of the ranked list, we measure the proportion of lexical mappings, P_{lexical}, and the coverage over all lexical mappings, C_{lexical}. As Fig. 2 (b) shows, when the Jaccard similarity is relatively high, most of the found mappings are actually lexical mappings. This proportion decreases with the Jaccard similarity. At the Jaccard similarity of 0.05, nearly half of the found mappings are non-lexical pairs. The coverage over all lexical mappings increases slowly up to around 20%. However, from the 273 lexical mappings that have a Jaccard similarity above zero (*i.e.*, there are joint instances) 95.6% are ranked among the top 1000 mappings.

We carried out a manual evaluation on the top 1000 Brinkman–GTAA mappings. The purpose is to check whether the precision decreases and how much it decreases with the increasing number of non-lexical mappings. Among the top 1000 mappings, there are 261 lexically equivalent pairs, which we consider as correct mappings. The remaining 739 non-lexically equivalent pairs were presented to a Dutch-speaker, who judged each pair to be a valid mapping or not.

The evaluation results are analysed as follows. For each 10th mapping in the list, we calculate the precision of all pairs within a window of size 40, 20 to the left and 20 to the right. This gives a local average precision, P_{local}, which is sensitive to its location in the list. It indicates, to some extent, the probability

---

[10] This Dutch language-specific lexical mapper makes use of the CELEX (http://www.ru.nl/celex/) morphology database, which allows to recognise lexicographic variants of a word-form, as well as its morphological components.

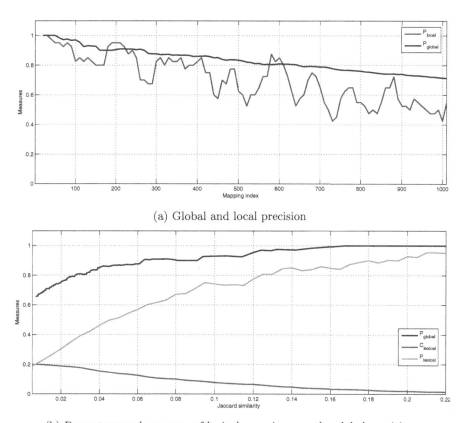

(a) Global and local precision

(b) Percentage and coverage of lexical mappings *vs.* the global precision

Fig. 2. Evaluation of mappings between Brinkman and GTAA

for a mapping to be correct in a certain neighbourhood when moving through the ranked list. The precision of all pairs from the top to the current pair, P_{global}, is also calculated, which indicates the precision from a more global view.

According to Fig. 2 (a), the global precision slowly decreases when moving further along the ranked mapping list. Although sensitive to its position in the ranked list, the local average precision also gradually decreases from 100% to 42.5%. A high precision of 71.6% at the 1000th mapping tell us that the quality of the returned non-lexical mappings is quite good. The local average precision at the 1000th mapping is still 42.5%,[11] which means that in this neighbourhood the mappings have an average probability of 42.5% to be correct. Depicted in Fig. 2 (b), with the increasing number of non-lexical mappings (from about 5% to nearly 80%), the precision does not decrease dramatically (from 100% to 71.6%). The nice results here illustrate the effectiveness of our method and its applicability to the heterogeneous case.

---

[11] In order to achieve this measure, the evaluator judged the 20 mappings after the 1000th one.

Fig. 2 to some extent indicates that lexically equivalent pairs often do not have similar extensional semantics, especially when used in different collections or across domains. This is an indication to consider the reliability or limitations of using lexical mappings in certain applications where extensional semantics play an important roles, such as retrieving or browsing across collections.

5   Related Work

Ontology matching, as a promising solution to the semantic heterogeneity problem, has recently become an interesting and important research problem. Many different matching techniques have been proposed. In order to make use of various properties of ontologies (*e.g.*, labels, structures, instances or related background knowledge), existing matching techniques adopt methods from different fields (*e.g.*, statistics and data analysis, machine learning, linguistics). These solutions share some techniques and attack similar problems, but differ in the way they combine and exploit their results. A detailed analysis of the different matching techniques has been given in [3]. Examples of individual approaches addressing the matching problem and latest development in this area can be found on www.OntologyMatching.org.

The most related matching technique to our work is the instance-based methods, also called extensional matching techniques. The idea behind such techniques is that similarity between the extensions of two concepts reflects the semantic similarity of these concepts. Many current mapping tools, such as [12], make very limited use of instances, where instance information are only used complementarity to other techniques. Instance-based method has not been very widely investigated until recently [4,5,6,13], where neural networks, machine learning or statistics were used to model the complex correlation between instances and the semantics of concepts. However, instances are in general simply used as literals and the instance-based similarity normally results from the set operations, such as in [7].

A simple instance based method requires the existence of common instances. However, the explicitly shared instances are often not available, as ontologies in different applications contain similar but different individuals. As a sufficient amount of instance data becomes available, it has been proposed to use machine learning and statistics to grasp the relations between instances themselves. The similarity between instances using their own information, such as the metadata of individuals, has recently been investigated in [8]. The method proposed in this paper is another way of using instance as informative individuals by themselves, instead of treating them only as simple literals.

6   Conclusion and Future Work

In this paper, we propose to use a lexical search engine to map instances from different ontologies. By exchanging concept classification information between these mapped instances, we can generate an artificial set of common instances shared by concepts from two ontologies, so that existing instance-based methods can apply. By comparing mappings between two thesauri, GTT and Brinkman, generated by explicit dually annotated instances and those by our method using singly annotated datasets, we have shown the feasibility of our method in a homogeneous case. Our experiments of mapping Brinkman and GTAA, using completely different and disjoint collections, have demonstrated this method to be an effective approach and applicable to a broad mapping context, *i.e.*, heterogeneous collections. To the best of our knowledge, this is new, and a very promising step towards effective semantic interoperability between different collections (*e.g.* in the Cultural Heritage domain).

In the future, we will further experiment with different query configurations in the instance mapping step, *e.g.*, the influence of single-field, multi-field and concatenated-field queries on the generated mappings, whether machine learning techniques can help map instances without many manual settings, *etc.*

In the GTT-Brinkman case, a threshold can be decided according to the optimal performance of the obtained mappings in a re-indexing scenario. However, it is not the case for the Brinkman-GTAA case. We will investigate more on how to find such optimisation tasks for deciding the relevant threshold parameters.

Finally, Lucene uses lexical information for answering queries. This hinders our method to be applied in a multi-lingual setting. In the future, we will explore the possibilities to increase the applicability of our method in this direction, such as using an automatic translation tool to reduce the language barrier.

References

1. Rahm, E., Bernstein, P.A.: A survey of approaches to automatic schema matching. VLDB J. 10(4) (2001)
2. Doan, A., Halevy, A.Y.: Semantic integration research in the database community: A brief survey. AI Magazine 26(1) (2005)
3. Euzenat, J., Shvaiko, P.: Ontology Matching. Springer, Heidelberg (2007)
4. Li, W.S., Clifton, C., Liu, S.Y.: Database integration using neural networks: Implementation and experiences. Knowledge and Information Systems 2, 73–96 (2000)
5. Doan, A.H., Madhavan, J., Domingos, P., Halevy, A.: Learning to map between ontologies on the semantic web. In: Proceedings of the 11th international conference on World Wide Web, pp. 662–673 (2002)
6. Ichise, R., Takeda, H., Honiden, S.: Integrating multiple internet directories by instance-based learning. In: Proceedings of the eighteenth International Joint Conference on Artificial Intelligence (2003)
7. Isaac, A., van der Meij, L., Schlobach, S., Wang, S.: An empirical study of instance-based ontology matching. In: Aberer, K., Choi, K.-S., Noy, N., Allemang, D., Lee, K.-I., Nixon, L., Golbeck, J., Mika, P., Maynard, D., Mizoguchi, R., Schreiber, G., Cudré-Mauroux, P. (eds.) ASWC 2007 and ISWC 2007. LNCS, vol. 4825, pp. 253–266. Springer, Heidelberg (2007)
8. Wang, S., Englebienne, G., Schlobach, S.: Learning concept mappings from instance similarity. In: Proceedings of the 7th International Semantic Web Conference (ISWC 2007), Karlsruhe, Germany (to appear, 2007)

9. Isaac, A., Matthezing, H., van der Meij, L., Schlobach, S., Wang, S., Zinn, C.: Putting ontology alignment in context: Usage scenarios, deployment and evaluation in a library case. In: Bechhofer, S., Hauswirth, M., Hoffmann, J., Koubarakis, M. (eds.) ESWC 2008. LNCS, vol. 5021, pp. 402–417. Springer, Heidelberg (2008)
10. Salton, G., Wong, A., Yang, C.S.: A vector space model for automatic indexing. Commun. ACM 18, 613–620 (1975)
11. Pirro, B., Talia, D.: An approach to ontology mapping based on the lucene search engine library. In: Proceedings of the 18th International Conference on Database and Expert Systems Applications (DEXA 2007), Regensburg, Germany, pp. 407–411 (September 2007)
12. Hu, W., Qu, Y.: Falcon-AO: A practical ontology matching system. Journal of Web Semantics (2007)
13. Dhamankar, R., Lee, Y., Doan, A., Halevy, A., Domingos, P.: iMAP: Discovering complex semantic matches between database schemas. In: Proceedings of the ACM International Conference on Management of Data (SIGMOD), pp. 383–394 (2004)

Deep Semantic Mapping between Functional Taxonomies for Interoperable Semantic Search

Yoshinobu Kitamura, Sho Segawa,
Munehiko Sasajima, Shinya Tarumi, and Riichiro Mizoguchi

The Institute of Scientific and Industrial Research, Osaka University
8-1, Mihogaoka, Ibaraki, Osaka, Japan
{kita,segawa,msasa,tarumi,miz}@ei.sanken.osaka-u.ac.jp

Abstract. This paper discusses ontology mapping between two taxonomies of functions of artifacts for the engineering knowledge management. The mapping is of two ways and has been manually established with deep semantic analysis based on a reference ontology of function for bridging the ontological gaps between the taxonomies. We report on the successful results thanks to such deep analysis not at the lexical level but at the ontological level. Using the mapping knowledge, we developed a semantic search system which can provide engineers with interoperable access to technical documents by searching for functional metadata based on either of functional taxonomies.

Keywords: Knowledge management, ontology, ontology mapping, metadata.

1 Introduction

Functionality is one of the key aspects of knowledge about artifacts [1,2]. The goal of this research is to manage engineering documents using semantic annotation about functionality of artifacts. Such function-oriented knowledge management is very useful in engineering design by finding previous design cases for the same required function or by finding related patents [2]. The semantic annotation about function is expected to solve the difficulty of the current document-based engineering knowledge management based on lexical expressions, that is, many terms (verbs) are used in documents for the same function (and vise versa) without clear semantics.

For this, we have proposed a framework of an ontology-based semantic annotation about functionality (we call *Funnotation* (abbreviation of FUNctional anNOTATION) hereafter) [3]. It includes a metadata schema in OWL for functional annotation. The schema is based on our functional ontologies [4,5,6] (we call FOCUS (abbreviation of Functional Ontology for Categorization, Utilization and Systematization)), which have been deployed successfully in industry [6]. Metadata in RDF based on the schema shows the function of the artifact mentioned in documents. Then, a document search system using the functional metadata as an engineering knowledge management system is designed to help engineers access technical documents in a web system on an intranet within a company by specifying "what they want to realize", i.e., function, independently of lexical terms in the documents.

J. Domingue and C. Anutariya (Eds.): ASWC 2008, LNCS 5367, pp. 137–151, 2008.
© Springer-Verlag Berlin Heidelberg 2008

Our aim in this paper is to realize interoperability between functional taxonomies in the functional annotation. Some taxonomies of verbs for generic functions have been proposed in the literature, e.g., [1,4,7,8]. Among others, we concentrate on (Reconciled) Functional Basis in the NIST Design Repository Project (hereafter FB) [8] and our functional concept ontology (hereafter FOCUS/Tx) [4]. Thus, our goal here is to search for documents using metadata based on either of these taxonomies.

The research issue here is to establish the two-way *mappings* (by which we here mean directed correspondence relations) between similar functional terms in those taxonomies. This is a problem so-called the *semantic integration* [9] or *ontology matching* [10]. General techniques for this problem can be categorized into 'automatic mapping discovery' [9] and 'manual mapping analysis'. The current majority of research efforts aim at 'automatic mapping discovery' which is to automatically determine which concepts in two ontologies represent similar notions [9]. Such techniques mainly use lexical information based on natural language processing techniques, the structural features of ontologies, and/or shared instances [9,10]. Although the automatic mapping discovery can be applicable to large-scale ontologies, it is difficult to get precise mappings reflecting the deep semantics[1] of the target concepts. Moreover, the automatic mapping discovery hardly contributes to revealing the underlying differences and in-depth investigation on the target concepts.

On the other hand, the manual mapping analysis can establish precise mappings based on deep analysis of the taxonomies and account for the ontological differences of taxonomies and the concepts. Of course, the manual analysis is a time-consuming task and then it is difficult to establish mappings between large-scale ontologies.

The crucial issue here is that the differences between those functional taxonomies are *not only* terminological *but also* ontological, because some functions are based on different conceptualizations. For example, "link" in FB implies not only "to couple flows together" [8] as the change at input and output but also "by means of an intermediary flow" [8] as how to realize it. Thus, it cannot be fully mapped onto "combine" in FOCUS/Tx which implies "to bring two operands into an operand" as the change at input and output, which corresponds to only the former part of the meaning of "link". This is not a terminological but an ontological difference, because "the change in the target object" and "how to realize the change" are ontologically different. One of the deep causes of such a confusion is the lack of clear understanding of the notion of function, though much research has been conducted on functionality in engineering design (e.g., [1,2]), in artificial intelligence (e.g., [12]) and in philosophy [13]. Our aims here include contribution to accounting for the notion of function by comparing those taxonomies as well.

On the basis of the above observation, this research adopts *not* the automatic mapping discovery *but* the manual mapping analysis based on a reference ontology of function. Its main reasons are the deep ontological gaps between taxonomies and our aim of investigating function ontologically discussed above. The small numbers of terms of the taxonomies (52 terms [8] and 89 terms [4]) enable us to analyze mappings manually. Although the numbers are small, FB is founded on a great number of empirical studies [7,8] and FOCUS/Tx has been successfully deployed in industry [6].

---

[1] Some matching methods use 'logical semantics' of axioms (e.g., [11]). The 'deep semantics' we would like to capture here is, however, not identical to those formal semantics.

These facts strongly suggest that these taxonomies cover wide-range of artifacts. So, it is worth to perform the labor-intensive and time-consuming manual process for precise mappings. The reference ontology of function [14] defines upper-categories of several kinds of function. It is utilized here for clarifying ontological differences between the taxonomies and for bridging the gaps for mappings between them. The mapping framework has been reported in [15,16]. This paper reports the concrete two-way mappings (only one-way has been reported in [16]), their analysis, and their use in interoperable semantic search for knowledge management.

This paper is organized as follows. Firstly, we overview the interoperable semantic search to be realized in this paper. Then, the taxonomies to be dealt with are introduced in Section 3. Section 4 discusses the reference ontology of function and the mapping process based on it. Section 5 reports the mappings obtained. Section 6 demonstrates the functional annotation and the interoperable search based on the mappings. Section 7 discusses related work followed by the conclusion.

2 Framework of Interoperable Semantic Search Based on Functional Annotation

Figure 1 shows an overview of the *Funnotation* framework. Its F-Core schema defines fundamental classes such as *device*, *stuff*, *energy*, *function* and *way* (of function achievement) together with properties such as *has-function* and *selected-way*. The *way* (of function achievement) represents how to achieve a function as discussed in Section 3.2. The F-Vocab schema defines generic functions based on the functional concept ontology; FOCUS/Tx [4]. Such schemata implemented in OWL enable us to describe metadata in RDF representing functionality of engineering devices in documents. For example, the metadata m_a in Fig. 1 shows that the device in annotated document d_a (a filter) can perform an instance of the *separating* function class defined in the schema. This is annotated to the term "extract" in d_a. The metadata m_b shows that the distiller in the document d_b has the same *separating* function, which is, however, annotated to the term "refine" in d_b. In this manner, functional metadata shows device's functions independently of the terms in documents and indicates URI to the original documents and/or terms. Moreover, the metadata show how to achieve a function, i.e., in this case, two different *ways* (i.e., the filtering way and the different-boiling-points (distilling) way) to achieve the *separating* function.

Moreover, the document d_c is annotated in terms of another functional taxonomy; Functional Basis (FB). The word "grinding" of a coffee grinder is annotated as the *branching* function in FB. As discussed in this paper, the authors prepare the mapping knowledge between FOCUS/Tx and FB based on the reference ontology of function. As discussed later, in this simple example, the *branching* function of FB has a direct mapping to the *separating* function of FOCUS/Tx.

Given a query in terms of functions of FOCUS/Tx, a semantic search system provides access to the annotated documents by searching for the functional metadata. In Fig. 1, if an engineer specifies the *separating* function as a goal-function (function to be achieved) as a query, the system provides links to the both documents d_a and d_b. Moreover, according to the mapping knowledge, the document d_c is also retrieved.

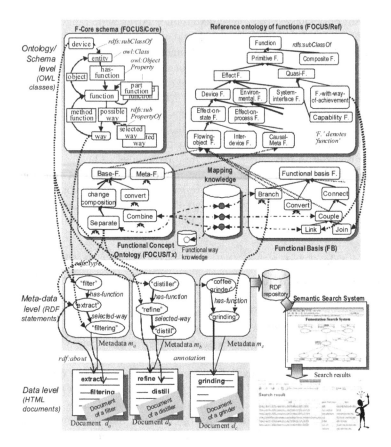

Fig. 1. Overview of *Funnotation*: A Framework for Semantic Annotation about Functionality for Engineering Documents

3   The Functional Taxonomies

3.1   Reconciled Functional Basis (FB)

Reconciled Functional Basis has been proposed by Hirtz et al. [8], which is a result of reconciliation of some previous taxonomies and empirical generalization based on a great number of empirical studies. A function of a device is expressed as a pair of an active verb and its (grammatical) object (called 'flow'). We call the taxonomy of function (verb) FB in this paper. FB consists of 52 terms in three levels of categorization. Table 1 shows its small portion [8]. Each of functional terms is defined in natural language with examples and correspondents (synonyms). For example, the *separating* function is defined as "to isolate a flow (material, energy or signal) into distinct components. The separated components are distinct from the original flow, as well as each other" [8].

Table 1. A Portion of Reconciled Functional Basis [8]

Class (Primary)	Secondary	Tertiary	Correspondents
Branch			
	Separate		Isolate, sever, disjoin
		Divide	Detach, isolate, release,
		Extract	Refine, filter, purify ....
		Remove	Cut, drill, lathe, polish,
	Distribute		Diffuse, dispel, disperse
Connect			
	Couple		Associate, connect
		Join	Assemble, fasten
		Link	Attach
	Mix		Add, blend, coalesce,

Such definitions in natural language are sometimes ambiguous and it is difficult to distinguish similar terms. Garbacz points out some problems of the classification of FB such as lack of principle of categorization and non-exhaustiveness from logical and ontological viewpoints [17]. Moreover, the concept of function is defined as "a description of an operation to be performed by a device or artifact" [7]. In this definition, the intention of a designer or a user is implicit, which is a crucial characteristic of function in comparison with objective *behavior* [1,2,4,6,12,13].

3.2   The Functional Concept Ontology (FOCUS/Tx)

In comparison with FB, the functional concept ontology (FOCUS/Tx) has an ontological foundation. It is based on a device-centered ontology; FOCUS/Core [6], which enables us to distinguish function from *behavior*. The behavior of a device is defined as temporal changes of things (called operands) as input-output relation in a black box. A (base-) function is defined as "a *role* played by such behavior in a specific *context of use*" [6]. The context of use depends on intentions of users or designers, or the system that the component embedded in. This definition is based on the notion of "role concept" in [18]. Much research has been conducted on "role" in Ontology Engineering [e.g., 19]. The concept of function satisfies fundamental characteristics in [19] as discussed in our paper [6].

FOCUS/Tx defines generic types of the base-functions (called functional concepts). Figure 2 shows its portion[2]. A functional concept (a class of function) is defined ontologically using constraints on the cardinality of operands, relationships among them and/or designer's intention to change (focus of intention). For example, a function, "to divide an operand", is defined by the following semantic constraints; (1) the cardinality of the input focused operand must be 1, (2) the cardinality of the output focused operands must be greater than 1, (3) there must be *material-product* relationship between the input operand and the output operands and (4) all the output operands are equally focused. The first three are inherited from the super-concepts such as 'separate'. The fourth one is the criterion of categorization at this level and enables us to distinguish the 'divide' function from the sibling function 'take_out'.

---

[2] The initial version of the ontology was organized in four *is-a* hierarchies [4]. It has been restructured into single *is-a* hierarchy based on the common definitions in the hierarchies.

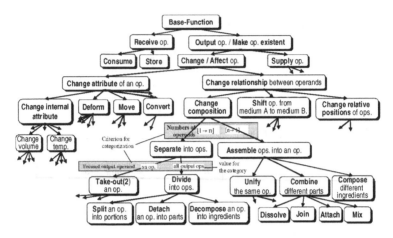

Fig. 2. A portion of FOCUS/Tx

We distinguish a function from a *way of function achievement* [5,6], which represents background knowledge such as physical principle in functional decomposition [1], in which part-functions achieve a whole-function. It enables us to distinguish "what to achieve"(function) from "how to achieve" (way of achievement).

FOCUS ontologies have been implemented in our role-centric ontology editor Hozo [18] (http://www.hozo.jp). Some portions of the implementation have been reported in [3,6]. Currently, we are rebuilding them and are implementing in OWL.

4   Mapping Process Based on a Reference Ontology

As discussed in Introduction and Section 2, the mappings are based on the reference ontology of function (FOCUS/Ref hereafter) [3], which defines *function categories,* that are, the upper types of functional terms defined functional taxonomies. By a reference ontology, we here mean that the ontology referred to for categorizing existing definitions of function and for defining the mappings between them (in comparison with "reference for system design" such as the ISO's OSI network reference model). Note that the set of the functional categories of FOCUS/Ref is *neither* a super-set *nor* a merged-set of those of functional taxonomies.

The upper-right part of Fig. 1 shows a portion of FOCUS/Ref. For example, an *effect function* implies changes of a target object (*operand*). It is categorized into a *device function*, an *environmental function*, and a *system-interface function*. These sub-categories imply changes of an operand *within* the system boundary, that *outside* of the boundary and that *on* the boundary, respectively. The *flowing-object function* as a sub-type of the *device function* represents input-output changes of an operand that flows through a device from the device-oriented viewpoint. The *function-with-way-of-achievement* category implies a specific *way of function achievement* (discussed above) as well as a function. Its examples include welding, washing, shearing and adhering. For example, the welding implies not only the *joining* function but also the *fusion way*. Because meaning of this type of function is impure, we regard this functional category as a subtype of the *quasi-function*. See [3] for the detail.

In the mapping process, the authors firstly analyzed the definitions of FB terms and gave them ontological definitions using Hozo. Then, the authors classified each functional term in the taxonomies into a *function category* of FOCUS/Ref. Because both FB and FOCUS/Tx adopt the device-centered viewpoint, all base-functions of FOCUS/Tx and many functional terms of FB are categorized into the *flowing-object function* category. The definition of *function* in FOCUS/Core also is based on the *flowing-object function*. Some functions of FB are, however, categorized into other categories of FOCUS/Ref. Then, according to such classification of functional terms, the mapping knowledge is described for each pair of two functional terms. If functions are categorized into the different categories, the mapping becomes complex for bridging the ontological gaps as discussed in the following section.

5 Mappings between Taxonomies

We have established two-way mappings (directed correspondence relations) between FB (52 terms) and FOCUS/Tx (89 terms) according to the mapping process discussed above. The statistics of the mappings is shown in Table 2. Figure 3 shows the types of the mappings. Table 2.1 shows statistics on the mappings from FB to FOCUS/Tx. If both functional terms are categorized into the same functional category of FOCUS/Ref, they are mapped onto each other directly. For example, 'couple' of FB and 'combine' of FOCUS/Tx are categorized onto the same *flowing-object function* category of FOCUS/Ref, and they are mapped onto each other (Table 2.1. (A)). In addition, we allow

Table 2. Statistics of the ontology mappings

Table 2.1. From FB to FOCUS/Tx

Mapping Type			Number of terms in FB	Ratio in FB terms
Mappings within the same category of function			31	60%
	1 to 1		17	33%
		(A) 1 to 1	15	29%
		(B) N (OR) to 1	2	4%
	(C) 1 to N (OR, selection)		8	15%
	(E₁) ⟳ is a func. type		6	12%
Mappings between different categories of function			21	40%
	(D) 1 to N (AND)		10	19%
	(E₂) ⟳ is a meta func.		5	9%
	(E₃) ⟳ is a way of function achievement.		4	8%
	(F) ⟳ is an operand.		2	4%

Table 2.2. From FOCUS/Tx to FB

Mapping Type			Number of terms in FOCUS/Tx	Ratio in FOCUS/Tx terms
Mappings with terms in the same grain-size			36	40%
Mappings within the same category of func.			32	36%
	1 to 1		23	26%
		(A) 1 to 1	15	17%
		(B) N (OR) to 1	8	9%
	(C) 1 to N (OR, selection)		4	4%
	(H₁) Mappings to a super-concept		5	6%
Mappings between different categories of function			4	4%
	(G) Partial mapping		4	4%
Mappings with terms in different grain-sizes			53	60%
	(H₂) Mappings to a super-concept		49	56%
	(I) Mappings to a sub-concept		4	4%

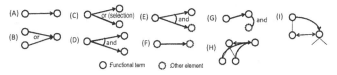

(A) ○ ⟶ ○   (C) ○ or (selection) ○/○   (E) ○ and ○/⟳   (G) ○ ○/⟳ and   (I) ○ ○/○⟶○

(B) ○/○ or ⟶ ○   (D) ○ and ○/○   (F) ○ ⟶ ⟳   (H) ⟳⟳⟳⟶○

○ Functional term   ⟳ Other element

Fig. 3. Mapping types

such mapping that several terms are mapped onto one term. For example, both 'extract' and 'remove' of FB are mapped onto 'take out' of FOCUS/Tx (Table 2.1. (B)). Next example is 'mix' in FB which is mapped onto 'unify' or 'compose' in FOCUS/Tx (Table 2.1. (C)). By "or" in (C), we here mean that the concrete corresponding term is selected according to the *context of use* (e.g., the whole system) in which the *mixing* function is used.

On the other hand, if two similar functional terms are classified into different categories of FOCUS/Ref, they are mapped in a complex manner. For example, 'guide' in FB is categorized into the *composite function* which consists of two primitive functional concepts. Thus, it is mapped onto 'supply motion' plus 'change direction of motion' functions (Table 2.1 (D)). The 'link' function of FB is categorized into the *function-with-way-of-achievement* category, because its definition implies "by means of an intermediary flow" [8] which represents *a way of achievement* as discussed in Introduction and Sections 3 and 4. Then 'link' of FB is mapped onto the 'combine' function of FOCUS/Tx plus the *intermediate-object way* for achieving the combining function (Table 2.1 (E_3)). The 'import' and 'export' of FB are categorized into the *system-interface* category of FOCUS/Ref. Because FOCUS/Tx is defined strictly based on the device-centered ontology, there is no corresponding functional concept in FOCUS/Tx. Thus, 'import' and 'export' of FB are mapped onto an operand from the outside of the system and an operand to the outside in a functional model of the FOCUS framework, respectively (Table 2.1 (F)).

Table 2.2 shows the statistics of the mappings from FOCUS/Tx to FB. Since the grain-sizes (granularity) of the functional concepts in FOCUS/Tx are finer than those of the FB terms, we took care of the difference of grain-sizes between two taxonomies in the mapping process. If the grain-size of each functional term is the same (the upper half of Table 2.2), the mappings have been established in the same manner of the mappings from FB to FOCUS/Tx[3]. If the grain sizes of functional terms are different (the lower part of Table 2.2), they are mapped to an upper-concept (H_2) or to a sub-concept (I). For example, the 'deform' (i.e., 'change shape') in FOCUS/Tx has subclasses such as 'change length' and 'change area', while 'shape' in FB has no subclass. In this case, 'deform' itself is mapped onto 'shape' in FB (Table 2.2 (A)), while those sub-concepts with finer granularity (e.g., 'change length') are mapped onto the coarser one; 'shape'. Table 2.2 (H_2) shows the numbers of such the mappings. The mapping type (I) indicates the reverse case of the type (H_2) for some highly abstracted concepts such as 'change an operand' in FOCUS/Tx. In both cases, those concepts are categorized into the same *flowing-object function* category.

We here compare the ratios of covering functional terms in the mappings. In the mappings from FB to FOCUS/Tx, the terms in FB cover (have mappings to) 33 terms in FOCUS/Tx out of the total of 89 (37%). In the mappings from FOCUS/Tx to FB, the terms in FOCUS/Tx cover 43 terms in FB out of the total of 52 (83%). Among the 9 terms of FB which are not covered, 6 terms are categorized into the *function-way-of-achievement* or the *system-interface* functions in FOCUS/Ref, both of which are not functional concepts from the device-oriented viewpoint, strictly speaking. The rest of 3 terms of FB are classified according to the quantitative difference (e.g., 'inhibit'

---

[3] In Tables 2.1 and 2.2, the numbers of the case (A) (the 1 to 1 mapping) are the same. The numbers of the case (B) are different, because they show the numbers of the source-side terms.

from 'prevent'), which is, we think, unnecessary classification of functional concepts. In other words, these 9 terms in FB are not target terms in the mapping from FOCUS/Tx to FB. Thus, we can say that FOCUS/Tx covers FB sufficiently.

Next, we discuss the ratios of successful mappings. There are, however, difficulties in accurate evaluation of their successfulness. Firstly, because the terms of FB are defined in natural language, it is difficult to calculate the similarity (or equality) between the meanings of the mapped terms. Secondly, we allow ambiguous mappings ('or'). Lastly, it is essentially difficult to determine criteria for evaluating differences between concepts in different categorizations. Considering these difficulties, in this article, we regard a mapping as successful if and only if that mapping is established only between the functional concepts with neither addition of extra information at either side nor heavy loss of information. For example, because the case (E_3) in Table 2.1 shows mappings to the way of function achievement (i.e., an element other than the functional concepts in FOCUS/Tx), those mappings in (E_3) are regarded as failure.

According to this criterion, in the mappings from FB to FOCUS/Tx, about 80% of the FB terms have been successfully mapped to the functional concepts of FOCUS/Tx (note that the covering ratio of those mappings is 37% as discussed above). On the other hand, in the mappings from FOCUS/Tx to FB, about 30% of the functional concepts of FOCUS/Tx have been successfully mapped to the terms in FB (the covering ratio is 83%). This low ratio is mainly due to the difference of granularity of the taxonomies, because we regard all the mappings between different grain-sizes (i.e., Table 2.2 (H_2) and (I)) as failure with the heavy information loss. The granularity of a taxonomy, however, heavily depends on an arbitrary decision made by its author and thus it is not essential to compare the core contents of different taxonomies. Thus, we can say that about 70% of the functional concepts of FOCUS/Tx successfully correspond to the terms in FB excluding the terms in the different grain-sizes. More accurate evaluation of the mappings remains as future work.

Even if we consider mappings only between terms in the same grain-size, their successful ratios in the two mapping directions (about 80% and 70%) are significantly different. One of its reasons is that FOCUS/Tx can represent the meanings of the terms of FB as combinations of the finely-categorized concepts such as the functional concepts and the meta-functions [4] as a system of the function-related concepts in the mappings from FB to FOCUS/Tx, while FB is single taxonomy of functional terms.

The mapping result discussed above can be regarded as very successful and interesting, considering the following backgrounds of the taxonomies. Firstly, they have been developed independently from each other. Secondly, the languages used for definition of the terms are different. FOCUS/Tx is designed firstly in Japanese, while FB is designed for (and defined by) English. Thirdly, the terms for describing functional knowledge have high diversity. The successful result has been gotten from the concentration *not* on lexical terms *but* on deep semantics of the functional concepts.

Consequently, the result strongly suggests the validity of the content of both FOCUS/Tx and FB from their commonality. The suggested validity is supported by their following applications as well. FOCUS/Tx has been deployed for modeling a real plant and knowledge management in manufacturing companies in Japan [6]. FB is widely used by researchers mainly in the United States. Furthermore the result suggests the adequacy of the mapping methodology in this paper.

6   Interoperable Semantic Search

6.1   Functional Metadata

The *Funnotation* schema [3] overviewed in Section 2 enables users to describe functional metadata with RDF which include (1) functions of the device/component (what is intended to achieve), (2) the used ways of function achievement (how to achieve a function), (3) a functional structure of the device (how to achieve the whole function), and (4) candidates (alternative) of ways of achievement. In the terminology in [20], (1) and (2) are "content descriptors" like keywords, while (3) and (4) are "logical structure" of "content representation" like a summary or an abstract.

Figure 4 shows an example of the metadata added to the document about a wire saw, which is a manufacturing machine to slice semiconductor ingots using moving wires. In Fig. 4, the wire-saw's function is described as an instance of the *splitting* function class (*Funnotation:split*) defined in FOCUS/Tx shown in Fig. 2. It is annotated to the term "cut" in the document. The wire-saw is annotated as the agent (performer) of the function using the *agent* property. The fact that the splitting function is achieved using frictional force is described using the *frictional_way* and the *selected_way* properties.

Much research has been conducted on automatically annotating web-documents with metadata elsewhere. Currently, we use two tools for functional annotation: one is to describe an instance model in Hozo and export it as a RDF file. The other is to use OntoMat-Annotizer with the schema in OWL exported from Hozo. Moreover, the authors and colleagues are currently investigating on automatic annotation of patent documents. It includes automatic identification of functional terms, semi-automatic mapping discovery from those terms to the functional concepts, and semi-automatic identification (mining) of functional structures as functional annotation.

6.2   Semantic Search System

In this section, firstly, we overview the basic usages and benefits of the *Funnotation* Semantic Search System [3]. Then, we will discuss interoperability with FB based on the mapping knowledge discussed thus far. This system consists of a user interface on

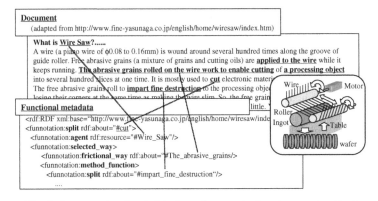

Fig. 4. An example of metadata for a document of a wire-saw (portion)

a web browser using JavaScript and a server module on a web server, which is implemented by Java and uses Tomcat with an HTTP server, Jena to operate the RDF repository, and SPARQL as a RDF query language.

As shown in Fig. 5, the users input the search condition such as a goal-function (i.e., function to be achieved). For example, let us consider a situation in which an engineer wants to know the possible ways to separate a semiconductor ingot. Firstly, the user selects '*separate*' from the functional terms defined in FOCUS/Tx as a goal-function. Giving such search conditions shown in Fig. 5, he/she gets the search results shown in Fig. 6. The center column indicated as "goal-function" shows the words in documents which are annotated both as the *separating* function class and as a subject of a *selected_way* (or *possible_way*) property. The rightmost column shows the terms annotated as a way.

This example shows that users can search for documents with a generic type of function independently of the lexical words in the documents (e.g., 'split' in the document (a) and 'cut off' in (b)). Using the *is-a* relations in FOCUS/Tx, the search result includes not only 'separate' but also its subclasses such as 'split'. Moreover, this search is based on semantic relations. If a 'separate' function is not a goal-function of a way in the metadata of a document, such documents are not retrieved.

Users can also search for possible *ways* for avoiding problems based on semantic relationships between functions. For example, in order to search for the ways to avoid problems caused by scrapings in a slicing machine, a user checks the "*supplementary*

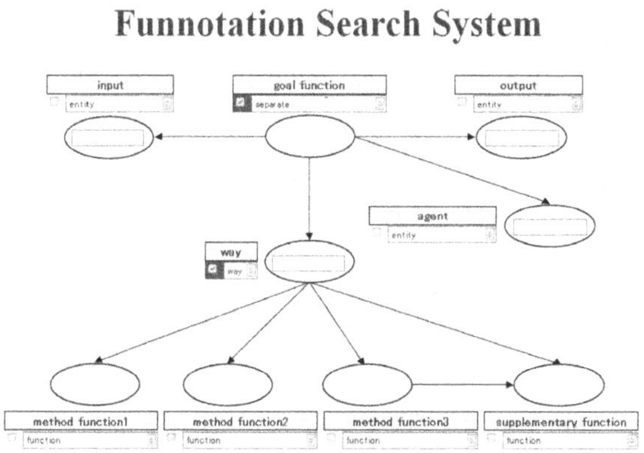

Funnotation Search System

Fig. 5. The interface of *Funnotation search system*

link	goal_function	way
(a) http://pc411:8083/search/c.vbs?t=A&n=2004-31639	split	rotatiing frictional way
http://pc411:8083/search/c.vbs?t=A&n=2003-318138	cutting	grind
http://www.fine-yasunaga.co.jp/english/home/wiresaw/indexc.htm	cut	frictional cutting way
http://pc411:8083/search/c.vbs?t=A&n=2004-221464	manufacture	manufacturing way
http://pc411:8083/search/c.vbs?t=A&n=2004-76053	make	electrolysis
(b) http://pc411:8083/search/c.vbs?t=A&n=H08-298250	cut off	A rotatiing friction way
	remove	Jet washing way
http://pc411:8083/search/c.vbs?t=A&n=2005-153005	comb out	rubbed up
	cut	cut

Fig. 6. Example of search result (1)

link	goal_function	way	supplementary function
(a) http://pc411:8083/search/c.vbs?t=A&n=2004-31639	split	rotatiing frictional way	harden
http://www.fine-yasunaga.co.jp/english/home/wiresaw/indexc.htm	cut	frictional cutting way	applied to the wire
http://pc411:8083/search/c.vbs?t=A&n=2004-221464	manufacture	manufacturing way	was aligned
http://pc411:8083/search/c.vbs?t=A&n=2004-76053	make	electrolysis	are along with
(b) http://pc411:8083/search/c.vbs?t=A&n=H08-298250	cut off	A rotatiing friction way	remove

Fig. 7. Example of search result (2) for supplementary functions

Link	Function	Taxonomy	Functional term
http://pc411:8083/search/c.vbs?t=A&n=2004-31639	split	FBRL	split
http://pc411:8083/search/c.vbs?t=A&n=2003-127058	manufacture	FBRL	split
http://pc411:8083/search/c.vbs?t=A&n=H11-58365	cut	FBRL	split
http://pc411:8083/search/c.vbs?t=A&n=2004-221464	split	FBRL	split
	making		split

Link	Function	Taxonomy	Function term
http://www.gti-usa.com/pages/semi_takatori_wiresaw_MWS_610SD.aspx	cuts	Functional basis	distribute
http://www.ctiattachments.com/overview_cutter.htm	cutting	Functional basis	distribute
			distribute
http://www.king-tool.com/drills.htm	cut	Functional basis	distribute

Fig. 8. Search result (3) for FOCUS/Tx (denoted as FBRL) and Functional Basis

function" box in Fig. 5 (by a supplementary function, we here mean a non-mandatory function that contributes to prevention of faults etc.) and sets the 'separate' function to cause the side-effect (e.g., scrapings). Figure 7 shows a result for this query which includes some possible solutions. The document (a) explains a way that hardens the target objects with ultraviolet rays before slicing to reduce the scrapings. In the document (b), to remove the scrapings by a fluid flow is a supplementary function.

The interoperability of the *Funnotation* framework with FB is enabled by the mapping knowledge discussed in Section 5. By translating the functional terms in the query and the metadata, the search system can access both documents that are annotated based on either of FB or FOCUS/Tx. Figure 8 shows a search result for a given goal-function 'split' of FOCUS/Tx. It includes not only documents annotated as 'split' of FOCUS/Tx but also those documents annotated as 'distribute' of FB which has a mapping to 'split' of FOCUS/Tx. In this manner, users can search for documents independently of the natural language of the documents and of the functional taxonomies used in the metadata. For example, an engineer can get ideas how to realize a function from both English documents annotated in FB from a US-based repository and Japanese documents annotated in FOCUS/Tx from a Japan-based repository. Even if he or she cannot read Japanese, the obtained metadata of the type 2, 3 or 4 (in Section 6.1) explain the possible way(s) for achieving the function.

7   Related Work and Discussion

As pointed out in [9], a "shared ontology" can facilitate semantic integration. The top-level generic ontologies such as DOLCE [21], SUMO [22] and PSL [23] can be used as the shared ontology [9]. Our FOCUS/Ref also can be regarded as a kind of such a shared ontology for matching concepts in ontologies, though a concept of the ontology is *not defined* as a subtype of a category of FOCUS/Ref in the ontology building process *but is classified* into a category in the mapping process. FOCUS/Ref is at the intermediate-level lower than those top-level ontologies. It is specific to the engineering domain, but it is applicable to wide-range of artifacts (see the discussion on limitation in [6]). It covers also several definitions of function that have been proposed in the literature [6]. We cannot claim its completeness in nature.

ONIONS methodology [24] is pioneering work to integrate terminologies based on formal and generic ontologies. It includes the "conceptual analysis" phase, in which the entities of a source terminology are represented in a formal way. Although our approach is *not* based on formal and generic (top-level) ontologies for integration, we described ontological descriptions of FB terms as a kind of the conceptual analysis and classified them into a category of FOCUS/Ref as an "intermediate ontology" [24].

A matching method based on an ontology that holds 'background knowledge' is proposed in [25]. A concept in source/target ontologies is connected to a concept in the background ontology ('anchoring matches' [25]) similar to the common ontology. Those anchoring matches are, however, produced automatically based on a simple lexical heuristic and the background knowledge (e.g., its semantic structure) is used to find semantic match between the source and target ontologies having few semantics.

Many methods for annotation-based semantic search have been proposed to date (e.g., [26]). Currently, our method simply shows the documents selected by the given query without ranking. More sophisticated search method remains as future work.

As mentioned in Introduction, there are many definitions of function (see [2,12,13]) and functional taxonomies [1,4,7,8]. Reconciled Functional Basis is a result of merging two existing taxonomies aiming at a 'standardized taxonomy' [8]. We aim at establishing mappings ('ontology matching' in the terminology of [10]) rather than merging ('ontology merging'), in order to allow the diversity of conceptualization of functions. Thus, FOCUS/Ref provides not a super-set (logical sum) of the existing taxonomies but generic upper categories of functions.

A functional modeling framework for the Semantic Web has been proposed in [27]. It is based on Functional Basis [8] and is used for reasoning tasks. It lacks an ontological foundation and interoperability with different taxonomies. For example, because it lacks the notion of "way of function achievement", the functional model in [27] is directly associated with components as a part of realization. Such direct association reduces flexibility in realization of functions.

The ontology-based integration and interoperability among engineering knowledge have been investigated from early 1990's such as PACT [28] and KIEF [29]. They mainly focus on generic mechanism for interoperability among engineering tools. The information integration of product data in the automobile industry is realized by ontology mapping [30]. Product data exchange based on a generic ontology has been proposed in [31]. The target knowledge in these papers is mainly the data level such as geometry rather than the conceptual knowledge level discussed in this paper. PhysSys ontology [32] is well-established ontology about physical objects. It, however, has no ontology for functions from the teleological viewpoint.

Our functional ontology is a domain knowledge and is different from "task" knowledge of designing or diagnosing, which has been discussed in the task ontology research (e.g., [33]).

8 Conclusion

In this paper, we have established two-way ontology mappings between two functional taxonomies by deep manual analysis based on the reference ontology of function for bridging the ontological gaps. Such ontological-level analysis has brought us the successful mappings, which suggest the validity of the taxonomies. Using the mapping knowledge, the semantic search system can provide users interoperable access to the technical documents by searching for functional metadata based on either of functional taxonomies.

In summary, the contributions of this paper includes (1) to show a successful application of ontological matching for interoperable annotation-based document management, (2) to demonstrate a successful case study of 'deep semantic mapping' based on an intermediate-level reference ontology rather than 'shallow matching', (3) to provide an interoperable engineering document management system and (4) to investigate ontological types of function by comparing the functional taxonomies.

Acknowledgements. The authors are most grateful to Robert B. Stone and his colleagues of Missouri University of Science and Technology for the discussion especially for clarifying some portions of the semantics of FB. The authors thank Masanori Ookubo and Naoya Washio for their contribution. Special thanks go to anonymous reviewers for their valuable comments.

References

1. Pahl, G., Beitz, W.: Engineering Design - a Systematic Approach. The design council, London (1988)
2. Stone, R.B., Chakrabarti, A.(eds.): Special Issues: Engineering applications of representations of function. AI EDAM 19(2/3) (2005)
3. Kitamura, Y., Washio, N., Koji, Y., Sasajima, M., Takafuji, S., Mizoguchi, R.: An Ontology-based Annotation Framework for Representing the Functionality of Engineering Devices. In: ASME International Design Engineering Technical Conferences (ASME IDETC 2006), DETC2006-99131. ASME (2006)
4. Kitamura, Y., Sano, T., Namba, K., Mizoguchi, R.: A Functional Concept Ontology and its Application to Automatic Identification of Functional Structures. Advanced Engineering Informatics 16(2), 145–163 (2002)
5. Kitamura, Y., Mizoguchi, R.: Ontology-based functional-knowledge modeling methodology and its deployment. In: Motta, E., Shadbolt, N.R., Stutt, A., Gibbins, N. (eds.) EKAW 2004. LNCS, vol. 3257, pp. 99–115. Springer, Heidelberg (2004)
6. Kitamura, Y., Koji, Y., Mizoguchi, R.: An Ontological Model of Device Function: Industrial Deployment and Lessons Learned. J. of Applied Ontology 1, 237–262 (2006)
7. Stone, R.B., Wood, K.L.: Development of a Functional Basis for Design. J. of Mechanical Design 122(4), 359–370 (2000)
8. Hirtz, J., Stone, R.B., McAdams, D.A., Szykman, S., Wood, K.L.: A Functional Basis for Engineering Design: Reconciling and Evolving Previous Efforts. Research in Engineering Design 13, 65–82 (2002)
9. Noy, N.: Semantic integration: A Survey of Ontology-based Approaches. ACM SIGMOD Record archive 33(4) (2004)
10. Euzenat, J., Shvaiko, P.: Ontology Matching. Springer, Heidelberg (2007)
11. Straccia, U., Troncy, R.: oMAP: Combining Classifiers for Aligning Automatically OWL Ontologies. In: Ngu, A.H.H., Kitsuregawa, M., Neuhold, E.J., Chung, J.-Y., Sheng, Q.Z. (eds.) WISE 2005. LNCS, vol. 3806, pp. 133–147. Springer, Heidelberg (2005)

12. Chandrasekaran, B., Josephson, J.R.: Function in Device Representation. Engineering with Computers 16(3/4), 162–177 (2000)
13. Perlman, M.: The Modern Philosophical Resurrection of Teleology. The Monist 87(1), 3–51 (2004)
14. Kitamura, Y., Takafuji, S., Mizoguchi, R.: Towards a Reference Ontology for Functional Knowledge Interoperability. In: ASME IDETC 2007, DETC2007-35373. ASME (2007)
15. Kitamura, Y., Washio, N., Ookubo, M., Koji, Y., Sasajima, M., Takafuji, S., Mizoguchi, R.: Towards a Reference Ontology of Functionality for Interoperable Annotation for Engineering Documents. In: Posters and Demos of ESWC 2006, pp. 75–76 (2006)
16. Ookubo, M., Koji, Y., Sasajima, M., Kitamura, Y., Mizoguchi, R.: Towards Interoperability between Functional Taxonomies using an Ontology-based Mapping. In: 16th International Conference on Engineering Design (ICED 2007) (2007)
17. Garbacz, P.: Towards a Standard Taxonomy of Artifact Functions. In: First Workshop FOMI 2005 - Formal Ontologies Meet Industry, CD-ROM (2005)
18. Mizoguchi, R., Sunagawa, E., Kozaki, K., Kitamura, Y.: The model of Roles within an Ontology Development Tool: Hozo. J. of Applied Ontology 2(2), 159–179 (2007)
19. Masolo, C., Vieu, L., Bottazzi, E., Catenacci, C., Ferrario, R., Gengami, A., Guarino, N.: Social Roles and Their Descriptions. In: 9th Int'l Conf. on the Principles of Knowledge Representation and Reasoning (KR 2004), pp. 267–277 (2004)
20. Euzenat, J.: Eight Questions about Semantic Web Annotations. IEEE Intelligent Systems, 55–62 (March/April 2002)
21. Masolo, C., Borgo, S., Gangemi, A., Guarino, N., Oltramari, A.: WonderWeb deliverable D18, ontology library (2003)
22. Suggested Upper Merged Ontology, http://ontology.teknowledge.com/
23. TC184/SC4/JWG8, I.S.O.: Process Specification Language (2003), http://www.tc184-sc4.org/SC4_Open/SC4_Work_Products_Documents/PSL_18629/
24. Gangemi, A., Pisanelli, D.M., Steve, G.: An overview of the ONIONS project: Applying Ontologies to the Integration of Medical Terminologies. Data and Knowledge Engineering 31 (1999)
25. Aleksovski, Z., Klein, M., ten Kate, W., van Harmelen, F.: Matching unstructured vocabularies using a background ontology. In: Staab, S., Svátek, V. (eds.) EKAW 2006. LNCS, vol. 4248, pp. 182–197. Springer, Heidelberg (2006)
26. Rocha, C., Schwabe, D., Poggi de Aragao, M.: A Hybrid Approach for Searching in the Semantic Web. In: WWW 2004, pp. 374–383 (2004)
27. Kopena, J.B., Regli, W.C.: Functional Modeling of Engineering Designs for the Semantic Web. IEEE Data Engineering Bulletin 26(4), 55–62 (2003)
28. Cutkosky, M.R., et al.: PACT: An experiment in integrating concurrent engineering systems. Computer, 28–37 (January 1993)
29. Yoshioka, M., Umeda, Y., Takeda, H., Shimomura, Y., Nomaguchi, Y., Tomiyama, T.: Physical Concept Ontology for the Knowledge Intensive Engineering Framework. Advanced Engineering Informatics 18(2), 95–113 (2004)
30. Maier, A., Schnurr, H.P., Sure, Y.: Ontology-based information integration in the automotive industry. In: Fensel, D., Sycara, K.P., Mylopoulos, J. (eds.) ISWC 2003. LNCS, vol. 2870, pp. 897–912. Springer, Heidelberg (2003)
31. Dartigues, C., Ghodous. P.: Product Data Exchange Using Ontologies. In: Artificial Intelligence in Design (AID 2002), pp. 617–637 (2002)
32. Borst, P., Akkermans, H., Top, J.: Engineering Ontologies. J of Human-Computer Studies 46(2/3), 365–406 (1997)
33. Schreiber, G., et al.: Knowledge Engineering and Management - The Common-KADS Methodology. The MIT Press, Cambridge (2000)

ROC: A Method for Proto-ontology Construction by Domain Experts

Nicole J.J.P. Koenderink[1], Mark van Assem[2], J. Lars Hulzebos[1],
Jeen Broekstra[1], and Jan L. Top[1,2]

[1] Agrotechnology & Food Innovations B.V., Wageningen UR
P.O. Box 17 6700 AA, Wageningen, The Netherlands
[2] Faculty of Sciences, Vrije Universiteit Amsterdam
The Netherlands

Abstract. Ontology construction is a labour-intensive and costly process. Even though many formal and semi-formal vocabularies are available, creating an ontology for a specific application is hindered in a number of ways. Firstly, the process of elicitating concepts is a time consuming and strenuous process. Secondly, it is difficult to keep focus. Thirdly, technical modelling constructs are hard to understand for the uninitiated. We propose ROC as a method to cope with these problems. ROC builds on well-known approaches for ontology construction. However, we reuse existing sources to generate a repository of proposed associations. ROC assists in efficiently putting forward all relevant concepts and relations by providing a large set of potential candidate associations. Secondly, rather than using intermediate representations of formal constructs we confront the domain expert with 'natural-language-like' statements generated from RDF-based triples. Moreover, we strictly separate the roles of problem owner, domain expert and knowledge engineer, each having his own responsibilities and skills. The domain expert and problem owner keep focus by monitoring a well-defined application purpose. We have implemented an initial set of tools to support ROC. This paper describes the ROC method and two application cases in which we evaluate the overall approach.

1 Introduction

Ontology construction is a laborious and expensive process. Even though many (semi-)formal vocabularies are available, creating one for a specific application context – in terms of domain and task – still requires considerable effort. Firstly, producing all relevant concepts and relations is time consuming and strenuous. Secondly, it is difficult to keep focus on the task for which the ontology is being developed. Thirdly, knowledge representation languages are hard to understand for those who are not trained as knowledge engineers.

We propose ROC (Rapid Ontology Construction) as a method to cope with these problems. ROC identifies three (idealized) roles in the ontology construction process, each role having its own responsibilities and skills: the problem

J. Domingue and C. Anutariya (Eds.): ASWC 2008, LNCS 5367, pp. 152–166, 2008.
© Springer-Verlag Berlin Heidelberg 2008

owner (PO), who explicates and monitors the purpose of the ontology; the domain expert (DE), who creates an informal version of the knowledge model – a so-called *proto-ontology*; and the knowledge engineer (KE), who is responsible for creating a formalized knowledge model from the proto-ontology. During the process, the roles interact frequently. A second specific feature of ROC is that it provides a repository of predefined associations to draw from and to stimulate the domain expert in making his knowledge explicit. The associations are stated in an informal way, taking away the necessity for the domain expert to understand knowledge engineering terminology. A third characteristic of the proposed method is that the application purpose of the model is stated in advance and monitored during the process by the problem owner. A unique characteristic is that the DE and PO interactively delineate the task and domain ranges.

The ROC approach is divided into three activities *identifying associations, creating a proto-ontology* and *constructing the final ontology.* In the first activity relevant sources are converted to triples, which are used in the second activity to construct the proto-ontology. A proto-ontology can be viewed as the starting material that is the basis for a full-fledged ontology. The activity to finalize the model in a formal representation is not considered here. The first activity is mainly performed by the KE, while the DE and PO perform the second activity. A tool allows the DE to select relevant triples from the repository. Another tool supports the PO in stating and maintaining the purpose of the ontology to be created. We provide the PO with a template to specify the purpose and monitor it during the construction activity.

This paper is organized as follows. In Section 2 we present methodologies and tools that we build our work on. Section 3 presents the activities ROC consists of. Then Section 4 presents two use cases – on geometry, resp. on supply chains – that we have performed to gain insight in the usefulness of the method, which is evaluated in Section 5. Section 6 provides a discussion and conclusions.

2   Related Work

Since the '90s, methodologies for ontology engineering have been developed, which cover all aspects of ontology development, management and support (see [1] for an overview). We are specifically interested in the specification, conceptualization and integration phases in which resp. the ontology's scope is specified, the ontology is edited, and existing sources are reused. Our methodology is roughly an adaptation and specialization of Methontology [2], where ROC emphasizes integration and reuse.[1]

In the past it was assumed that the knowledge engineer would gather information from the domain expert (knowledge acquisition), and would then use a formal language to represent it. One of the reasons for this workflow is that generally domain experts have no background in formal representation languages and find the associated tools (*e.g.* Protégé [3]) hard to use. One way to involve domain experts more in the knowledge creation process is to let them use a

---

[1] A full comparison is beyond the scope of this paper.

controlled natural language such as ACE, Rabbit or CLoNe [4,5,6]. Sentences in the controlled language (*e.g.* "River is a type of Waterway") are parsed and the appropriate OWL statements are generated. The DE has to be trained to use highly constrained sentence structures, because only then is it possible to unambiguously translate sentences to OWL constructs. Instead, in ROC we explicitly do not target a specific set of OWL constructs, but allow the DE as much freedom as possible to express knowledge. The only prescribed structure is that knowledge has to be entered in a subject-predicate-object format. There are no restrictions on the triple's content. In this our work is closer in spirit to traditional knowledge acquisition methods such as mentioned in [7] (*e.g.* laddering, concept sorting).

The integration phase is traditionally also handled by the KE. Semi-automatic support for reuse activities such as mapping can be supported by tools such as Falcon-AO [8]. A novel idea in ROC is to combine the conceptualization and (partial) reuse of existing sources by offering the DE possibly relevant concepts from a preprocessed repository. Thus we explicitly support the associative process of knowledge elicitation (for example, by helping the expert in remembering to include related concepts). It is also expected that association speeds up the acquisition process. Some work in this direction is also done by a team from KMi. It is developing a plugin for both the Protégé and NeOn toolkits which enables to search for related triples on the Web using the Watson semantic search engine[2]. The user can select relevant triples for inclusion in the current project. We provide similar functionality but instead base ourselves on an information repository focused on the domain at hand that contains triples gathered from RDF-sources but also semi-structured web documents.

3   Rapid Ontology Construction (ROC) Methodology

3.1   Overview

The ROC method aims to support the process of constructing a purpose-specific proto-ontology. Proto-ontologies solely consist of concepts and relations; formal term definitions, knowledge rules and logical constructs are not part of the proto-ontology. The notion of a proto-ontology has been introduced to provide the DE with a means to easily gather relevant knowledge, while keeping the process close to the domain expert's frame of reference. In other words, we do not require the DE to use 'good modelling practices', as for example defined in [9], since concepts like 'subclassOf', 'datatypeProperty', or 'inverseProperty' as used in the ontology language OWL are mostly meaningless to the domain expert and may even hinder the knowledge identification process. Instead, we stay close to natural language to better support the domain expert. The resulting proto-ontology is a useful intermediate product consisting of RDF-triples that the KE can work on; hence the KE has to spend significantly less time in the knowledge acquisition trajectory.

---

[2] http://watson.kmi.open.ac.uk/editor_plugins.html

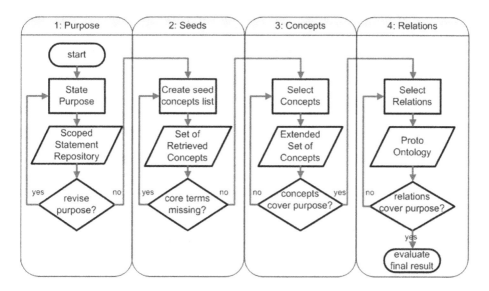

Fig. 1. Workflow of the proto-ontology creation process

By using the ROC method, the domain expert is supported in identifying relevant knowledge, for ROC incorporates a prompting process that offers the DE terms associated with the terms that have already been selected by the DE. This prompting process uses existing knowledge sources that have been processed into simple natural language statements in a 'subject – predicate – object' format. Such statements can easily be mapped onto an RDF/OWL expression, but are, at the same time, understandable to the domain expert.

3.2   Building the Information Repository

The ROC method requires the presence of semi-structured knowledge in an information repository. In most cases, web-based sources are not specified in a semantically structured format. In such cases, the two steps below are needed to obtain the knowledge and store it in a triple format. In some cases though, web-based content is already in triple format, e.g. in the form of SKOS thesauri. In those cases, the triples are directly stored in the information repository.

Step 1.1. Source identification. The first step is to identify sources that may contain relevant information for the proto-ontology to be developed. We identify the following types of sources: (i) *semi-structured sources*, like web pages in a structured layout (tables, pages with rigid section structures, etc.), and (ii) *existing ontologies and thesauri*, typically formalised in OWL or RDFS.

To effectively use semi-structured sources, an additional *triple extraction* step is required. Existing ontologies and thesauri are already formalized and structured as triples. Informal alternatives of the formal relations are required to ensure that triples can be presented to the DE in a format that is intuitive to him and that does not burden him with formal knowledge representation terminology. Hereto, the KE has to review the selected sources and provide such mappings. For example, the formal 'rdfs:subClassOf' relation could be mapped to the natural language expression 'is a'.

Step 1.2. Triple Extraction. In the second step we parse semi-structured sources and extract triples from them. Various techniques can be employed to do so. The possibility used in this paper is to create a custom parser and triple extractor for each identified semi-structured source. This 'tailor-made' approach yields relatively high-quality triples, but is labor-intensive for the KE, since it requires adaptation of the extraction tool for each new source. Alternative approaches using more generic and robust tooling are supported as well, for example in the form of more generic parsers and crawlers, but also the integration of text mining and named entity recognition software (*e.g.* Calais[3]). Note that our claim is not to have created a particularly novel triple extraction technique, but that when a source is harvested for triples, those triples can be efficiently used to support the proto-ontology creation process.

Whichever extraction method is used, the quality of the extracted triples can not be taken for granted and an editorial filtering step is necessary. The triples that are retrieved from the information sources are stored in an information repository, in our case realised using the Sesame RDF framework[4].

3.3   Creating the Proto-ontology

The proto-ontology creation process is divided into four main parts; stating (1) purpose, (2) seed terms, (3) concepts, and (4) relations, not unlike existing methods such as Methontology [2]. The progression through these tasks is presented in the workflow in Figure 1. However, the DE is free to switch back and forth between different steps as he sees fit. Below, these steps are presented in more detail and an example in the plant domain is provided.

Orthogonal to these four steps is the process of evaluation. While each step has a decision criterion to decide whether or not to progress to the next step, Step 4 finishes with an overall evaluation of the resulting proto-ontology. The DE and the PO evaluate the created proto-ontology with respect to the specified purpose. They decide whether the result is satisfactory. If this is not so, any step in the proto-ontology construction workflow can be revisited.

Step 2.1. Defining the purpose. Although ontologies are typically considered purpose-independent artefacts, in ROC we take the purpose-as defined a priori by a problem owner-into account as a crucial aspect during development, since it helps the DE to keep focus and to decide on issues such as coverage and depth of the proto-ontology. The scope of the proto-ontology is constrained in two dimensions: the application perspective, represented by the problem owner, and the domain perspective, represented by the domain expert (see Figure 2).

---

[3] http://www.opencalais.com/

[4] http://www.openrdf.org/

The PO is the main stakeholder for the application that is to be supported by the ontology. Hence, he takes the lead in defining the application's scope and purpose. During the proto-ontology construction process, the DE and PO interactively refine the purpose and adapt the proto-ontology to converge to a knowlege model that is well suited to support the envisioned application. A domain expert selects concepts from the domain; the problem owner may reject some of them and indicate that other concepts are to be explored further.

Aspects that help to set the scope of the proto-ontology are, for instance:

- **Description of the application:** a description in natural language of the application for which the proto-ontology is to be used.
- **Proto-ontology domain:** a list of available domain(s) from which the PO and DE can choose. Examples are: food, agriculture, mathematics, *etc.*
- **Type of end users:** the type of end users of the application influences the knowledge needed for the proto-ontology. In ROC we distinguish 'general public', 'professional', and 'expert'.
- **Level of detail:** the level of detail that is appropriate for the proto-ontology can be specified by offering examples of concepts that are or are not to be considered for inclusion.

For example, we imagine an application in the plant domain. The description of the application could be: *A computer vision application that uses expert knowledge to automatically determine the quality of recorded tomato plants.* The system will typically be used by *Experts.* The domains of relevance that are chosen by the problem owner may be *Agriculture* and *Plants.* During the process, the PO identifies concepts that are to be explored in more detail, such as *Cotyledon* and *True leaf,* and rejects concepts that are too detailed for inclusion such as *Petiolule* and *Stem hair.* The domain expert follows through by finding related concepts to *True leaf* such as *Terminal leaflet* and *Vein structure.* The problem owner and domain expert work together to find the concepts that best suit the purpose of the proto-ontology.

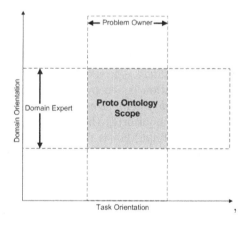

Fig. 2. Scoping of the proto-ontology

Step 2.2. Seeding the Proto-ontology. The DE is asked to compile a list of terms that are relevant to the proto-ontology domain, the so-called *seed concepts.* The PO has the possibility to identify concepts that are *not* to be included in the proto-ontology. This list of non-concepts serves to restrict the proto-ontology to its intended scope. The DE and PO can revisit the seeding step whenever they think of concepts that should or should not be included in the proto-ontology.

For the example in the plant domain, the domain expert may choose the concepts *Stem* and *Leaf* as initial seed concepts.

Step 2.3. Extending the Set of Concepts. The purpose of this task is to identify relevant terms from a pool of terms associated with the previously defined seed concepts. The method uses either the seed concepts or previously associated and approved concepts to *automatically look up associated terms in the information repository*. The identified terms are offered to the DE and the PO for inspection. If a new term is relevant for the domain and task of the proto-ontology, both the DE and PO accept the term; otherwise, the term is rejected. Accepted terms may be adapted to better reflect the domain knowledge.

If a term from the seed list is not found in the information repository, it is still added to the proto-ontology as a single concept. In Step 2.4, the DE can link this concept to the rest of the proto-ontology by defining appropriate relations.

For the example, the information repository may yield the terms *Stem hair*, *Plug* and *Lobe* as related terms for *Stem*. The domain expert may accept all of these terms; the problem owner may reject the term *Stem hair* as too detailed, accept the term *Plug* and change the *Lobe* into the preferred term *Cotyledon*.

Step 2.4. Adding Relations. In this task, the DE identifies relevant relations and labels them properly. Hereto, *relations between approved concepts are looked up in the information repository and offered for review* to the DE. The DE is asked to either approve or reject the relations. If a relation is approved, the DE can change the label of the relation. The DE can also add new relations to the proto-ontology.

With respect to the plant example, the system may present the expert with the relations *Stem – grows from – Plug* and *Stem – develops into – Cotyledon*. The first relation is accepted by the expert, but the relation *Stem – develops into – Cotyledon* is replaced by the relation *Cotyledon – is connected to – Stem*.

Note that monitoring the purpose of the proto-ontology is important for the process of defining labels for the relations. Depending on the type of task, the extent to which the labels have to be specified differs. In some cases, it suffices to know that a relation exists, but the type of relation is irrelevant. In other cases, the precise specification of the type of relation is required for a useful deployment of the proto-ontology.

4   Use Cases

4.1   Case Study 1: The Geometric Proto-ontology

As a first test case for the ROC-method, a proto-ontology containing geometric concepts has been created. The problem owner had indicated that such a knowledge model has to be created that could be used to support an imaginary 3D drawing program. In this case study, we used a preliminary implementation of the ROC method.

Building the information repository. To prepare the ROC repository, we have asked the expert to indicate semi-structured sources relevant for a future 3D drawing application. The expert mentioned the Geometrical Classroom on Mathworld[5]. Besides this source, we used the sources that were already present in the information repository. Below we give a short description of the used sources.

- The CABI thesaurus[6], consisting of terms related to applied life sciences.
- The NAL thesaurus[7], containing agricultural and biological terminology.
- OUM, the ontology of units and measures[8], containing units of measure, quantities, dimensions, and systems of units.
- The OpenCyc thesaurus, a generic knowledge base[9].
- Mathworld Geometrical Classroom, containing an overview of geometrical terms, their definitions and the categories to which they belong.

The Mathworld Geometrical Classroom is a semi-structured source for which we have created a tailor-made parser. We have used the structure of the Mathworld page to find triples like <term> is_part_of_category <category name>, <term> has_definition <definition> and <word> is_related_to <term>. Examples of identified triples are: *Triangle – is part of category – Polygon*, *Triangle – has definition – A three-sided (and three-angled) polygon*, and *Hypothenuse – is related to – Triangle*. The process of creating the parser and harvesting the triples took the KE approximately 0.5 days. The other sources were harvested in a similar way. The retrieved triples have been added to the information repository.

Creating the proto-ontology. For this early test of the ROC method, both the KE and the DE were present at the proto-ontology creation process. The KE operated the ROC system, and the DE provided the input. In this first implementation of the ROC method, the purpose was only defined in terms of a global description of the application and of the domain of interest.

The DE started the seeding process with two concepts in the seed list: *cubes* and *cylinders*. In the 'concept step' these terms were looked up in the repository. In this first implementation of ROC, we did not distinguish between the 'concept step' and the 'relation step'. Therefore, the DE was asked to assess the statements and to adjust relations and concepts when necessary. At certain points in time, the created intermediate proto-ontology was visualised; the KE manually mapped the concepts in the IR to classes and the relations to properties and used the TGViz plug-in of Protégé to show the intermediate proto-ontology to the DE. The DE has used thirteen iterations to reach a satisfactory proto-ontology. For this process, approximately 20 hours have been used by the DE and 25 by the KE. The resulting proto-ontology contains 453 triples.

---

[5] http://mathworld.wolfram.com/classroom/classes/Geometry.html
[6] http://www.cabthesaurus.info
[7] http://agclass.nal.usda.gov.agt/agt.shtml
[8] http://www.afsg.nl/foodinformatics/index.asp
[9] http://www.opencyc.org

4.2   Case Study 2: The Supply Chain Proto-ontology

For a university, it is important that the expertises of its employees are known to properly answer questions from *e.g.* journalists. We have developed a prototype system for Wageningen UR in which a search term can be entered to find the corresponding expert.

For the prototype of the expert finder system, we have performed a pilot study that focussed on the areas of expertise of 'agrifood supply chains'. Hereto, we have invited a DE to participate in a ROC session. In these sessions, two KEs were present: one to guide the DE through the ROC process, the other to operate the preliminary ROC tools.

Building the information repository. To prepare the information repository, the DE was asked to identify relevant Web-based sources on 'supply chains'. The expert indicated that 'chain logistics' and 'supply chain management' are more appropriate terms. Below we give a short description of the identified sources.

- The CABI and NAL thesaurus as in the first case study.
- The MeSH vocabulary[10]: a controlled vocabulary in the area of life sciences.
- The UMLS vocabulary[11]: controlled vocabularies in the biomedical sciences.
- The AGROVOC thesaurus[12]: covering concepts in the agrifood domain.
- Agrologistics list[13]: a structured list with terms in agrologistics.
- Sustainability list[14]: a structured list with terms in sustainability.
- Expertise list: a structured list with the expertises of Wageningen UR.
- Wikipedia supply chain management[15]: containing information on supply chain management.

A tailor-made parser for each of these sources was used to harvest triples. This process resulted in triples of the form <term> is_subcategory_of <chain logistics term>, <term> is_subcategory_of <supply chain management term> , and <term> is_related_to <term>. Examples are *Cost-benefit analysis – is related to – costs, Chain integration – is a subcategory of – organisation*, and *Food safety – is a subcategory of – chain transparency.*

The process of identifying appropriate additional knowledge sources and writing parsers for these sources, took the domain expert 0.75 days. The harvested statements were added to the information repository.

Creating the proto-ontology. The purpose of the proto-ontology was defined as *being useful for expert identification within Wageningen UR*. The proto-ontology domain was *supply chains* and *food domain*, the expert type of end users was defined as *general public*.

---

[10] http://www.nlm.nih.gov/mesh/filelist.html

[11] http://www.nlm.nih.gov/research/umls/documentation.html

[12] http://www.fao.org/aims/faq_aos#30.htm

[13] Internal reports 'kenniskaart agrologistiek en visie agrologistiek' (in Dutch), and 'platform agrologistiek' (in Dutch).

[14] Internal report 'Vitaal en samen' (in Dutch).

[15] http://en.wikipedia.org/wiki/Supply_chain_management

The DE started with a seed concepts list of 49 terms specified in Dutch and translated into English. The translation to English terms was needed since the used sources were partly in English. The remainder of the first session was used for the 'concept step'. All terms that were automatically looked up in the information repository were presented to the DE in separate sets; each set centered around a seed concept. The advantage of this set-based way of presentation is that the list of retrieved terms is presented to the DE in manageable chunks instead of in an overwhelmingly large list. The DE checked for each set all terms and indicated whether they had to be included or not in the proto-ontology.

The proto-ontology construction step was concluded after a second iteration of the 'concept step'. Since the purpose did not require any further specification of the relations – a simple 'has-relation-with' label sufficed – the 'relation step' was not entered. In total the process of creating the proto-ontology took 5 hours for the domain expert and a little less than 16 hours for the knowledge engineers. The resulting proto-ontology contains 248 triples.

4.3 Experiences and Lessons Learned

One of the goals of the ROC method is to support the domain expert in optimally performing his task, while staying in his frame of reference. The domain experts indicated that the use of 'term – relation – term' statements was clear to them. Both experts were well capable of performing the knowledge specification activity within the predefined knowledge format. They indicated that the separation of roles of DE and KE was satisfactory, since it made sure that the DE was responsible for the knowledge specification.

With respect to the first case study, we noted that the combination of the concept step and the relation step was cumbersome for the domain expert. She had to look at the visualisation of the intermediate proto-ontologies to recall what earlier decisions had already been made. In the second implementation of ROC, the separation of these steps was embedded. The domain expert in the second case study showed signs of irritation when already discarded concepts, showed up in relation to other concepts in other sets. As a result, we have added to the ROC method a filtering module to remove such concepts from all sets.

5 Evaluation

We evaluate ROC with respect to its original objectives and the chosen solutions. Firstly, separated roles ensure that the DE and PO are not confronted with technical KE constructs. The quality of the ontology construction process can be assessed by looking at the time spent at ontology construction by DE, PO and KE and by inspecting the structure of the resulting proto-ontologies. This evaluation is presented in the next two subsections. Secondly, the developed proto-ontology should focus on the task for which it is intended. The degree to which an ontology is task-oriented can be estimated by assessing the performance of an application using this ontology. This evaluation can be found in the

Table 1. In this table, an overview is given of the properties of three manually developed proto-ontologies and the two proto-ontologies covered in the case studies

proto-ontology	# interviews	$\sum$ time DE	$\sum$ time KE	total time	# concepts
Plants	13 (13 DEs)	58 hrs	88.5 hrs	146.5 hrs	37
Food Components	4 (4 DEs)	10 hrs	30 hrs	40 hrs	120
Potato	8 (1 DE)	8 hrs	32 hrs	40 hrs	279
Geometry	12 (1 DE)	20 hrs	25 hrs	45 hrs	208
Chains	2 (1 DE)	5 hrs	16.5 hrs	21 hrs	236

system-based evaluation subsection. Thirdly, existing knowledge supports the association process of the DE through the repository of predefined associations. The usefulness of the information repository can be measured by identifying how often the automatically proposed statements are adopted by the domain expert. The subsection on the use of the repository provides some clues. Finally, we note that the full evaluation of ROC with respect to other methods awaits the development of the final phase of the method on ontology construction.

5.1   Cost-Benefit Analysis

To obtain an idea of the costs and benefits of the ROC method, we compared the method to interview-based proto-ontology creation. Within our group, we have developed for example the Plant Ontology [10], the Healthy Food Components Ontology[16], and the Potato Ontology [11]. In Table 1, an overview of the creation processes is given. The problem owner was not explicitly included in the reference processes. We see that both the KE and the DE are involved in the knowledge acquisition process, be it in the interview-based or in the ROC-based method. The potato ontology and the supply chain ontology (see Section 4) are most comparable, since in both cases only one DE was interviewed and the proto-ontologies are of comparable size. The geometric ontology was hampered by the immature character of the initial ROC tools.

When we compare the supply chain ontology and the potato ontology, we see that for the supply chain ontology, the KE needs three times the time of the DE, whereas in the potato proto-ontology this ratio is 1 to 4. Although these results are far from statistically conclusive, they suggest that the ROC method could reduce the amount of time required by the KE to develop the proto-ontology. It is indeed one of the design criteria of ROC to not require the KE to study the domain of the proto-ontology. The numbers for the DE are more difficult to interpret. In the potato and chain proto-ontologies, the DE is involved in fewer sessions of the ROC method than for traditional interview-based knowledge acquisition methods. Whether this is a generic trait of the ROC-method is still to be seen. The total time used to develop both ontologies differs a factor of two in favour of the ROC method.

---

[16] The application based on this ontology can be found at www.afsg.nl/icgv

Table 2. Network analysis measures on the two cases. The top 3 scoring concepts are shown.

Case	Degree Centr.	Betweenness Centr.	PageRank
Geo	2d concept (0.13) 3d concept (0.11) polygon (0.07)	polygon (791) triangle (187) square (129)	3d-geometry (59) 2d-geometry (10) 2d-concept (7)
Chain	supply chain man'ment (0.20) transport (0.09) models (0.02)	transport (371) models (19.5) risk (11)	agro-industrial-chains (21) food-chains (21) transport (7)

5.2 Structure Analysis of the Resulting Proto-ontologies

Following [12], we applied some network measures to gain insight in structural features of the developed proto-ontologies: (1) *Degree Centrality* (which nodes have a relative high number of in/out links) (2) *Betweenness Centrality* (which nodes are central to the network), and (3) *PageRank* (which nodes are important). The results in Table 2 show the top 3 scoring concepts. The average degree is lower for the chain case (2.21) than for the geometric case (3.75), which is not surprising since the purpose of the geometric ontology implies that it has to provide many interlinked concepts to better find appropriate experts. The top concepts for degree and betweenness centrality are in both cases considered as 'important' concepts by the DEs. The top 3 for PageRank are root concepts, which makes sense given that most relations in the network point 'upward'. Distribution of all three measures show a steep curve towards the high end of the value range. This skewed distribution indicates an unbalanced network. We are considering the possibility to provide the DE with these measures to give an indication of which parts of the network might need more attention.

We did an additional qualitative analysis of the hierarchical structure of the geometrical proto-ontology (only for this case, since the supply chain case has no hierarchy as this was not needed for its purpose). We identified the hierarchical relations and analyzed the structure. The geometric case has a maximum depth of 5 nodes, a mean number of subconcepts of 2.9 and a median of 1. There are three cases of overly many subconcepts (14, 20, 25).

The first objective of the ROC method is to enable the DE to gather relevant concepts in his domain. Part of our future research is to guide the DE in adding not too few or too many concepts in a specific area of the domain. Hereto, we will enrich ROC with a 'proto-ontology dashboard' that indicates to the DE which parts of the proto-ontology require attention.

5.3 System-Based Analysis

One of the evaluation measures of a proto-ontology is to see how well it supports an envisaged application. For the geometrical case study no specific application was created. For the second case study, we developed an expert recommender system that can be used to identify experts in the field of supply chains.

For this system, we assumed that experts publish on subjects that are within their area of expertise. These publications can be used to create individual 'fingerprints'[17], containing the characteristic terms of his expertise. To link a search term to a fingerprint, we need ontologies for a number of expert domains. If, for example, a journalist needs information about avian influenza, the appropriate expert will probably be known as expert on bird diseases. When only text based search is used, the journalist will not find the desired expert. When a (lightweight) ontology is used, the link between avian influenza and bird disease is made, and the expert can be contacted.

To find an expert, a user enters a free text string indicating the topic for which an expert is required. This term is matched with the terms in the proto-ontology. If a matching term is found, its *related terms in the proto-ontology* are identified. The application uses these terms scans the publications' fingerprints for the original search term and its related terms. The authors from the publications are identified and ranked. The user can see which related terms have been found, and which experts best match the original query.

To evaluate the supply chain proto-ontology, we randomly selected ten terms from the proto-ontology for which we queried the system for experts. Next, we checked with the involved DE whether the Top 3 of scientists returned are indeed experts in the indicated areas. For the terms *food supply*, *supply chains*, *supply chain management* and *food production*, the identified experts were the persons expected by the DE (*i.e.* high precision). For the terms *logistics* and *food safety* the DE did not know all identified experts. After looking into the experts' affiliations, though, the DE concluded that it was reasonable to assume that the unknown persons were indeed experts in the indicated fields. The terms *trade barrier* and *chain governance* were on the border of the expertise of the DE; he could not give an indication of the correctness of the selection. The terms *quality* and *networks* are important for the food supply field of expertise, but also have a meaning in other areas. The found experts indeed related to these terms, but were not specifically linked to the area of supply chain quality or networks. Overall, the expert finder tool seems to have indicated the expected experts or related persons in the expected departments. This suggests that the proto-ontology supporting the expert finder tool fulfills its expectations sufficiently well.

5.4   Use of the Information Repository

In this section we look at the amount of information from the repository that is actually reused in the case studies. We see in Table 3 that in the geometrical case, the DE has mainly taken concepts and associations from Mathworld, and has added many new statements (*e.g.* equations, parameters, etc.). The skewed ratio between reused and new concepts was caused by (1) spontaneous associations at the presented concepts – a desirable effect – and (2) a limited amount of available dedicated statements in the information repository. This shows that

---

[17] http://www.collexis.com

Table 3. Contribution of sources to total triple size of proto-ontology

(a) Case 1: Geo

Source	# Triples
CABI	2
Mathworld	164
OUM	8
Expert	279

(b) Case 2: Chains

Source	# Triples	Source	# Triples
CABI	81	NALT	28
AGROVOC	61	Intranet agrologistics	2
Wikipedia	42	Intranet sustainability	3
MeSH	30	UMLS	1

identifying sufficient sources is important to profit optimally from ROC. For the supply chain case more sources were available. The ratio between triples from these sources is more balanced and no new concepts were added by the DE.

6 Conclusion

We have observed that the ROC method may accelerate proto-ontology construction by supporting different players in the process. First, the problem owner is assisted in defining the application context. Second, the domain expert specifies a proto-ontology without being hindered by technical modelling details. Third, the time spent by the knowledge engineer to get to know the domain is minimized. With ROC, association rate, focus and readability during ontology development is enhanced. Existing knowledge sources are used from the start of the construction process and the purpose of the proto-ontology is continuously monitored. This purpose determines the scope, level of detail and level of expertise for the proto-ontology to be developed. Furthermore, we use natural language statements generated from a triple format as intermediate representation.

Even though ROC and its tools are still under development, we have already used them successfully. In the two case studies we have constructed proto-ontologies in relatively short time. The evaluation of the results shows that combining multiple sources works well, as they all appear in the resulting proto-ontology. Not many additional triples need to be added by the DEs when sufficient reusable sources were available. We also analyzed the costs and benefits of developing ontologies with and without ROC for five cases. It shows that the time needed by the KE and to a lesser extent the DE is reduced. Measuring the quality of the proto-ontology, though, remains difficult and is ultimately expressed by the effectiveness by applying the model in some context.

An aspect of ROC that needs further attention is ensuring that domain experts stay motivated during the process. This can be achieved by certifying that the domain expert is committed to the intended application and by minimizing the amount of manual editing. Another issue we will attend is to include the selection of appropriate domain experts as a step in the ROC method, not unlike choosing appropriate text sources. A third extension that we presently investigate is the inclusion of existing triple extraction tools in the ROC toolkit. The fourth proposed addition to the ROC method is a proto-ontology structure dashboard, to guide the PO and DE in creating balanced, high-quality proto-ontologies. Lastly, the steps for refinement of a proto-ontology towards *e.g.* a full-blown OWL model – if needed – requires additional work.

Acknowledgements

This work was carried out within the VL-e project, subprogram Food Informatics. This project is supported by a BSIK grant from the Dutch Ministry of Education, Culture and Science and is part of the ICT innovation program of the Ministry of Economic Affairs. Additional funding was obtained from the Ministry of Agriculture. We would like to thank the DEs S. Tromp and M. Vollebregt. We are also grateful to L. Gazendam for providing the PageRank calculations.

References

1. Gómez-Pérez, A., Fernández-López, M., Corcho, O.: Ontological Engineering. In: Advanced Information and Knowledge Processing. Springer, Heidelberg (2003)
2. Fernández-López, M., Gómez-Pérez, A., Juristo, N.: Methontology: From ontological art towards ontological engineering. In: AAAI Ontological Engineering: Papers from the 1997 Spring Symposium. AAAI Press, Menlo Park (1997)
3. Knublauch, H., Fergerson, R.W., Noy, N.F., Musen, M.A.: The protégé owl plugin: An open development environment for semantic web applications. In: 3rd International Semantic Web Conference (2004)
4. Kaljurand, K., Fuchs, N.E.: Bidirectional mapping between OWL DL and attempto controlled english. In: Alferes, J.J., Bailey, J., May, W., Schwertel, U. (eds.) PPSWR 2006. LNCS, vol. 4187, pp. 179–189. Springer, Heidelberg (2006)
5. Hart, G., Johnson, M., Dolbear, C.: Rabbit: Developing a control natural language for authoring ontologies. In: 5th European Semantic Web Conference (2008)
6. Funk, A., Tablan, V., Bontcheva, K., Cunningham, H., Davis, B., Handschuh, S.: CLOnE: Controlled Language for Ontology Editing. In: Aberer, K., Choi, K.-S., Noy, N., Allemang, D., Lee, K.-I., Nixon, L., Golbeck, J., Mika, P., Maynard, D., Mizoguchi, R., Schreiber, G., Cudré-Mauroux, P. (eds.) ASWC 2007 and ISWC 2007. LNCS, vol. 4825, pp. 142–155. Springer, Heidelberg (2007)
7. Schreiber, G., Akkermans, H., Anjewierden, A., De Hoog, R., Shadbolt, N., Van de Velde, W., Wielinga, B.: Knowledge Engineering and Management – The CommonKADS Methodology. MIT Press, Cambridge (2000)
8. Hu, W., Qu, Y.: Discovering simple mappings between relational database schemas and ontologies. In: International Semantic Web Conference, pp. 225–238 (2007)
9. Noy, N., McGuinness, D.L.: Ontology development 101: A guide to creating your first ontology (2001)
10. Koenderink, N., Top, J., van Vliet, L.: Expert-based ontology construction: a case study in horticulture. In: TAKMA workshop at the DEXA conference (2005)
11. Haverkort, A., Top, J.L., Verdenius, F.: Organizing data in arable farming: towards an ontology of processing potato. Potato Research 49, 177–201 (2006)
12. Hoser, B., Hotho, A., Jäschke, R., Schmitz, C., Stumme, G.: Semantic network analysis of ontologies. In: Sure, Y., Domingue, J. (eds.) ESWC 2006. LNCS, vol. 4011, pp. 514–529. Springer, Heidelberg (2006)

A Pattern Based Approach for Re-engineering Non-Ontological Resources into Ontologies

Andrés García-Silva, Asunción Gómez-Pérez, Mari Carmen Suárez-Figueroa, and Boris Villazón-Terrazas

Ontology Engineering Group, Departamento de Inteligencia Artificial, Facultad de Informática, Universidad Politécnica de Madrid, Spain
hagarcia@delicias.dia.fi.upm.es,{asun,mcsuarez,bvillazon}@fi.upm.es

Abstract. With the goal of speeding up the ontology development process, ontology engineers are starting to reuse as much as possible available ontologies and non-ontological resources such as classification schemes, thesauri, lexicons and folksonomies, that already have some degree of consensus. The reuse of such non-ontological resources necessarily involves their re-engineering into ontologies. Non-ontological resources are highly heterogeneous in their data model and contents: they encode different types of knowledge, and they can be modeled and implemented in different ways. In this paper we present (1) a typology for non-ontological resources, (2) a pattern based approach for re-engineering non-ontological resources into ontologies, and (3) a use case of the proposed approach.

Keywords: Patterns for Re-engineering, Ontologies, Non-Ontological Resources.

1 Introduction

Research on Ontology Engineering methodologies has provided methods and techniques for developing ontologies from scratch. Well-recognized methodological approaches such as METHONTOLOGY [6], On-To-Knowledge [21], and DILIGENT [17] provide guidelines to help researchers in the development of ontologies. However, they have one important limitation: the lack of guidelines for building ontologies by reusing and re-engineering existing knowledge-aware resources widely used in a particular domain.

There are some initial works related to the re-engineering of non-ontological resources (NORs). Examples of projects that perform re-engineering are: (1) the NeOn Project[1], in which Fisheries Ontologies were developed for their use within the Fish Stock Depletion Assessment System (FSDAS) [4], by reusing resources available for the fisheries domain; and (2) the SEEMP[2] project in which a Reference Ontology has been built by reusing human resources management

---

[1] http://www.neon-project.org
[2] http://www.seemp.org

J. Domingue and C. Anutariya (Eds.): ASWC 2008, LNCS 5367, pp. 167–181, 2008.
© Springer-Verlag Berlin Heidelberg 2008

standards. However, none of these projects propose any guidelines about how to carry out that re-engineering process of NORs.

Within the context of the NeOn project, we are proposing a novel scenario-based methodology for builing ontology networks[3]. One of the scenarios in the NeOn methodology is *Building Ontology Networks by Reusing and Re-engineering Non-Ontological Resources*. For such scenario we propose methodological guidelines for reusing and re-engineering NORs. In this paper we present our approach for re-engineering NORs, which refers to the process of taking an existing non-ontological resource and transforming it into an ontology. The rest of the paper is organized as follows: Section 2 depicts the proposed typology of NORs. Section 3 presents the state of the art on re-engineering NORs. Section 4 presents our approach for re-engineering NORs. Section 5 presents a particular use case of our approach. Finally, section 6 concludes the paper and proposes future lines of work.

2   Types of Non-Ontological Resources

Non-Ontological Resources are existing knowledge-aware resources whose semantics have not been formalized yet by means of an ontology.

There is a big amount of NORs that embody knowledge about some particular domains, and that represent some degree of consensus for a user comunity. These resources present the form of free texts, textual corpora, web pages, standards, catalogues, web directories, classifications, thesauri, lexicons and folksonomies, among others. NORs have related semantics which allow to interpret the knowledge they contain. Regardless of whether the semantic is explicit or not, the main problem is that the semantics of NORs are not always formalized, and this lack of formalization avoids the use of them as ontologies.

The analysis of the literature has revealed that there are different ways of categorizing NORs [14,20,7,13]. Maedche et al. [14] and Sabou et al. [20] classify NORs into unstructured (e.g. free text), semi-structured (e.g. folksonomies) and structured (e.g. databases) resources. Gangemi et al. [7] distinguish catalogues of normalized terms, glossed catalogues, and taxonomies. Hodge [13] proposes characteristics such as structure, complexity, relationships among terms, and historical functions for classifying them. However, an accepted typology of NORs does not exist yet. Additionally, the existing NOR categorizations do not take into account the NOR data model, an important artifact the re-engineering process.

In this paper we propose a new categorization of NORs according to three different features: (1) the type of NOR, which refers to the type of knowledge encoded by the resource; (2) the data model, that is, the design data model used to represent the knowledge encoded by the resource; and (3) the resource implementation. Below we explain in more detail the proposed classification.

---

[3] An ontology network or a network of ontologies is a collection of ontologies together through a variety of different relationships such as mapping, modularization, version, and dependency relationships [10].

1. According to the **type of NOR** we classify them into:
   - *Glossaries*: A glossary is a terminological dictionary that contains designations and definitions from one or more specific subject fields. The vocabulary may be monolingual, bilingual or multilingual. As an example we mention the FAO Fisheries Glossary[4].
   - *Lexicons*: In a restricted sense, a computational lexicon is considered as a list of words or lexemes hierarchically organized and normally accompanied by meaning and linguistic behaviour information. An example is WordNet[5], the best known computational lexicon of English.
   - *Classification schemes*: A classification scheme is the descriptive information for an arrangement or division of objects into groups based on characteristics the objects have in common. For example, the Fishery International Standard Statistical Classification of Aquatic Animals and Plants (ISSCAAP)[6].
   - *Thesauri*: Thesauri are controlled vocabularies of terms in a particular domain with hierarchical, associative and equivalence relations between terms. Thesauri are mainly used for indexing and retrieval of articles in large databases. As an example we can mention the AGROVOC[7] thesaurus.
   - *Folksonomies*: A folksonomy is the result of personal free tagging of information and objects (anything with a URI) for one's own retrieval. An example of the use of folksonomies is the *del.icio.us*[8] website.
2. There are different ways for representing the knowledge encoded by the resource. In the following we present several **data models** for classification schemes, which are shown in Fig. 1.
   - *Path Enumeration* [2]: A path enumeration model is a recursive structure for hierarchy representations defined as a model which stores for each node the path (as a string) from the root to the node. This string is the concatenation of the nodes code in the path from the root to the node. Fig. 1-a) shows this model.
   - *Adjacency List* [2]: An adjacency list model is a recursive structure for hierarchy representations comprising a list of nodes with a linking column to their parent nodes. Fig. 1-b) shows this model.
   - *Snowflake* [15]: An snowflake model is a normalized structure for hierarchy representations. For each hierarchy level a table is created. In this model each hierarchy node has a linked column to its parent node. Fig. 1-c) shows this model.
   - *Flattened* [15]: A flattened model is a denormalized structure for hierarchy representations. The hierarchy is represented using one table where each hierarchy level is stored on a different column. Fig. 1-d) shows this model.

---

[4] http://www.fao.org/fi/glossary/default.asp
[5] http://wordnet.princeton.edu/
[6] http://www.fao.org/figis/servlet/RefServlet
[7] http://www.fao.org/agrovoc/
[8] http://del.icio.us/

Path Enumeration	Category Name	Category Description
1	Category1	Category1Desc
11	Category11	Category11Desc
111	Category111	Category111Desc
12	Category12	Category12Desc
121	Category121	Category121Desc
2	Category2	Category2Desc
...	...	...

Category Code	Category Name	Parent Category Code
1	Category1	Null
2	Category2	Null
3	Category3	1
4	Category4	1
5	Category6	3
6	Category7	4
...	...	...

a) Path Enumeration   b) Adjacency List

First level categories entity

Category Code	Category Name	Category Description
1	Category1Level1	Category1Level1Desc
2	Category2Level1	Category2Level1Desc
..		

Second level categories entity

Category Code	First Level Category	Category Name	Category Description
1	1	Category1Level2	Category1Level2Desc
2	1	Category2Level2	Category2Level2Desc
..			

Third level categories entity

Category Code	Second Level Category	Category Name	Category Description
1	1	Category1Level3	Category1Level3Desc
2	2	Category2Level3	Category2Level3Desc
..			

c) Snowflake

Flattened entity

First level		Second level		Third level		...
Category Code	Category Name	Category Code	Category Name	Category Code	Category Name	...
1	Category1Level1	1	Category1Level2	1	Category1Level3	...
1	Category1Level1	2	Category2Level2	2	Category2Level3	...
2	Category2Level1	...	...	...	...	...
..	..	..	...	...	...	...

d) Flattened

Fig. 1. Classification Schemes Data Models

3. According to the **implementation** we classify NORs into:
   - *Databases*: A collection of logically related data stored together in one or more files.
   - *XML file*: eXtensible Markup Language is a simple, open, and flexible format used to exchange a wide variety of data on and off the Web. XML is a tree structure of nodes and nested nodes of information, in which the user defines the names of the nodes.
   - *Flat file*: A flat file is a file that is usually read or written sequentially. In general, a flat file is a file containing records that have no structured inter-relationships.
   - *Spreadsheets*: An electronic spreadsheet consists of an array of cells into which a user can enter formulas and values.

Fig. 2 shows how a given type of NOR can be modeled following one or more data models, each of which could be implemented in different ways at the implementation layer. As an example, Fig. 2 shows a classification scheme

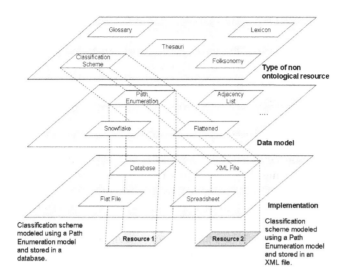

Fig. 2. Non-Ontological Resources (NORs) Categorization

modeled following a path enumeration model. In this case, the classification scheme is implemented in a database and in an XML file.

3 Related Work

In this section we present an overview of sofware re-engineering and a review of the state of the art on NOR re-engineering.

3.1 Software Re-engineering

Software re-engineering [5] is defined as the (1) examination of the design and implementation of an existing legacy system, and (2) application of the different techniques and methods to redesign and reshape that system into hopefully better and more suitable sofware.

Software re-engineering main activities are:

1. *Reverse engineering* [5] is the process of analyzing a subject system to identify the system components and their interrelationships, and create representations of the system in another form or at a higher level of abstraction.
2. *Alteration*, also called restructuring [5], is the transformation from one representation form to another at the same relative abstraction level, while preserving the subject system's external behaviour.
3. *Forward engineering* [5] is the traditional process of moving from high level abstractions and logical, implementation-independent designs to the physical implementation of a system.

Re-engineering patterns [18] are patterns that describe how to change a legacy system into a new, refactored system that fits current conditions and requirements. Their main goal is to offer a solution for re-engineering problems. They are also on a specific level of abstraction. They describe a process of re-engineering without proposing a complete methodology, and they can sometimes suggest a type of tool that one could use.

3.2   Non-Ontological Resource Re-engineering

Non-ontolgical resource re-engineering, defined in the Glossary of Activities in Ontology Engineering [24], refers to the process of taking an existing non-ontological resource and transforms it into an ontology.

The research in NOR re-engineering has been mainly centered on the transformation of standards [16,12], thesauri and lexicons [12,20,25], XML files [8], hierarchical classifications [9,12], folksonomies [20], relational databases [1,22], and spreadsheets [11]. These works only concentrate on the re-engineering process of the type and implementation of NOR.

In [20] Sabou et al. two approaches for the non-ontological resource transformation are distinguished. The first one consists in transforming resource schema into an ontology schema, and then resource content into instances of the ontology (Approach 1). The second one transforms resource content into an ontology schema (Approach 2). We add a third transformation approach which consists in transforming the resource content into instances of an existing ontology (Approach 3).

Table 1 shows a summary of the analyzed research works which have been focused on NOR type. Table 2 shows a summary of the research works which have been focused on the implementation of NORs. Both tables show the transformation approach, and also, if available, the name of the tool which supports the transformation approach. These research works just include *ad-hoc* methods and techniques for the transformation, i.e. the research works are specific of the NOR type or NOR implementation.

Re-engineering patterns are defined in [19] as transformation rules applied in order to create a new ontology (target model) from elements of a source model. The target model is an ontology, while the source model can either be an ontology or a NOR, e.g., a thesaurus concept, a data model pattern, a UML model, a linguistic structure, etc. In fact, [19] presents a unique example of a schema re-engineering pattern, which includes four rules to transform a knowledge organization system into SKOS[9]. These rules just identify the elements of the source model that are mapped to their corresponding elements of the target model, but the rules do not provide information about how to carry out the mapping. Re-engineering patterns are not integrated within a method to carry out the re-engineering process. Moreover, a template to describe re-engineering patterns in a unified way is not proposed.

---

[9] http://www.w3.org/2004/02/skos/

Table 1. Research works centered in the NOR type

Research Work	NOR Type	Transformation approach	Tool
Hepp et al. [12]	Classification schemes, thesauri, taxonomies	2	SKOS2GenTax
Mochol et al. [16]	Classification schemes	2	-
Sabou et al. [20]	Folksonomies	2	-
Sabou et al. [20]	Lexica	1,2	-
van Assem et al. [25]	Thesauri	1	-

Table 2. Research works centered in the NOR implementation

Research Work	NOR Implementation	Transformation approach	Tool
Stojanovic et al. [22]	Relational Database	1	KAON REVERSE
Barrasa et al. [1]	Relational Database	3	R_2O, ODEMapster
Garcia et al. [8]	XML files	1	XSD2OWL, XML2RDF
Han et al. [11]	SpreadSheet	3	RDF123

After having analyzed the state of the art on NORs re-engineering, we conclude that research efforts have been mainly devoted to the implementation and the type of NOR. It has also been analyzed how to map NORs content and schema into ontology instances and schema, but none of the analyzed research works have taken advantage from the data model which underlies the NOR to guide the re-engineering process. Finally, it is left to say that none of the analyzed re-engineering approaches propose a set of re-engineering patterns to guide the re-engineering process, and that there is also a lack of re-engineering methods.

4 Approach for Non-Ontological Resource Re-engineering

In this section we present our approach for NOR re-engineering. We describe a proposal for carrying out the NOR re-engineering process. Then, we present an example of the patterns for re-engineering NORs.

4.1 General Model for Non-Ontological Resource Re-engineering

In a nutshell, our approach for NOR re-engineering considers as input a pool of NORs and patterns for re-engineering NORs. NORs, as we mentioned in section 3, include lexica, classification schemes, thesauri, etc. Regarding patterns for

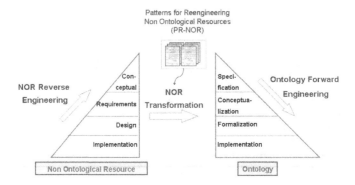

Fig. 3. Re-engineering Model for Non-Ontological Resources

re-engineering NORs, they provide solutions to the problem of transforming NORs into ontologies. These patterns will be included in the NeOn project patterns library[10].

Based on the software re-engineering model presented in [3] we propose our re-engineering model for NOR re-engineering in Fig.3.

The NOR re-engineering process consists of the following activities, which are defined in a Glossary of Activities in the Ontology Engineering[24]:

1. *Non-Ontological Resource Reverse Engineering*, whose goal is to analyze a NOR to identify its underlying components and create representations of the resource at the different levels of abstraction (design, requirements and conceptual). Since NORs can be implemented as XML files, databases or spreadsheet among others, we can consider them as software resources, and therefore, we use the software abstraction levels shown in Fig. 3 within this activity. Here the requirements and the essential design, structure and content of the NOR must be recaptured.

2. *Non-Ontological Resource Transformation*, whose goal is to generate a conceptual model from the NOR. We propose the use of Patterns for Re-engineering Non-Ontological Resources (PR-NOR) to guide the transformation process. First, the transformation approach has to be selected: (1) transforming resource schema into an ontology schema, and then resource content into instances of the ontology, (2) transforming resource content into an ontology schema, or (3) transforming the resource content into instances of an existing ontology. Second, the semantics of the relations between the NOR entities have to be identified, these semantics can be a)*subClassOf*, b)an *ad-hoc* relation like *partOf* or c)a mix of *subClassOf* and *ad-hoc* relations. Finally a pattern for re-engineering NORs according to the type of NOR, as well as the selected transformation approach, and the semantics of the relations between the NOR entities, has to be searched.

---

[10] http://www.ontologydesignpatterns.org

3. *Ontology Forward Engineering*, whose goal is to output a new implementation of the ontology on the basis of the new conceptual model. We use the ontology levels of abstraction to depict this activity because they are directly related to the ontology development process.

4.2   Patterns for Re-engineering Non-Ontological Resources

Patterns for re-engineering non-ontological resources (PR-NOR) define a procedure to transform the NOR components into ontology representational primitives. To this end, patterns take advantage of the NOR underlying data model. The data model defines how the different components of the NOR are represented.

According to the NOR categorization presented in section 3, the data model can be different even for the same type of NOR. For every data model we can define a process with a well-defined sequence of activities to extract the NORs components and then map them to the conceptual model of an ontology. Each process can be expressed as a pattern for re-engineering NORs.

The resultant ontologies proposed by the patterns for re-engineering NORs are modeled following the recommendations provided by some other ontological patterns such as logical and architectural patterns [23]. The current inventory of *NeOn Ontology Modelling Components* considered as Architectural Patterns includes the following ones: taxonomy, lightweight ontology and modular architecture. A taxonomy is the way of organizing an ontology as a hierarchical structure of classes only related by subsumption relations. A lightweight ontology adds the following features to the taxonomy structure: (a) a class can be related to other classes through the *disjointWith* relation, (b) object and datatype properties can be defined and used to relate classes, (c) a specific domain and range can be associated with defined object and datatype properties. Finally, the modular architecture consists in structuring an ontology as a configuration of components, each having its own identity based on some design criteria.

Moreover, the patterns for re-engineering NORs define the transformation process but they do not provide either an algorithm or an implementation of the process. We plan to include the algorithms and implementations later on in a framework which will implement the transformation process.

We have created eight patterns for re-engineering classifications schemes into taxonomies and lightweight ontologies, two for each data model identified (path enumeration, adjacency list, snowflake and flattened). We plan to extend this pool of patterns with more patterns for the rest of transformation approaches. Also we plan to include patterns for re-engineering the other types of NORs.

Next, we present an example of a re-engineering pattern identified in our ongoing research work on transforming classification schemes into ontologies. To present the patterns for re-engineering NORs we adapted the tabular template for ontology design patterns used in [23].

The pattern for re-engineering NOR shown in Table 4.2 suggests a guide to transform a classification scheme into a lightweight ontology. The classification scheme is modeled with a snowflake data model. This pattern aims at creating a lightweight ontology from the classification scheme.

Table 3. Pattern for Re-engineering a Classification Scheme

Slot	Value
General Information	
Name	Classification scheme to Lightweight Ontology (Snowflake model)
Identifier	PR-NOR-CLLO-01
Type of Component	Pattern for Re-engineering Non-Ontological Resources (PR-NOR)
Use Case	
General	Re-engineering a classification scheme which follows the snowflake model to design a Lightweight Ontology.
Example	Suppose that someone wants to build a lightweight ontology based on the ISO 3166 standard for the representation of names of countries and their subdivisions. This standard is divided in ISO 3166-1 for countries, and ISO 3166-2 for subdivisions (regions).
Pattern for Re-engineering Non-Ontological Resources	
Resource to be Re-engineered	
General	A NOR holds a classification scheme which follows the snowflake model. A classification scheme is a rooted tree of concepts, in which each concept groups entities by some particular degree of similarity. The semantics of the hierarchical relation between parents and children concepts may vary depending on the context. The snowflake model for hierarchical classifications proposes to create a fixed but separated entity (table, file) for each level of the hierarchy.
Example	The ISO 3166 standard (codes for the representation of names of countries and their subdivisions) is divided in ISO 3166-1 for countries, and ISO 3166-2 for country subdivisions (regions). For the example, ISO 3166-1 and ISO 3166-2 are hold on different entities. The relation semantics between the sub-ordinate and the super-ordinate concepts is *partOf*.
Graphical Representation	
General	First level categories entity 	Category Code	Category Name	Category Description
1	Category1Level1	Category1Level1Desc
2	Category2Level1	Category2Level1Desc
...		| Second level categories entity | Category Code	First Level Category	Category Name	Category Description
1	1	Category1Level2	Category1Level2Desc
2	1	Category2Level2	Category2Level2Desc
...			| Third level categories entity | Category Code	Second Level Category	Category Name	Category Description
1	1	Category1Level3	Category1Level3Desc
2	2	Category2Level3	Category2Level3Desc
...				
Example	ISO 3166-1 Country 	Code	Name
GB	UNITED KINGDOM
ES	SPAIN
...		 ISO 3166-2 Subdivision 	Code	Name	ISO 3166-1 Code
GB-NI	Northern Ireland	GB
GB-EA	East Anglia	GB
...		
Designed Ontology	

Table 3. (*continued*)

Slot	Value
General	The generated ontology will be based on the lightweight ontology architectural pattern (AP-LW-01)[23]. Each snowflake entity is mapped to a class. An *ad-hoc* binary relation is defined between the new classes according to the semantics of the relation between super-ordinate and sub-ordinate categories. Each data included on an entity is mapped to an instance of the entity class. The semantics of the relationship between sub-ordinate and super-ordinate instances is mapped to an *ad-hoc* binary relation instance.
	Graphical Representation
(UML)General Solution Ontology	
(UML)Example Solution Ontology	
	How to Re-engineer
General	1. Create a class for each entity in the snowflake model. 2. If there is a relationship between the entity classes then create it as an *ad-hoc* binary relation. 3. If there is a super-class for the new entity related classes then create it and set the appropriate *subClassOf* relation between the entity classes and the super-class. 4. For each record on each entity of the snowflake model, create an instance of the appropriate entity class. 5. If you have created an *ad-hoc* binary relation between the entity classes then you have to create the relation instance between the entity class instance.

Table 3. (*continued*)

Slot	Value
Example	1. Create a COUNTRY class for the ISO 3166-1 Countries entity and a REGION class for the ISO 3166-2 Subdivisions entity. 2. Create the *Has_region* binary relation with COUNTRY as domain and REGION as range. 3. Create a LOCATION class and assert that COUNTRY and REGION are *subClassOf* LOCATION. 4. For each record on the ISO 3166-1 Countries entity create an instance of the COUNTRY class. 5. For each COUNTRY instance look for its REGION on the ISO 3166-2 Subdivisions entity and create an instance of REGION for each subdivision found. Also create an instance of the *Has_region* relation associated to the current country instance and related to the current region instance.
Relationships	
Relations	Use the Architectural Pattern: AP-LW-01 [23]

5   SEEMP Use Case

A preliminary experimentation of our approach was done within the SEEMP project, in which NORs of the human resources domain were transformed into ontologies. We re-engineered four classification schemes using the overall set of patterns. We obtained the following ontologies:

- Occupation, Education, Economic activity ontologies. We applied the pattern *classification scheme (path enumeration) to lightweight ontology* (PR-NOR-CLTX-01), to re-engineer the ISCO-88 (COM), FOET, and NACE standards. These standards are classification schemes modeled following a path enumeration data model and they are stored in a MS Access database.
- Geography ontology. We applied the pattern *Classification scheme (adjacency list) to lightweight ontology* (PR-NOR-CLLO-02), to re-engineer the ISTAT[11] geography italian standard. This standard is a classification scheme modeled following an adjcency list data model and it is stored in a MS Excel spreadsheet.

In this section we present the activities carried out to re-engineer the ISTAT standard. This standard contains information about the divisions, regions and provinces of Italy. It is available in MS Excel spreadsheet format.

- *Non-Ontological Resource Reverse Engineering.* Within this activity we gathered documentation about ISTAT from domain web sites such as ISTAT web site itself and Eurostat. From this documentation we extracted the schema of the classification scheme which consists of 4 divisions, 20 regions and 106 provinces. Since the data model was not available in the documentation, it was necessary to extract it for the resource implementation itself.

---

[11] http://www.istat.it/

ISTAT is modeled following the adjacency list data model, i.e. each row of the spreadsheet contains the information related to a province, its region and its division.

- *Non-Ontological Resource Transformation.* Within this activity we carried out the following tasks:
  1. We followed approach 1, described in section 3, to carry out the transformation. This approach consists in transforming resource schema into an ontology schema, and then resource content into instances of the ontology.
  2. We identified the semantic of the relations between the NOR entities. In this case the relation was identified as *part Of*.
  3. Then, we looked in our local pattern repository for a suitable pattern to re-engineer NORs taking into account the selected transformation approach, the semantics of the relations between the NOR entities, and the data model of the resource.
  4. The most appropriate pattern for this case is the PR-NOR-CLLO-02 pattern. This pattern takes as input a classification scheme modeled with an adjacency list data model and produces a lightweight ontology.
  5. The selected pattern suggests to create a class for each one of the columns related to the main entities of the ISTAT standard. With this information we outlined the conceptual model for the ontology.
     (a) Create the DIVISION, REGION, and PROVINCE classes according to the ISTAT entities.
     (b) Create the *has_region* binary relation with DIVISION as domain and REGION as range.
     (c) Create the *has_province* binary relation with REGION as domain and PROVINCE as range.
     (d) Create a LOCATION class and assert that DIVISION, REGION and PROVINCE are *subClassOf* LOCATION.
     (e) Create an instance of the DIVISION class for each distinct ISTAT division .
     (f) Look for the REGIONS of each DIVISION instance in the ISTAT regions and create an instance of REGION for each distinct region. Create an instance of the *has_region* relation associated to the current division instance and related to the current region instance.
     (g) Look for the PROVINCES of each REGIONS instance in the ISTAT provinces and create an instance of PROVINCE for each distinct province. Create an instance of the *has_province* relation associated to the current region instance and related to the current province instance.
- *Ontology Forward Engineering.* WSML[12] is the ontology implementation language used in the SEEMP project. Because of the number of divisions, regions and provinces of the ISTAT standard, it was not practical to create the ontology manually. Therefore, we created an *ad-hoc* wrapper, implemented

---

[12] http://www.wsmo.org/wsml/

in Java, that reads the data from the resource implementation and automatically creates the corresponding classes, attributes and relations of the new ontology following the suggestion given by the pattern for re-engineering NORs and the conceptual model. The resultant ontology is available at http://droz.dia.fi.upm.es/ontologies/.

6 Conclusions and Future Work

In this paper we have introduced a three level categorization of NORs according to three different features: type of NOR, data model and implementation. Moreover, we present a pattern based approach for re-engineering NORs into ontologies. We take advantage of the NOR data model to define patterns for re-engineering NORs. We also describe a pattern for re-engineering a classification scheme into an ontology. Additionally, we present a use case of the proposed approach. Further work needs to be done to consider data models of the other NORs. If we can identify data models as we made for classification schemes we will be able to create more patterns to guide the re-engineering process. This approach will be extended for creating richer and more complex ontologies. We also need to calculate how much effort do we save re-engineering NORs using patterns compared with re-engineering NORs without them.

Acknowledgments. This work has been partially supported by the European Comission projects NeOn(FP6-027595) and SEEMP(FP6-027347), as well as by a UPM-BSCH grant, and an I+D grant from the UPM.

References

1. Barrasa, J., Corcho, O., Gómez-Pérez, A.: R2O, an Extensible and Semantically Based Database-to-Ontology Mapping Language. In: Bussler, C.J., Tannen, V., Fundulaki, I. (eds.) SWDB 2004. LNCS, vol. 3372. Springer, Heidelberg (2005)
2. Brandon, D.: Recursive database structures. Journal of Computing Sciences in Colleges (2005)
3. Byrne, E.J.: A conceptual foundation for software re-engineering. In: Proceedings of the International Conference on Software Maintenance and Reengineering. IEEE Computer Society Press, Los Alamitos (1992)
4. Caracciolo, C., Gangemi, A.: Revised and Enhanced Fisheries Ontologies. Technical report, NeOn project deliverable D7.2.2 (2007)
5. Chikofsky, E.J., Cross, J.H.: Reverse engineering and design recovery: a taxonomy. In: IEEE Software (1990)
6. Gómez-Pérez, A., Fernández-López, M., Corcho, O.: Ontological Engineering. In: Advanced Information and Knowledge Processing. Springer, Heidelberg (2003)
7. Gangemi, A., Pisanelli, D., Steve, G.: Ontology integration: Experiences with medical terminologies. Ontology in Information Systems, 163–178 (1998)
8. García, R., Celma, O.: Semantic Integration and Retrieval of Multimedia Metadata. In: Proceedings of the ISWC 2005 Workshop on Knowledge Markup and Semantic Annotation, Semannot 2005 (2005)

9. Giunchiglia, F., Marchese, M., Zaihrayeu, I.: Encoding Classifications into Lightweight Ontologies.. In: The Semantic Web: Research and Applications. Springer, Heidelberg (2006)
10. Haase, P., Rudolph, S., Wang, Y., Brockmans, S.: Networked Ontology Model. Technical report, NeOn project deliverable D1.1.1 (2006)
11. Han, L., Finin, T., Parr, C., Sachs, J., Joshi, A.: RDF123: a mechanism to transform spreadsheets to RDF. In: Proceedings of the Twenty-First National Conference on Artificial Intelligence (AAAI 2006). AAAI Press, Menlo Park (2006)
12. Hepp, M., de Bruijn, J.: GenTax: A Generic Methodology for Deriving OWL and RDF-S Ontologies from Hierarchical Classifications, Thesauri, and Inconsistent Taxonomies. In: Franconi, E., Kifer, M., May, W. (eds.) ESWC 2007. LNCS, vol. 4519, pp. 129–144. Springer, Heidelberg (2007)
13. Hodge, G.: Systems of Knowledge Organization for Digital Libraries: Beyond Traditional Authority Files (2000),
    http://www.clir.org/pubs/reports/pub91/contents.html
14. Maedche, A., Staab, S.: Ontology learning for the semantic web. IEEE Intelligent Systems (2001)
15. Malinowski, E., Zimányi, E.: Hierarchies in a multidimensional model: From conceptual modeling to logical representation. Data and Knowledge Engineering (2006)
16. Mochol, M., Paslaru, E.: Practical Guidelines for Building Semantic eRecruitment Applications. In: International Conference on Knowledge Management (iKnow 2006), Special Track: Advanced Semantic Technologies (2006)
17. Pinto, H.S., Tempich, C., Staab, S.: DILIGENT: Towards a fine-grained methodology for DIstributed, Loosely-controlled and evolvInG Engineering of oNTologies. In: Proceedings of the 16th European Conference on Artificial Intelligence (ECAI 2004), pp. 393–397. IOS Press, Amsterdam (2004)
18. Pooley, R., Stevens, P.: Software reengineering patterns. Technical report (1998)
19. Presutti, V., Gangemi, A., David, S., Aguado de Cea, G., Suárez-Figueroa, M.C., Montiel-Ponsoda, E., Poveda, M.: NeOn Deliverable D2.5.1. A Library of Ontology Design Patterns: reusable solutions for collaborative design of networked ontologies. In: NeOn Project (2008), http://www.neon-project.org
20. Sabou, M., Angeletou, S., dAquin, M., Barrasa, J., Dellschaft, K., Gangemi, A., Lehman, J., Lewen, H., Maynard, D., Mladenic, D., Nissim, M., Peters, W., Presutti, V., Villazón, B.: Selection and integration of reusable components from formal or informal specifications. Technical report, NeOn project deliverable D2.2.1 (2007)
21. Staab, S., Schnurr, H.P., Studer, R., Sure, Y.: Knowledge processes and ontologies. IEEE Intelligent Systems (16), 26–34 (2001)
22. Stojanovic, L., Stojanovic, N., Volz, R.: A Reverse Engineering Approach for Migrating Data-intensive Web Sites to the Semantic Web. In: Proceedings of the Conference on Intelligent Information Processing (2002)
23. Suárez-Figueroa, M.C., Brockmans, S., Gangemi, A., Gómez-Pérez, A., Lehmann, J., Lewen, H., Presutti, V., Sabou, M.: Neon modelling components. Technical report, NeOn project deliverable D5.1.1 (2007)
24. Suárez-Figueroa, M.C., Gómez-Pérez, A.: Towards a Glossary of Activities in the Ontology Engineering Field. In: Proceedings of the 6th Language Resources and Evaluation Conference, LREC 2008 (2008)
25. van Assem, M., Menken, M., Schreiber, G., Wielemaker, J.: A method for converting thesauri to RDF/OWL. In: McIlraith, S.A., Plexousakis, D., van Harmelen, F. (eds.) ISWC 2004. LNCS, vol. 3298, pp. 17–31. Springer, Heidelberg (2004)

Efficient Index Maintenance for Frequently Updated Semantic Data

Yan Liang[1], Haofen Wang[1], Qiaoling Liu[1], Thanh Tran[2], Thomas Penin[1], and Yong Yu[1]

[1] Department of Computer Science & Engineering
Shanghai Jiao Tong University, Shanghai, China
{yliang,whfcarter,lql,tpenin,yyu}@apex.sjtu.edu.cn
[2] Institute AIFB, Universität Karlsruhe, Germany
{dtr}@aifb.uni-karlsruhe.de

Abstract. Nowadays, the demand on querying and searching the Semantic Web is increasing. Some systems have adopted IR (Information Retrieval) approaches to index and search the Semantic Web data due to its capability to handle the Web-scale data and efficiency on query answering. Additionally, the huge volumes of data on the Semantic Web are frequently updated. Thus, it further requires effective update mechanisms for these systems to handle the data change. However, the existing update approaches only focus on document. It still remains a big challenge to update IR index specially designed for semantic data in the form of finer grained structured objects rather than unstructured documents. In this paper, we present a well-designed update mechanism on the IR index for triples. Our approach provides a flexible and effective update mechanism by dividing the index into blocks. It reduces the number of update operations during the insertion of triples. At the same time, it preserves the efficiency on query processing and the capability to handle large scale semantic data. Experimental results show that the index update time is a fraction of that by complete reconstruction w.r.t. the portion of the inserted triples. Moreover, the query response time is not notably affected. Thus, it is capable to make newly arrived semantic data immediately searchable for users.

1 Introduction

Nowadays, indexing and retrieving the Semantic Web data is drawing an increasing attention. Some systems such as Swoogle [1] and Watson [2] have adopted IR (Information Retrieval) approaches for indexing these data. In particular, the semantic search engine [1] has indexed over 1.4 million Semantic Web documents and began to provide search services in the Semantic Web community similar to Google. The success is due to the fact that IR is proved to handle Web-scale data and be efficient on query answering. Moreover, these systems are benefited from the IR approaches to exploit huge amounts of textual information on the Semantic Web by keyword queries.

J. Domingue and C. Anutariya (Eds.): ASWC 2008, LNCS 5367, pp. 182–196, 2008.
© Springer-Verlag Berlin Heidelberg 2008

Additionally, the huge volumes of semantic web data are frequently updated. Thus it requires semantic search engines not only to be scalable but also have flexible update mechanisms to make the newly arrived data immediately searchable for users. For example, with a large number of indexed Semantic Documents, Swoogle has to update its index on a regular basis.

However, to keep the Semantic Web data up-to-date in an IR index is a difficult task. The IR-based approaches index the Semantic Web data by reusing the existing structure of inverted index. Although there are many discussions on the index update for traditional IR search engines [3,4,5,6,7], current update approaches are just suitable for semantic documents. However, the index updating is more difficult for an IR index which is designed as a repository of triples, since during the update it should also keep the original relations between existing individuals. Although some methods (such as [8]) have presented to speed up the index construction, frequently index rebuilding is still costly.

In this paper, we propose an efficient updating mechanism on top of Semplore [9], which is the state-of-art of the current IR approaches to index and retrieve the large scaled semantic instances (RDF triples). It extends the IR engine's index structure and functions to provide efficient query processing. Moreover, it supports both the structured queries for semantic web data and the keyword queries for textual information.

Our approach is based on the idea of dividing the index into blocks, which reduce the number of update operations during the insertion of triples. Our index mechanism can also be used for the index update based on an incremental crawler. Experimental results show that the index update time is a fraction of complete reconstruction w.r.t. the portion of the inserted triples. Thus it's capable to make the newly arrived semantic data immediately searchable by users. At the same time it preserves both the efficiency on query processing and the scalability to handle the large-scaled semantic data. Moreover, the reuse of IR search engine not only can index the structural Semantic Web data but also the textual information. Thus it supports the hybrid query capability for both structured queries and keyword queries.

The paper is organized as follows. Section 2 introduces the related work. Section 3 describes the basic index structure we are based on. Section 4 discusses the extended block index structure, along with an update mechanism. Moreover, a comprehensive analysis on the performance of our update mechanism is presented in Section 5. Section 6 shows the experimental results and we will give a conclusion in Section 7.

2   Related Work

The update mechanism for inverted index is a well studied field. Work in [5] presents a hybrid approach in which long posting lists are updated in-place, while short lists are updated using a merge strategy. The method proposed in [3] maintains a dual inverted list which stores short lists in the memory and long lists on the disk. When the area for the short lists is full, the longest short list will be merged into a long list. Work in [6] improves the in-place update by saving the short posting lists within the vocabulary and over-allocating the long lists. [4] uses overflow 'buckets' to handle the new arriving postings. Work in [7] presents a method to update previously indexed documents whose content have changed. The idea is based on blocking together with the diff algorithm. [10] presents a just-in-time indexing component which invests less in the preprocessing of arriving data, at the expense of a tolerable latency in query response time. Index update can also be achieved by reconstruction. Method in [8] is presented to speed up the index construction.

However, there are few work on the index update for Semantic Web data. [11] enables incremental update of index for XML documents part of which are changed.

Querying and searching semantic web data using IR-based approaches are emerging areas. Work in [1] presents a crawler based indexing and retrieval system for semantic web data. It uses an IR engine to index the crawled semantic web documents by using the n-gram and taking URIrefs as terms. [2] provides an interface for searching ontologies and Semantic Documents using keywords. However, these works are designed to index Semantic Documents and they do not index triples. [9] is designed as a repository of RDF triples based on the IR engine's index structure. It supports both structured query and keyword query. [12] uses keyword search results to do spread activation on semantic networks. But it does not support structured queries for Semantic Web data. While [13] combines full text and ontology search based on an IR engine. Work in [14] is presented as a lookup index over sources crawled on the Semantic Web. But it acts as locator of RDF resources and not as a query engine. [15] borrows XML Fragment query language to search semantically annotated text corpora but not for semantic web data such as RDF triples. Moreover, the inverted index also can be used in DBMSs to support containment queries in XML documents [16,17].

3   Overview of Semplore

Our work is based on [9], which indexes and retrieves RDF triples using the existing index structure and functions of current IR engines. It provides the hybrid query capability by combining both structured queries and keyword queries. In this section, we will give a brief introduction to its query capability, index structure and query evaluation algorithm.

3.1   Hybrid Query Capability

Here the hybrid query is an extension of the DL-based conjunctive query which was introduced in [18] and can be presented by the SPARQL query language. To support the keyword queries, an extension of the ordinary conjunctive query is made by taking keyword as a virtual concept. An individual is an instance of a certain virtual concept if the textual content of its properties contain the corresponding keyword. Then the users can input a conjunctive query containing keyword constraints. For example, to find all films which are about "romantic" and directed by some Chinese director, the query is:

$\{f \mid \texttt{"romantic"}(f) \land \texttt{directs}(d, f) \land \texttt{ChineseDirector}(d)\}$

Here the queries are restricted as tree-shaped unary queries, whose query graphs are trees. The detailed definition can be found in [9].

3.2   Index Structure

The index structure of traditional IR search engine is the inverted index which is based on fields, documents and terms. Work in [9] uses the inverted index structure to index triples and provides searching and querying based on the functions of IR search engine. Its main idea is to translate semantic web data into documents, fields and terms which can be indexed and retrieved by traditional IR engine. The translation is shown in Table 1.

After the translation, semantic web data can then be indexed by the IR engine. The IR engine's retrieval functions can also be used over these indexed data. For example, for each relation, the IR engine can find all its super relations by inputting the relation name and the field "superRelOf" as a query. For each concept, the IR engine can also returns all its individuals by taking the concept name and the field "type" as a query.

For relation triples, the index saves relation names as terms and the subject individuals as the documents. As it is shown in Fig.1, for each subject in a certain relation's posting list, its position list stores all its corresponding objects in this relation. As an example in Fig.1, i_2 and i_3 are corresponding objects of i_1, then i_2 and i_3 are stored in i_1's position list in relation R_1's posting list. Thus the IR engine can find all the objects of a certain relation with a given set of subjects by return the union of the corresponding position lists. The index structure is symmetric, for the objects of a relation can be taken as the subjects of the inverse relation. So in the similar way, objects of triples are stored in the posting list of inverse relations and they also have position lists to stored the corresponding subjects.

To force the search engine to save the object individuals as position information, the actual contents stored in the position lists are the object's local

Table 1. Translation from semantic web data to fields, documents, and terms

Document	Field	Term
concept C	subConOf	super-concepts of C
	superConOf	sub-concepts of C
	text	tokens in textual properties of C
relation R	subRelOf	super-relations of R
	superRelOf	sub-relations of R
	text	tokens in textual properties of R
individual i	type	all concepts that i belongs to
	subjOf	all relations R that $(i, R, ?)$ is a triple in data
	objOf	all relations R that $(?, R, i)$ is a triple in data
	text	tokens in textual properties of i

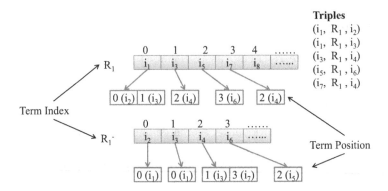

Fig. 1. Triples stored in the inverted index

positions in the inverse relation's posting list. For example in Fig.1, in i_1's position list stores 0 and 1, which are i_2 and i_3's local position in the posting list of R^-. By reading the subject's position list the search engine can quickly skip to the corresponding objects in the inverse position list. Based on this index structure, the search engine can provide efficient query evaluation algorithm which will be discussed in Section 3.3.

IDs are used throughout the indexing process to uniquely represent a resource (individual). Individuals in the posting lists or position lists are sorted in ascending order according to their IDs in order to provide fast query evaluation.

3.3   Query Evaluation

Basic Operations. In modern IR engines, two basic operations can be efficiently achieved, which are the Basic Retrieval and the Merge Sort. Given a field f and a term t, Basic Retrieval (f, t) returns the corresponding posting list from inverted index. The result is sorted by individual IDs in ascending order. The input of Merge Sort are two sorted lists of individual IDs S_1 and S_2 and a binary operator op which can be $\cap$, $\cup$ or $-$. The Merge Sort operation $m(S_1, op, S_2)$ computes $S_1 \ op \ S_2$ by merging the lists S_1 and S_2 and returns the result as a new sorted list of individual IDs. According to the index structure mentioned in Section 3.2, works in [9] reuses and extends these basic functions of IR engine to support its own query evaluation algorithm.

(1) Concept Constraints

The input of this operation is a boolean combination of concepts and keyword concepts. It's output is a sorted list of individual IDs which match the constraints. This operation can be implemented using basic retrieval and merge-sort operation mentioned above. For example, for the input Film $\sqcap$ "romantic", the Concept Constraints can be achieved through two Basic Retrievals and one Merge Sort: $m(\,(\texttt{type}, \texttt{Film}),\ \cap,\ (\texttt{text}, \text{"romantic"})\,)$.

(2) Relation Expansion

The input of this operation is a relation R and two sets S_1 and S_2 of individual IDs. The operation computes the set $\{y \mid \exists x : x \in S_1 \wedge (x, R, y) \wedge y \in S_2\}$ and returns it as a sorted list of individual IDs. The Relation Expansion is not directly supported by traditional IR engines. This operation needs to find all the objects of a certain relation with a given set of subjects. According to the index structure in 3.2, these objects can be obtained by computing the union of the subjects' position lists. Since in these position lists it stores the objects' local position in the inverse position list, the union can be computed based on a bit vector which has the same length as the inverse relation's posting list.

Query Evaluation Algorithm. From these basic operations, a tree-shaped hybrid query can be evaluated using a bottom-up method. At first for each leaf nodes in the query graph, it uses the Concept Constraints operation to obtain the satisfied individuals. When all of the children nodes are evaluated, it moves forward to the parent node using the Relation Expansion to filter the results. Then these children nodes are removed. Doing this procedure iteratively then the final result is obtained when finish visiting all the edges in the query graph.

4   Index Update Mechanisms

Based on the index structure in Section 3.2, for concept names, relation names or concept individuals which are indexed without using position lists, we can update the index by adopting the optimizations of traditional index maintenance([3,6]). However, for relation triples, updating the index is time consuming. During the index update, newly arrived triples need to be added into the index. Their subjects and objects are inserted into the posting lists of corresponding relations and inverse relations. However, inserting new individuals into a posting list would make some of the original individuals' local position moved behind. These affected local positions are stored in the position lists in the inverse relation's posting lists. As a result, these position lists which store the updated local positions should be updated. It is certainly a heavy cost, since inserting one individual may sometimes leads to the reconstruction of the whole posting list. In this section, we present a Block Index structure based on Section 3.2. It reduces the cost of inserting new relation triples into the index.

4.1   Block Index Structure

The purpose of our Block Index structure is to minimize the changes in the position lists when inserting relation triples. The basic idea is to split posting lists into blocks. The first individual of each block is taking as landmark. All individuals in the same block have their offsets comparing their local position to that of the block's landmark. Then the local position of each individual in the posting list can be presented as a $< Landmark\_ID, offset >$ pair. Take Fig.2 as an example, in the posting list of relation R_1, individuals i_1 and i_7

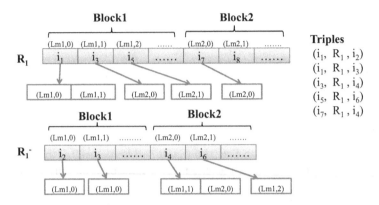

Fig. 2. Structure of block based index

are landmarks of $Block_1$ and $Block_2$ respectively. Then the local position of i_5 can be represented by the pair $< Lm1, 2 >$ where $Lm1$ is the landmark_ID of $Block_1$ and 2 is the offset. An auxiliary landmark table is needed to store all the landmarks and their real position in the posting lists. Thus real positions of individuals in a posting list can be obtained by getting the landmark's real position from landmark table and adding the offset value.

Note that the Block Index structure is only for the storage of relation triples. For concept names, relation names or concept individuals, which are only stored in posting lists, the index structure is the same as it defines in Section 3.2.

4.2   Single Update Operation

In this section, we discuss the insertion of a single relation triple into the index, which is shown as Algorithm 1. First we need to insert the subject into the relation's posting list if necessary. Second is to insert the object into the inverse relation's posting list in a similar way if necessary. After the insertion, we will get the local positions of the subject and the object in corresponding posting lists. Then we can add their local positions into each other's position list.

Algorithm 1. Single Update Algorithm

    Input: An inserted triple (s, R, o)
1 **if** $s \notin Posting(R)$ **then** $Insert(s, R)$;
2 **if** $o \notin Posting(R^-)$ **then** $Insert(o, R^-)$;
3 Add $LocalPosition(R, s)$ to $PositionList(o, R^-)$;
4 Add $LocalPosition(R^-, o)$ to $PositionList(s, R)$;

The procedure of $Insert(s, R)$ is to insert an individual s into the posting list of a relation R. For each following individual i in the Block, we read every local position $(< p_l, p_o >)$ in it's position list to find the corresponding object o in the inverse-relation list. Then we update the old local position of i which is stored in o's position list by increasing the offset value by one. We should also maintain

---

Procedure. Insert(s, R)

---

1 Find block B that s should inserted into;
2 **foreach** *instance* i *that* $i \in B \wedge i > s$ **do**
3   **foreach** $< p_l, p_o > \in PositionList(i, R)$ **do**
4          $o = Skip\_To(< p_l, p_o >, Posting(R^-))$;
5   Find $< n_l, n_o >$ in $PositionList(o, R^-)$ that
            $Skip\_To(< n_l, n_o >, Posting(R)) == i$;
6          $n_o = n_o + 1$;
7 Add s to Posting(R);
8 Update the Landmark Table;

---

the landmark table after the insertion. Since the insertion of s makes the real positions of all landmarks of the following blocks moved backward for one space.

Fig.3 shows the procedure of $Insert(i_2, R_1)$ when inserting a single triple (i_2, R_1, i_7) into the index. Since the subject i_2 does not exist in the posting list of R_1, i_2 is then inserted into $Block_1$ of R_1's posting list. The local position of all following individuals in this block (i_3 and i_5) should be updated by increasing the offset value by one. The old local positions of i_3 and i_5 should be updated, which are stored in the position list of their corresponding objects. By reading the position list of i_3 and i_5 in R_1's posting list, we can easily get these corresponding objects (i_4 and i_6) and skip to their positions in the posting list of R_1^-. Then we can find in i_4 and i_6' position lists and update i_3 and i_5's old local position by increasing the offset value by one. In the similar way, the object i_7 is inserted into $Block_2$ of R_1^-'s posting list and all the local positions of following individuals in this block have to be updated. As a result, their old local positions which are stored in the position list of the corresponding subjects in R_1 should be changed.

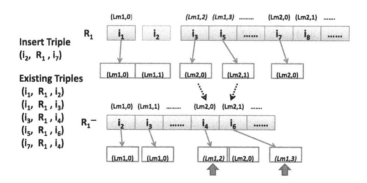

Fig. 3. Procedure of inserting a single individual into posting list

The delete operation is much easier, for we only need to delete the local positions of the subject and the object in each other's position lists. Considering the expenses of inserting an individual in the posting list, we do not delete an individual with empty position list for further insertion.

4.3 Batch Update Operation

In this section we present a batch update operation, which can reduce the number of update operations in position lists for multi-triple's insertion. In the batch update operation, every time we insert all individuals which belong to the same block into the posting list. It will avoid the redundant update in position lists. The Algorithm 3 shows how it works.

Algorithm 3. Batch Update Algorithm

    Input: Inserted triples $(s_1, R, o_1), (s_2, R, o_2) \ldots (s_n, R, o_n)$

1 **foreach** $block\ B_i \sqsubseteq Posting(R)$ **do**

2     $S_{sub} =$
    $\{s_t \mid s_t \notin Posting(R) \wedge s_t \geq Landmark(B_i) \wedge s_t < Landmark(B_{i+1})\}$;

3     $BatchInsert(S_{sub}, B_i, R)$;

4 **foreach** $block\ B_i \in Posting(R^-)$ **do**

5     $S_{obj} =$
    $\{s_t \mid s_t \notin Posting(R^-) \wedge s_t \geq Landmark(B_i) \wedge s_t < Landmark(B_{i+1})\}$;

6     $BatchInsert(S_{obj}, B_i, R^-)$;

7 **foreach** (s_i, R, o_i) **do**

8   Add $LocalPosition(R, s_i)$ to $PositionList(o_i, R^-)$;

9   Add $LocalPosition(R^-, o_i)$ to $PositionList(s_i, R)$;

Procedure. BatchInsert(S,B,R)

---

1 **foreach** $instance\ i \in B \wedge i > min\{S\}$ **do**

2   **foreach** $< p_l, p_o > \in PositionList(i, R)$ **do**

3         $o = Skip\_To(< p_l, p_o >, Posting(R^-))$;

4   Find $< n_l, n_o >$ in $PositionList(o, R^-)$ that
        $Skip\_To(< n_l, n_o >, Posting(R)) == i$;

5         $n_o = n_o + |\{s \mid s \in S \wedge s < i\}|$;

6 Add each $s \in S$ to $Posting(R)$;

7 Update the Landmark Table;

When more triples of a relation are inserted into the index, each time we select all the individuals which belong to the same block and insert them into the posting list at one time. The operation of $BatchInsert(S, B, R)$ is an expansion of $Insert(s, R)$ presented in Section 4.2. Once inserting these individuals in the posting list, the local position of every original individual in the block which behind the minimum inserted individual should be moved backward. These local positions are stored in the position lists of the corresponding objects in the inverse relation's posting list. The offset value are updated due to the number of individuals which will be inserted in front of it. The batch insert operation is more efficient since it reduce the number of position update when inserting individuals belong to the same block.

In order to be efficient for the update operation, the block size must be chosen in a certain range, which will be discussed in Section 5.3. After the batch update, if a block size exceeds the threshold, it will be split into two smaller blocks.

5   Performance Analysis

5.1   Space Requirement

According to [7], the landmark-offset encoding for local position does not increase the space requirement of the index. Suppose k bits are allocated for a location position in the posting list, then the same k bits can be used to encode a $<$ $landmark\_ID, offset >$ pair with $b < k$ bits for the landmark ID and the rest $k - b$ bits for the offset. However, an extra landmark table will be stored on disk and loaded in memory during index update and query processing. The size of landmark table is usually small. For a index with average block size B, the total number of landmarks is $\Sigma(\lceil L_R/B \rceil)$,where L_R is the length of R's posting list and R is every relation or inverse relation in the index.

5.2   Query Performance

In essence, query evaluation time with block index is not significantly affected comparing to the index structure mentioned in 3.2. For concept individuals, which are stored in traditional inverted index without blocks, the IR engine provides fast processing time. For relation triples, the main difference is that the individual's real positions in the posting lists needs to be computed by seeking the real position of the landmark in the landmark table and then adding the individual's offset. In the landmark table, all landmarks in a certain relation's posting list is sorted by their real positions. Using the binary search, the seek operation takes $O(log(L/B))$ time, where L is the length of posting list and B is the average block size.

5.3   Index Update Time

In the block based index, contents in the same block are stored continuously. Based on the optimization of traditional index maintenance([3,6]), inserting individuals into a block is not time consuming. Moreover, with the help of landmark table, seeking in the posting lists can be finished efficiently. During the index update, the main cost is to look up and update the offset values in the position lists. Since contents of a position list are physically stored continuously in modern IR engines, the look up operations in the position lists enjoys the benefit of spatial locality for fast access.

Using the batch update introduced in Section 4.3, the total index update time for all newly arrived triples depends on the block size and the number of blocks which will be inserted with new individuals. So our update mechanism is especially efficient when the index size is large and the number of update triples is small. Blocks with larger size lead to more individuals whose local positions are affected during the insertion. Thus it increases the number of update operations in the position list. While block with too small size will decrease the efficiency of both index updating and query processing. That's because the individuals' real positions are computed by looking up the landmark table. Small blocks will increase the size of landmark table and thus slow down the look up operation.

Small blocks also produce many fragments on disk which will affect the disk access time. In our experiment in Section 6.2, we will demonstrate and further discuss the impact of block size to the index update.

For concept names, relation names or concept individuals, which only stored in posting lists, the index structure is the same as it defines in Section 3.2 and we update the index by using the existing optimizations for traditional IR index maintenance([3,6]). The update in these posting lists are infrequent and less time costing comparing to that of relation triples.

6   Evaluation

6.1   Experiment Setup

We use both the real world data and the artificial semantic data in our experiment. In order to simulate the data change on the Semantic Web, a representative NTriple file (persons.nt) from DBpedia is used as the real world data. Table 2 shows its content update during a time interval of five months [1].

Table 2. Triples of real world dataset

Dataset	Version 2.0	Version 3.0	Percentage
No. of triples	557,126	569,051	-
Inserted triples	-	157,127	28.2%
Deleted triples	-	145,280	26.1%

In order to test the efficiency of our index update mechanism and its impact on query answering, we also use the LUBM [19] benchmark data. In the LUBM dataset, data is randomly generated and can be scaled to an arbitrary size. For each dataset from LUBM(1,0) to LUBM(20,0), we treat its content as triples to be inserted into an existing block index which is build for the LUBM(50,0) dataset. Table 3 shows the number of triples from LUBM(1,0) to LUBM(50,0). In Section 6.3 we also evaluate the query processing time under LUBM(20,0) and LUBM(50,0).

Table 3. Triples of artificial datasets

Dataset	LUBM(1,0)	LUBM(5,0)	LUBM(10,0)	LUBM(15,0)	LUBM(20,0)	LUBM(50,0)
No. of triples	102,737	643,435	1,311,787	2,014,462	2,772,017	6,865,225

The proposed experiments are carried out on a desktop PC with Pentium 4 CPU of 3.2 GHz and 2Gb memory, running Microsoft Windows Server 2003 with Sun Java JRE 1.5.0. Note that single indexing thread is used in our experiment to obtain a raw indexing speed for ease of comparison.

---

[1] The DBPedia 2.0 version was launched in 09/2007 while the 3.0 version was launched in 02/2008.

6.2 Index Update Performance

In this section, we evaluate the efficiency and scalability of our index update mechanism. Table 4 shows both the index construction time and index space size for persons.nt v2.0 using the two different index structures. Semplore [9] is based on the index structure already introduced in Section 3.2. Note that our block index only slightly increases the index construction time. Moreover, the same conclusion can be drawn on the size of index space, which indicates that it would not lead to the space overhead.

Table 4. Index construction for persons.nt v2.0

Person.nt v2.0 from DBpedia	Semplore	Block Index
Index Construction Time (s)	218	231
Index Space (MB)	37	38

When updating the index to persons.nt v3.0, we choose different block size to test the performance. As shown in Fig. 4, we can see that when the block size is increasing, the update time increases. Blocks with larger size lead to more individuals whose local positions are affected during the insertion and thus increase the number of update operations. However, when the block size is chosen as 50, it took more time to update the index than with block size equals to 100. The main reason is that getting individuals' real position needs to look up the landmark table. When the block size is small, the size of landmark table is increased and thus the index update time is slow down. Since Semplore does not provide the index update mechanism, it's index can only be updated by complete reconstruction, which takes 225 seconds.

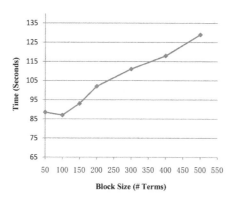

Fig. 4. Index update time with different block sizes

For the LUBM dataset, we first build the index based on the Block Index structure under the dataset of LUBM(50,0). Here we chose the block size equals to 1000. For every dataset from LUBM(1,0) to LUBM(20,0), we take it as the set of triples be to inserted into the existing Block Index. Then we insert each of them into the original index of LUBM(50,0) to evaluate the index update time. For Semplore, we rebuild the index for the original dataset LUBM(50,0) together with the new inserted dataset. The results are shown in Fig.5. Together with Table 3, we can find that the index update time is a fraction of complete rebuild due to the portion of the inserted triples.

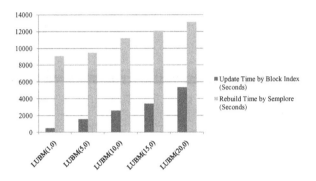

Fig. 5. Index update time vs. index rebuild time

6.3   Query Response Time

In additional to the efficiency of index update performance, it is also important to test whether the block index structure would largely influence the time of query answering. We choose 8 of 14 LUBM benchmark queries mentioned in [19] for the evaluation, which is shown in Table 5. The excluded queries are either cyclic or with multiple variables which are out of the query capability of Semplore (i.e. unary tree-shaped conjunctive query). Here we only focus on testing the efficiency on retrieval but not the reasoning capability.

Table 6 shows the query response time under LUBM(20,0) and LUBM(50,0) by the two different indices. The block index is built by setting the block size as 1000. The response time of the block index is slightly slower than that of

Table 5. LUBM benchmark queries

Q1	(type GraduateStudent ?X) (?X takesCourse Department0.University0.GraduateCourse0)
Q3	(type Publication ?X) (?X publicationAuthor Department0.University0.AssistantProfessor0)
Q5	(type Person ?X) (?X memberOf Department0.University0)
Q6	(type Student ?X)
Q10	(type Student ?X) (?X takesCourse Department0.University0.GraduateCourse0)
Q11	(type ResearchGroup ?X) (?X subOrganizationOf University0)
Q13	(type Person ?X) (University0 hasAlumnus ?X)
Q14	(type UndergraduateStudent ?X)

Table 6. Query Response Time for LUBM Datasets (ms)

Query	LUBM(20,0)		LUBM(50,0)	
	Semplore	Block Index	Semplore	Block Index
Q1	14	63	14	84
Q3	2	43	3	51
Q5	1	32	2	43
Q6	16	19	31	34
Q10	2	43	2	47
Q11	1	1	1	2
Q13	1	39	13	59
Q14	1	2	32	36

Semplore as it needs to lookup the landmark table stored in the main memory when returning the individuals local position. Moreover, the retrieval time is almost the same as Semplore when queries are tend to find individuals of concepts or keywords. This is due to the fact that the corresponding posting lists do not use the position lists thus are not stored in the block index. This way, the block index is proved to provide much more flexibility for index update mechanisms while preserving the efficiency of query answering.

7   Conclusion

In this paper, we present a well-designed update mechanism on the state of the art IR index (Semplore) for triples. Benefited from the basic idea of dividing the index into blocks, it reduces the number of update operations during inserting triples, which results in several orders of magnitude decrease on the index update time compared to that by complete reconstruction. Moreover, both the size of index space and query response time are not notably effected. Thus, our proposed mechanism makes it possible for Semplore to handle frequent semantic data update while preserving efficient hybrid query answering. One future work we are considering is to offer more suitable block sizes and update strategies to meet the requirements of different situations in order for self-tuning.

References

1. Ding, L., Finin, T., Joshi, A., Pan, R., Cost, R.S., Peng, Y., Reddivari, P., Doshi, V.C., Sachs, J.: Swoogle: A Search and Metadata Engine for the Semantic Web. In: Proceedings of the Thirteenth ACM Conference on Information and Knowledge Management. ACM Press, New York (2004)
2. d'Aquin, M., Baldassarre, C., Gridinoc, L., Sabou, M., Angeletou, S.: Watson: Supporting next generation semantic web applications. In: WWW 2007 (2007)
3. Tomasic, A., García-Molina, H., Shoens, K.: Incremental updates of inverted lists for text document retrieval, pp. 289–300 (1994)
4. Brown, E., Callan, J., Croft, W.: Fast incremental indexing for full-text information retrieval. In: Proceedings of the 20th International Conference on Very Large Databases (VLDB), Santiago, Chille, pp. 192–202 (1994)

5. Büttcher, S., Clarke, C.L.A., Lushman, B.: Hybrid index maintenance for growing text collections. In: Proceedings of SIGIR 2006, New York, NY, USA, pp. 356–363. ACM, New York (2006)

6. Lester, N., Zobel, J., Williams, H.: Efficient online index maintenance for contiguous inverted lists. Inf. Process. Manage. 42(4), 916–933 (2006)

7. Lim, L., Wang, M., Padmanabhan, S., Vitter, J.S., Agarwal, R.C.: Efficient update of indexes for dynamically changing web documents. In: World Wide Web, pp. 37–69 (2007)

8. Brin, S., Page, L.: The anatomy of a large-scale hypertextual Web search engine. Computer Networks and ISDN Systems 30(1–7), 107–117 (1998)

9. Zhang, L., Liu, Q., Zhang, J., Wang, H., Pan, Y., Yu, Y.: Semplore: An ir approach to scalable hybrid query of semantic web data. In: Proceedings of ISWC/ASWC 2007, pp. 652–665 (2007)

10. Lempel, R., Mass, Y., Ofek-Koifman, S., Sheinwald, D., Petruschka, Y., Sivan, R.: Just in time indexing for up to the second search. In: CIKM, pp. 97–106 (2007)

11. Jang, H., Kim, Y., Shin, D.: An effective mechanism for index update in structured documents. In: CIKM, pp. 383–390 (1999)

12. Rocha, C., Schwabe, D., Aragao, M.P.: A hybrid approach for searching in the semantic web. In: Proceedings of the 13th international conference on World Wide Web, pp. 374–383. ACM Press, New York (2004)

13. Bast, H., Chitea, A., Suchanek, F.M., Weber, I.: ESTER: efficient search on Text, Entities, and Relations. In: Proceedings of SIGIR 2007, Amsterdam, Netherlands, pp. 671–678. ACM, New York (2007)

14. Tummarello, G., Oren, E., Delbru, R.: Sindice.com: Weaving the open linked data. In: Aberer, K., Choi, K.-S., Noy, N., Allemang, D., Lee, K.-I., Nixon, L., Golbeck, J., Mika, P., Maynard, D., Mizoguchi, R., Schreiber, G., Cudré-Mauroux, P. (eds.) ASWC 2007 and ISWC 2007. LNCS, vol. 4825, pp. 552–565. Springer, Heidelberg (2007)

15. Chu-Carroll, J., Prager, J.M., Czuba, K., Ferrucci, D.A., Duboué, P.A.: Semantic search via XML fragments: a high-precision approach to ir. In: Proceddings of SIGIR, pp. 445–452 (2006)

16. Li, Q., Moon, B.: Indexing and querying XML data for regular path expressions. In: The VLDB Journal, pp. 361–370 (2001)

17. Tatarinov, I., Viglas, S., Beyer, K.S., Shanmugasundaram, J., Shekita, E.J., Zhang, C.: Storing and querying ordered XML using a relational database system. In: SIGMOD Conference (2002)

18. Horrocks, I., Tessaris, S.: Querying the semantic web: A formal approach. In: Horrocks, I., Hendler, J. (eds.) ISWC 2002. LNCS, vol. 2342, pp. 177–191. Springer, Heidelberg (2002)

19. Guo, Y., Pan, Z., Heflin, J.: Lubm: A benchmark for owl knowledge base systems. J. Web Sem. 3(2-3), 158–182 (2005)

Towards a Component-Based Framework for Developing Semantic Web Applications

Raúl García-Castro[1], Asunción Gómez-Pérez[1], Óscar Muñoz-García[1], and Lyndon J.B. Nixon[2]

[1] Ontology Engineering Group, Departamento de Inteligencia Artificial
Facultad de Informática, Universidad Politécnica de Madrid, Spain
{rgarcia,asun,omunoz}@fi.upm.es
[2]AG Netzbasierte Informationssysteme, Freie Universität Berlin, Berlin, Germany
nixon@inf.fu-berlin.de

Abstract. For those outside the research community, to develop Semantic Web applications entails real difficulty. This difficulty is due in part to the lack of usable approaches for planning Semantic Web solutions, even though Semantic Web tools have already reached industrial maturity. We propose here the Semantic Web Framework, a component-based framework for analysing rapidly the required components, the dependencies between them, and selecting existing solutions. This approach has been tested with a number of industrial partners, which justifies the effort made in this direction.

1 Introduction

Semantic Web technologies are slowly but surely moving out of the borders of the research community and reaching all types of business users, ranging from large multinational companies to individuals. These users, when convinced of the benefits that the Semantic Web technology provides to their problems and processes, may want to switch from being technology consumers to technology producers, by building their own Semantic Web-based solutions on top of existing tools and methodologies. However, when non-expert users try to plan and develop Semantic Web solutions they currently face several obstacles:

- They do not know the types of technologies now existing nor the functionalities that these provide, nor do they know what are the dependencies between the different technologies.
- They do not know how to use the Semantic Web technology, so they cannot reuse or include this technology into their own applications.
- They do not know whether these technologies can interoperate either between themselves or with their own technologies and, if so, how this interoperability can be achieved.
- They cannot accurately make decisions, such as cost or resource estimations, when including semantic capabilities into their applications or when building Semantic Web applications from scratch.

J. Domingue and C. Anutariya (Eds.): ASWC 2008, LNCS 5367, pp. 197–211, 2008.
© Springer-Verlag Berlin Heidelberg 2008

Although reaching a universal agreement on how to develop a Semantic Web application is almost impossible, facilitating the understanding and development of Semantic Web applications by giving design guidelines through software patterns and exploiting software reuse techniques is really feasible. Nowadays, to construct applications from a collection of reusable components and frameworks is a popular approach to software development. Components provide a number of benefits because they simplify application development and maintenance, and thus, they allow systems to be more adaptive and to respond rapidly to changing requirements [1].

The Semantic Web Framework is intended to help Semantic Web application developers design and build Semantic Web applications. This framework can be a first step to solve the above problems, though later on it should be extended with interface descriptions, benchmarking, interoperability tests and cost models. The framework is a reference framework that currently provides descriptions of the existing types of Semantic Web technologies and their functionalities, and of the dependencies between these technologies.

Our approach involves classifying the different Semantic Web technologies according to their functionalities and representing them as independent components grouped under a smaller set of component groups. For each component, we give a description of the functionalities that the component provides and then we identify the dependencies between the different components. The level of the descriptions is understandable enough to non-experts; additionally, with our industry partners we have validated through use case analysis the accessibility of the framework to non-experts, enabling them to identify rapidly the required components with their planned Semantic Web application, thus ensuring a viable final concept through taking component dependencies into account.

With the appropriate extensions to the framework, we expect to facilitate the use and reuse of this technology and to avoid inconsistencies when developing Semantic Web applications by providing further specifications and guidelines for components.

This paper is structured as follows: Section 2 presents a brief explanation of component-based software development, software architectures and frameworks. Section 3 describes the commonalities of Semantic Web applications and the related work that supports application development in this context. Section 4 focuses on the Semantic Web Framework, the components involved in it and in the dependencies between such components. Section 5 shows how the Semantic Web Framework is used to support real industrial use cases and to determine their component needs and dependencies. Finally, Section 6 draws the conclusions of this work and proposes future lines of research.

2   Background

2.1   Component-Based Software Engineering

Reuse-based software engineering is becoming the main development approach for business and commercial systems. One of this reuse-based approaches is

Component-Based Software Engineering (CBSE), which is the process of defining, implementing and composing loosely coupled independent components into systems [2]. In CBSE, application developers reuse components already developed and tested to build their applications in a robust and rapid way, only knowing the component interface or contract and not knowing the details of the component implementation or the way the component was conceived to be used.

CBSE relies on *independent components* that are completely specified by their interfaces, *component standards* that facilitate the integration of components, *middleware* that provides software support for component integration and *a development process* that is geared to CBSE. According to this, the Semantic Web Framework provides the skeleton for a specification of the independent components needed.

A software component is a software composition unit that specifies a set of interfaces and a set of requirements; and that can be composed with other components independently in time and space [3].

Component-based systems have the following characteristics:

- *Interoperability*. Components cooperate despite differences in language, interface, and execution platform.
- *Distribution*. Components can be hosted in different machines in a network.
- *Heterogeneity*. Components can be executed in different platforms or operating systems and written in different languages by different developers.
- *Extensibility independence*. The applications are modifiable and extensible adding new components.
- *Dynamism*. Applications can evolve by component extension, extinction, substitution, or by reconfiguring the relationships between components.

The Semantic Web Framework has been defined as a component-based framework because Semantic Web applications possess similar characteristics to component-based systems above presented. Furthermore, component-based frameworks provide the features that facilitate software reuse [4]: *abstraction*, to reduce and factor out details; *selection*, to help developers locate, compare and select reusable software artifacts; *specialisation*, to particularize generic artifacts; and *integration*, to combine a collection of artifacts.

2.2   Software Architectures and Frameworks

A **software architecture** is defined as the fundamental organization of a system embodied in its components, their relationships to each other and to the environment, and the principles guiding its design and evolution [5].

The objectives of software architectures are to understand and improve complex application structures; to reuse application structures so as to solve similar problems; to plan the application evolution; to analyse the application correction and the compliance degree with respect to the initial requirements; and to allow the study of some domain specific parts.

Software architectures are described by a) the *components* that realise the computational and data storage aspects; b) the *interaction* between components during the execution; c) the *patterns* that describe the component composition; and d) the *restrictions* imposed when applying those patterns.

Frameworks are a kind of domain-specific software architecture [6], which define the architectural style relating the components inside a system. Furthermore, they define a set of components and their interfaces in an abstract way, establishing the interaction rules and mechanisms between them.

Depending on the framework applicability, frameworks can be classified into horizontal and vertical frameworks [3]. *Horizontal frameworks* are valid for every application domain relative to a concrete aspect of the system (e.g., communication infrastructures, user interfaces, visual environments, etc.). *Vertical frameworks* are developed specifically for a concrete application domain such as telecommunications, manufacturing, multimedia services, etc.

In this paper, the Semantic Web Framework is a horizontal framework that constitutes an abstract reusable design represented by the components commonly involved in the architecture of semantic applications as well as the dependencies between these components.

3   Semantic Web Applications

The Semantic Web is an extension of the current web, in which information is given well-defined meaning, better enabling computers and people to work in cooperation [7]. In this context, in which the web is a network of application-usable information, we can define a Semantic Web application as a software application that uses or produces information for the Semantic Web.

As companies begin to perceive the benefits of semantic technologies, they will explore how to apply this technology to build Semantic Web applications. These applications have been characterised by different authors [8] and by events such as the Semantic Web Challenge[1] with the following features:

- Data has semantics and is represented using formal descriptions.
- Semantic data is reused, manipulated and processed.
- Data sources are heterogeneous and are owned or controlled by different organisations.
- Applications assume an open world (i.e., the information is never complete).
- Multiple natural languages are supported.
- RDF(S) and OWL, the open standards recommended by the W3C, are used.

In the Semantic Web, the term reuse appears not only at the data level, as shown above, but also at the application level, because nowadays there exist many open software from a wide range of sources that can be reused when building Semantic Web applications. At the application level, reuse follows three different approaches: a distributed services approach, which integrates web service technology into their architectures; a shared memory approach, which composes components using a shared space of common memory to communicate, as is the case of libraries being reused inside an application; and a mixed approach, which combines the two approaches explained before.

---

[1] http://iswc2007.semanticweb.org/callfor/SemanticWebChallenge.asp

3.1   Semantic Web Application Architectures

Only a few architectures for Semantic Web applications have been proposed so far.

Mika et al. sketch a generic architecture of ontology-based applications grounded in a call-and-return style and structured in hierarchical layers [9]. The layers involved from bottom to top are the following: ontology, middleware and application. The ontology layer contains the components concerned with the creation and maintenance of the model of the application; the middleware layer supplies common ontology-related services; and the application layer rests on the ontology and on related services to provide some kind of ontology functionality to an end user.

Tran et al. [10] present a service-oriented architecture also structured in hierarchical layers: the data layer hosts any kind of data sources, including sources different from ontological ones; the logic layer includes application-specific services that are implemented for a particular use case and that operate on specific object models; finally, the presentation layer hosts presentation components that the user interacts with. These authors also classify the components inside the logic layer into ontology services, ontology engineering services and ontology usage services.

By contrast, the framework described in this paper is an open system and is not divided in layers. Layered approaches, on the other hand, present several disadvantages, such as the difficulty in structuring some systems in a layered fashion; performance considerations when high level functions require close coupling to low level implementations; and the difficulty in finding the right level of abstraction, especially if existing systems cross several layers [11].

The two architectures presented above identify some example components that illustrate their approaches. However, in the Semantic Web Framework we have tried to identify exhaustively the existing semantic components of Semantic Web applications. The 32 components we have identified in the Semantic Web Framework cover the 16 and 21 components identified in the previous approaches.

4   The Semantic Web Framework

In this paper, the Semantic Web Framework is defined as a structure in which Semantic Web applications can be organised and developed. The Semantic Web Framework is guided by some general design principles that state that the Semantic Web Framework should be

- *Developer-oriented.* Different audiences such as developers with low expertise in Semantic Web technologies or ontology practitioners should be considered.
- *Easy to understand.* To facilitate the understanding and use of the Semantic Web Framework, its components have been organised in dimensions according to the major properties of the problem space that have significant variation over Semantic Web technology.

- *Inexpensive to adopt.* To develop a Semantic Web application or to upgrade an existing application with semantic capabilities should be easy and thus, the impact on legacy systems is minimised.
- *Semantics focused.* To describe only the components that provide semantic functionalities and functionalities to manage semantics. Other components that deal with communication, distribution, etc. have not been taken into account to ease the integration of the components of the Semantic Web Framework into other software architectures.
- *Component based.* To define some specifications of these components that allow different implementations of them, providing each of these components a basic functionality.
- *Evolving.* To extend easily the Semantic Web Framework by inserting new components or by modifying existing ones because the Semantic Web, and its technology, is continuously evolving.

According to the definition of software architecture presented in Section 2, if we want to define the architecture of the Semantic Web Framework, we need to identify its components, the interaction between them, the patterns that describe their composition, and the restrictions to impose when applying those patterns.

Therefore, this paper is focused on the identification of the components of the Semantic Web Framework; on their classification, as stated below; and on the main interfaces of the Semantic Web Framework components with other components and with the environment. In a future work, we will define a concrete specification of the interfaces and the different patterns that can be used in Semantic Web applications.

4.1   Definition and Classification of Components

We follow the definition of component given by Szyperski [3] since a Semantic Web Framework component is as an autonomous and modular unit with well defined interfaces that describes a service and performs a specific functionality. These components can be used either independently or together to develop applications for the Semantic Web; and they can be implemented using services, program libraries or applications.

Components are usually defined by specifying some *general information* about them, such as a natural language description; their *interfaces*, including the functionalities that the component implements and those that it uses; and their *contracts*, which are specifications added to the interface that establish use and implementation conditions [3].

Within the Semantic Web Framework, we do not describe the component contracts, since these will be defined in future work, but we explicitly classify the interfaces into the functionalities that a component implements and those that it uses. Therefore, each component is defined by the following:

- *Name.* The name of the component.
- *Description.* A high-level description of the component.

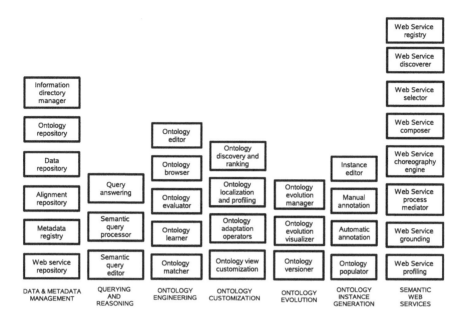

Fig. 1. Components of the Semantic Web Framework

- *Functionalities provided.* An enumeration of the functionalities that the component provides, specifying for each functionality the type or types of interface that it has (user interface, programming interface, service interface, hardware interface, etc.).
- *Component dependencies.* These include an enumeration of the functionalities required by the component to work correctly and that are provided by other components.

To classify the components of the Semantic Web Framework, we have considered the dimensions of an architecture as the major properties of the problem space that have significant variation over Semantic Web systems, in other words, the groups of components that provide some specific support to the architecture. These dimensions, however, are not exhaustive; we have classified the different components according to the main functionalities that they provide, as stated in previous Semantic Web technology classifications [12,13].

Figure 1 presents the components that have been identified from software currently available or under construction. The enumeration of components is neither exhaustive nor complete, and is open to improvements and extensions. The current components have been identified by members of the Knowledge Web[2] Network of Excellence who have great expertise in each of the dimensions.

In Figure 1, each dimension of the architecture is represented as a column and reflects those components that provide a particular functionality to the architecture. It should be noted that the order of the components or of the

---

[2] http://knowledgeweb.semanticweb.org/

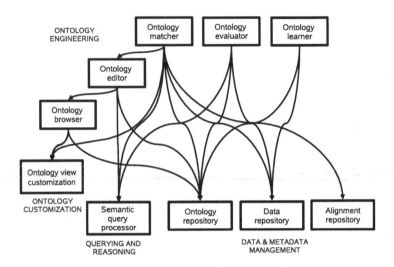

Fig. 2. Dependencies of the components on the *Ontology engineering* dimension

dimensions in the figure does not imply any precedence or relation between them.

The dependencies of each of these components on other components of the framework were identified. Figure 2 shows the basic dependencies of the components on the *Ontology engineering* dimension. Component dependencies are represented graphically in the following way: when one component depends on the functionalities of another, it is then represented with an arrow going from the first component to the component that provides the functionalities.

Existing software that implements the components was also identified. It must be observed that existing implementations may include the functionalities of multiple components. This is clearly seen in ontology engineering platforms, which give support to different tasks of the ontology development process and cover multiple components. In total, we identified 200 component implementations: 43 in the *Data and Metadata Management* dimension, 10 in the *Querying and Reasoning* dimension, 78 in the *Ontology Engineering* dimension, 25 in the *Ontology Customization* dimension, 10 in the *Ontology Evolution* dimension, 15 in the *Ontology Instance Generation* dimension, and 19 in the *Semantic Web Services* dimension.

On the other hand, even if there is a dependency between two components (e.g., an *Ontology editor* requires an *Ontology repository*), in the real world all the implementations of a certain component will not be compatible with all the implementations of the dependent component.

Next, a description of the dimensions of the Semantic Web Framework and of the components included inside each dimension is given. The full description of the Semantic Web Framework components, dependencies and implementations can be found in [14].

Data and Metadata Management. This dimension includes those components that manage knowledge and data sources, such as:

- *Information directory manager.* It handles query distribution, manages provider directories, identifies information providers from a query, and handles the storage and access to distributed ontologies and data.
- *Ontology repository.* It locally stores and accesses ontologies and instances.
- *Data repository.* It locally stores and accesses data and ontology annotated data.
- *Alignment repository.* It handles the storage and access to distributed alignments.
- *Metadata registry.* It locally stores and accesses metadata information.

Querying and Reasoning. This dimension includes those components that generate and process queries, such as:

- *Query answering.* It takes care of the logical processing of a query by providing reasoning functionalities to search results from a knowledge base.
- *Semantic query processor.* It takes care of the physical processing of a query by providing functionalities to manage query answering over ontologies in distributed sources.
- *Semantic query editor.* It takes care of the user interface for editing queries.

Ontology Engineering. This dimension includes those components that provide functionalities to develop and manage ontologies, such as:

- *Ontology editor.* It allows creating and modifying ontologies, ontology elements, and ontology documentation. These functionalities include single ontology component editing or more advanced editing, such as ontology pruning, extension or specialization.
- *Ontology browser.* It allows visually browsing an ontology.
- *Ontology evaluator.* It evaluates ontologies, either their formal model or their content, during the different phases of the ontology life cycle.
- *Ontology learner.* It acquires knowledge and generates ontologies of a given domain through some kind of (semi)-automatic process.
- *Ontology matcher.* It matches two ontologies or an ontology and another data source and outputs some alignments. Two types of ontology matchers can be distinguished, one that generates matchings and one that uses matchings for other tasks (merging, mediating, etc.).

Ontology Customisation. This dimension includes the components that customize and tailor ontologies, such as:

- *Ontology localization and profiling.* It adapts an ontology according to some context or some user profile.
- *Ontology discovery and ranking.* It finds appropriate views, versions or subsets of ontologies, and ranks them according to some criterion.

– *Ontology adaptation operators.* It is in charge of applying appropriate operators to the ontology in question, resulting in an ontology customized according to some criterion.
– *Ontology view customisation.* It enables the user to change or amend a view on a particular ontology to fit a particular purpose.

Ontology Evolution. This dimension includes those components that manage the ontology evolution, such as:

– *Ontology versioner.* It maintains, stores and manages different versions of an ontology.
– *Ontology evolution visualizer.* It visualises different versions of an ontology.
– *Ontology evolution manager.* It is in charge of the timely adaptation of an ontology to the changes undergone and of the propagation of such changes to dependent artifacts.

Ontology Instance Generation. This dimension includes those components that generate ontology instances, such as:

– *Instance editor.* It allows creating and modifying manually instances of concepts and of relations between such concepts in existing ontologies.
– *Manual annotation.* It allows the manual and the semi-automatic annotation of digital content documents (e.g. web pages) with concepts in the ontology. This annotation process may be assisted or guided by a machine (semiautomatic annotation).
– *Automatic annotation.* It allows the automatic annotation of digital content (e.g., web pages) with concepts in the ontology. Occurrences in the considered content of concept instances are automatically detected and subsequently annotated.
– *Ontology populator.* It automatically generates new instances in a given ontology from a data source.

Semantic Web Services. This dimension includes those components that discover, select, mediate, compose, choreograph, ground, and profile semantic web services, such as:

– *Web service discoverer.* It publishes and searches service registries, controls access to registries, and distributes and delegates requests to other registries.
– *Web service selector.* After discovering a set of potentially useful services, this component checks whether the services can actually fulfil the user's concrete goal and under what conditions.
– *Web service composer.* It automatically composes web services to provide new value-added web services.
– *Web service choreography engine.* It uses the choreography descriptions of the service requester and provider to drive their conversation.
– *Web service process mediator.* It reconciles the public process heterogeneity that can appear during the invocation of web services.

– *Web service grounding.* It is responsible for web service communication.
– *Web service profiling.* It creates web service profiles based on their execution history.
– *Web service registry.* It registers semantic web services.

5   Use Cases

In order to check the viability of use of the Semantic Web Framework by non-experts from the industry, we selected some of the use cases from Knowledge Web and carried out face-to-face interviews with industry members. Then, a few days before the meeting, we sent them a copy of the Semantic Web Framework specification to read. When the meeting was held, they had the opportunity to raise any questions about the framework they had encountered. Then, their use case was analysed according to the required components. This analysis was led by the industry partner while the Knowledge Web researcher's function was to help the industry partner understand the functioning of the components.

We found out that even before being prompted by the researcher, the industry partners were able to identify most of the components required by their use case and were able to intuitively understand the dependency diagrams, leading to avoidance of inconsistencies (e.g., recognizing that they had forgotten to explicitly add a certain component). In total, 8 use cases were analysed with the Semantic Web Framework. Here we show only one of those use cases, but the reader can find them all in [14].

5.1   Semantic Aggregation of News Stories

We chose a use case from the technology provider Neofonie GmbH[3]. Neofonie represents the typical case of a small company with an interest in deploying semantic solutions to improve their technology offer and better their competitiveness. They have a general knowledge of what semantic technologies are, but lack expert knowledge to successfully evaluate and deploy the technology. We illustrate the framework with their use case as we consider this an ideal scenario for our work to support industry in better modelling of semantic solutions for their needs, the necessary first step before further evaluation and deployment of the technology.

The selected use case deals with the provision of an aggregated news service able to provide business clients with accurate search, thematic clustering, classification of news stories, and e-mail notification of stories of interest. The news sources used are not just the main news feeds and media outlets but also press releases, announcements on websites and other "alternative" sources.

The result of the analysis is shown in Figure 3, which presents the components of the Semantic Web Framework that can support this use case and their dependencies. This analysis could be performed within the company based on a reading of the component descriptions and dependencies, with the final diagram resulting from a briefer meeting with an expert to clarify open issues.

---

[3] http://www.neofonie.de

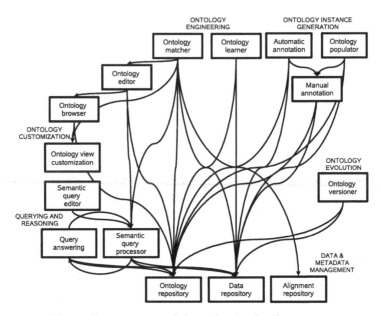

Fig. 3. Components and dependencies for the use case

In order to achieve all of the goals proposed in the business use case, the system could use the following Semantic Web Framework components:

- The *Ontology repository*, the *Data repository*, the *Alignment repository* and the *Metadata registry* store all the data necessary for the use case: the ontologies used for each source, the instance data extracted from these sources and the alignments that have been created between each source ontology.
- The *Query answering*, the *Semantic query processor* and the *Semantic query editor* provide both the user interface support for formulating the query and displaying the results and the system-intern support for performing the query across the aligned instance data and extracting the results.
- The ontologies for representing the data of each source are semi-automatically created using ontology learning techniques through the *Ontology learner* component. The initial ontology extraction is refined with the *Ontology browser* component to view the ontology and the *Ontology editor* component to complete the ontology manually.
- It is possible that with the use of the system over time, the ontologies will need to be revised as new concepts or properties gain relevance. Hence, the *Ontology versioner* component may be employed at a later stage in the system. Likewise, in the ontology extraction part, extracted terms may overlap with those of existing ontologies for related domains such as politics, sport etc. Given the existence of an ontology that represents terms from a certain source, knowledge extraction can take place. Instance data is generated through semi-automatic annotation approaches with the *Automatic annotation* component, the *Manual annotation* component for adding semantic data to news sources, and the *Ontology population* component.

– Finally, two approaches to searching can be considered. In one, queries are expressed in terms of one ontology and, at run time, they are mapped into the other ontologies of the sources; then they are executed across the different source data and the results are combined at the end. However, this approach is very resource intensive at query time. The other approach considered is that, given that we update the source data only periodically, it makes better sense to transform all source data into a core ontology, which can be built from the merge of all source ontologies. Then, we first generate alignments between the source ontologies and a core ontology using the *Ontology matcher* component. These alignments need manual proofing and correction. The alignments also help refine the core ontology. Given now a core ontology and alignments to the individual source ontologies, mediators can be generated for the transformation of instance data from any source in terms of the core ontology. Hence a core ontology is maintained against which the queries are executed.

5.2 Results from Use Cases

The findings of the eight selected use cases reveal that some of the components, namely, the *Ontology repository*, the *Data repository* and the *Metadata registry*, are used in all the use cases. Other components, such as the *Alignment repository*, the *Query answering*, the *Semantic query processor*, the *Ontology editor*, the *Ontology browser*, the *Ontology view customization*, the four components of the *Ontology evolution* dimension, and the *Ontology matcher* are used in almost all the use cases. On the other hand, some other components, namely, the *Information directory manager*, the *Ontology evaluator*, the *Ontology discovery and ranking*, the *Ontology adaptation operators*, the *Instance editor* and all the components of the *Semantic Web Service* dimension are not used in the use cases or almost not used. These findings can serve as an indicator of those fields of research that should be focused on to meet more readily industrial requirements on Semantic Web applications.

Another benefit of this analysis is that the industry members had a basis for choosing which existing Semantic Web tools could be directly re-used in their applications. For each identified component, we provide a list of existing implementations.

Our dependency diagrams are a first step towards a formal analysis of the overall design, where the industry partner can prove whether all dependencies between components are taken into account. In future work, this will be supported further by specifications of component interfaces and reports on component interoperability.

6 Conclusion and Future Work

The Semantic Web Framework is intended to help developers build Semantic Web applications and to diminish development costs. This work is a first step to provide the foundation for large-scale development of Semantic Web applications; it presents a first definition of the Semantic Web Framework and describes the existing types of Semantic Web technology, their functionalities, and the dependencies between these technologies.

Although the Semantic Web Framework is useful as a reference and helps reusing existing technology, Semantic Web application developers will still have to develop their applications and their functionalities.

Immediate uses of the Semantic Web Framework include the identification of the components needed for a Semantic Web application in the software design phase or the identification of existing implementations of components to be reused. In these cases, having descriptions of the Semantic Web Framework components and their implementations in a machine-processable form can help automate these tasks.

Future work includes providing sets of compatible tools from the components which are already implemented by existing tools. Therefore, the Semantic Web Framework will provide not just single component implementations but also groups of already-interoperable implementations.

We will extend the usability of the framework by providing evaluations and benchmarks of component implementations, interoperability testing between components and cost/benefit models for Semantic Web application development.

Another line of work is to realise the Semantic Web Framework as an infrastructure of semantic focused services so they can be used in the context of a Service Oriented Architecture when semantic functionalities are needed. This will require to develop specifications of the component interfaces, of their interactions, and to develop the middleware needed to adapt the interface specifications to the concrete implementations API. These developments will allow utility computing for semantic resources, i.e., to organise semantic resources so that they may be accessed when needed, just like traditional utilities such as gas, water, or electricity [15].

Within the NeOn project (IST-2005-027595) we are creating a methodology to support the rapid prototyping and development of a new generation of large scale, complex, semantic applications. The overall goal of this methodology is to ensure that economically viable solutions will appear on the market and help application developers to build Semantic Web applications from scratch or by including semantic components into traditional information systems. In this context, the Semantic Web Framework constitutes the starting point of the NeOn methodology that will take into account the existing methods for building component-based software as for example the described in [16].

Acknowledgements

Thanks to the collaborators in the definition of the Semantic Web Framework components: S. Costache, S. Dasipoulou, Y. Ding, M. Dzbor, J. Euzenat, M. Kaczmarek, F. Lécué, D. Maynard, V. Novacek, R. Palma, R. Piskac, M.C. Suárez-Figueroa, and D. Zyskowski. This work is partially supported by a FPI grant from the Spanish Ministry of Education (BES-2005-8024), by the IST project Knowledge Web (FP6-507482), by the CICYT project Infraestructura tecnológica de servicios semánticos para la web semántica (TIN2004-02660), and by the InnoProfile-Corporate Semantic Web project funded by the German Federal Ministry of Education and Research (BMBF) and the BMBF Innovation Initiative for the New German Länder - Entrepreneurial Regions. Thanks to Rosario Plaza for reviewing the grammar of this paper.

References

1. Oberle, D.: Semantic Management of Middleware. Semantic Web and Beyond (2006)
2. Sommerville, I.: Software Engineering, 8th edn. International Computer Science Series. Addison-Wesley, Reading (2007)
3. Szyperski, C.: Component Software, Beyond Object Oriented Programming. Addison-Wesley, Reading (1998)
4. Krueger, C.W.: Software Reuse. ACM Comput. Surveys 24, 131–183 (1992)
5. IEEE: IEEE Std 1471-2000. IEEE Recommended Practice for Architectural Description of Software-Intensive Systems. IEEE (2000)
6. Traz, W.: DSSA frequently asked questions. ACM Software Engineering Notes 19, 52–56 (1994)
7. Berners-Lee, T., Handler, J., Lassila, O.: The Semantic Web. Scientific American (2001)
8. Motta, E., Sabou, M.: Next generation semantic web applications. In: Mizoguchi, R., Shi, Z.-Z., Giunchiglia, F. (eds.) ASWC 2006. LNCS, vol. 4185, pp. 24–29. Springer, Heidelberg (2006)
9. Mika, P., Akkermans, H.: D1.2 Analysis of the State-of-the-Art in Ontology-based Knowledge Management. Technical report, SWAP Project (2003)
10. Tran, T., Haase, P., Lewen, H., Muñoz-García, Ó., Gómez-Pérez, A., Studer, R.: Lifecycle-Support in Architectures for Ontology-Based Information Systems. In: Proceedings of the 6th International Semantic Web Conference, pp. 508–522 (2007)
11. Shaw, M., Garlan, D.: Software Architecture: Perspectives on an Emerging Discipline, 1st edn. Prentice Hall, Englewood Cliffs (1996)
12. Gómez-Pérez, A., Fernández-López, M., Corcho, O.: Ontological Engineering. Springer, Heidelberg (2003)
13. Davies, J., Studer, R., Warren, P. (eds.): Semantic Web Technologies - trends and research in ontology-based systems. John Wiley & Sons, Chichester (2006)
14. García-Castro, R., Muñoz-García, O., Suárez-Figueroa, M., Gómez-Pérez, A., Costache, S., Maynard, D., Dasiopoulou, S., Palma, R., Novacek, V., Lécué, F., Ding, Y., Kaczmarek, M., Piskac, R., Zyskowski, D., Euzenat, J., Dzbor, M., Nixon, L., Léger, A., Vitvar, T., Zaremba, M., Hartmann, J.: D1.2.5 Architecture of the Semantic Web Framework v2. Technical report, Knowledge Web (2007)
15. Pulier, E., Taylor, H.: Understanding Enterprise SOA. Manning (2006)
16. Cheesman, J., Daniels, J.: UML Components. A Simple Process for Specifying Component-Based Software. Component Software Series. Addison-Wesley, Reading (2001)

Bounded Ontological Consistency
for Scalable Dynamic
Knowledge Infrastructures

Maciej Zurawski, Alan Smaill, and Dave Robertson

Centre for Intelligent Systems and their Applications (CISA), School of Informatics,
University of Edinburgh, Informatics Forum
10 Crichton Street, EH8 9AB Edinburgh, Scotland
m.zurawski@sms.ed.ac.uk, A.Smaill@ed.ac.uk, dr@inf.ed.ac.uk

Abstract. Both semantic web applications and individuals are in need of
knowledge infrastructures that can be used in dynamic and distributed environments where different autonomous entities create knowledge and build their
own view of a domain. Our framework represents this using evolving simple
contextual ontologies and mappings between them, at the same time as incremental logical coherence is maintained. The definition of semantic autonomy
includes these aspects. Our earlier research has shown that a knowledge infrastructure can have semantic autonomy that maintains global consistency, if the
knowledge representation is kept simple. We generalize that research by investigating what happens if the consistency of a knowledge infrastructure is
bounded 1) within certain regions called spheres of consistency, and 2) by allowing a limited variable degree of inconsistency. Our experiments show that a
phase transition can occur in this kind of system, beyond which constant-time
and constant-memory complexity is approached.

Keywords: Semantic autonomy, ontology evolution, ontology management,
bounded consistency, rule-based process modelling, phase transition.

1 Introduction

Distributed knowledge infrastructures are becoming more important in a networked
society where it is difficult for a single authority to provide a perfect model of how
knowledge should be represented using ontologies. This is for example true for decentralized organizations that in order to operate efficiently have to make decisions that
are adapted to their local needs and where every division develops an ontology that
reflects how they conceive reality [1]. Practically, this means that many different
ontologies will evolve - but to give up the task of maintaining their coherence is to
give up the idea of *one* knowledge infrastructure being used by such an organization.
Similar arguments can be made about distributed semantic-web applications where
ontologies are needed and used.

In our earlier work [3], we assumed that a system maintaining these ontologies has to
enforce global consistency in every state of evolution, but we now relax and generalize

J. Domingue and C. Anutariya (Eds.): ASWC 2008, LNCS 5367, pp. 212–226, 2008.
© Springer-Verlag Berlin Heidelberg 2008

that assumption (see table 1). By doing so we wish to make our framework applicable to other scenarios, e.g. modelling *two* organizations where there is a higher logical coherence between the divisions *within* one organization than the coherence *between* the divisions of two different organizations. Also, this is useful in semantic-web related applications where certain local environments (that all have their own ontology) must be kept perfectly consistent because their interaction is of great importance whereas not that of others. Full logical consistency would sometimes be impractical or unnecessary because of different reasons, e.g. such as:

1) The application domain that is modelled as such does not assume such perfect consistency, and
2) Application domains that require extensive or many ontologies, and where better computational scalability is required and even though full consistency would be beneficial that is seen as a less important requirement than a highly scalable infrastructure with a short response time
3) Even though from an epistemological point of view we assume there is an objective reality (but no single language) that in theory would facilitate a perfectly consistent mapping between all ontologies, perhaps

Table 1. The generalised definition of semantic autonomy

The definition of *Semantic autonomy* requires these properties to hold:
1. The local contexts have the freedom to propose a change in their local ontology (i.e. the ontology of the local context) or in the mappings from their ontology. All the possible request types, operation types and explicit change process that manage them, are explicated and formalised
2. The system does "in some way" maintain full or bounded consistency as defined within every sphere of consistency.
3. The ontological language is dynamic and open-ended (i.e. not confined by a pre-defined set) but there is a knowledge source that can provide knowledge about this language.

in practice that knowledge cannot be acquired immediately so then *some* disagreement and inconsistency are accepted as provisional phenomena.

We are interested in properties of the whole dynamic system as such that are incrementally sustained when the system is evolving – this time two types of consistency. We are interested in scalability of the whole system for managing evolving ontologies, the autonomy of the various divisions of the system (formalized as contexts) and in formalizing an *explicit change process* that manages the initiatives and decisions of the autonomous units and consequently can change both the individual ontologies and the mappings between them. The framework is a specification of a knowledge infrastructure (that could include other elements as well that will not be analyzed here, e.g. authorization and a graphical user interface).

Consider several organizations with many divisions (every having its own ontology using the subsets of RDFS or OWL mentioned in 4.1.3) and then the engineers define the required levels of consistency between the divisions. When engineers from any division (in any organization) feel they need to propose a change in their ontology or mappings to other ontologies, they can do so, while this infrastructure mechanism automatically guarantees to maintain all the defined consistency levels in the whole infrastructure – so humans are freed from this hard but crucial task that preserves the integrity of the infrastructure, the organizations and the relationships between them.

Motivation of the Logical Languages Used. Instead of using an expressive Description Logic (DL) this time, we have chosen another much more scalable RDFS-like logic (see the end of section 4.1.3). Therefore, the ontologies and mappings can be visualized as graphs. That simple logic has been extended with explicit notions of context and temporal states, because an explicit notion of state is useful for a dynamic system and the ontologies express several cognitive contexts describing a domain. One could develop a framework and knowledge infrastructure that use a different logic than the one we use and we hope that our work will be repeated using another scalable logic (even without both these explicit features - temporal states and context). We also use a rule-based language for modelling the change processes, and it is independent of the languages for expressing ontologies and mappings.

1.1 Novel Contribution

The novel contribution is that we have both formalized and empirically evaluated **a notion of spheres of consistency, that can contain several ontologies and the mappings between them, inside which the framework and its reasoning process automatically maintains full or proof-bounded consistency (as defined by the user).**

This novelty utilizes an underlying ontology-based framework specifying a knowledge infrastructure and formalizing semantic autonomy (see Table 1) by means of:

Distributed multi-contextual state-based semantics (i.e. a particular logic).
Distributed ontology evolution (governed by an explicit *change process*).
Distributed mapping evolution (governed by an explicit *change process*).
The distributed explicit *process of initiating* change.

all in one framework. We will express the explicit change processes using a rule-based language.

2   Related Research

Some researchers [6] distinguish between structural consistency (that the ontology obeys the constraints of the language) and logical consistency (the ontology is satisfiable). A model-theoretic inconsistency measure is presented by [7], whereas ours is proof based. The question how to make ontologies autonomous is investigated by [8]. Their theoretical model uses unidirectional relationships (bindings) for borrowing entities from other ontologies. We instead use ontology mappings that are directed relationships. They formalize two kinds of reasoning: cautious and brave. The first one uses a local ontology and its neighbours whereas the second one uses the transitive closure. In our own work, a sphere of consistency is a symmetric relationship that can include an arbitrary amount of ontologies and we do explicit modelling of change processes. Research related to the NEON Project [9] compares four different description-logic based formalisms for modular ontologies – all of them have better expressivity than our current logical language but at the cost of exponential worst-time time complexity. The authors mention the interesting distinction between two different approaches: 1) linking/mapping between ontologies and 2) importing (parts of) ontologies into other ontologies. We use the first approach. Some researchers [10] motivate why the social process of creating meaning is important. An interesting application that is also using an ontology-based *layered approach* is described by the authors of [11]. However, their system is designed for the particular purpose of automating system administration and not for the purpose of maintaining distributed ontologies.

3 The Notions Used and an Introductory Example

Before defining our framework we will present the basic assumptions and show a simple example. We assume that there are multiple different contexts and every different context has a potentially unique *ontology* that expresses the point of view of that context (the formal side of this will be explained later). The ontologies that the current system supports are actually simple because that benefits the system's ability to maintain coherence among distributed ontologies while keeping the system scalable. These simple ontologies can be visualized as *graphs* where every node corresponds to a logical *concept* and every edge to an *ontological relation* (currently we define four different ones). Every *ontology mapping* between two ontologies (currently we define five different ones in section 4.1.3) can be visualized as an edge that connects the nodes of two different graphs. The reason for this simplicity of the ontology language is that instead of studying a complex knowledge representation

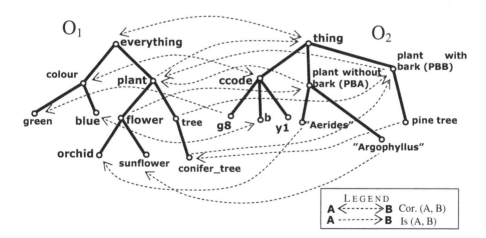

Fig. 1. The ontologies of two local contexts and some mappings between them

language we want to study a whole
dynamic system where this complexity
is observed at the macroscopic level
instead, e.g. many ontologies that are
changing while being constrained – this
complexity will be clear in expression
(7) and the behaviour exhibited in fig 5.
Once the logic is defined, inconsistency
can be defined as a contradiction of the
logical entities within a sphere of consistency, whose degree of contradiction
surpasses that of the defined level inside that consistency sphere (see sec-

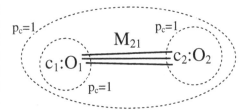

Fig. 2. An abstract illustration that shows two
contexts having an ontology each, a set of
mapping connecting them, and three spheres of
consistency having full consistency ($p_c=1$)

tion 4.1.3 for details). To make things more clear, assuming this simple logic we can
always visualise a *contradiction* as a closed loop (or in special cases several connected loops) in the graph. This means that given two specific concepts and then traversing one path (connecting them) and combining the relationships transitively to a
new single relationship gives different results depending on the path taken and that
these different results cannot coexist – they contradict each other when the meaning
of the relationships is taken into account (see fig. 4 in section 4.1.2.). An application-
oriented interpretation of these notions is to view every context as a division within an
organization and the ontologies as models of the division's local understanding of a
certain domain (e.g. in fig. 1 the first context could represent a customer-facing division and the other context a product division).

Figure 1 shows an example of two local contexts and their ontologies. The dark
edges within the ontologies are subsumption relations, whereas the dotted lines visualize the ontology mappings (two ontology mapping types are shown here: correspondence and the IS-mapping). Let us now conceive that the first local context *initiates*
the proposal to add a new concept, e.g. **yellow** that actually is a type of **colour**. The
framework mechanism should then consider this proposal and *formally investigate* its
consequences – more concretely, it has to investigate if this operation would induce
contradiction or redundancy *in any of the spheres of consistency* that ontology 1 belongs to and if that the levels of contradiction or redundancy surpass the accepted rate.
Fig. 2 illustrates the spheres of consistency. As the next step, the framework accepts
these changes (but it could have rejected them in other situations) and then the first
local context initiates a proposal to add an ontology mapping between **yellow** and e.g.
ccode in the other ontology. Then the framework mechanism has to formally investigate the consequences and the *opinions* of both local contexts, before a potential
change is made. Let's now look at the framework in general.

4   The Framework and Its Layers

We are proposing a solution that will have semantic autonomy as we have defined it.

The solution is a framework consisting of five layers (see fig. 3). The two bottom
layers represent the epistemological and logical assumptions whereas the three top
layers constitute the executable system itself (they are the main focus of this paper).
We will now describe the whole framework (the logical formalization used at the

bottom is mentioned in section 4.1.3). Also, we are interested in the process of pro-posing and reconciling ontological changes and therefore we define a rule-based process language (for the purpose of evolving the ontologies and mappings) that al-lows these three types of statements:

- *entity*: *Predicate* (*parameters*)   where

$entity = c_i \mid$ F   (c_i is a local context i and F the framework mechanism)

(1)

- *Predicate* (*parameters*)
(this purely declarative statement can be true or false)

(2)

- *statement1* $\Rightarrow$ *statement2* (this is the definition of *rule*) where

(3)

statement1 = *entity*: *Predicate* (*parameters*) $\mid$

CNF$_j$(*entity*$_j$: *Predicate* (*parameters*$_j$)) $\wedge$ *Predicate*(*parameters*)

statement2 = *rule*$_1$ $\mid$ *rule*$_1$ or *rule*$_2$ $\mid$ *rule*$_1$ or *rule*$_2$ or *rule*$_3$ $\mid$ *statement1*

We will call the three types of statement type 1, 2 and 3 respectively. The first type of statement means that *entity* makes *Predicate* (*parameters*) true. The second statement type means that *Predicate*(*parameters*) returns its global truth value. The third type of statement is a rule where if *statement1* has been made true, then the rule fires. **CNF**$_j$(exp$_j$) means conjunctive normal form that can contain exp$_1$, exp$_2$, ... etc. If *statement2* is a disjunction of several rules, then they are investigated sequentially until one of them fires – when that happens then the remaining ones are not executed. This formalism has been in-spired by [2] but our formalism is simpler because it does not use message-passing, except in one case (when F: COMM(*message, recipient*) is used to explic-itly communicate something. The whole system S

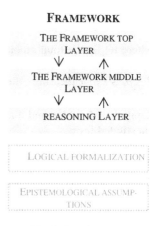

FRAMEWORK

THE FRAMEWORK TOP
LAYER
$\vee$ $\wedge$
THE FRAMEWORK MIDDLE
LAYER
$\vee$ $\wedge$
REASONING LAYER

LOGICAL FORMALIZATION

EPISTEMOLOGICAL ASSUMP-
TIONS

Fig. 3. Our framework

consists of *n* different local contexts c_i ($i=1...n$), their ontologies and the mappings between them. Every local context possesses its own ontology.

4.1   The Reasoning Layer and Spheres of Consistency

4.1.1   Defining Spheres of Consistency
Whereas our framework has in the past [3] supported a notion of semantic autonomy that requires full consistency of the whole system, we now relax that assumption and thereby generalize this notion. A *proof tree* is analogous to a tableau or resolution derivation, where the reasoning mechanism tries to prove a new statement by refuting its negation. In our case the proof tree uses a type of breadth-first search when adding new facts (see end of section 4.1.3.) and it has a certain depth.

Now we define spheres of consistency, in the following way:

Given a set of contexts $\{c_1, c_2, c_3, ... \}$
where every context c_i has an ontology O_i

and given a set of mapping sets $\{m_{12}, m_{13}, m_{23}, \ldots\}$
where every m_{ij} is the set of all mappings connecting contexts i and j where $i{<}j$,
a sphere of consistency is defined as

$$Cons(\{c_i, c_j, \ldots\}, \{m_{ij}, m_{ik}, \ldots\}, p_c) \qquad (4)$$

where $\{c_i \ldots\}$ and $\{m_{ij} \ldots\}$ are defined as above and p_c is a continuous consistency
parameter that can vary between

$p_c{=}1$ which means full consistency, and
$p_c{=}2$ which means that inconsistencies of all depths are fully allowed

When $1{<}p_c{<}2$ then the sphere of consistency defines a proof-bounded consistency,
where there is no proof of contradiction where the proof tree has a smaller depth than
d (that must be an integer), and p_c and d are related through the following formula:

$$p_c = 2 - \frac{d-2}{tot_s - 1} \qquad (5)$$

where tot_s is the total amount of relationships in all the ontologies and mapping sets
that are included in the sphere of consistency. In an analogous way we define
$1 \leq p_r \leq 2$ that measures the amount of bounded redundancy within a sphere of consistency where there is no proof of redundancy where the proof tree has a smaller
depth than d, and p_r and d are related through the same formula as (5) but p_r is simply
substituted for p_c.

4.1.2 Explaining and Motivating the Definition of Spheres of Consistency

The reasoning layer detects if a proposal
would cause a contradiction or redundancy
within some specified spheres of consistency.
If consistency has to be maintained within a
reasoning space that has ontologies and/or
mappings (that in total contain tot_s relationships) then in the worst case the depth of the
reasoning proof tree will be $d{=}tot_s{+}1$ because
if there is no contradiction smaller than
$d{=}tot_s{+}1$, complete consistency can be guar-
anteed. The reason for this is that this particular reasoning algorithm uses breath-first

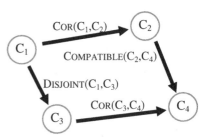

Fig. 4. An example of contradiction
between concepts C_1 and C_4

search. If we use $d{=}tot_s{+}1$ in formula (5) then $p_c{=}1$, i.e. full consistency, but $d{=}2$ will
give the result $p_c{=}2$, i.e. all inconsistencies are possible. The smallest contradiction
that our reasoning recognizes has size 2 [e.g. fig.4 shows a contradiction of size 4]
and if contradictions of that size are accepted then all operations that would introduce
contradictions are accepted.

The reason why we have chosen to define consistency in terms of the depth of the
proof tree is that we can make sure that the system holds this property *incrementally*
when it moves to the next state. Before the whole system S starts to evolve the user
has to *define* all spheres of consistency, their regions (i.e. the sets in formula (4)) and

their individual degrees p_c and p_r. Then there must be some expectation of the size the system will reach, and using that expectation (as tot_s) the reasoning layer will then calculate d in every sphere using formula (5) but solving for d. Then a proposed change that would introduce a contradiction of length d or smaller would be discovered by the reasoning mechanism that always investigates all possible contradictions (that the proposed change would create) starting with the *small ones*.

Finally, one should note that a inconsistency that has a proof tree that is more shallow, is a more serious one because it is very direct (e.g. when the size is 2), whereas if it requires a more extensive proof then there are more choices for how to resolve it and a smaller proportion of the relationships have to be removed. One could also claim that in organizational policies the "obvious" contradictions first have to removed, whereas the more subtle ones are discovered later. Also, in the agent-related theory of "bounded-rationality" agents are required only to be able to achieve tasks that require a limited amount of reasoning – and that is true for smaller contradictions.

4.1.3 The Reasoning Layer
The vocabulary of this layer is:

Expression	Meaning
F: C_CONTRA(*sp*, *ont_op*)	returns true if *ont_op* would have introduced a contradiction in sphere *sp* of degree d_c that is higher than the defined degree p_c in that sphere.
F: IS_INFERABLE(*sp*, *ont_op*)	returns true if *ont_op* would have introduced a redundancy in sphere *sp* of degree d_r that is higher than the defined degree p_r in that sphere.
F: IS_NEW(*sp*, *ont_op*)	returns true if neither F: C_CONTRA(*sp*, *ont_op*) or F: IS_INFERABLE(*sp*, *ont_op*) are true.
F: CREASON(*sp*, *ont_op*)	returns a subset of *sp* that creates a contradiction if *ont_op* is performed.

The first 3 statements return either true or false, the fourth returns a subset. Notice that "higher" degree means worse. In the case when C_CONTRA(S, *ont_op*) is true, CREASON(S, *ont_op*) returns one of the contradiction reasons, i.e. one of the minimal subsets in the whole system S that show that S with the *ont_op* performed creates a contradiction (the best solution would be for the user to select one of these).

Function	Return values
Spheres(c_i, c_j)	All spheres of consistency that contain *both* contexts *i* and *j*.
Spheres(c_i)	All spheres of consistency that contain context *i*.

The paper title uses the word "dynamic" because we define several ontology operations that change ontologies and mappings. The current list of operations that evolve ontologies or mappings between them is the following (and the second mentions a special ontology operation):

ont_op = add_mapping(m, c_j, c_k) $\mid$ add_ontorel(m, c_j, d_j) $\mid$

delete_mapping(m, c_j, c_k) $\mid$ delete_ontorel(m, c_j, d_j) $\mid$ ε

$spec\_ont\_op$= RC(sp, P) [where P $\subseteq$ sp,]

RC() is the whole sphere sp that remains after one of the inconsistent subsets P has been removed. If there are several alternative such inconsistent subsets, several of them might have to be removed in order to make the whole sphere sp consistent.

The Reasoning Method. Because of limited space we will not investigate the reasoning algorithms in this paper. However, we have adopted the approach from [4] in order to do efficient and complete reasoning using this inexpressive language. The logical meaning of every ontology mapping uses a combined temporal and contextual logic that defines relationships between concepts in different local context and how it will persist in future states. That representation is transformed to one only using propositions, and C_CONTRA () and IS_INFERABLE() are implemented building refutation proof trees (analogous to using resolution by refutation) in a breadth-first fashion that use caching, loop-prevention and proof-pruning (automatically cutting of branches that cannot be successful). Our algorithm has worst-case linear time and memory complexity for computing C_CONTRA () and IS_INFERABLE() in spheres where $p_c=1$, based on an analysis of the reasoning algorithm.

The Logical Formalization. This will only be briefly discussed here, but [3] provides a full formalization. From a logical point of view a "context" is something that adds an index to the logical language, domain and interpretation functions (i.e. these are multiplied), and prohibits direct import of and access to concepts from other ontologies, because only ontology mappings can mediate such access. Informally, contexts model several cognitive points of view of a reality. They are expressed by different ontologies, and we define ontology mappings between them. We have been informally inspired by the five ontology mappings proposed by [5] as a part of C-OWL. However, the *actual* semantics are different and are defined in [3]. There is no single standard for ontology mappings. These are the five ontology mappings types between a concept A in ontology i and a concept B in ontology j that we use in the contextual state-based logic we have chosen for this scenario (and is used in the implemented prototype): COR(A_i, B_j), IS (A_i, B_j), IS2 (A_i, B_j), DISJOINT (A_i, B_j), and COMPATIBLE(A_i, B_j). They approximately correspond to (but C-OWL doesn't have a temporal notion of state, and our logic doesn't have directionality) and could be created by importing the following 5 C-OWL bridge rules [5]:

$$i: A \xrightarrow{\equiv} j: B, i: A \xrightarrow{\sqsubseteq} j: B, i: A \xrightarrow{\sqsupseteq} j: B, i: A \xrightarrow{\perp} j: B, i: A \xrightarrow{*} j: B$$

The formal meaning of our mappings and ontology language is defined in [3] and also the epistemological assumptions. Here is our *ontology language* (given an ontology j) and how axioms can be imported from OWL (the first three axioms can be represented in RDFS and could therefore be imported from RDFS to our language).

OWL Axiom	Our ontology language	OWL Axiom	Our ontology language
$C \doteq D$	COR(C_j, D_j)	$C \sqsupseteq D$	IS2 (C_j, D_j)
$C \sqsubseteq D$	IS (C_j, D_j)	$C \sqsubseteq \neg D$	DISJOINT (C_j, D_j)

4.2 The Framework Middle Layer

This vocabulary will be used for defining this layer (and also some of the functionality of the reasoning layer):

Expression	Meaning
c_i : CONFIRM (ont_op)	c_i accepts that ont_op should be performed
c_i : REFUSE (ont_op)	c_i rejects the performance of ont_op
F: DO (ont_op)	The framework mechanism performs ont_op
F: COMM($message$, recipient)	The framework mechanism sends a $message$ to recipient (that must be a context).
F:MCHOICE($\{c_j, c_k, \ldots\}$, ont_op_1, ont_op_2)	The framework mechanism chooses between ont_op_1 and ont_op_2 on behalf of several contexts.
contexts_of(sp)	A conjunction of all contexts included in the spheres sp.

The first rule available at this framework level is the following.

$$F: MCHOICE(\{c_j, c_k,...\}, stat_1, stat_2) \Rightarrow$$

$$\left(\left(\bigwedge_{n=\{j,k,...\}} c_n: CONFIRM(stat_i)\right) \wedge i \in \{1,2\} \Rightarrow stat_i\right) or \tag{6}$$

$$\left(i \in \{1,2\} \wedge \bigvee_{n=\{j,k,...\}} c_n: REFUSE(stat_i) \wedge \bigvee_{n=\{j,k,...\}} c_n: CONFIRM(stat_i) \Rightarrow DO(\varepsilon)\right)$$

MCHOICE() is a choice between two mutually exclusive statements (e.g. *do* an ontology operation or not) that is done involving a set of local contexts (at least two) and sent back to the framework mechanism. The formalization (6) says that if all involved local contexts choose one of the ontology operations, then that becomes their joint choice, and if there is some disagreement then the joint choice is to do nothing.

The DO() statement is important and can only be performed by the framework mechanism itself when it actually performs an ontology operation. **It is actually the DO() statement that moves the whole system S to the next state** – simply because it changes S. Before then, the system only does hypothetical reasoning ("what would have happen if an operation ont_op_1 would be performed?").

We will now investigate what happens when a local context initiates a proposal to add a mapping to another local context.

Case 1. Proposing to add an ontology mapping

$$c_j: PROPOSE(add\_mapping(m, c_j, c_k)) \Rightarrow$$

$$\left(\begin{array}{l} (\bigvee_{sp \in spheres(c_j,c_k)} F:C\_CONTRA(sp, add\_mapping(m, c_j, c_k))) \Rightarrow \\ (\bigwedge_{sp \in spheres(c_j,c_k)} F: COMM("contradicted:"+ \\ F: CREASON(sp, add\_mapping(m, c_j, c_k)),contexts\_of(sp))) \\ \wedge F: MCHOICE(contexts\_of(spheres(c_j, c_k)),DO(\varepsilon), \\ \bigwedge_{sp \in spheres(c_j,c_k)} F:C\_CONTRA(sp, add\_mapping(m, c_j, c_k)) \Rightarrow \\ DO(RC(sp,CREASON(sp, add\_mapping(m, c_j, c_k))))) \end{array}\right) or \tag{7}$$

$$\left(\begin{array}{l} (\bigvee_{sp \in spheres(c_j, c_k)} \text{F: IS\_INFERABLE}(sp, \text{add\_mapping}(m, c_j, c_k))) \Rightarrow \\ \text{F: COMM("already\_known", contexts\_of(spheres}(c_j, c_k)) \land \\ \text{F: MCHOICE}(\{c_j, c_k\}, DO(\varepsilon), DO(\text{add\_mapping}(m, c_j, c_k))) \end{array} \right) or$$

$$\left(\begin{array}{l} (\bigwedge_{sp \in spheres(c_j, c_k)} \text{F: IS\_NEW}(sp, \text{add\_mapping}(m, c_j, c_k))) \Rightarrow \\ \text{F: COMM("new", contexts\_of(spheres}(c_j, c_k)) \land \\ \text{F: MCHOICE}(\{c_j, c_k\}, DO(\varepsilon), DO(\text{add\_mapping}(m, c_j, c_k))) \end{array} \right)$$

This statement in the beginning of the rule (7) means: a local context c_j is proposing to add a mapping from its ontology to another local context c_k. This rule uses some statements from the reasoning layer. Only one of the three sub-rules (inside this big rule) can actually be activated. Intuitively, this formalization (7) then says that if the proposed change would introduce a forbidden contradiction in *any* sphere containing both c_j and c_k, then one of the contradiction reasons is communicated to them and the involved contexts either do nothing or one of the reasons for the contradiction is removed in every sphere where a contradiction occurs (but without adding the proposed mapping within the same step). Notice that all local contexts that have ontologies in spheres where one of the contradictions resides, have to participate in making this decision (this situation is referred to as "All involved inside the sphere..." below). The second and third sub-rule in (7) express that if the mapping can already be inferred or is new, the change is allowed but the two local contexts involved have to decide if they actually want to have it performed (this situation is referred to as "Pair" below). E.g. if, in fig.1, we invent the concept **yellow** in O_1 and then propose to add Is(**yellow, colour**) then it will be classified as IS_NEW by the process above, because this proposal neither creates a contradiction nor redundancy.

Table 2. The table shows how decisions are made if c_i: PROPOSE (*ont_op*) is proposed

ont_op=	C_CONTRA() is true	IS_INFERABLE() is true	IS_NEW() is true
add_mapping(m, c_j, c_k)	All involved inside the sphere(s) of consistency	Pair	Pair
delete_mapping(m, c_j, c_k)	Contradiction cannot happen.	Pair (i.e. mappings was already deleted)	Pair (i.e. mapping existed)
add_ontorel(m, c_j, c_j)	All involved inside the sphere(s) of consistency	Individual	Individual
delete_ontorel(m, cj, cj)	Contradiction cannot happen.	Individual (i.e. relationship didn't exist)	Pair (i.e. rel. existed)

Now we have investigated the case (case 1 above) when the proposal is to add a mapping between two ontologies, and the formalization showed what happens in the three cases. Table 2 above summarizes how this formalization would look like to for the cases when the proposal is to delete an ontology mapping, or add or delete an ontology relation. In all these cases formulas similar to (7) would be defined, but using the appropriate decision-making entities. The term "individual" in the table means that the local context that created the proposal, can decide itself if it wants the logically allowed change to *actually* be performed. If the proposal is to delete a mapping or delete a relation within an ontology this always logically allowed (because that is safe in this particular logic), so the individual local context or the pair of contexts decide whether to actually perform this act. This policy is allowing for individual ontologies that are not always singly connected.

4.3 The Framework Top Layer

We now formalize the top layer. This layer's vocabulary is:

Expression	Meaning
c_i: PROPOSE (ont_op), where $i \in \{1,..,n\}$ **(8)**	context c_i proposes operation ont_op
NEWCONCEPT(d_j)	that is true iff concept d_j was created in the previous state
F: REQUEST(c_i : PROPOSE (ont_op))	The framework requests context c_i to propose operation ont_op

The top layer governs the general system because all action is initiated there.

Firstly, any local context can *initiate* the synchronization processes of the whole framework by activating statement (8) above assuming the framework mechanism is in waiting mode (and doesn't process another proposal then, e.g. is in "busy mode"). So this is the formal sense in which the local contexts can exercise their semantic autonomy. After this statement is invoked, the framework mechanism invokes the corresponding procedural rules of the other layers. The top layer contains this rule:

$$F: DO (add\_ontorel(m, c_j, d_j)) \wedge NEWCONCEPT(d_j) \Rightarrow$$
$$F: REQUEST(c_j : PROPOSE(add\_mapping(m, d_j, c_k)))$$
$$\text{where } k \in \{1,...,n\} \wedge k \neq j \tag{9}$$

This means that if an ontology relation m has actually been created within the ontology of a local context j and it connects a new concept d_j (to an existing concept c_j) then that local context is requested to "try" to generate proposals that would map this new concept to the other local contexts. It means that it has to ask the knowledge source to generate knowledge that fits that pattern, and sometimes that will actually result in this knowledge being generated. The rules are acting here as performative statements in a multi-agent system.

5 Experiments

We now present an experimental evaluation where our implementation (built in Java) tests the core part of the framework. We assume two contexts c_1 and c_2 and they have one ontology each (that evolves) and an evolving set of mappings that connect these.

We then define spheres of consistency like in fig. 2, with the exception that the largest sphere has a variable consistency we shall call v_c. So the three defined spheres are $Cons(\{c_1\},\{\}, 1)$, $Cons(\{c_2\},\{\}, 1)$, $Cons(\{c_1, c_2\},\{m_{12}\}, v_c)$.

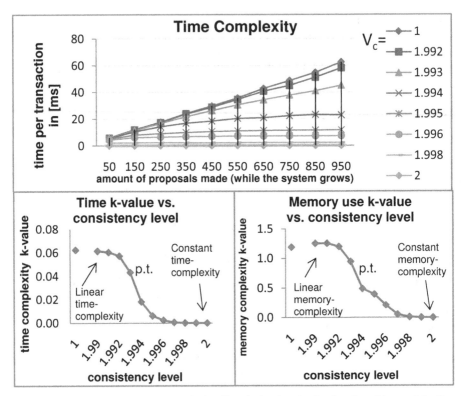

Fig. 5. Visualization of the time-complexity directly (top) and using k-values (slope of the lines in the top graph) and memory/consistency k-values. The phase transition (in both time and memory use) is marked with "p.t."

We now think of an application scenario where the contexts represent divisions within an organization, and all the "proposals" express organizational needs to adapt to a changing business environment. In this evaluation we assume that both contexts evolve ontologies having tree-like structures with a recursive branching factor 3, but it is randomized how quickly they grow. These things can happen:

- c_i : PROPOSE (Is ($C_{new,i}$, $C_{old, i}$)) - a proposal to change(grow) the ontology i where the concept C_{new} is invented and added as a sub-concept of C_{old}.
- c_i : PROPOSE (add_mapping(m, c_i, c_k)) - a proposal to add a mapping m from ontology i to k. One of the 5 mapping types is randomly chosen.

I.e. either one of the ontologies decides to grow or it proposes a random mapping from a recently created concept. So the rules (5), (6) and (7) are obeyed but we don't yet simulate the removal of contradictions (in (7)). During this simulation the system grows while obeying the rules and the sphere of consistency constraints. We have evaluated this implementation looking at the scalability and more precisely the incremental effort of the system to respond to a proposal and do the required reasoning. The total system size is 2000 in the experiment so e.g. d=2001 if v_c=1 but d=13 if v_c=1.994 (we solve formula (5) for d that is defined just before that formula). Our results are seen in fig. 5. The experiments also showed that the system's memory use has a linear upper bound. These figures show the effort of the system to process proposals for adding the ontology mappings while maintaining all the above-mentioned constraints – so some proposals are rejected but others accepted. Every marked data point is the average for 100 consecutive proposals (x-values) for 420 different runs of the system (for every fixed v_c). We see that the difference between v_c=1 and v_c=1.994 is small, because the proofs of contradiction that occurs in an application domain having *this* structure have a depth that is small compared to the overall system size (but they could be very wide). One could therefore re-normalize parameter p_c (here v_c=p_c) depending on the application. However, around v_c=1.993, v_c=1.994 we observe a **phase transition** between two states with *different behaviour*: to the left of the transition the system has linear time- and memory-complexity but to the right of it the system is approaching constant time- and memory-complexity.

Validation of the algorithm. We have implemented a module that use a slow brute-force method for measuring the smallest inconsistency in a set of ontologies and mappings – for any pair of variables it investigates if it can prove and disprove that any of the five mappings hold. We have generated 170 times a system of size 100 and validated that the measured inconsistency is never higher than the promised one.

In fact, often it is much lower (especially in a small system) because we have measured that the probability of a single proposal (in a system growing to the size of 2000) creating a contradiction is less than 0.6% for p_c=1 (it does decrease when p_c increases). But because of the potential *risk* that a contradiction could occur reasoning about the existence of a contradiction still must be done.

6 Conclusions

The semantic web relevance. Ideally, many ontology-based applications should maintain perfect consistency. In reality, many distributed information systems have no notion of consistency at all and sometimes this could nourish disorder. A framework for maintaining bounded consistency could help in bridging this gap – the system would *safeguard a chosen level of consistency* in every sphere of consistency. In the introduction (section 1) we mentioned three reasons when it is not necessary or possible to provide full consistency, which we now refer to. In the first case our framework can be used for specifying a knowledge infrastructure that maintains incremental bounded consistency. As regards the second argument, we have seen from fig. 5 that if consistency is bounded enough (i.e. parameter p_c is increased bey-ond the phase transition) the system become very scalable, i.e. its behaviour is altered. As regards the third argument, our framework supports the maintenance of bounded-consistency, but we have not yet evaluated the effort to reduce inconsistency. Future work to be done is investigating what happens when the spheres change their scope and parameters during run-time in cases when the application domain exhibits this behaviour and requires the knowledge infrastructure to follow in its footsteps.

References

1. Zurawski, M.: Towards a context-sensitive distributed knowledge management system for the knowledge organization. In: Workshop on Knowledge Management and the Semantic Web, 14th International Conference on Knowledge Engineering and Knowledge Management (EKAW 2004), UK (2004)
2. Robertson, D.: Multi-agent Coordination as Distributed Logic Programming. In: Demoen, B., Lifschitz, V. (eds.) ICLP 2004. LNCS, vol. 3132, pp. 416–430. Springer, Heidelberg (2004)
3. Zurawski, M.: Distributed multi-contextual ontology evolution – A step towards semantic autonomy. In: Staab, S., Svátek, V. (eds.) EKAW 2006. LNCS, vol. 4248, pp. 198–213. Springer, Heidelberg (2006)
4. Zurawski, M.: Reasoning about multi-contextual ontology evolution. In: The First International Workshop on Context and Ontologies: Theories, Practice and Applications, The Twentieth National Conference on Artificial Intelligence (AAAI 2005), Pittsburgh, PA, USA, July 9-13 (2005)
5. Bouquet, P., Giunchiglia, F., van Harmelen, F., Serafini, L., Stuckenschmidt, H.: C-OWL: Contextualizing Ontologies. In: Sekara, K., Mylopoulis, J. (eds.) Proceedings of the Second International Semantic Web Conference. LNCS, pp. 164–179. Springer, Heidelberg (2003)
6. Haase, P., Stojanovic, L.: Consistent evolution of OWL ontologies. In: Gómez-Pérez, A., Euzenat, J. (eds.) ESWC 2005. LNCS, vol. 3532, pp. 182–197. Springer, Heidelberg (2005)
7. Ma, Y., Qi, G., Hitzler, P., Lin, Z.: An algorithm for computing inconsistency measurement by paraconsistent semantics. In: Mellouli, K. (ed.) ECSQARU 2007. LNCS, vol. 4724, pp. 91–102. Springer, Heidelberg (2007)
8. Zhao, Y., Wang, K., Topor, R., Pan, J.Z., Giunchiglia, F.: Semantic cooperation and knowledge reuse by using autonomous ontologies. In: Aberer, K., Choi, K.-S., Noy, N., Allemang, D., Lee, K.-I., Nixon, L., Golbeck, J., Mika, P., Maynard, D., Mizoguchi, R., Schreiber, G., Cudré-Mauroux, P. (eds.) ASWC 2007 and ISWC 2007. LNCS, vol. 4825, pp. 666–679. Springer, Heidelberg (2007)
9. Wang, Y., Bao, J., Haase, P., Qi, G.: Evaluating Formalisms for Modular Ontologies in Distributed Information Systems. In: Marchiori, M., Pan, J.Z., Marie, C.d.S. (eds.) RR 2007. LNCS, vol. 4524, pp. 178–193. Springer, Heidelberg (2007)
10. Froehner, T., Nickles, M., Weiß, G.: Towards modeling the social layer of emergent knowledge using open ontologies. In: ECAI Workshop on Agent-Mediated Knowledge Management (AMKM), pp. 10–19 (2004)
11. Stojanovic, L., Schneider, J., Maedche, A., Libischer, S., Studer, R., Lumpp, T., Abecker, A., Breiter, G., Dinger, J.: The role of ontologies in autonomic computing systems. IBM Systems Journal 43(3), 598–616 (2004)

An Editorial Workflow Approach For Collaborative Ontology Development

Raúl Palma[1], Peter Haase[2], Oscar Corcho[1], Asunción Gómez-Pérez[1], and Qiu Ji[2]

[1]Ontology Engineering Group, Laboratorio de Inteligencia Artificial
Facultad de Informática, Universidad Politécnica de Madrid, Spain
{rpalma,ocorcho,asun}@fi.upm.es
[2]Institute AIFB, University of Karlsruhe, Germany
{pha,qiji}@aifb.uni-karlsruhe.de

Abstract. The widespread use of ontologies in the last years has raised new challenges for their development and maintenance. Ontology development has transformed from a process normally performed by one ontology engineer into a process performed collaboratively by a team of ontology engineers, who may be geographically distributed and play different roles. For example, editors may propose changes, while authoritative users approve or reject them following a well defined process. This process, however, has only been partially addressed by existing ontology development methods, methodologies, and tool support. Furthermore, in a distributed environment where ontology editors may be working on local copies of the same ontology, strategies should be in place to ensure that changes in one copy are reflected in all of them. In this paper, we propose a workflow-based model for the collaborative development of ontologies in distributed environments and describe the components required to support them. We illustrate our model with a test case in the fishery domain from the United Nations Food and Agriculture Organisation (FAO).

1 Introduction

The growing use and application of ontologies in the last years has lead to an increased interest of researchers in the development of ontologies, either from scratch or by reusing existing ones. Ontology development and maintenance activities are addressed by many different methodologies (e.g. Methontology, On-To-Knowledge, DILIGENT, etc.). However, most of them only consider the development of ontologies by single users or a small group of ontology engineers placed in the same location. More important is that even though they address the methodological aspects, in general they focus less on the process followed by organisations to coordinate the collaborative ontology development. In practice ontologies may be distributed, and a whole team of ontology engineers with different roles may collaborate in the development and maintenance, usually following a well defined process. Examples of such collaborative development processes can be found in international institutions like the United Nations Food and Agriculture Organisation (FAO), who are developing and maintaining large ontologies in the fishery domain [8]. Other similar examples are those of the Gene Ontology (GO)

J. Domingue and C. Anutariya (Eds.): ASWC 2008, LNCS 5367, pp. 227–241, 2008.
© Springer-Verlag Berlin Heidelberg 2008

project[1], which addresses the need for consistent descriptions of gene products in different databases, the caGrid project[2], which aims at providing a virtual informatics infrastructure that connects data, research tools, scientists, and organizations, etc.

Consequently, in this collaborative organisational setting, existing approaches are not enough to support all ontology development and maintenance needs. Furthermore, although recently some proposals and tools have been designed specifically to support collaborative ontology development (e.g. client-server mode in Protégé along with the PROMPT and change-management plugins), they generally only address parts of the overall problem (see section 4). Most of the existing advanced ontology tools (e.g. Protégé core system, SWOOP, etc.) support only the single-user scenario, where there is just one user involved in the development and later modification of the ontologies. With such tools, a typical scenario of collaborative ontology development would look as follows: An editor changes an ontology using his ontology editor system and then sends (e.g. using email or uploading it to an ontology repository) his locally changed ontology to other users (i.e. to add more changes using their own Protégé system, or review current changes). Even in the scenario where all users are editing the same ontology stored in a central server (e.g. using client-server mode in Protégé), the coordination of the actions of the editors (e.g. when editors want their changes to be reviewed or what kind of actions they can perform) is not yet fully supported.

As we can see from the previous discussion, in this type of collaborative scenario, change management is central. Hence, we need appropriate procedures (and corresponding infrastructure) to control and support the management of ontology changes. This procedure can be modelled as a collaborative workflow, which according to [2], is a special case of epistemic workflow characterized by the ultimate goal of designing networked ontologies and by specific relations among designers, ontology elements, and collaborative tasks. The need for such workflows has also been acknowledged in the past by other related works (e.g. [17]). An example of such workflow is that followed by the FAO (described in [8]), which we take as a use case in our work, in order to derive a generic set of required activities to support it.

Following this workflow, the development process starts with proposals for ontology changes. These proposals are discussed by multiple users (with different roles) in a collaborative way. For instance, if a change is made by an ontology editor, it has to be approved by a validator. After that, the change will be considered definitive and permanently added to the structure. Once changes are definitive, we will have a new stable version of the ontology, which requires the appropriate support to manage different ontology versions. Of course one could think of other kinds of workflows in different situations.

In this paper we present our approach for the **management of collaborative ontology development in a distributed scenario by means of an editorial workflow**. We analyse the collaborative development process using as illustrating scenario the case study at FAO and derive a set of functional requirements to support the process. We then introduce our proposal to support the collaborative ontology development where we address the identified problems. In particular, we propose a formal model for the representation of the workflow and describe the relationship with other models and

---

[1] http://www.geneontology.org/
[2] http://www.cagrid.org/

methods required in the management of ontology changes in distributed environments. Our contribution also includes the implementation of the proposed approach. The remainder of this paper is organised as follows: In section 2 we analyse the collaborative workflow scenario at FAO and derive a set of requirements based on the editorial workflow. In section 3 we introduce our approach for the collaborative ontology development based on the requirements derived in the previous section and present our implementation that provides the technological support to the presented models and methods. Section 4 provides a brief summary of related approaches to collaborative ontology development. Finally we conclude with a discussion in section 5.

2   Requirements for Collaborative Ontology Development

In this section we present the most relevant requirements to support the collaborative ontology development based on the analysis of the process (i.e. workflow) typically followed by organisations in the development and maintenance of ontologies[3]. For our analysis we considered existing processes for collaborative ontology development, also taking similar works in the state of the art into account (e.g. [10]). We use a case study of the NeOn project[4] for illustration: Specifically, we consider the editorial workflow of the fisheries ontologies lifecycle from FAO [8].

2.1   Overview of the Fisheries Ontologies Lifecycle

Within this case study, NeOn partners are developing an ontology-based information system to facilitate the assessment of fisheries stock depletion by integrating the variety of information sources available. In this context, the goal of the case study is to to implement an ontology-based Fishery Stock Depletion Assessment System (FSDAS) as well as an application to manage the fishery ontologies and their lifecycle.

The full lifecycle of the fisheries ontologies is introduced in [9], we here focus on the ontology engineering phase: In a nutshell, there are several actors involved in the engineering phase of the fishery ontology lifecycle, including experts in ontology modeling, that are in charge of defining the original skeleton of the ontology, ontology editors, that are in charge of the everyday editing and maintenance of the ontologies, and subject matter experts who know about the domain to be modeled. Finally, validators are subject experts who can move a change to production status for external availability. Ontology development follows a well defined collaborative workflow, which needs to be supported in the engineering environment. The editorial workflow allows ontology editors to consult, validate and modify the ontology keeping track of all changes in a controlled manner. Finally, once editors in charge of validation consider the ontology final, they are authorized to release it and make it available to end users and systems.

2.2   Functional Requirements

The functional requirements for the collaborative editorial workflow specify the specific functionality of the workflow, including the specification of the workflow behavior.

---

[3] In the remainder of this paper we refer to this process as the collaborative editorial workflow.
[4] http://www.neon-project.org/

Some of the requirements that we introduce in the following are similar to the ones that have been already identified in the past (see the analysis of [10] in section 4). However, in our work, we further identified additional requirements to support the process followed typically by many organisations to coordinate the collaborative ontology development. When appropriate, we illustrate the requirement using FAO scenario.

Lifecycle Requirements. A collaborative editorial workflow should implement the necessary mechanisms to allow ontology editors to: **consult, modify** and **validate** ontologies. In some cases, ontologies may also need to be **published** on the internet once they are fully validated. Furthermore, the process should ensure that the aforementioned activities are carried out in a controlled and coherent manner. Hence, the editorial workflow is responsible for the coordination of who (depending on the user *role*) can do what (i.e. what kind of *actions*) and when (depending on the *status* of the ontology elements – classes, properties and individuals – and the role of the user).

The activities of the editorial workflow are being done by users that are ontology editors in charge of the everyday editing and maintenance work of the ontologies. Each user is assigned a specific role (which has associated permissions) by the organisation based on his expertise and responsibilities. Depending on the user permissions, he can be in charge of developing specific fragments of ontologies, revising work done by others, or developing new versions of ontologies. Ontology editors know about the ontologies domain, but usually know little or nothing about ontology software or design issues. For instance, one approach for assigning the user role can be driven by the module of the ontology the user is responsible for (e.g. [6]). As another example, in FAO, an ontology editor can be assigned one of the following roles:

- *Subject experts* (SE) know about specific aspects of the ontology domain and are in charge of adding or modifying ontology content.
- *Validators* (V) revise, approve or reject changes made by subject experts, and they are the only ones who can copy changes into the production environment for external availability. They have a broader knowledge of the ontology domain and have at least some knowledge about design issues.

To enforce permissions, it is required that (i) the system supports the different user roles and (ii) users identify themselves to the system before using it. Furthermore, to control when the ontology editors are allowed to work with an ontology element, in addition to the user roles, every ontology element is required to have a *status*. Ontology editors can change the status depending on their role. For instance, the possible status ontology elements can have in FAO include: *Draft* for the proposed additions or updates, *To Be Approved* for the proposed changes that are ready to be reviewed by a validator, *Approved* for the accepted changes, *To Be Deleted* for the proposed removals and *Published* for the changes released to the internet.

Workflow Activities. Activities required to support the editorial workflow include the operations (or possible *actions*) the ontology editors are allowed to perform depending on their roles and the status of the ontology elements.

Edit ontology element
Insert an ontology element. This operation triggers the start of the editorial workflow.

Update an ontology element. Editors can update ontology elements. Depending on their role and the status of the element, this operation could trigger also the start of the editorial workflow. For instance, in our illustrative scenario, a SE can only update elements in "Draft" or "Approved" status. In both cases the status of the element is automatically reset by the system to "Draft", and the element will need to pass through the whole workflow again.

Delete an ontology element. Editors propose elements for deletion. In general this is not a definitive action, and it has to be authorized by an appropriate editor.

Change status of ontology element. While inserting, updating and deleting elements, their status is automatically changed by the system. There are other cases where a specific action from editors is required to move an element from one status to another and make the editorial workflow to function (e.g. SE's need to explicitly send elements in "Draft" status to the "To Be Approved" status).

Publish ontology. In some organisations, authorized editors are allowed to copy an ontology from the test and validation environment (editorial workflow in the Intranet) to the production environment (Internet). By doing so, the system automatically assigns the right version to the published ontology following a versioning scheme.

Visualization Requirements

View change history. Editors need to be able to view the logs of ontology changes and their related information including the history notes e.g. argumentation of the tracked changes.

View based on status and user role. The interface should be able to provide different data views based on the user role.

View use statistics. Editors can view information about an ontology regarding how the ontology has been used or evolved throughout the time e.g. provenance, editors, frequency of changes, the fragment/domain of the ontology changed most rapidly, etc.

View ontology statistics. Authorized users can view statistics of the ontology being edited e.g. depth of the class hierarchy; number of child nodes; number of relationships and properties; number of concepts per branch.

Change Management

Representation of changes. A main requirement is the explicit representation of the changes that editors are able to perform to ontologies. The representation should ensure the accessibility and interoperability with other components (e.g. workflow, ontology metadata, etc.) and the maintenance of the chronological order of the changes to support e.g. undo/rollback operations or reconstruction of performed operations (e.g. when syncrhonizing/propagating changes). Additionally, to facilitate the previous tasks and provide an efficient link between what the user sees (e.g. ontology elements) and what the system manage internally (e.g. axioms), the representation should provide a flexible classification of changes that considers the actual "atomic" operations that can be performed over ontologies in addition to operations at the element level (e.g. to support the different status that each ontology element can have during the editorial workflow) or the complex operations that have been considered in the past (see section 4). Information about changes should include e.g. the operation performed, the time of the operation, the user, the element associated, the previous change and the description.

Capture ontology changes. The system should automatically log ontology changes.

Change Propagation and notification. After new changes are submitted to the ontology, editors involved in this workflow process should be informed when they log into the system. Each author (or the coordinator) should be able to view changes made by other authors, even without editing permission.

Versioning. An additional requirement is the management of ontology versions. The first modification to an approved/published ontology automatically changes the current version. This modified version of the ontology will either become a new version (i.e. with a different version information) or if specified by the editor remain the same version. In any case, versions need to be uniquely identified.

Concurrency Control and Conflict Resolution. An important issue that has to be addressed in this collaborative scenario is to ensure the integrity of the ontology via concurrency control mechanisms and appropriate means for the resolution of conflicts whenever two or more editors submit changes to the same element concurrently.

3   A Workflow-Based Collaborative Ontology Development Approach

In this section we present our solution to support the collaborative ontology development and describe how it tackles the aforementioned requirements. We first present the conceptual models[5] that provide the foundations to represent the required information in our solution and then we present the implementation support.

3.1   Conceptual Models

Change Representation. A core element in our approach is the representation of changes (c.f. *change management requirement*). In [13] we presented our proposal for the representation of changes which integrates many of the features of the existing approaches (e.g. [15], [5]) in a consistent layered manner. In this paper we highlight only the most relevant parts of our representation of changes: We refine and extend existing work and propose a layered approach for the representation of changes that consists of a generic ontology that models generic operations in a taxonomy of changes that are expected to be supported by any ontology language and that can be specialized for specific ontology languages (e.g. OWL) while still providing a common, independent model for the representation of ontology changes. It comprises three levels for the classification of changes: Atomic (i.e. the smallest and indivisible operation that can be performed in a specific ontology model), Entity (i.e. basic operations that can be performed over ontology elements usually from an ontology editor) and Composite (i.e. group of changes applied together that constitute a logical entity). It also provides the link to capture the argumentation of changes and it relies and uses some of the knowledge defined in our early work, the Ontology Metadata Vocabulary (OMV) [3] to refer to ontologies and users. OMV is a metadata schema that captures relevant information about ontologies such as provenance, availability, statistics, etc. Besides the main class

---

[5] Our conceptual models are available in OWL at http://omv.ontoware.org

Ontology, OMV also models additional classes and properties required to support the reuse of ontologies, such as Organisation, Person, LicenseModel, OntologyLanguage and OntologyTask among others. Our change ontology has been implemented as an OMV extension because it models specific ontology metadata (i.e. ontology changes).

Furthermore, the change ontology provides the means to support not only the tracking of changes but also the information that identifies the original and the current version of the ontology after applying the changes (*versioning requirement*). This is not a trivial issue: even though ontologies are in general identified by an URI, in practice it is not enough to identify a particular ontology version (i.e. different versions of the same ontology have the same URI). Hence, the management of ontology versions requires a clear definition of the ontology identification. In our solution, we rely on the identification of ontologies that we presented in [3], which consists of a tripartite identifier: the ontology URI, the ontology version (if present), and the ontology location.

Finally, to keep track of the actual sequence of changes (i.e. the order in which changes were performed), our ontology relies on two elements: each change is linked to its predecessor via the "hasPreviousChange" object property and a "Log" class provides the pointer to the last change in the ontology history.

Workflow Model. Based on the analysis presented in section 2 we found that some of the possible actions and states in the editorial workflow apply at different levels of abstraction. Therefore our solution considers the editorial workflow at two levels: ontology level and ontology element level. Although the workflows can be used independently of the underlying ontology model, the specific set of ontology elements depend on the ontology model. In our approach we are mainly considering the OWL ontology model, in which an OWL ontology consists of a set of axioms and facts[6]. Facts and axioms can relate to classes, properties or individuals, and hence that is the set of ontology elements we are considering.

As previously discussed, the workflow details (e.g. the specific roles, actions, etc.) depend on the organisation setting. To exemplify, in the rest of this section we discuss our solution for the particular scenario in FAO. Figures 1 and 2 show the two different workflow levels (i.e. element and ontology level). States are denoted by rectangles and actions by arrows. The information in parenthesis specifies the actions that an editor can perform depending on its role, where "SE" denotes Subject Expert, "V" denotes Validator and "-" denotes that the action is performed automatically by the system.

The possible states (see Figure 1) that can be assigned to ontology elements are:

- *Draft*: This is the status assigned to any element when it passes first into the editorial workflow, or when it was approved and then updated by a subject expert.
- *To be approved*: Once a "SE" is confident with a change in draft status the element is passed to the "To Be Approved" status, and remains there until a "V" approves/rejects it.
- *Approved*: If a "V" approves a change in an element in the "To Be Approved" status, it passes to the "Approved" status. Additionally, this is the default state for every element of the initial version of a stable ontology.

---

[6] In our current implementation we support the upcoming OWL 2 language. See
http://www.w3.org/TR/owl2-syntax/

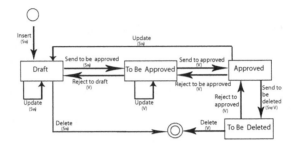

Fig. 1. Editorial workflow at the element level

Fig. 2. Editorial workflow at the ontology level

– *To be deleted*: If a "SE" considers that an element needs to be deleted, the item will be flagged with the "To Be Deleted" status and removed from the ontology, although only a "V" will be able to definitively delete it.

The ontology has a state (see Figure 2) that is automatically assigned by the system (denoted with "-" in Figure 2), except from the "published" state as described below:

– *Draft*: Any change to an ontology in any state automatically sends it into draft state.
– *To be approved*: When all changes to an ontology version are in "To Be Approved" state (or deleted) the ontology is automatically send to "To Be Approved" state.
– *Approved*: When all changes to an ontology version are in "Approved" state (or deleted) the ontology is automatically send to "Approved" state. Additionally, this is the default state of the initial version of a stable ontology.
– *Published*: Only when the ontology is in "Approved" state, it can be sent by a validator to "Published" state.

As described in section 2.1, the editorial workflow starts after getting a stable populated ontology that satisfies all the organizational requirements. Hence, we assume that the initial state of this stable ontology (and all its elements) is "Approved"[7].

Note that during the editorial workflow, actions are performed either implicitly or explicitly. For instance, when a user updates an element he does not explicitly perform an update action. In this case the action has to be captured from the user interface and recorded when the ontology is saved. In contrast, Validators explicitly approve/reject proposed changes and the action is recorded immediately when performed.

Similarly to our change ontology, we decided to model the workflow elements (i.e. roles, status, actions) using an (OWL-Lite) ontology (i.e. a workflow ontology) that allows the formal and explicit representation of knowledge in a machine-understandable

---

[7] In a different scenario, the workflow could start with an empty ontology (without elements), which we could assume that will be by default in "Approved" state.

format. Furthermore, having both models (i.e. ontology changes and workflow) formalized as ontologies will facilitate the representation of the tight relationship that exists between both of them. For instance, consider a user with role "subject expert" that "inserts" a new ontology "class" to the ontology. That "class" will receive automatically the "draft" state. All the information related to the process of inserting a new ontology element will be captured by the workflow ontology, while the information related to the particular element inserted, along with the information about the ontology before and after the change is captured by the change ontology. Additionally, the workflow process also relies on OMV to refer to ontologies and users.

Workflow ontology. The main classes and properties of the workflow ontology and its relationships with the other ontologies in our approach are shown in Figure 3.

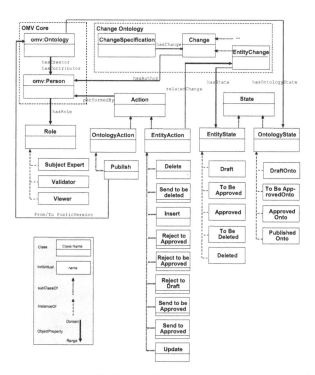

Fig. 3. Workflow ontology

The different roles of the ontology editors are modelled as individuals of the `Role` class that is related to the `Person` class of the OMV core ontology (i.e. a person has a role). To explicitly model the separation between the possible states of ontology elements (i.e. classes, properties and individuals) and the possible states of the ontology itself, the `State` class is specialized in two subclasses (i.e. `EntityState` and `OntologyState`. Similarly to the roles, the possible values of the states are modelled as individuals of their respective subclass. Furthermore, the two subclasses of `State` allow to represent the appropriate relationships at the element and ontology level: To specify that an ontology element has a particular state we rely on the class

`EntityChange` from the change ontology which is associated to a particular ontology element (as described in section [13]) and associate it with subclass `EntityState`, and to specify that an ontology has a particular state we rely on the class `Ontology` from the OMV core and associate it with the subclass `OntologyState`.

Finally, for the actions there is also a separation between the possible actions at the element level and actions at the ontology level. Hence, the `Action` class is specialized in two subclasses (i.e. `EntityAction` and `OntologyAction`. To track the whole process (and keep the history) of the workflow, the possible actions are modelled as subclasses of the appropriate `Action` subclass. Similar to the states, the two subclasses of `Action` also allow to represent the appropriate relationships at the element and ontology level: to specify that an action was performed over a particular ontology element, the subclass `EntityAction` is associated with class `EntityChange`. As we explained before, actions at the ontology level are performed automatically by the system except from publish which changes the public version of the ontology. Therefore, the only subclass of `OntologyAction` is `Publish` that is associated to the class `Ontology` to specify the previous and next public version of the ontology.

3.2 Implementation Support

Our approach has been implemented within the NeOn Toolkit[8], an extensible ontology engineering environment based on Eclipse, by means of a set of plugins and extensions. A high level conceptual architectural diagram of the involved components is shown in Figure 4. We present in the following, first the change capturing related components (i.e. left side of the figure), then the workflow management related components (right side of the figure), next the user related components for editing and visualizing ontologies (and related information) in the editorial workflow (upper part of the figure) and finally our distributed registry implementation (bottom part of the figure).

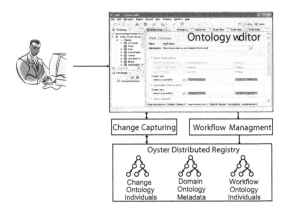

Fig. 4. Conceptual architecture for the collaborative ontology development support

Change Capturing Components. Once the ontology editor specifies that he wants to monitor an ontology, changes are automatically captured (*change management*

---

[8] http://www.neon-toolkit.org/

requirement) from the ontology editor by a change capturing plugin. This plugin is notified about events that consist of ontology changes performed by the user in the ontology editor. For each of these events, the change is represented according to the change ontology by creating the appropriate individual. For example, adding a class individual in the ontology editor creates the entity change "Add Individual" and the two corresponding atomic changes (OWL 2 axioms): "Add Declaration" and "Add ClassMember". As described by the change ontology, each individual includes relevant information such as the author, the time, the related ontology, etc. The individuals are stored into the Oyster distributed registry [12]. This plugin is also in charge of applying changes received from other clients to the same ontology after Oyster synchronizes the changes in the distributed environment (see last subsection). Finally, this plugin extends the NeOn Toolkit with a view to display the history of ontology changes (*visualisation requirements*).

Workflow Management Components. In our implementation, the workflow management component (i) takes care of enforcing the constraints imposed by the collaborative workflow, (ii) creates the appropriate action individuals of the workflow ontology and (iii) registers them into the distributed registry. Hence, whenever a new workflow action is performed, the component performs the following tasks:

– It gets the identity and role of the user performing the action (if it is an explicit action) e.g. send to approve, or the associated change (if it is an implicit action) e.g. adding a new class implicitly creates an insert action.
– It gets the status of the ontology element associated to the action/change.
– It verifies that the role associated to the user can perform the requested action when the ontology element is in that particular status.
– If the verification succeeds, it creates the workflow action and registers it.
– If the verification fails, it undoes the associated change(s) for the implicit actions because the complete operation (e.g. adding a new class) failed.

Ontology Editing and Visualization Components. To support the workflow activities (*workflow activities requirements*) we rely on the NeOn Toolkit which comes with an ontology editor that allows the editing of ontology elements. Additionally, according to the *visualisation requirements* the NeOn Toolkit is extended with a set of views that allow editors to (i) see the appropriate information of ontologies in the editorial workflow and (ii) perform (as described in 3.1) the applicable workflow actions (*approve, reject*, etc.), depending on their role. There are four views[9]:

– *Draft view*: Shows all proposed changes (from all editors) to that ontology version. In accordance to FAO scenario the changes of the current editor are editable while changes from other editors are non editable (see Figure 5).
– *Approved view*: Shows the approved changes.
– *To Be Approved view*: Shows all changes (from all editors) pending to be approved.
– *To Be Deleted view*: Shows all proposed deletions (from all editors).

---

[9] Subject experts see the first two views, validators see the latter three.

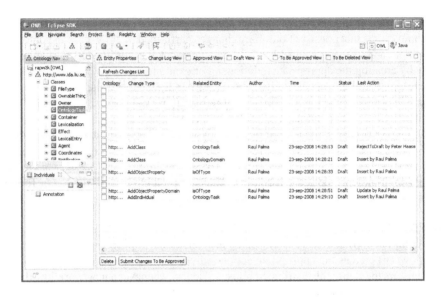

Fig. 5. Draft View in the NeOn Toolkit

Distributed Registry. Ontologies are stored within a repository and their metadata is managed by the Oyster distributed registry[10] (*change management requirement*). The metadata includes information about ontologies and users (represented using OMV), the changes to the ontology (represented using the change ontology) and about the actions performed (represented using the workflow ontology). For each change the status is also kept to support the editorial workflow. When a new change is registered into an Oyster node, Oyster automatically updates the log history keeping track of the chronological order of changes: It gets the last registered change (using the "Log" class) and adds it as the previous change of the current one. Then it updates the "Log" class to point to the current change.

The local Oyster nodes contact each other creating a distributed ontology registry. In this distributed environment, Oyster also propagates the ontology changes, thus allowing the notification of new changes to ontology editors (*change management requirement*). That is, once we have the required changes in a machine-understandable format, the system propagates them to the distributed copies of the ontology. For this task, we follow a synchronization approach that is a combination of a push and pull mechanism. During the synchronization, nodes periodically contact other nodes in the network to exchange updated information (pull changes) and optionally they can push their changes to a specific node (called the super node) such that if a node goes offline before all other nodes pull the new changes, the node changes are not lost. In this way, Oyster minimizes the conflicts or inconsistencies due to concurrent editing as it automatically synchronizes changes periodically (and it allows to force the synchronization immediately) in the distributed environment such that every editor will have an up-to-date copy of the ontology with the proposed changes (*concurrency control and conflict*

---

[10] http://ontoware.org/projects/oyster2/

resolution requirement). Nevertheless, conflicts in the collaborative workflow could still occur as logical conflicts in the form of inconsistencies or conflicts due to concurrent editing of an ontology. The strategies to deal with those potential problems are out of the scope of this paper, and we refer the reader to [13] for additional information.

4 Related Work on Collaborative Ontology Development

The problem of collaborative ontology development has been partially addressed in the literature with methodological and technological results, which are not necessarily aligned. In the remainder of this chapter we mainly focus on the existing works of ontology development where some kind of reviewing process has been acknowledged.

In [16], [14] the authors introduce DILIGENT, an ontology engineering process for decentralized cases of knowledge sharing. It identifies several key roles involved in collaboratively building the same ontology. The process entails different users in the creation of a shared ontology and adaptation to local needs and a control board in charge of deciding how the shared ontology will be changed based on the user requests. DILIGENT also considers the provision of arguments for the requested changes and design decisions in a semi-formal way (similar to [15]).

A related work – DOGMA-MESS – is presented in [1] (and [7]). The authors propose a generic model for understanding the interorganisational ontology engineering process where the knowledge moves in an upward spiral starting at the individual level, moving up to the organisational level, and finally up to the interorganisational level.

In [10] the authors present the Change and Annotation Ontology (CHAO) to represent changes between two versions of an ontology and user annotations related to these changes, and two Protégé plugins: The Change-management plugin that provides access to a list of changes (i.e. instances of CHAO) and enables users to add annotations to changes and the PROMPT plugin (also introduced in [11]) that provides comparisons between two versions of an ontology, allowing to examine the list of users who performed changes and to accept and reject changes. Finally they introduce the client/server mode in Protégé for synchronous editing by multiple users.

Another similar tool support is presented in [17]. The authors introduce an extension of the existing Protégé system that supports collaborative ontology development (i.e. Collaborative Protégé). The extension enables (1) the annotation of ontology components and ontology changes and (2) the searching and filtering of user annotations based on simple or complex criteria. The authors also propose two types of mechanisms for voting change proposals (i.e. a *5-star* voting or a *Agree/Disagree* type of voting).

Although [16] and [1] consider the collaborative development of ontologies in a distributed setting, it is not clear how change requests are represented, there is no explicit tracking of the change operations in the shared ontology that would be useful for local users to identify the approved changes or compare it with the local copy and there is no history of the rejected changes. Moreover, local users are not notified automatically of changes and consequently they could be working with different versions of the ontology which might hamper the interoperability. Also, [1] does not consider how users interact in the process depending on their role.

A main difference in [10] and [17] with respect to our solution is that the tracking of changes and curator actions (i.e. accept/reject changes) is done in a centralized manner i.e. either in a local copy or in a centralized server. Additionally, although the approaches consider the reviewing of changes (e.g. acceptance/rejection of changes), it is not clear what kind of roles (and related permissions) are considered, how those actions are traced or how is the process flow for the reviewing.

In our solution, we rely on a formal representation of changes which provides the basis for the creation of change logs that support e.g. the comparison of ontologies or the synchronization of distributed copies of the ontology. Moreover, we identify different user roles involved in the process of the ontology development and propose to formalize the reviewing process in a machine-understandable format such that all taken actions can be tracked and exchanged. In our solution, users are able to use the same version of the ontology (i.e. a local copy) and work in a decentralized manner given that changes and curator actions are maintained in a distributed registry which is in charge of synchronizing the information automatically.

5 Discussion

The need for a systematic approach to ontology development in a highly distributed environment has been emphasized many times in the past in the ontology engineering community. As a result, different solutions have been proposed ranging from informal or lightweight strategies (e.g. [4]) to semi-formal approaches like in the Gene Ontology project to formal methodologies (e.g. [16]). In this paper we have presented our solution to support the collaborative ontology development in distributed environments. It consists of a formal strategy where the ontology development process is explicitly represented. The criteria for choosing this strategy was based on the analysis of the requirements of large organisations using as test case the FAO scenario.

Hence, we proposed two generic workflows specialised at different levels of abstraction (i.e. ontology element level and ontology level). Our proposal includes the definition of a workflow ontology for the formal representation of the workflow process. Additionally, we introduced the role of the workflow in the infrastructure required for the management and control of ontology changes and describe its relationships with other components and activities (i.e. change representation, versioning, etc.). We illustrated in a simple scenario how the workflow ontology supports the collaborative ontology development and its tight relationship with the ontology for the representation of changes. Finally, we introduced our implementation to support the proposed model.

Although we are already evaluating individual components of our approach, our next step is the complete evaluation of our approach within FAO and other scenarios. In the future, we plan to provide additional features to the NeOn toolkit such as a *Change View* to shows the changes (diff) between two versions of an ontology according to the change ontology or a full undo/redo support. Further, we are working on the integration of our work with other threads of related work, such as argumentation support.

Acknowledgments. Research reported in this paper was partially supported by the EU in the IST project NeOn (IST-2006-027595, http://www.neon-project.org/.

References

1. de Moor, A., De Leenheer, P., Meersman, R.: DOGMA-MESS: A meaning evolution support system for interorganizational ontology engineering. In: Proc. of the International Conference on Conceptual Structures (ICCS 2006), Aalborg, Denmark. Springer, Heidelberg (2006)
2. Gangemi, A., Lehmann, J., Presutti, V., Nissim, M., Catenacci, C.: C-ODO: an OWL metamodel for collaborative ontology design. In: Workshop on Social and Collaborative Construction of Structured Knowledge (CKC 2007) at WWW 2007, Banff, Canada (2007)
3. Hartmann, J., Palma, R.: OMV - Ontology Metadata Vocabulary for the Semantic Web, vol. 1.0 (2005), http://omv.ontoware.org/
4. Hepp, M., Bachlechner, D., Siorpaes, K.: Ontowiki: Community-driven ontology engineering and ontology usage based on wikis (2005)
5. Klein, M.: Change Management for Distributed Ontologies. PhD thesis, Vrije Universiteit, Amsterdam (2004)
6. Kozaki, K., Sunagawa, E., Kitamura, Y., Mizoguchi, R.: A framework for cooperative ontology construction based on dependency management of modules. In: Proceedings of the International Workshop on Emergent Semantics and Ontology Evolution (ESOE2007) at ISWC/ASWC2007, Busan, South Korea (November 2007)
7. De Leenheer, P., Mens, T.: Ontology Evolution. State-of-the-art and Future Directions. In: Ontology Management. Semantic Web, Semantic Web Services, and Business Applications. Springer, Heidelberg (2007)
8. Muñoz-García, Ó., Gómez-Pérez, A., Iglesias-Sucasas, M., Kim, S.: A workflow for the networked ontologies lifecycle. A case study in FAO of the UN. In: Proceedings of the CAEPIA-TTIA 2007, Spain. Springer, Heidelberg (2007)
9. Muñoz-García, O., Kim, S., Iglesias Sucasas, M., Caracciolo, C., Bagdanov, A., Wang, Y., Haase, P., Suarez-Figueroa, M., Gomez-Perez, A.: Software architecture for managing the fisheries ontologies lifecycle. Technical Report D7.4.1, NeOn Consortium (October 2007)
10. Noy, N., Chugh, A., Liu, W., Musen, M.: A framework for ontology evolution in collaborative environments. In: International Semantic Web Conference, pp. 544–558 (2006)
11. Noy, N., Kunnatur, S., Klein, M., Musen, M.: Tracking changes during ontology evolution. In: International Semantic Web Conference (2004)
12. Palma, R., Haase, P.: Oyster - sharing and re-using ontologies in a peer-to-peer community. In: International Semantic Web Conference, pp. 1059–1062 (2005)
13. Palma, R., Haase, P., Wang, Y., d'Aquin, M.: D1.3.1 propagation models and strategies. Technical Report D1.3.1, UPM; NeOn Deliverable, November (2007)
14. Pinto, S.: Ontoedit empowering swap: a case study in supporting distributed, loosely-controlled and evolving engineering of ontologies (diligent) (2004)
15. Stojanovic, L.: Methods and Tools for Ontology Evolution. PhD thesis, University of Karlsruhe (TH), Germany (August 2004)
16. Tempich, C.: Ontology Engineering and Routing in Distributed Knowledge Management Applications. PhD thesis, University of Karlsruhe (TH), Germany (2006)
17. Tudorache, T., Noy, N.: Collaborative protege. In: Workshop on Social and Collaborative Construction of Structured Knowledge (CKC 2007) at WWW 2007, Banff, Canada (2007)

Identifying Key Concepts in an Ontology, through the Integration of Cognitive Principles with Statistical and Topological Measures

Silvio Peroni, Enrico Motta, and Mathieu d'Aquin

Knowledge Media Institute
The Open University
Milton Keynes, United Kingdom
{s.peroni,e.motta,m.daquin}@open.ac.uk

Abstract. In this paper we address the issue of identifying the concepts in an ontology, which best summarize what the ontology is about. Our approach combines a number of criteria, drawn from cognitive science, network topology, and lexical statistics. In the paper we show two versions of our algorithm, which have been evaluated against the results produced by human experts. We report that the latest version of the algorithm performs very well, exhibiting an excellent degree of correlation with the choices of the experts. While the generation of automatic methods for ontology summarization is an interesting research issue in itself, the work described here also provides a basis for novel approaches to a variety of ontology engineering tasks, including ontology matching, automatic classification, ontology modularization, and ontology evaluation.

Keywords: Ontology, semantic web, key concepts, ontology summarization, natural categories, cognitive science.

1 Introduction

The Semantic Web is growing fast and already contains a large amount of data, measured in millions of semantic documents and billions of triples. According to our own estimates, which are based on our experience with the Watson ontology search engine [1], at least seven thousand[1] ontologies[2] exist on the Semantic Web, providing an unprecedented set of resources for developers of semantic applications. Thus, consistently with Mark Stefik's vision of a *knowledge medium* [2], the Semantic Web is rapidly emerging as a large scale platform for publishing and sharing formalized knowledge models. Given this context, for the past two years we have been working on a new generation of knowledge-based applications, which are able to exploit the

---

[1] This number refers only to ontologies which are formalised in either OWL, RDFS, or DAML+OIL and are also publicly available on the web.

[2] In this context we use the term 'ontology' to refer to a semantic web document, which contains class and relation specifications, rather than simply data about individuals.

J. Domingue and C. Anutariya (Eds.): ASWC 2008, LNCS 5367, pp. 242–256, 2008.
© Springer-Verlag Berlin Heidelberg 2008

Semantic Web as a source of background knowledge, e.g., to provide new solutions to tasks such as ontology matching, or to add semantics to tag spaces [3].

In addition, we have also developed tools, such as the Watson Plug-in, which exposes the functionalities provided by Watson within ontology engineering editors, such as Protégé (http://protege.stanford.edu) and the NeOn Toolkit (http://neon-toolkit.org), thus making it possible for ontology developers to locate relevant semantic web entities, and integrate them with the ontology under construction.

While the vision of a large scale reuse of semantic resources available on the web is in principle very exciting, in reality the current level of tool support for the process of ontology development by reuse is rather limited. For example, while the aforementioned Watson Plug-in makes it possible to locate entities on the Semantic Web and import them into an ontology, it actually provides only limited support for navigating and making sense of the ontologies in which these entities reside. Indeed, a key problem faced by an ontology engineer when considering the reuse of an ontology is *ontology understanding*: how to make sense speedily of the content and organization of an ontology, in order to make decisions about the suitability of the ontology in question for the current ontology engineering development project.

A number of people have partially tackled this problem from different angles. For example, the ontology engineering environments available today, such as Protégé, TopBraid Composer (http://www.topbraidcomposer.com/), or The NeOn Toolkit, all provide functionalities for exploring and visualizing an ontology, to facilitate ontology understanding. Nevertheless, formal evaluations of these tools [4] indicate that these environments do not actually do a particularly good job in helping a user to deal with multiple ontologies, to make sense of an ontology, or in general to develop ontologies by reuse. In particular the aforementioned study reported on the lack of *abstraction mechanisms* in these tools, both at the micro-level (notation) and at the macro-level (providing high level ontology summaries).

In this paper we focus on the latter problem and we present an approach to identifying the *key concepts* in an ontology, to generate a meaningful snapshot of an ontology and facilitate the process of ontology understanding. In contrast with other approaches to ontology summarization [5, 6] our work integrates criteria from both cognitive science, lexical statistics, and graph analysis, to try and come up with the same kind of summaries as human experts.

We will start the discussion in the next section by illustrating both the high-level criteria, which inform our approach, and their initial computational realization. We will then discuss the results obtained from an empirical evaluation of this initial version of our method, which unfortunately showed a low degree of correlation with the choices made by human experts. This negative result led to a revision of our algorithm, which is described in section 3. Among other things, this new version introduces an additional criterion, which attempts to estimate the *popularity,* determined using lexical statistics, of a concept in the ontology. As discussed in section 3.3, the revised version of the algorithm shows an excellent degree of correlation with human experts. Finally, in sections 4-6, we discuss related work, reiterate the key contributions of this work, and outline a number of new opportunities for research and development made possible by it.

2 Our Approach

Our aim is to design a method that, given an ontology and an integer n, extracts the n concepts, which can be considered as 'best descriptors' of the ontology: the *key concepts*. Obviously there is no formal definition of what is a key concept and, especially if we take a task-independent stance, it is unlikely that such a formal definition can be produced. For this reason, our work is empirically grounded and specifically our goal is to define a method able to generate results that match as closely as possible those produced by human experts. Support for such empirical stance is given by some initial evidence in the literature, indicating that some degree of convergence exists when multiple experts are asked to identify the 'important' concepts in an ontology [6].

Consistently with the stated empirical grounding of our work, we consider both criteria drawn from cognitive science as well as others based on the topological structure of the ontology. Specifically, in the initial version of our method we used both the notion of *natural categories* [7], which aims to identify concepts that are information-rich in a psycho-linguistic sense, and the notion of *density*, which highlights concepts which are information-rich in an ontological sense. In addition we also used a *coverage* criterion, to ensure that no important part of the ontology is ignored in the resulting selection. In what follows we define these criteria more precisely and present the first implementation of these ideas.

2.1 Natural Categories

Let's consider as an example the AKT Reference Ontology (AKT-RO)[3], which has been extensively analysed in a number of applications – e.g., see [5]. This ontology has been defined primarily to characterise computer science departments in academia, and would be briefly summarized by its main designer (who happens to be also one of the authors of this paper) by stating that it provides concepts to describe projects, categories of staff and students, organizations, events (in particular, academic events), technologies, publications, etc. Now, if we look at the analysis presented in [5], we can see that it indicates that, out of about 70K queries which had been posted to the AKT-RO, all but twelve focused on only four classes: Technology, Organization, Research-Area and Person. An interesting feature that links these four classes to the informal summary of the AKT-RO given by its designer is that both selections of concepts appear to be pitched at a level of abstraction akin to what Eleanor Rosch termed *natural categories* [7]. Specifically, in her seminal work, Rosch showed that people characterise the world primarily in terms of *basic objects*, such as chair or car, rather than more abstract concepts, such as furniture or vehicle, or more specific ones, such as sportscar or kitchen chair. Hence, an initial hypothesis underlying our approach was that this notion of natural categories could provide a useful basis to identify good descriptors of an ontology[4].

---

[3] http://www.aktors.org/publications/ontology/

[4] It is important to emphasise that we are by no means the first researchers to highlight the value of natural categories in identifying good descriptors of an ontology. In particular, the advantages of a *middle-out* approach to ontology design, where basic concepts are identified first and used to drive the ontology development process, have long been recognized in ontology engineering [8].

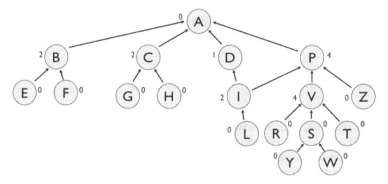

Fig. 1. Basic levels of nodes in a taxonomy – please note that measures are not normalised

Unfortunately, to our knowledge there is no available repository of natural categories and for this reason we had to approximate this notion by devising mechanisms which operationalize it for our scenario. Specifically we have devised two measures, which we use to try and identify concepts that may play the role of 'natural categories' in ontologies.

> **Name simplicity.** The *name simplicity*, $NS(C) \in [0..1]$, of a concept C favours concepts that are labelled with simple names, while penalizing compounds. The rationale for this criterion is that natural categories normally have relatively simple labels, such as chair or dog. In other words, they are unlikely to be compound terms. Accordingly, the name simplicity of a concept is 1 if its label is made of only one word. It decreases following the number of compounds in the label, in accordance with the following formula: $NS(C) = 1 - c(nc\text{-}1)$, nc being the number of compounds in the label and c a constant —in our experiments, we use $c = 0.3$. For example, the name simplicity of the concept *Artist* is 1, while that of *MusicalArtist* is 0.7.

> **Basic level.** The *Basic Level*, $BL(C)$, of a concept C is a measure between 0 and 1, which indicates how 'central' C is in the taxonomy of the ontology. It is computed by counting, for each branch of the ontology containing C, how many times C can be found in the middle of a path from the root to a leaf of the branch (see Figure 1) and then normalising the value.

Given these two measures, there are two steps needed to decide the set of concepts corresponding to natural categories in a given ontology. First, the basic level and name simplicity scores are used to generate a set of candidate concepts, by choosing the ones for which $w_{BL}*BL(C) + w_{NS}*NS(C)$ is greater than a given threshold T_{nc} —in our experiments, we used $T_{nc} = 0.5$, $w_{BL} = 0.8$, $w_{NS} = 0.2$. Then, this set of candidates is filtered, by giving priority to the concepts which are neither roots or leaves of the branch, and also by assuming that only one natural category exists on a given branch of the hierarchy. If a branch contains more than one candidate concept, the one which maximizes $w_{BL}*BL(C) + w_{NS}*NS(C)$ is chosen. The output is a set of concepts, $NC(O)$, which are considered as corresponding to natural categories in the context of the ontology O.

As shown by the above definitions, while natural categories in Rosch have a universal connotation, our operationalization takes into account the design of the ontology and therefore somewhat contextualises this notion with respect to the granularity of the ontology.

2.2 Topology-Based Criteria: Density and Coverage

While natural categories provide a criterion to decide what type of concepts ought to be part of an ontology summary, such a criterion is not sufficient on its own as a basis for an algorithm. We also need structuring criteria, which take into account the overall organization of the ontology. These criteria are meant to ensure that the chosen concepts embed enough information and that no important part of the ontology is left out in the 'summary'. To this purpose we also use two criteria defined on the basis of the structure of an ontology, *density* and *coverage*.

2.2.1 Density

The *density(C)* $\in$ *[0..1]* of a concept C is a measure of how richly described the concept is in the ontology and is computed on the basis of its number of direct sub-concepts, properties and instances. When computing the overall density of a concept, we use two sub-measures, *global* and *local* density. The former measures density in relation to the entire ontology, the latter only considers the neighborhood of a concept.

The global density, *globalDensity(C)* $\in$ *[0..1]*, of a concept C is computed by a simple, weighted aggregation on the number of direct sub-concepts, properties and instances of C:

$$globalDensity(C, O) = \frac{aGlobalDensity(C)}{max(\{\forall N_i \in O \rightarrow aGlobalDensity(N_i)\})}$$

$$aGlobalDensity(C) = n.SubClasses(C) * w_S + n.Properties(C) * w_P + n.Instances(C) * w_I$$

In our experiments, we used $w_S = 0.8$, $w_P = 0.1$, $w_I = 0.1$.

The local density, *localDensity(C)* $\in$ *[0..1]*, of a concept C refers to a density value which is relative to those of the surrounding concepts. The rationale for this measure is that, even within the same ontology, the richness of the description of concepts can vary dramatically: some areas of an ontology may contain many dense concepts, which will all be picked-up by the global density measure, while some other areas may only contain shallow concepts. For instance, the 'triangle' concept in Figure 2 is locally dense, but has a low global density, at least compared to some of the other concepts in the ontology. Hence, the local density criterion favours the densest concept in a local area, for being potentially the most important for this particular part of the ontology. It is computed using the formula below, where by "nearest concepts" to C, we refer to the set which includes sub- and super-concepts reachable through a path of maximum length 2 in the hierarchy from C, as well as C itself.

$$localDensity(C) = \frac{globalDensity(C)}{maxGlobalDensityNearestClasses(C)}$$

Finally, the overall density is computed by combining the local and global densities, each of these sub-measures being associated with a particular weight:

$$density\,(C) = globalDensity\,(C) * w_G + localDensity\,(C) * w_L$$

In our experiments, we used $w_G = 0.2$, $w_L = 0.8$.

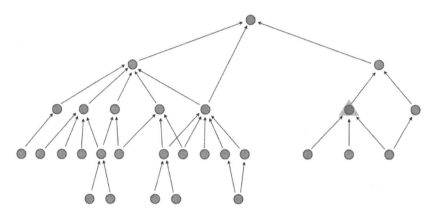

Fig. 2. Example of a locally dense concept

2.2.2 Coverage

The coverage criterion states that the set of key concepts identified by our algorithm should maximise the coverage of the ontology with respect to its is-a hierarchy. More precisely, if $C = \{C_1,.....C_n\}$ is the set of concepts returned by the algorithm and D_i is a concept in the ontology, there should be a $C_k \in C$ such that either $D_i \subseteq C_k$ or $C_k \subseteq D_i$ holds. The rationale for this criterion is that not only we want the right type of concepts to be returned by our method, but also the right spread of concepts must be achieved, to provide the best possible illustration of the ontology.

Let $Covered(C)$ be the set of concepts covered by a concept C, i.e., $Covered(C) = C \cup allSubClasses(C) \cup allSuperClasses(C)$. We define $Coverage(S)$ as the measure of the level of coverage of a set of concepts S in a given ontology. Specifically, $Coverage(\{C_1,...,C_n\})$ is computed using the following formula (with $|O|$ being the size of the ontology O given as the number of concepts included in O):

$$Coverage\,(\{C_1,...,C_n\}) = \frac{\left|\{Covered\,(C_1) \cup ... \cup Covered\,(C_n)\}\right|}{|O|}$$

Another useful measure related to coverage indicates how balanced a set of concepts is, i.e., the degree to which each concept contributes to the overall coverage of the set. This measure, called $Balance(S)$, where S is a set of concepts, is equal to the standard deviation of the elements in S, computed with respect to the cardinality of $Covered(C_k)$, for each $C_k \in S$.

The algorithm presented in the next section requires a procedure *able to complete a set of concepts according to coverage*. That is, considering a set S of concepts of size

$m<n$, we want to complete this set with additional concepts such that the resulting set is of size n, while maximizing coverage. This is realized by first computing the set S' of all the concepts not covered by S, and then generating all the possible sets, with cardinality equal to n, obtained by merging S with concepts in S'.

2.3 Key Concepts Extraction: First Version

Our algorithm takes as input an ontology, O and an integer n, with $n \leq |O|$, and returns as output n concepts in O, which best summarize it. Below we describe the algorithm in detail:

1. Using the procedure described in section 2.1, compute the set $NC(O)$ of natural categories in O.
2. If the size m of $NC(O)$ is
   - equal to n, then return $NC(O)$ and stop.
   - greater than n, then generate the set *CandidateSets* of all the possible subsets of $NC(O)$ of size n.
   - smaller than n, then generate the set *CandidateSets* of all the completed sets of concepts from $NC(O)$, according to the procedure described in section 2.2.2.
3. Select the set of key concepts to return in *CandidateSets* by applying successively the following criteria, until only one candidate set is left:
   a. Restrict *CandidateSets* to the sets of concepts $S \in$ *CandidateSets*, which maximise *Coverage(S)*
   b. Restrict *CandidateSets* to the sets of concepts $S \in$ *CandidateSets*, which minimise *Balance(S)*
   c. Restrict *CandidateSets* to the sets of concepts $S \in$ *CandidateSets*, which maximise the average of $w_{BL}*BL(C_k) + w_{NS}*NS(C_k)$, where $C_k \in S$
   d. Restrict *CandidateSets* to the sets of concepts $S \in$ *CandidateSets*, which maximise the average of *density(C_k)*, where $C_k \in S$
   e. Randomly choose one set S in *CandidateSets* and return it.

Essentially, this algorithm returns a set of size n of concepts from O, which is computed by selecting concepts that appear to be 'natural categories', then taking into account how this set of concepts covers the ontology, and finally using the density of the concepts to discriminate between possible alternatives.

2.4 Evaluation of the First Version of the Algorithm

In order to evaluate the level of similarity between the output produced by our method and human experts, we performed an evaluation using four different ontologies: *biosphere*[5], *music*[6], *financial*[7], *aktors portal*[8].

---

[5] http://sweet.jpl.nasa.gov/ontology/biosphere.owl
[6] http://pingthesemanticweb.com/ontology/mo/musicontology.rdfs
[7] http://www.larflast.bas.bg/ontology
[8] http://www.aktors.org/ontology/portal

We asked eight people with good experience in ontology engineering to select up to 20 concepts they considered the most representative for summarizing the contents of the ontologies. We also told them that if possible they should try and achieve a good coverage of the various parts of the ontology, rather than simply selecting all concepts from one particular branch in a taxonomy and ignore the others. In other words, we explained to them that achieving a good coverage was a desirable feature, but of course we did not give any formal guidance on how to apply this criterion, nor we mentioned the other criteria used by our approach.

Table 1. The concepts shared by more than half of the experts

Ontology	Number of concepts in O	Concepts shared by the experts
biosphere	87	Animal, Bird, Fungi, Insect, Mammal, MarineAnimal, Microbiota, Plant, Reptile, Vegetation
music	91	Event, Genre, Instrument, Medium, MusicArtist, MusicGroup, MusicalExpression, Record, Sound
financial	188	Bank, Bond, Broker, Capital, Contract, Dealer, Financial_Market, Order, Stock
aktors portal	247	Computing-Technology, Geopolitical-Entity, Event, Organization, Person, Publication, Publication-Reference, Software-Technology

Table 2. Average proportion of the concepts in Table 1 selected by each expert

Ontology	mean agreement among experts
biosphere	73.75%
music	76,39%
financial	75%
aktors portal	73,61%

Table 1 shows the concepts that were chosen by at least 50% of the experts, while Table 2 measures the level of agreement on the concepts shown in Table 1. Hence, the tables show that a consensus emerged on a number of concepts in each ontology and, for these concepts, the level of agreement among experts was good, with a mean value of *74.68%*. Indeed, it is important to emphasise that the ontologies used in our study are significantly larger than those used in [6], hence our experiments show that not just in small ontologies but also in medium sized ones, a degree of consensus emerges when experts are asked to identify key concepts.

Unfortunately the results for our method were disappointing. As shown in Table 3, our method only exhibits an average 42.56% level of agreement with the experts, much lower than the measure of inter-expert agreement shown above.

Table 3. Correlation between the first version of our method and the experts

Ontology	Common choices between the testers and the algorithm	%
biosphere	Animal, MarineAnimal, Plant	30
music	Event, Genre, Instrument, MusicalExpression	44,44
financial	Broker, Dealer, Order	33,33
aktors portal	Computing-Technology, Event, Organization, Person, Publication-Reference	62,5

3   Revised Approach

3.1   What Went Wrong? How Experts Select Key Concepts

The analysis of the results we obtained from the experts shows that while people may employ the three criteria used by our algorithm, their application is different from the way the algorithm combines them. Our subjects did not apply coverage as strictly as our algorithm and moreover they seemed to use density ahead of natural categories. In addition, our approximation of the notion of natural category, with its emphasis on centrality and name simplicity, did not work well. Many concepts which are not natural categories may have a very simple label and, given that different ontologies have different degrees of structure and depth, centrality turned out not to be crucial, especially when it did not correlate with density. In other words, we did not find any evidence that contextualizing the notion of natural categories to the granularity of a specific ontology correlates with expert choices. Let's clarify this point with an example. Figure 3 shows some of the subclasses of the class *Animal* in the *biosphere* ontology. These have all very simple labels and have no children. However, several experts selected *Bird* and *Insect* as key concepts, even though none of the criteria we use is able to select them ahead of their siblings: they are neither dense nor central and their labels is not lexically simpler than any of the other subclasses *of Animal*.

To deal with these cases we introduced a new criterion, called *popularity*, to try and identify concepts that are particularly common, such as *Bird* and *Insect*. The advantage of this approach is that it allows us both to pick many natural categories (such as *Bird* and *Insect*) and also to identify *best exemplars* of a concept, in those cases in which we are not dealing with natural categories. Operationally, we measure the popularity of a concept, C, as the number of results returned by querying Yahoo with the name of *C* as keyword. Compound names are transformed to a sequence of lower case keywords separated by a space. For instance, *Marine-Animal, MarineAnimal, marineAnimal, marine_animal* are all transformed in *"marine animal"*.

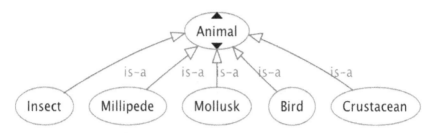

Fig. 3. Subclasses of class *Animal* in the *biosphere* ontology

3.2 Revising the Algorithm

On the basis of the considerations discussed in the previous section, we revised our method and implemented and tested two new versions of the algorithm for key concept extraction, which include the popularity criterion as well as those used in the first version of our method. For the sake of conciseness, in what follows we will focus on the third and final version of our system, which is the one exhibiting the best overall performance –i.e., the highest degree of correlation with the choices of the experts.

3.2.1 New Concepts and Formulas

In order to understand the new version of the algorithm, we need to introduce a number of new concepts and formulas. First of all, we want to improve the way we compute local densities, to obtain a more continuous spread of values. This is achieved by means of the following formula:

$$localDensity(C) = \frac{globalDensity(C)}{max(\{\forall N_i \in nearest_k(C) \rightarrow weightedGD(C, N_i)\})} + w_{GDL} * globalDensity(C)$$

$$weightedGD(C, N) = (1 - (ratio_D * distance(C, N))) * globalDensity(N)$$

The function *nearest_k(C)* returns the class *C* and its sub- and super-classes, which are reachable through a path of maximum length *k* in the hierarchy. In our experiments we used $k = 2$, $ratio_D = 0.1$ and $w_{GDL} = 0.5$.

The function *weightedGD(C,N)* is used to ensure a more continuous distribution of the local density values, compared to the definition given in section 2.2.1. To this purpose, when we calculate the maximum global density value of the set *nearest_k(C)* we take into consideration a weighted global density value for the classes $N \in nearest_k(C)$. In a nutshell, as the distance from *N* to *C* increases, the weighted global density of *N* with respect to *C* decreases.

As in the case of density, we also want to take into consideration both the *global* and *local* popularity of a concept, and we compute these analogously to the way we derive global and local densities:

$$globalPopularity(C,O)=\frac{hits(C)}{max(\{\forall N_i\in O\rightarrow hits(N_i)\})}$$

$$localPopularity(C)=\frac{globalPopularity(C)}{max(\{\forall N_i\in nearest_k(C)\rightarrow weightedGP(C,N_i)\})}+w_{GPL}*globalPopularity(C)$$

$$weightedGP(C,N)=(1-(ratio_p*distance(C,N)))*globalPopularity(N)$$

The function *hits(C)* returns the number of hits that we obtain querying Yahoo with the name of C as keyword. In our experiments we used $k = 1$, $ratio_P = 0.1$ and $w_{GPL} = 0.5$.

The new version of the algorithm is based on the calculation, for each class C of an ontology O, of its local and global density, local and global popularity and its natural category value, *NCValue*, which is the normalized value of $w_{BL}*BL(C) + w_{NS}*NS(C)$, as described in section 2.1. All these measures are aggregated in a new overall value associated with a concept, called *score*, which corresponds to a weighted sum of all the above measures, as shown by the following formulas:

$$score(C)=D(C)+P(C)+NCValue(C)$$

$$D(C)=w_{LD}*localDensity(C)+w_{GD}*globalDensity(C)$$

$$P(C)=w_{LP}*localPopularity(C)+w_{GP}*globalPopularity(C)$$

$$NCValue(C)=w_{BL}*BL(C)+w_{NS}*NS(C)$$

In our experiments we used $w_{LD} = 0.32$, $w_{GD} = 0.08$, $w_{LP} = 0.1$, $w_{GP} = 0.2$, $w_{BL} = 0.66$, $w_{NS} = 0.33$.

We also extended the coverage criterion with a new function called *contribution*, which aims to measure the actual 'contribution' of a class C_i to the coverage of a set of classes $\{C_1,..., C_i, ..., C_n\}$ in O, by counting the classes of O covered only by C_i in this set. This value is computed as follows:

$$contribution(C_i, \{C_1,..., C_i,..., C_n\}) = |Covered(C_i) - \cup_{1\leq k\leq n\wedge k\neq i}Covered(C_k)|$$

Finally, we define the *optimal coverage* for an ontology O as a set $S = \{C_1,..., C_n\}$, where *Coverage(S)* = 1, and each $C_i \in S$ provides the same *contribution* with respect to S as the other concepts in S.

3.2.2 Specification of the Revised Algorithm

As in the first version, our revised algorithm takes as input an ontology O and an integer n, with $n \leq |O|$, and returns as output n classes in O, which best summarize it. In our experiments we used $n = 20$.

Below we describe the algorithm in detail:

1. For each class C in O we compute its global and local density, global and local popularity and the natural category value.
2. For each class C in O we compute *score(C)*, as described in section 3.2.1.

Table 4. Correlation between the final version of our algorithms and the experts. Concepts in *italic* in the second column are the ones also picked by more than half of the experts.

Ontology	Algorithm choices	% matches with experts' choices
biosphere	*Animal*, Bacteria, *Bird*, Crown, Fish, *Fungi*, FungyTaxonomy, Human, Litter, LivingThing, *Mammal*, *MarineAnimal*, Marine-Plant, *Microbiota*, MicrobiotaTaxonomy, Mold, Mushroom, *Plant*, *Vegetation*, Yeast	80
music	Agent, CorporateBody, Document, *Event*, Expression, *Genre*, Group, *Instrument*, Item, *Medium*, *MusicalExpression*, MusicalManifestation, MusicalWork, OriginMap, Person, *Record*, Show, Signal, TimeLine, Work	66.67
financial	Agent, *Bond*, *Capital*, Card, Cost, *Dealer*, Financial_Asset, Financial_Instrument, *Financial_Market*, Money, *Order*, Organization, Payment, Price, Quality, Security, *Stock*, Supplier, Transaction, Value	66.67
aktors portal	Educational-Organization-Unit, Employee, *Event*, Information-Bearing-Object, Intangible-Thing, Integer, Legal-Agent, Location, Message, Month, Number, *Organization*, *Person*, *Publication*, *Publication-Reference*, Set, *Software-Technology*, Technology, University, Working-Person	75

3. Given a number $k \leq n$ (in our experiments $k = 15$), let S be the set of k classes in O with the best *score* and let T be the set of n-k classes in $\{O \ — \ S\}$ with the best *score*. If T is empty, we return S and we stop.
4. Otherwise, let c be the average of all the values obtained by invoking the function *contribution*$(C_i, \{S \cup T\})$, for each $C_i \in \{S \cup T\}$. And let a be the average of all the values obtained by invoking the function *overallScore*$(C_i, \{S \cup T\})$, again for each $C_i \in \{S \cup T\}$. The function *overallScore* is defined as follows.

$$overallScore\left(C_i, \{C_1, ..., C_i, ..., C_n\}\right) = w_{CO} * \frac{contribution(C_i, \{C_1, ..., C_i, ..., C_n\})}{maxContribution(\{C_1, ..., C_i, ..., C_n\})} + w_{CR} * score(C_i)$$

In our experiments we have used $w_{CO} = 0.6$ and $w_{CR} = 0.4$.

5. Let W be the class in T with the worst *overallScore*$(W, \{S \cup T\})$ of all the classes in $\{S \cup T\}$, and let R be the set $\{\{S \cup T\} \ — \ \{W\}\}$. If there is a class $B \in \{O \ — \ \{S \cup T\}\}$, such that

(a) the average a' of all the values obtained by invoking *overallScore*$(C, \{R \cup \{B\}\})$, computed for each $C \in \{R \cup \{B\}\}$, is greater than a,

(b) the average c' of all the values obtained by invoking *contribution*$(C, \{R \cup \{B\}\})$, computed for each $C \in \{R \cup \{B\}\}$, is greater than or equal to c,

we swap W with B in $\{S \cup T\}$ and we go back to step 4. Otherwise we return $\{S \cup T\}$ and we stop.

3.3 Evaluation of the Revised Version of the Method

The tests performed with the new version of the algorithm produced much better results than the previous version. In particular, as shown in table 4 the average measure of agreement between our algorithm and the human experts is now *72.08%*, only 1.5 points lesser than the inter-expert agreement (*74.68%*). In practice the final version of our method, at least on the current benchmark, is indistinguishable in its output from human experts.

4 Related Work

As already mentioned, a few papers have addressed the topic of ontology summarization. In particular, in [6] the authors describe a family of algorithms to select the *salient RDF sentences* from a RDF graph. These algorithms work primarily on the basis of the topological structure of the graph. The paper shows that while there is a relatively low correlation between experts at the sentence level, there is a much better degree of agreement with respect to vocabulary overlap. In addition, they also show that the results produced by their method exhibit a good degree of correlation with the experts. However, the ontologies used in their case studies are much smaller than the ones used here, so those results are potentially less significant than those presented here, even though no firm conclusion can be stated without trying out both approaches on a common benchmark. The work described in [5] focus on *winnowing* an ontology – i.e., reducing the size of an ontology to facilitate its reuse. Hence, in this work the focus is on a different type of summarization, which aims to make the ontology more easily reusable, rather than facilitating ontology understanding in a context in which the user wishes to quickly get a snapshot of what an ontology is about. The same consideration applies to work on *ontology customization* [9], which provides mechanisms to enable particular views over an ontology. While this work can be seen as a particular kind of ontology summarization, it differs from our work both with respect to the output of these techniques (a particular *cut* over an ontology) and also because it expects the user to specify which part of an ontology she is interested in.

5 Discussion

While the generation of automatic methods, able to extract ontology summaries in a way which correlates with human experts, is an interesting research issue in itself, the work described here also provides a potentially useful basis for a number of novel contributions to ontology engineering and semantic web research. In section 1, we have already pointed out that a key motivation for this work was to facilitate the process of ontology understanding for users of the Watson ontology search engine and the Watson Plug-in. In particular, by providing quick snapshots of an ontology as part of the results returned by Watson, we hypothesise that it will be easier for users to quickly home in on the ontology most relevant to her needs. We also plan to use this work as the basis for a novel visualization algorithm, to complement and to address the weaknesses of the traditional taxonomic-centric support for navigating ontologies, which is provided by current ontology engineering editors. As discussed in [10], classic hierarchical views of ontologies are not very helpful for supporting tasks related to understanding the general structure of an ontology. In particular, consistently with the experiments carried out here, the concepts that experts select to describe an ontology tend to be on different branches of the hierarchy at various levels of depth. Hence, they cannot be easily identified with standard top-down taxonomy browsers.

Initial presentations of this work to a number of audiences have also elicited interesting suggestions for applying the work described here to a number of ontology-centric scenarios. In particular, colleagues have suggested the use of our summarization technique in scenarios where an ontology is used to support automatic data classification, but it is too expensive to try and classify large quantities of data against a large number of classes. In these scenarios, our method could be used to identify the most useful concepts in an ontology, so that these can be tried first. Similar ideas have been suggested by colleagues working on ontology matching and evolution, where the ability to prioritize which concepts the system ought to focus on could also be useful. Analogously, key concept selection could also be used as the basis for a new family of ontology modularization algorithms. For example, modules could be built around each key concept, so that the resulting partitioning of the ontology would identify 'key areas' of the ontology, consistently with the criteria presented in this paper. Finally, we also intend to use this method in the context of the work on 'cautious knowledge sharing', which we are carrying out in the OpenKnowledge project (http://www.openk.org). This work is concerned with scenarios where the content of an ontology is proprietary or otherwise restricted, and cannot be made publicly available. In these scenarios, automatic ontology summaries can be useful as a way to advertise an ontology while disclosing as little content as possible.

6 Conclusions

In this paper we have introduced a user-independent approach to identifying automatically the key concepts in an ontology. The approach integrates both topological measures, such as density (both global and local) and coverage, as well as statistical lexical measures (popularity), and cognitive criteria (natural categories). The approach has been validated empirically, by showing that the revised version of our implementation shows an excellent degree of correlation with human experts. However, we should stress that these results, although promising, are still preliminary. A more extensive evaluation study will be needed, to determine more conclusively both the extent to which experts are able to agree on what are the best concepts to describe an ontology and also the extent to which this approach can emulate expert concept selection in a variety of domains. It will also be interesting to extend the algorithm, so to be able to add also 'key properties' and even 'key individuals' to the ontology summaries. In particular, adding key properties introduces interesting issues, as some degree of coherence needs to be ensured between the set of concepts and the set of properties identified by the algorithm. Hence, a possible strategy could be to focus on concepts first, using the approach presented in this paper, and then extend such selection by identifying the most important properties associated with the selected concepts, rather than with the ontology as a whole.

As already mentioned, we also intend to apply these ideas to a number of ontology engineering tasks, e.g., to explore new approaches to ontology visualization and navigation, ontology evolution, and ontology modularization. Finally, we plan to make our system available as a resource for the ontology engineering and semantic web communities, by exposing it as a web application.

Acknowledgments. This work has been partially funded by the OpenKnowledge IST-FP6-027253 and NeOn IST- FP6-027595 projects. The authors would like to thank an anonymous referee for his numerous insightful suggestions.

References

1. d'Aquin, M., Baldassarre, C., Gridinoc, L., Angeletou, S., Sabou, M., Motta, E.: Characterizing Knowledge on the Semantic Web with Watson. In: Workshop on Evaluation of Ontologies and Ontology-based tools, 5th International EON Workshop, collocated with the International Semantic Web Conference (ISWC 2007), Busan, Korea (2007)
2. Stefik, M.: The next knowledge medium. AI Magazine 7(1), 34–46 (1986)
3. d'Aquin, M., Motta, E., Sabou, M., Angeletou, S., Gridinoc, L., Lopez, V., Guidi, D.: Towards a New Generation of Semantic Web Applications. IEEE Intelligent Systems 23(3), 20–28 (2008)
4. Dzbor, M., Motta, E., Buil Aranda, C., Gomez, J.M., Goerlitz, O., Lewen, H.: Developing ontologies in OWL: An observational study. In: Workshop on OWL: Experiences & Directions, Georgia, US (November 2006)
5. Alani, H., Harris, S., O'Neil, B.: Winnowing ontologies based on application use. In: Sure, Y., Domingue, J. (eds.) ESWC 2006. LNCS, vol. 4011, pp. 185–199. Springer, Heidelberg (2006)
6. Xiang, Z., Cheng, G., Qu, Y.: Ontology Summarization Based on RDF Sentence Graph. In: 16th International World Wide Web Conference (WWW2007), Banff, Alberta, Canada, May 8-12 (2007)
7. Rosch, E.: Principles of Categorization, Cognition and Categorization. Lawrence Erlbaum, Hillsdale, New Jersey, Mahwah (1978)
8. Uschold, M., Gruninger, M.: Ontologies: Principles, Methods and Applications. Knowledge Engineering Review 11(2), 93–136 (1996)
9. Bercovici, N., Gröner, G., Schenk, S., Kubias, A., Dzbor, M.: Ontology customization and module creation: query-based customization operators and model. NeOn Project Deliverable D4.2.2 (February 2008)
10. Katifori, A., Halatsis, C., Lepouras, G., Vassilakis, C., Giannopoulou, E.: Ontology Visualization Methods - A Survey. ACM Computing Surveys 39(4) (2007)

The Art of Tagging:
Measuring the Quality of Tags

R. Krestel and L. Chen

L3S Research Center
Universität Hannover, Germany
{krestel,lchen}@L3S.de

Abstract. Collaborative tagging, supported by many social networking websites, is currently enjoying an increasing popularity. The usefulness of this largely available tag data has been explored in many applications including web resources categorization,deriving emergent semantics, web search etc. However, since tags are supplied by users *freely*, not all of them are useful and reliable, especially when they are generated by spammers with malicious intent. Therefore, identifying tags of high quality is crucial in improving the performance of applications based on tags. In this paper, we propose TRP-Rank (Tag-Resource Pair Rank), an algorithm to measure the quality of tags by manually assessing a seed set and *propagating the quality* through a graph. The three dimensional relationship among users, tags and web resources is firstly represented by a graph structure. A set of seed nodes, where each node represents a tag annotating a resource, is then selected and their quality is assessed. The quality of the remaining nodes is calculated by propagating the known quality of the seeds through the graph structure. We evaluate our approach on a public data set where tags generated by suspicious spammers were manually labelled. The experimental results demonstrate the effectiveness of this approach in measuring the quality of tags.

1 Introduction

With the recent rise of Web 2.0 technologies, many social media applications like *Flickr*, *Del.ici.ous*, and *Last.fm* provide features which allow users to assign tags [1] to a piece of information such as a picture, blog entry, video clip etc. Web users from different backgrounds annotate (tag) resources on the Web at an incredible speed, which results in a large volume of tag data obtainable from the Web today. The hidden value of tag data has been explored in many applications. For example, Tso-Sutter et al [2] incorporated tags into collaborative filtering algorithms to enhance recommendation accuracy. In [3], the authors discussed using tags to lighten the limitation of the amount and quality of anchor text to improve enterprise search. The usage of tags in Web search has also been investigated in Bao et al [4].

One notable reason which supports the increasing popularity of collaborative tagging is that users are permitted to enter tags at will, without referring to

J. Domingue and C. Anutariya (Eds.): ASWC 2008, LNCS 5367, pp. 257–271, 2008.
© Springer-Verlag Berlin Heidelberg 2008

any pre-specified taxonomy or ontology. On the one hand, this easy and flexible utility boosts the spreading of collaborative tagging systems. On the other hand, allowing users to *freely* choose tags sometimes leads to poor quality of the tag data. For example, ambiguity and synonymy are two frequently cited problems. The tag "XP" is used to annotate both web pages about "Extreme Programming" and pages about "Windows XP". Synonymous tags, like "RnB" and "R&B", are also widely used. Such problems hamper the applications built upon tags. Another problem which even damages the performance of applications using tags is tag spam, which refers to misleading tags generated maliciously in order to increase the visibility of some resources or simply to confuse users. *Therefore, measuring the quality of tags is an important issue and discriminating high quality from low quality tags improves the effectiveness of different tag-based applications.*

In [5], the authors discussed some properties a good tag combination (e.g., the set of tags annotating a common resource) should possess. For example, a good tag combination should cover multiple facets of the tagged resource; the set of tags should be used by a large number of people; and the number of resources identified by the tag combination should be small etc. They further proposed a tag suggestion algorithm based on these properties. In contrast to suggesting new tags to users based on existing tags so that a good tag combination can be achieved, our objective here is to assess the quality of tags assigned by users. Koutrika et al [6] proposed to combat tag spam by ranking the results returned from a query tag, based on the co-occurrence frequency between the tag and each resource. Thus, their approach is specially designed for tag based search. Our research objective is more general so that the results can be used in various applications of tags.

Note that, whether a tag is good or bad can only be assessed with respect to a particular resource. Hence, our investigation is based on the unit of a tag-resource pair. We aim to measure the quality of each individual pair of tag and resource. For this purpose, we firstly construct a graph which models tag-resource pairs as nodes and co-user relationship as edges. We then select a set of seed nodes whose qualities are assessed manually. The qualities of the remaining nodes are calculated by propagating the qualities of seed nodes through the graph. In order to improve the performance of this approach, a set of various seed selection strategies are employed. We evaluate the effectiveness of our approach on a bibsonomy data set[1] labelled manually.

The rest of this paper is organized as follows. We discuss the background knowledge by reviewing related work in Section 2. In Section 3, we describe the approach which propagates the quality of tag-resource pairs and discuss improving the performance by employing different strategies to select a set of seeds. The evaluation results conducted on a public data set are presented and analyzed in Section 4. Finally, Section 5 concludes this paper with some summary remarks and future work discussions.

---

[1] http://www.kde.cs.uni-kassel.de/ws/rsdc08/dataset.html

2   Related Work

In this section, we review related work in two areas, collaborative tagging systems and spam detection.

A collaborative tagging system allows users of a web site to freely attach to a particular resource arbitrary tags which, in the opinion of the user, are somehow associated with the resource in question. The commonly noted structure of collaborative filtering systems is a tripartite model consisting of users, tags and resources. This model is developed as a theoretical extension of the bipartite structure of ontologies with an added "social dimension" in [7]. The dynamics of collaborative systems are examined in [8] using the tag data at the bookmarking site Del.ici.ous. According to this work, tag distributions tend to stabilize over time. Halpin et al. confirm these results in [9] and show additionally that tags follow a power law distribution. Considering the structure and stable dynamics of collaborative tagging systems, it seems likely that tag data would be a reliable source of semantic information reflecting the cultural consensus of a particular system's users. As a result, various applications of tag data have been researched. Mika [7] investigates the automatic extraction of ontological relationships from tag data and proposes the use of such emergent ontologies to improve currently existing ontologies which are less capable of responding to ontological evolution. Dmitriev et al. [3] explore the use of "annotations" for enterprise search to compensate for the lack of sufficient anchor text in intranet environments. In [4], tag data is exploited for the purpose of web search through the use of two tag based algorithms: one exploiting similarity between tag data and search queries, and the other utilizes tagging frequencies to determine the quality of web pages. Tso et al [2] incorporate the tag data into the collaborative filtering systems. Berendt and Hanser [10] demonstrate the benefits of using tag data for weblog classification by treating it as content instead of meta data. For searching and ranking within tagging systems, [11] proposes the exploitation of co-ocurrence of users, resources, and tags. This is done using a graph model to represent the *folksonomy*.

Everywhere in the internet where information is exchanged, malicious individuals try to take advantage of the information exchange structure and use it for their own benefit. The largest amount of spam and historically the first field where spam was generated is the electronic communication system (e-mail). Afterwards, various internet applications were attacked by spammers such as search engine spam, blog spam, wiki spam etc, which triggered numerous research efforts in spam combating. For example, TrustRank [12] separates spam pages from non-spam pages based on the intuition that trustworthy pages usually link to also trustworthy pages and so on. They select a seed set of highly trusted pages first and then propagate the trust score of seed pages by following the links from these pages through the Web. A survey of approaches fighting spam on social web sites can be found in [13]. Comparing to spam detection from other web applications, studies on detecting spam from collaborative tagging systems are very limited. Koutrika et al [6] propose to combat spam in the particular situation when users query for resources annotated with certain

tags. Their method ranks a resource higher if more users annotated it with the
queried tags, based on the assumption that tag spam may not be used by the
majority. Our work is different in the way that our approach is not designed for
a particular application. Consequently, the output of our algorithm can be used
by any application based on tags. Xu et al [5] assign authority scores to users,
and measure the quality of each tag with respect to a resource by the sum of the
authority scores of all users who have tagged the resource with the tag. Then,
the authority scores of users are computed via an iterative algorithm similar to
HITs [14]. Their approach treats every tag-resource pair used by a user equally
even if a spam user may use good tag-resource pairs frequently and bad ones
occasionally. Our approach addresses this problem by measuring the quality of
a tag-resource pair more independently from a particular user.

3   Measure Tag Quality

The hidden value of tag data has been explored by a wide range of applications.
However, as mentioned before, since there is no limitation on the vocabulary
users are allowed to use for taggging, the quality of tags varies. In other words,
tags are not equally useful for a particular application. For example, recovery
and discovery of resources on the web is one of the main uses of tags. Although
tags describing the general topics of resources might be useful for search engines, *personal* or *subjective* (see [15,16] for a taxonomy of tags) tags such as
"myFavorite", "funny", "home" do not seem to be promising for this task. Furthermore, it is common that tags which describe one resource very well may not
be suitable for another resource. Consequently, measuring the quality of tags is
critical for applications to exploit the positive usage of tag data. The quality of
a tag should be measured with respect to the resource to which it is assigned.

In this section, we first formally define the problem we focus on in this paper.
Then, the data structure which models the relationship among tags, resources,
and users is described. Next, we illustrate our algorithm, called *TRP-Rank* (Tag-
Resource Pair Rank), which iteratively assesses the quality of each pair of tag-
resource in the data set. Finally, several strategies which select various sets of
seed nodes, serving as the input of TRP-Rank, are discussed.

3.1   Problem Specification

Let $\mathcal{T}$ be a set of tags, $\mathcal{R}$ be a set of resources, and $\mathcal{U}$ be a set of users. We denote
a tag assignment of a tag $t \in \mathcal{T}$ to a resource $r \in \mathcal{R}$ as a tag-resource pair tr.
All tag assignments in the data $\mathcal{T} \times \mathcal{R}$ is a set of tag-resource pairs denoted
as $\mathcal{TR} = \{tr | t \in \mathcal{T}, r \in \mathcal{R}\}$. Each tag-resource pair is assigned by at least one
user $u \in \mathcal{U}$. We define the function $getU(tr)$ to retrieve the set of users who
assigned t to r. Note that, $getU(tr) \neq \emptyset$. Then, given the complete set of tag-
resource pairs $\mathcal{TR} = \{tr_1, \cdots, tr_n\}$, and associated users of each tag-resource
pair $getU(tr_i) \subseteq \mathcal{U}$, our goal is to find a function $Q(tr_i)$ which assigns a score
to each tag-resource pair tr_i such that the higher the value of $Q(tr_i)$, the better

the quality of the pair tr_i. The value of $Q(tr_i)$ ranges in $[-1, 1]$ (the reason why negative values are involved will be explained later in Section 3.3).

3.2 Tagging System Model

Given a set of data including tags $\mathcal{T}$, resources $\mathcal{R}$ and users $\mathcal{U}$, we model the data as a bidirected weighted graph $\mathcal{G} = \{\mathcal{V}, \mathcal{E}\}$, where $\mathcal{V}$ is a set of vertices with each $v \in \mathcal{V}$ represents a $tr \in \mathcal{TR}$. $\mathcal{E}$ is a set of edges such that each edge (v_i, v_j) indicates that the two corresponding tag-resource pairs tr_i and tr_j are assigned by at least one common user. That is, $|getU(tr_i) \cap getU(tr_j)| \geq 1$. Additionally, we associate a weight to each edge so that the weight of an edge is the number of common users who assigned the tag-resource pairs corresponding to the two end nodes of this edge, $W(v_i, v_j) = W(tr_i, tr_j) = |getU(tr_i) \cap getU(tr_j)|$.

In Figure 1 (a), we present a very simple tagging scenario: Suppose we have three users $\mathcal{U} = \{u_1, u_2, u_3\}$, three different tags $\mathcal{T} = \{t_1, t_2, t_3\}$ and two resources $\mathcal{R} = \{r_1, r_2\}$. Each user has annotated the resources with certain tags. For example, the leftmost link in Figure 1 (a) indicates that both users u_1 and u_2 have supplied the tag t_1 with the resource r_1. Observing the tag assignments in this figure, we notice that there are a total 5 tag-resource pairs $\mathcal{TR} = \{t_1r_1, t_2r_1, t_3r_1, t_1r_2, t_3r_2\}$. Hence, as shown in Figure 1 (b), there are five nodes involved in the data model where each node represents a particular tag-resource pair. An edge connects two nodes if the two corresponding tag-resource pairs are supplied by at least one common user. For example, there is an edge between $v_1 : t_1r_1$ and $v_3 : t_3r_1$ because they are supplied by the common user u_2. Accordingly, the weight of this edge, as shown in the figure, is $|\{u_2\}| = 1$.

Based on this graph model, we introduce a right stochastic transition matrix T, which is defined as:

$$T(i, j) = \begin{cases} 0 & \text{if } (v_i, v_j) \notin \mathcal{E} \\ \frac{W(v_i, v_j)}{\sum_{v_k \in \mathcal{V}} W(v_i, v_k)} & \text{if } (v_i, v_j) \in \mathcal{E} \end{cases}$$

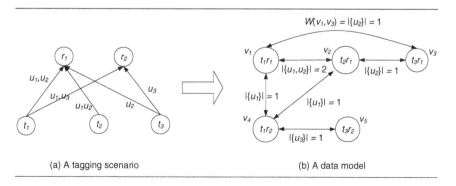

(a) A tagging scenario   (b) A data model

Fig. 1. A tagging scenario and its data model

Figure 2 shows the adjacency matrix and the transition matrix for the example in Figure 1. Note that, the adjacency matrix is symmetric since the graph model is bidirected, while the transition matrix is asymmetric.

	v_1	v_2	v_3	v_4	v_5
v_1		2	1	1	
v_2	2		1	1	
v_3	1	1			
v_4	1	1			1
v_5				1	

$$T = \begin{pmatrix} 0 & \frac{1}{2} & \frac{1}{4} & \frac{1}{4} & 0 \\ \frac{1}{2} & 0 & \frac{1}{4} & \frac{1}{4} & 0 \\ \frac{1}{2} & \frac{1}{2} & 0 & 0 & 0 \\ \frac{1}{3} & \frac{1}{3} & 0 & 0 & \frac{1}{3} \\ 0 & 0 & 0 & 1 & 0 \end{pmatrix}$$

Fig. 2. Adjacency (left) and transition (right) matrixes of the example in Figure 1

3.3 Quality Propagation

Similar to TrustRank [12], which semi-automatically separates web pages from spam, the basic idea of TRP-Rank is to manually assign quality scores to a subset of $\mathcal{TR}$ first, and propagate these quality values through the graph. As the TrustRank algorithm is based on the well-known PageRank [17] algorithm, we briefly review PageRank and TrustRank in the following before illustrate TRP-Rank.

PageRank. PageRank is an algorithm that assigns scores to web pages based on link information. When important pages point to a particular page, this page should also be considered important as well. Thus importance information is propagated through the web graph via an iterative process:

$$\text{p-rank}_{i+1} = \alpha \cdot T \cdot \text{p-rank}_i + (1 - \alpha) \cdot \frac{1}{N} \cdot 1_N. \tag{1}$$

where α is a decay factor, T is the transition matrix and N is the number of web pages. The transition matrix is not weighted and all web pages get the same initial value of p-rank. The iteration process goes on until the difference between two consecutive runs' results is below a certain threshold.

TrustRank. TrustRank extends the Equation (1) to identify web spam. Therefore the original PageRank algorithm was altered to be biased towards a seed set of high quality sites, where each site x was manually assessed with an oracle function $O(x)$. Then, the column vector $\frac{1}{N} \cdot 1_N$ in Equation (1) is replaced with a vector d, such that elements corresponding to manually assessed sites are set as $O(x)$ and the remaining elements are set as 0. d is then normalized, $\mathsf{d} = \mathsf{d}/|\mathsf{d}|$, and feed as t-rank$_0$.

$$\text{t-rank}_{i+1} = \alpha \cdot T \cdot \text{t-rank}_i + (1 - \alpha) \cdot \mathsf{d}. \tag{2}$$

The set of seed sites is selected using an inverse PageRank algorithm. Particularly, nodes from where lots of other nodes can be reached are identified and ranked accordingly, similar to the idea of Hubs [14]. Then, the top-k nodes are manually assigned values 1, or 0 in case of a spam web site, and these initial values are stored in d.

TRP-Rank. For TRP-Rank, the quality of each tag-resource pair, $Q(tr)$, is computed similarly as the Equation (2) in TrustRank. That is, we propagate initial quality scores of seed tag-resource pairs through the graph. In addition to TrustRank which propagates only trust information, we adopt the distrust propagation idea described in [18] to allow the propagation of scores for not only good tag assignments but also explicitly bad ones. Consequently, in TRP-Rank, we extend the manual seed set assessment to include both tag-resource pairs of high quality and those of low quality. We populate the initial vector d with:

$$d(tr_i) = \begin{cases} O(tr_i) & \text{if } tr_i \in SEED \\ 0 & \text{if } tr_i \notin SEED \end{cases} \tag{3}$$

where $O(tr_i) \in \{-1, 0, 1\}$ is the oracle function which assigns initial quality score 1 to good tag-resource pairs, -1 to bad ones and 0 to the rest. $SEED \subseteq T\mathcal{R}$ is a set of seed nodes, which will be defined in Section 3.4.

Consider the running example shown in Figures 1 and 2, the results of TRP-Rank (i.e. quality of tag-resource pairs) after 10 iterations are shown in Figure 3, where v_3 and v_4 are selected as seed nodes and the decay factor α is set as 0.85.

$$\text{trp-rank}_{i+1} = 0.85 \cdot \begin{pmatrix} 0 & \frac{1}{2} & \frac{1}{4} & \frac{1}{4} & 0 \\ \frac{1}{2} & 0 & \frac{1}{4} & \frac{1}{4} & 0 \\ \frac{1}{2} & \frac{1}{2} & 0 & 0 & 0 \\ \frac{1}{3} & \frac{1}{3} & 0 & 0 & \frac{1}{3} \\ 0 & 0 & 0 & 1 & 0 \end{pmatrix} \cdot \text{trp-rank}_i + (1 - 0.85) \cdot \begin{pmatrix} 0 \\ 0 \\ -1 \\ 1 \\ 0 \end{pmatrix}$$

$i = 10$	v_1	v_2	v_3	v_4	v_5
trp-rank(10)	-0.03341879	-0.03341879	-0.16368952	0.180295	0.05023218

Fig. 3. TRP-Rank computation and results for the example in Figure 1

3.4   Seed Selection Strategies

In our approach, we experiment with three different seed selection strategies, whose performance will be presented and discussed in Section 4.3. The two main challenges for seed set selections are: 1) finding an appropriate size for the seed set. A small seed set may not be enough to reach most nodes in the graph, while a large seed set means an expensive manual assessment process; 2) picking the *right* set of tag-resource pairs as seeds. On the one hand, the seed set should contain not only good tag-resource pairs but also pairs of low quality, so that explicit information of both good and bad quality can be propagated. On the other hand, the seed set should contain nodes from which many of the remaining nodes can be reached.

Algorithm 1. Different Seed Selection Strategies

Input:
    N: a set of graph nodes, K $(K < |N|)$: the number of seeds
Output:
    $SEED$: A set of selected seed nodes

1: order N as $\hat{N} = <v_1, v_2, \cdots, v_n>$ such that $PR(v_i) \geq PR(v_{i+1})$
2: **for** each $v_i \in \hat{N}$ **do**
3:    **if** Top-K Seed Selection **then**
4:        **if** $|SEED| < K$ **then**
5:            $SEED = SEED \cup \{v_i\}$
6:        **end if**
7:    **end if**
8:    **if** Exponential Base Seed Selection **then**
9:        **if** $i \in \{a_n\}$; $a_n = n + \lfloor b^n \rfloor$; $b = e^{\frac{\ln\,(|N|-K-1)}{K-1}}$; $\forall n \in \{0,\dots,K-1\}$ **then**
10:            $SEED = SEED \cup \{x_i\}$
11:        **end if**
12:    **end if**
13:    **if** Constant Base Seed Selection **then**
14:        **if** $\exists n \in \mathbb{N} \mid \lfloor an = i \rfloor$; $a = \frac{|N|}{K}$ **then**
15:            $SEED = SEED \cup \{x_i\}$
16:        **end if**
17:    **end if**
18: **end for**

We first compute PageRank scores for each tag-resource node to examine the connectivity of each node in the graph. The resulting list, with the nodes ordered according to PageRank, is the starting point for the three strategies we evaluated. Algorithm 1 shows the three seed selection processes.

1. **Top-k seed set.** TrustRank also employed the (inverse) top-k PageRank selection to find highly connected nodes whose quality influences a lot of neighboring nodes. However, since our data model is a bidirected graph, we consider the top-k PageRank directly without computing the inverse PageRank scores. This strategy can be easily adjusted to satisfy the first requirement of the seed set size, while it may not be able to select the right seed set which includes both good and bad tag-resource nodes. The reason is that, as will be shown in the next section, bad tag-resource nodes usually have lower PageRank values.

2. **Exponential base seed set.** Motivated by the observation that the top-k strategy mainly select the good tag-resource nodes, this strategy aims to include more bad tag-resource nodes in the seed set. However, in order to propagate quality scores through the graph as far as possible, nodes with high PageRank values (i.e., high connectivity) are favored. Hence, after ordering nodes based on their PageRank scores, seed nodes are selected with an increasing interval, such as $\{v_1, v_2, v_4, v_8, \cdots\}$.

3. **Constant base seed set.** In contrast to exponential base seed selection which favors nodes with high connectivity to those less connected, so that more good tag-resource nodes are selected, this strategy selects good and bad tag-resource nodes with equal chances. For example, let the constant base be 10, then every 10th node will be selected. The inclusion of more bad

tag-resource nodes may be able to discover more tag-resource nodes with inferior quality, while the propagation may not be as extensive as before.

4   Evaluation

Since there is no manually annotated corpus – of which we are aware of – that could be used to compare our results for the quality of tags with a gold standard, we have to resort to an indirect approach. Particularly, we use the tag data compiled for a competition[2] to detect spam users. In this section, we first describe the data set. Then, an indirect approach to evaluate TRP-Rank is discussed. Next, we evaluate the performance of TRP-Rank, with different seed selection strategies. Finally, we examine the performance of our approach when applied to a larger dataset.

4.1   Data Set

The data set used by us consists of $221,354$ tag assignments by $1,328$ users of the BibSonomy[3] system for publications. Out of these users, 118 were marked manually as spammers and $1,210$ as non-spammers. The size of the set of unique tag-resource pairs $\mathcal{TR}$ is $195,198$. We discarded tag-resource pairs which were made by users having only one tag assignment (these tag-resource pairs would be disconnected nodes in our data model). And we only picked the first 1000 tag assignments of users whose number of tag assignments exceed this threshold. The remaining set has $132,520$ trs.

In order to show the connectivity of tag-resource nodes, Table 1 summarizes the numbers of pairs of tag-resource nodes, $\{tr_i, tr_j\}$, and their associated common users. For example, the second column of the table indicates that there are $175,619$ pairs of trs that are used by only one common user. In other words, in the adjacency matrix of our data model, there are $2 * 175,619$ elements with value 1. Although these numbers seem to imply that the graph is not highly connected, as we will show in Section 4.3, a rather small seed set is sufficient to reach most of the nodes in the graph.

Table 1. Number of pairs of trs assigned by common users

Number of pairs of trs	175619	15767	2664	641	197	115	55	41	24	75
Shared by # of Users	1	2	3	4	5	6	7	8	9	$\geq$10

Some necessary preprocessing has been done before using the data. For example, since the data set consists of the raw BibSonomy data, we have to give IDs to each individual tr. To identify the semantic relationship between certain tags, we use stemming and ignore capital letters to assign one ID to a group of tags (e.g. "Book", "book", or "Books").

---

[2] http://www.kde.cs.uni-kassel.de/ws/rsdc08/
[3] http://www.bibsonomy.org

4.2   Indirect Evaluation Method

The TRP-Rank algorithm aims to measure the quality of each tag-resource pair, while the data set contains only the spammer information. Hence, an indirect evaluation method needs to be used. Basically, we need to consider the following two issues: 1) The input of TRP-Rank needs manually assessed quality scores of a set of seed tag-resource nodes. How to assign the initial quality using the spammer information in the data? 2) The output of TRP-Rank is the converged quality scores of all tag-resource pairs. How to map the quality scores of tag-resource pairs to some score which could reflect whether a user is a spammer or not? We discuss the solutions of the two problems respectively as follows.

For assigning the initial quality scores to seed tag-resource nodes, we make use of the available spammer information in the dataset by defining a function $notSpammer(u) \in \{1, -1\}$. When a user u is not a spammer, the function returns value 1; otherwise, it returns value -1. Thus, the oracle function $O(tr)$ assigns the scores to each $tr \in SEED$ as:

$$O(tr) = \begin{cases} 1 & \text{if } \frac{1}{|getU(tr)|} \sum_{u \in getU(tr)} notSpammer(u) > 0 \\ -1 & \text{if } \frac{1}{|getU(tr)|} \sum_{u \in getU(tr)} notSpammer(u) < 0 \\ 0 & \text{otherwise} \end{cases} \quad (4)$$

That is, when a tag-resource pair is assigned by more normal users than spammers, it is deemed as a good tag-resource node and assigned a positive quality score. Otherwise, a negative score is given to reflect the inferior quality of the tag-resource node.

For mapping the result quality scores $Q(tr)$ of all tag-resource pairs, returned by TRP-Rank, to the scores indicating whether a user is a spammer or not, we aggregate the quality of all tag-resource pairs assigned by the user. Let $getTR(u)$ return the set of tag-resource pairs used by u, $getTR(u) = \{tr_1, \cdots, tr_n\}$. We define the function $isSpammer(u)$ as:

$$isSpammer(u) = \begin{cases} 1 & \text{if } \frac{1}{|getTR(u)|} \sum_{tr_i \in getTR(u)} Q(tr_i) < 0 \\ 0 & \text{otherwise} \end{cases} \quad (5)$$

4.3   TRP-Rank Performance

We first examine the maximum performance which can be achieved theoretically with our approach. Namely, the performance generated when the complete set of tag-resource nodes are used as seeds. As shown by the top confusion matrix in Table 2, the accuracy is approximately 97.66% (1210/1239). It is actually promising considering that our algorithm is not designed for spammer detection. We further investigate the theoretically achievable maximum by using all nodes with positive initial quality scores and all nodes with negative initial quality scores as seeds respectively. The middle and bottom confusion matrixes in Table 2 show the results. We notice that, compared with using only the nodes

Table 2. Confusion matrixes for theoretically achievable maximum using different seeds

Positive and Negative spread information	
True Positives: 1210	True Negatives 89
False Positives: 29	False Neagatives 0
Only positive spread information	
True Positives: 1079	True Negatives 114
False Positives: 4	False Neagatives 131
Only negative spread information	
True Positives: 1210	True Negatives 91
False Positives: 27	False Neagatives 0

with positive initial scores as seeds, using all nodes with negative initial quality scores is able to detect more spammers correctly.

Then, we investigate the performance of TRP-Rank which uses a combination of good and bad nodes as seeds. We conduct the experiments by varying the size of seed sets. As discussed in Section 3.4, the PageRank of nodes is used as the starting point to select seeds. Figure 4 shows the PageRank scores for all trs in our data set. By examining the PageRank scores of nodes, we notice that nodes

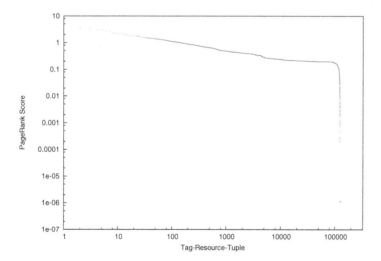

Fig. 4. Log-log graph of PageRank scores for the whole data set

related to spammers usually have lower PageRank values. That observation implies that the *top-k* method probably will not include as many negative nodes related to spammers as seeds as the exponential base and constant base seed selections do. The results for the different selection strategies are shown in Table 3, which verify the previous hypothesis. The top-k approach is not comparable to the other two seed selection approaches. It also could not outperform the method

Table 3. Accuracy for different seed set selection strategies with seed set size 10000/20000

Strategy	Accuracy Seed Set Size	
	10000	20000
Top-k	91.11 %	91.11 %
ExponentialBase	94.58 %	96.39 %
ConstantBase	94.88 %	96.31 %

which uses all nodes with negative initial scores as seeds. In contrast, the other two seed selection methods exhibit similar good performance.

We further investigate how the performance of TRP-Rank varies with respect to the seed set size. The seed set size is a crucial factor for the algorithm. Since we need an oracle function that gives us $O(tr)$ $\forall tr \in SEED$, and the oracle function usually invokes human assessing procedures, a large seed set could be expensive. However, a smaller seed set may be not able to propagate the quality through the graph wide enough. As shown in Table 4, which are the performance of TRP-Rank with constant base seed selection running on seed sets with different size, we notice that our approach can achieve an accuracy as good as 93.75% even if only 3.7% (5000/132, 250) of the nodes are selected as seeds, which equals roughly the manual assessment of 50 users.

Table 4. Results for different sized seed sets using constant base TP=true positives, TN=true negatives, FP=false positives, FN=false negatives

Seed Set Size	TP	FP	TN	FN	Accuracy
132520	1210	29	89	0	97.82 %
50000	1210	29	89	0	97.82 %
20000	1210	49	69	0	96.31 %
10000	1210	68	50	0	94.88 %
5000	1210	87	31	0	93,45 %

4.4   Data Reduction

For large data sets the matrix of our algorithm can become very large. To reduce the amount of data to process, we examine the effect of considering only trs where tags were used by at least x ($x > 1$) users. This seems to be justifiable at least for the case of measuring the quality of a certain tag for a certain resource. For detecting spam users, this filtering scheme is also an option. We examine the performance by using the whole data set as seed set and setting the parameter x as 3 and 10 respectively. The results are shown as below. We observe that the performance drops by only 2.94 % when considering only tags that were used by at least 10 users (compared with the performance where $x = 1$), while the transition matrix size is reduced by more than 50%.

- Minimum 10 Users → Accuracy 94.80 %
- Minimum 3 Users → Accuracy 95.63 %

- Minimum 1 Users → Accuracy 97.67 %

4.5   Discussion

The experimental results demonstrate that our algorithm performs quite well on distinguishing spammers from normal users based on the quality of their tag-resource pairs. After looking at the data into more detail, it seems that our approach could perform even better when modifying the notion of "spammer". For example, users with only one "test" tag assignment are considered as non spammers in the data set. Since they are not malicious users, this might be an acceptable classification. Nevertheless, from the tag quality point of view, these users would be considered unreliable because they use bad quality tag-resource pairs.

As observed from the experiments, an appropriate seed set should be well representative so that it contains not only good tag-resource pairs but also bad one. However, in a real-world tagging system, the majority are usually good/non-spam tags. Thus, the negative seeds are ranked rather low by PageRank which makes them hard to be found. The constant base seed selection method is generally applicable and has shown to be effective.

Regarding the size of the whole data set, we saw that the accuracy drops only little when putting some restrictions on the tags which are allowed for valid tag-resource pairs. Filtering out tag-resource pairs with tags used by few users is useful under the assumption that tags that are regarded valuable are used by a lot of users.

5   Conclusions and Future Work

In this paper, we focus on the problem of measuring the quality of tags which are supplied by users to annotate resources on the Web. Due to the intrinsic feature of existing collaborative tagging systems that users are allowed to supply tags freely, the resulting tags can have great disparity in quality. Consequently, measuring the quality of tags appropriately is important towards effectively exploiting the usefulness of tags in many applications. The main characteristics of our algorithm are represented by the data model we adopt and the seed selection functions we investigate. By decoupling the relationship between users and tag-resource pairs, we model the tag-resource pairs as nodes and co-user relationship as edges of a graph. Different from existing models, this structure allows every two tag-resource pairs used by the same user to have different quality, which complies with the practical situation better. Our algorithm, which propagates quality scores iteratively through the graph, needs to be initialized with the scores of a set of seed nodes. We investigate various seed selection strategies with the aim to not only minimize the size of the seed set but also minimize the error of the resulting quality scores. The effectiveness of our algorithm is evaluated on a manually labelled data set and demonstrated by the promising experimental results.

For future work, we are interested in pursuing the following problems:

- We currently assign the three distinct values $\{-1, 0, 1\}$ to the set of seeds. However, finer initial quality scores such as 0.2, 0.5 might be able to dissect the quality of tag assignments better.
- The manually assessment of the quality of seed nodes is expensive. How to make use of Web 2.0 and let users generate the seed set is an interesting issue which is worthwhile to consider.
- Since TRP-Rank demonstrated good performance of detecting spammers in tagging systems, we are considering to revise our approach to specifically address combatting tag spam. For example, our current model represents tag-resource pairs as nodes in order to measure the quality of tag-resource pairs. We can alternatively model users as nodes and common tag-resource pairs as edges to directly find spam users.

Acknowledgements

This work is supported by the EU project IST 45035 - Platform for searcH of Audiovisual Resources across Online Spaces (PHAROS).

References

1. Marlow, C., Naaman, M., Boyd, D., Davis, M.: Ht06, tagging paper, taxonomy, flickr, academic article, to read. In: Wiil, U.K., Nürnberg, P.J., Rubart, J. (eds.) Hypertext, pp. 31–40. ACM, New York (2006)
2. Tso-Sutter, K.H.L., Marinho, L.B., Schmidt-Thieme, L.: Tag-aware recommender systems by fusion of collaborative filtering algorithms. In: Wainwright, R.L., Haddad, H. (eds.) SAC, pp. 1995–1999. ACM, New York (2008)
3. Dmitriev, P.A., Eiron, N., Fontoura, M., Shekita, E.J.: Using annotations in enterprise search. In: [19], pp. 811–817 (2006)
4. Bao, S., Xue, G.R., Wu, X., Yu, Y., Fei, B., Su, Z.: Optimizing web search using social annotations. In: [20], pp. 501–510 (2007)
5. Xu, Z., Fu, Y., Mao, J., Su, D.: Towards the semantic web: Collaborative tag suggestions. In: WWW2006: Proceedings of the Collaborative Web Tagging Workshop, Edinburgh, Scotland (2006)
6. Koutrika, G., Effendi, F., Gyöngyi, Z., Heymann, P., Garcia-Molina, H.: Combating spam in tagging systems. In: AIRWeb (2007)
7. Mika, P.: Ontologies are us: A unified model of social networks and semantics. In: Gil, Y., Motta, E., Benjamins, V.R., Musen, M.A. (eds.) ISWC 2005. LNCS, vol. 3729, pp. 522–536. Springer, Heidelberg (2005)
8. Golder, S.A., Huberman, B.A.: The structure of collaborative tagging systems. CoRR abs/cs/0508082 (2005)
9. Halpin, H., Robu, V., Shepherd, H.: The complex dynamics of collaborative tagging. In: [20], pp. 211–220 (2007)

10. Berendt, B., Hanser, C.: Tags are not metadata, but just more content - to some people. In: ICWSM (2007)
11. Hotho, A., Jäschke, R., Schmitz, C., Stumme, G.: Information retrieval in folksonomies: Search and ranking. In: Sure, Y., Domingue, J. (eds.) ESWC 2006. LNCS, vol. 4011, pp. 411–426. Springer, Heidelberg (2006)
12. Gyöngyi, Z., Garcia-Molina, H., Pedersen, J.O.: Combating web spam with trustrank. In: Nascimento, M.A., Özsu, M.T., Kossmann, D., Miller, R.J., Blakeley, J.A., Schiefer, K.B. (eds.) VLDB, pp. 576–587. Morgan Kaufmann, San Francisco (2004)
13. Heymann, P., Koutrika, G., Garcia-Molina, H.: Fighting spam on social web sites: A survey of approaches and future challenges. IEEE Internet Computing 11(6), 36–45 (2007)
14. Kleinberg, J.M.: Authoritative sources in a hyperlinked environment. J. ACM 46(5), 604–632 (1999)
15. Sen, S., Lam, S.K., Rashid, A.M., Cosley, D., Frankowski, D., Osterhouse, J., Harper, F.M., Riedl, J.: Tagging, communities, vocabulary, evolution. In: Proceedings CSCW, New York, NY, USA, pp. 181–190. ACM, New York (2006)
16. Golder, S.A., Huberman, B.A.: Usage patterns of collaborative tagging systems. Journal of Information Science 32, 198–208 (2006)
17. Page, L., Brin, S., Motwani, R., Winograd, T.: The pagerank citation ranking: Bringing order to the web. techreport (1998)
18. Wu, B., Goel, V., Davison, B.D.: Propagating trust and distrust to demote web spam. In: Finin, T., Kagal, L., Olmedilla, D. (eds.) MTW of CEUR Workshop Proceedings, vol. 190 (2006), `CEUR-WS.org`
19. Carr, L., Roure, D.D., Iyengar, A., Goble, C.A., Dahlin, M. (eds.): Proceedings of the 15th international conference on World Wide Web. In: Carr, L., Roure, D.D., Iyengar, A., Goble, C.A., Dahlin, M. (eds.) WWW 2006, Edinburgh, Scotland, UK, May 23-26, 2006. ACM Press, New York (2006)
20. Williamson, C.L., Zurko, M.E., Patel-Schneider, P.F., Shenoy, P.J. (eds.): Proceedings of the 16th International Conference on World Wide Web, WWW 2007, Banff, Alberta, Canada, May 8-12, 2007. ACM, New York (2007)

STAN: Social, Trusted Annotation Network

Hyun Namgoong[1], Kyoung-Mo Yang[2], Sung-Kwon Yang[1],
Charles Borchert[1], and Hong-Gee Kim[1]

[1] Seoul National University, Biomedical Knowledge Engineering Lab,
Seoul, Republic of Korea
{ngh,sungkwon.yang,cborchert,hgkim}@snu.ac.kr
[2] SunMoon University, Department of Computer Science Frank.Holzwarth, Asan,
Chungnam, Republic of Korea
{kyoungmo.yang}@sunmoon.ac.kr

Abstract. Annotated data play an important role in enhancing the usability of information resources. Single users can be easily frustrated by the task of annotating. Collaborative approaches to annotation have been applied to web resources, but have not yet been applied to the task of local documents, due in part to the lack of a uniform identification method. In this paper, we use hash-based virtual URIs for identifying documents, and introduce the concept of a STAN (Social, Trusted Annotation Network), which enables collaborative annotation of documents through their URIs. STAN also incorporates quantitative trust rates between users in social networks based on their interactions with each other. The STAN framework is described, demonstrating how these trust networks are constructed through collaborative annotation. Finally, we evaluate the usefulness of collaborative annotation and the feasibility of the resulting trust rates through empirical experiment.

Keywords: Semantic Desktop, Social, Trusted Annotation Network, Trust Network, Hash algorithm, Virtual URI.

1 Introduction

Annotated data play an important role in enhancing the usability of information resources. Annotation with either plain-text tags or semantically richer RDF or OWL is essential in Semantic Web and more specifically Semantic Desktop applications [1][2]. It facilitates the information access and management by utilizing meaningfully closer contingents of resources or documents. However, annotating every resource is sometimes bothersome and time-consuming for individual users. Some recent works have been committed to enable the collaborative annotation of web resources [6][7]. In spite of these works, it is still not a simple task to collaboratively annotate local desktop resources, i.e. PDF documents, copies of which are difficult to uniquely identify due to lack of explicit URIs (universal resource identifiers).

In this paper we introduce STAN (Social, Trusted Annotation Network) which enables collaborative annotation for local documents. Firstly, STAN provides hash-based virtual URIs for local documents for the uniform identification. Secondly,

J. Domingue and C. Anutariya (Eds.): ASWC 2008, LNCS 5367, pp. 272–286, 2008.
© Springer-Verlag Berlin Heidelberg 2008

STAN can identify annotators in a socially connected network, recognize others who have the same copy of a document, and group them accordingly. Finally, it incorporates quantitative trust ratings measured through multiple users' interactive annotations for the effective collaboration. A social network represented in STAN can be extended to a trust network since it comprises a group of connected annotators, whose meta-data can be recommended and shared based on their quantified trust. The network allows users to know and to collaborate with others who might have the same interests and views through the use of quantitative trust rates.

After describing the concept of STAN, we describe framework of the implemented prototype, which allows for the sharing of annotated data for documents, particularly PDF files intended for distribution without manipulation of the content. Each user's annotation of local documents can be aided by the delivery of others' annotated data, perhaps filling in gaps a user has left in their annotations. Also, the framework defines a specific algorithm for quantitative trust measuring based on their shared annotated data. The framework mainly consists of individuals' repositories (for publishing their data), STAN applications (for collaborative annotation and searching), and the STAN mediator (which manages communication through the network). These components together demonstrate how a STAN can be used to improve cooperative annotation for users.

2   State of the Art

Annotated data can enrich and add value to a document, to the benefit of understanding and expanding its content. They can come in the form of plain text tags or semantic annotations. This annotated data has the potential to be useful in many areas -not only for the web, but also for semantic desktop and social semantic desktop applications [1][2]. Works like NEPOMUK and Haystack are devoted to using this potential for enhancement of desktop resource usability [3][4]. SALT also provides techniques for encapsulating this annotated data in a document for easy distribution as a meaningful addition to the document [5]. These works show the potential that can be harnessed from the annotated data of local resources.

Obviously annotated data is meant for more than a single person's benefit, so collaborative annotating is an important aspect of annotation as a whole. Many sites and research projects manage multiple users' annotated data on web resources for the purpose of sharing. CiteULike is a web-based social bookmarking service for academic research papers, storing optionally 'tagged' references to papers in users' online profiles [6]. Web Discussions is a tool for annotating parts of websites, which shares annotated data among users connected to the same server [7]. However, the above mentioned services focus on web resources which all implicitly have URLs or URIs that provide universal identification. The sharing of annotated data for local resources has still not seen significant consideration.

In the study of trust networks, determining quantitative trust rates and reputations has become a big issue. These networks' focus can range from matters of taste, like FilmTrust, to the source of claims of facts, like IWTrust [11][12]. The above implementations show human-driven rating approaches, however users' assignments are difficult to formalize without explicit context or history of their neighbors, as Marsh addresses in [9]. According to Jennifer Golbeck's definition of trust in [8], 'trust in a person is a commitment to an action based on a belief that the future actions of that person will lead to a good outcome', the quantitative trust rate gives many use-cases for future behavior, e.g., email filtering [13].

Our research provides the concept of and framework for STAN, intended for ensuring collaborative annotation, primarily by addressing the identification problem with hash-based virtual URIs for each document, and further providing a network for users' exchange of document annotations. Additionally, we believe that users' annotation sharing and selection behavior can be assessed for further benefit. To that end, the framework delivers quantitative trust rates between users, giving an indication of reliability of future annotation contributions based on their previous actions.

3 Grouping Users with Document Identification

Desktop users currently engage with the flow of information not only on the Internet, but also on their desktops. Numerous documents clutter desktop computers, and can number from the hundreds to even the thousands. Most of these are simply downloaded from the Internet for browsing, e.g. research papers. In these cases, the user is only the reader or carrier, rather than an author. Annotation and tagging of the documents can be utilized to easily and effectively exploit their contents. Also, there can be some benefit from knowing other users and their annotations, and searching documents based on those annotations. Users could simply download and apply the annotations of others to their documents, so they need not attach every semantic annotation by themselves. The first task of this work is to group people who have the same copy of a given document.

In order to group people for annotation sharing, local documents firstly require a universal identification strategy, since local documents do not have a URI or URL representing them. To enable such universal identification, we use a hash algorithm to digest the whole content of the document as an array of codes [14]. The algorithm was originally invented to guarantee security problems in file or data transmission. The hash algorithm turns some series of data into a small set of integer codes, called the hash value. The hash value is unique for each unique document copy, so that we can detect duplicates [15].

In the proposed approach, hash values are employed both to test for equality of copied documents and as part of a URI indicating a virtual document that represents all copies of that document. As Figure 1 shows, the hash-based virtual URI also acts as the URL of a web resource. When we share annotated data about a document, we can treat every copy of the document as the same document through this URI. When the algorithm is executed, only the actual content of the document is used to determine the hash value, since including the annotated meta-data would modify the hash value of the whole document. This also means that any local annotation and addition to meta-data do not modify the file's identity as a copy of the original.

For the sharing of annotations, a social network based on FOAF [16] (the Friend of a Friend ontology) is used, associating users with their 'repository addresses.' As a basic who-knows network, it can allow users to find people around them and check their repositories for annotated data on their shared documents.

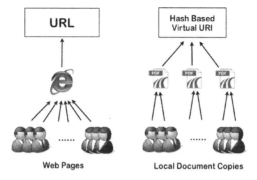

Fig. 1. The rule of hash based virtual URI

For publishing annotated data to the web, each user has a repository address described in FOAF. For each user's document annotated data, this address is combined with the document's hash value for publishing on the web. For example, suppose there is a paper titled 'Social Semantic Desktop' by Stefan Decker. We can calculate the hash value from the content of the paper, resulting in something like this: 'bbf2e6a7ed78a2e3857f4e1a6aede12f9f712a4f.' Then, the document URI for virtually indicating the document would be 'http://stan-project.org/bbf2e6a7ed78a 2e3857f4e1a6aede12f9f712a4f.' The URLs for each user's annotated data are derived from their designated repository address and the file's hash value. If my FOAF has 'http://blog.stan-project.org/nghyun' as my private repository, the driven URL for my annotated data for the paper would be 'http://blog.stan-project.org/nghyun/bbf2e6a7e d78a2e3857f4e1a6aede12f9f712a4f.rdf.'

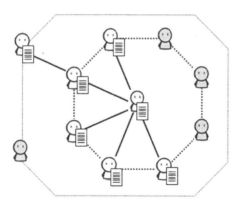

Fig. 2. An example of an annotating group

The described approach enables user to gather separate annotated data, and store it in generated URLs. Figure 2, a subset of a social network, shows a group of people who have the same document. For each document, groups can be generated dynamically, and used for data sharing. STAN and the STAN framework, the details of which will be explained later, enable the sharing of annotated data through this social network.

4 Social, Trusted Annotation Network (STAN)

We define STAN as a derived trust network based on a social network that is constructed from users' collaborative annotation behavior. Within a group of people, a user can cooperate with those connected persons. Then, continuous acceptance and rejection of annotated data suggested by others will inform a quantitative scaling of trust. For example, If one accepted some person's semantic annotations and tags frequently, and frequently denied another person's, the actions involving them will be assessed as a measure of trust between that user and each of the others. Also, similarities between a user's own annotations and other users' will be another basis for measurement, as we can assume similar annotation behavior will carry on into the future. The trust rates derived from those acceptances and comparisons will refine and extend the social network as a STAN.

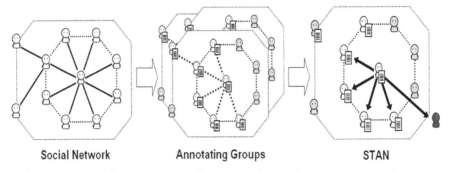

| Social Network | Annotating Groups | STAN |

Fig. 3. Social, Trusted Annotation Network from social network and temporal annotating group

For a more clear explanation, see Figure 3. The central user's social network gives him connections to his friends, and friends of his friends. From this, an annotating group can be derived, consisting of the subset of the social network's users who have copies of the same file. This temporal annotation group can publish and accept annotated data amongst others in the group, which can lessen the annotating workload between them. However from this group the central user will still selectively accept only the annotated data that he judges useful. This acceptance frequency and the similarity of his and other users' annotated data will be used to measure his trust of other users in the group. Based on all of such measured trust rates, a trust network, STAN, will be constructed for the user.

We claim that this network is a kind of trust network, following Jennifer Golbeck's definition in [8]. The network includes people committed to a purposed action, in this case annotating, and it includes the beliefs of users in each other for future actions. The network is constructed with personalized actions, so the directions between users can be distinct. Therefore, this network satisfies the properties of trust, asymmetry and personalization. Also, we assume the trust can be transmitted and believed along edges, which entails another property: transitivity.

In this proposed annotation network, the authorship of tags and semantic annotated data are preserved during data sharing. Every instance of annotated data will include the first writer's URI (or FOAF URL). When an annotation is 'accepted' by another user, his network will be refined by discovering new relations with indirect people. At the right-most network graph of Figure 3, originally the gray-colored man was out of range of the central man's social network, but after some transmission of trust, he can eventually enter the central man's network. This happens due to the gray man's annotations being made, then spread by other people, and finally reaching the central man for acceptance. With further acceptance, the gray man will become a person in the central man's trust network. STAN does not remain simply a sub-set of the original social network. Rather, it is extended by discovering new relations with previously unfamiliar people.

In the STAN, person A's rate of trust for person B is decided based on their similarity and A's acceptance frequency. Annotated data similarity means the shared annotations A and B have for the same document, while the acceptance frequency means the number of annotations that A accepted from B's data.

$$Tr(P_a, P_b) = Similarity(P_a, P_b) + AcceptRate(P_a, P_b)$$

For DI_x

$$Similarity(P_a, P_b) = \sum_{x=1}^{z} \frac{n(As_{ax} \cap As_{bx})}{n(As_{bx})}$$

$$AcceptRate(P_a, P_b) = \sum_{x=1}^{z} \frac{n(Acc(i, As_{bx}))}{n(As_{bx})} \tag{1}$$

When $DI_x = D_a \cap D_b$

i = person, D_i=Documents of i, As_{ij} =Annotated data by Person i on Document j

The equation shows the trust measurement using acceptance rate and similarity. In the equation i means a person, D_i means the set of document held by i, and As_{ij} is all annotated data sets attached by i to document j. DI_x is the intersected document set of A and B. The trust rate is calculated from the summated values, B's accepted frequency and the number of the same data divided by the total number of the shared data, where As_{bx} is not empty. The function $Acc(i, As_{ij})$ returns the total number of accepted data by i in As_{ij}.

5 STAN Framework

Here we introduce the STAN Framework, an infrastructure enabling users, and semantic desktops to construct annotating groups, and build STANs through sharing and collaboration in a social network. Figure 4 shows the architecture of the STAN framework and its main components. It consists of STAN applications, the STAN Mediator, and each user's private repositories. Also, other social semantic desktop applications can be incorporated.

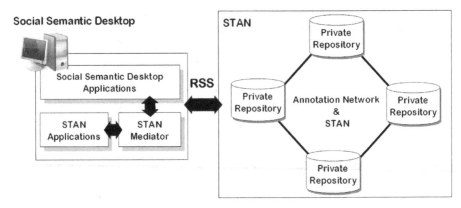

Fig. 4. STAN Framework as a bridge between social semantic desktop and STAN

To enable data sharing among grouped people, the STAN Mediator stores anno-tated data into each person's private repository. It deals with users' requests for pub-lishing, updating, and deleting of annotated data in their private repositories. Every page storing the published data is provided to other users through driven URLs based on the virtual document URIs. The STAN Mediator gathers the annotated data from the given URL, and manages the data as a data source, enabling desktop applications, including STAN applications, to access the annotated data.

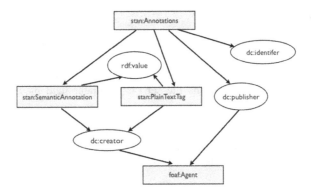

Fig. 5. STAN RDF scheme for representation of Annotations

In this framework, the associated applications and components handling the anno-tated data share a simple RDF scheme, as shown in Figure 5. As the figure shows, a set of annotated data (stan:Annotation) has a publisher who is storing the annotations. Each single annotation has a creator who initially created the annotation, an identify-ing hash value for the document, and one of two values; a plain text tag or a semantic annotation represented as an N-Triple. The set of annotated data is delivered and stored in the STAN framework according to this scheme.

For the purpose of publishing users' trust networks to the web, this framework uses FOAF with some relationship properties added as shown in Figure 6. These published values could also be used for reputation or TidalTrust matrix calculations [8]. As explained above, the user's acceptance and similarity measures are used to calculate an overall trust rate. These values are all incorporated as a new relationship, 'Acceptance'. The defined properties for Acceptance include the user's acceptance rate and annotation similarity, and the derived trust rate. An agent (or user) can have an acceptance instance which defines an 'acceptee' (the target user of the acceptance), and the agent's trust rate, acceptance rate, and similarity associated with that acceptee.

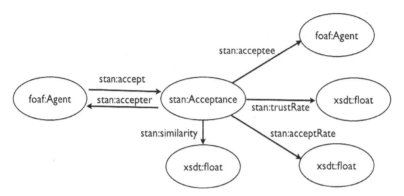

Fig. 6. Extended FOAF properties for STAN

In the following subsections, we explain the components of the STAN Framework in detail; STAN Mediator and individuals' Private Repositories. STAN applications will be described later.

5.1 STAN Mediator

The STAN Mediator takes the main role allowing a social semantic desktop to communicate with a STAN. Positioned between the network and the semantic desktop, it delegates each application's web communications with other networked nodes. The responsibilities are listed below:

1)   Annotation Publishing/Subscribing

The STAN Mediator manages annotated data publishing and subscribing processes. Publishing of annotated data is done by a call from a desktop application. For example, an annotation tool can submit new data or an update. Then, the mediator publishes this annotated data to the private repository. The mediator continually gathers others' annotated data using driven URIs and stores them into the local repository for other applications.

2)   Annotated data Storing & Management

Other users' downloaded annotated data are stored and managed by the mediator as files. When there are modifications and updates detected of other users' annotated data, the changes are applied to the stored data. It also performs the similarity calculation between the desktop user's annotations and others in the network, and correspondingly updates the user's FOAF file. The desktop applications can obtain the annotated data from the data storage, which has a shared XML scheme as described.

3) Social Network Handling

The Mediator handles and updates the social network expressed in the FOAF file, allowing the location of related persons, and storage of their corresponding trust rates. It expands the original user's FOAF to a limited depth and obtains each person's repository address for data acquisition. Upon a change in some values due to other users' actions, the mediator updates and republishes the FOAF with new values.

5.2 Private Repositories

As we described, the STAN Mediator publishes annotated data and user profiles to the web. Therefore, to use the STAN, every user should have his own repository, e.g., web folder. This allows publishing and updating of annotated data, document lists, and optionally a FOAF file, for other users to access. The FOAF file can be preserved in another location, if it has a link to the user's repository.

First, private repositories should contain a list of annotated data in RSS format. A user agent, such as the STAN Mediator, can subscribe to that list and observe updates by checking documents or publishing dates, as RSS typically is used. Second, the documents' annotated data that are created or accepted by the user have to be stored in the proper location though the driven URL from the hash based virtual URI.

In our implementation, the private repositories are equipped with SVN and HTTP technology to meet these requirements. SVN enables agents to access and publish their data remotely without any server side implementation, and makes it easy to handle the authorization for the server. HTTP allows us to provide a universal standard for accessing the data.

6 STAN Applications

In this chapter, we show two applications of STAN. The applications are implemented to demonstrate our framework. The Collaborative Annotator provides an environment for annotating desktop PDF files with collaborative features. Also, the Collaborative Document Finder shows improved document searching approach using tag sharing in the desktop environment.

6.1 Collaborative Annotator

The Collaborative Annotator is a tool for semantic annotating and tagging of PDF documents, and it also supports collaborative annotating as a STAN application. With this tool, a user attaches plain-text tags or semantic annotations (referencing domain ontology) while reading the document. The tool also provides view of multiple users' annotations for the purpose of collaboration. Each user can see the other users' annotations in a single file, and accept particular ones as his annotation. Then, the described annotated data are encoded as metadata in the XMP (eXtensible Metadata Platform) standard, for storage in the document file and publishing to the web. The published data is delivered to the STAN mediator and uploaded to the user's repository.

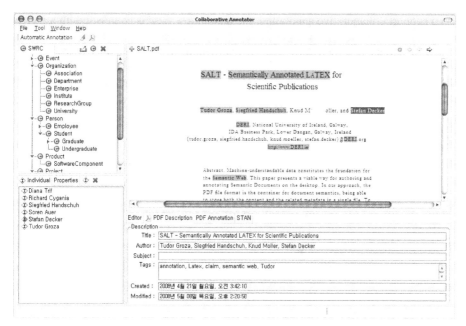

Fig. 7. Screenshot of Collaborative Annotator

As shown in Figure 7 above, the annotator provides three main view frames, the ontology view, document browser view, and annotation view frames. The ontology view shows ontology resources for the semantic annotating, with several tabs like class, individual, and properties. A user can browse the chosen ontology through this view to make a semantic annotation or RDF triple expression. Also, the browser view displays the content of the currently opened PDF document, so that users can select a specific entity in the document as the target of the annotation. The Annotation view part includes the PDF Description tab, PDF Annotation Information tab and Social-Trusted Annotation Network (STAN) tab. The PDF Description tab shows the basic meta-data of the opened PDF document, such as title, author, subject, and keywords. Also, as shown in Figure 8, the PDF Annotation Information tab lists all of the entities with annotations in the PDF document alongside their related ontology resources.

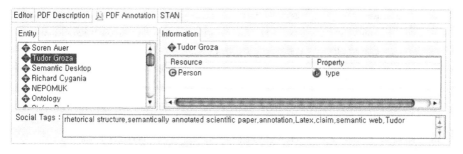

Fig. 8. Screenshot of PDF Annotation tab

To support collaborative annotating, the annotation view is equipped with a STAN tab. In this tab, a user can view the people in his social network and STAN, along with their annotated data. Seeing others' annotations, a user can select particular semantic annotations or tags to be included in his PDF file. This selection, which we call acceptance, will invoke the update of the FOAF information, the acceptance rate for the annotation publisher, and the author.

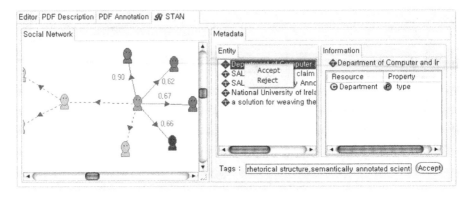

Fig. 9. Screenshot of STAN tab

6.2 Collaborative Document Finder

The Collaborative Document Finder is another kind of STAN application for searching desktop documents. The Finder uses gathered tags as criteria for document searching. Even if a user did not attach any tags to a document, this tool performs tag based searching through other users' tags. Communicating through the STAN mediator, the Finder can gather others' tags for a desktop document. We expect this finder can provide improved document search, particularly for files that a user has not read or tagged yet.

Location	FileName	...	Tags
/Users/ngh/Documents/My	757.pdf		Alfil Gliozzo,Aldo Gangemi,Legacy System,ISWC2007,TBox,Semantic Web,Semant
/Users/ngh/Documents/My	785.pdf		BBN Tech,BBN Technologies,Dave Kolas,ISWC2007,Semantic Web,Spatially-Augn
/Users/ngh/Documents/My	823.pdf		Application,Ontology Translation,James Ressler,Mike Dean,ISWC2007,Semantic W
/Users/ngh/Documents/My	910.pdf		Amanda Bouffier,Data Structuazation,GEM,Mike Dean,ISWC2007,Semantic Web,
/Users/ngh/Documents/My	920.pdf		Data Menagement,Ontology,Reto Krummenacheer,Semantic Space,ISWC2007,Sem
/Users/ngh/Documents/My	ArnetMiner.pdf		ArnetMiner,Expertise System,Jie Tang,Jing Zhang,ISWC2007,Semantic Web,Searc
/Users/ngh/Documents/My	CHIP.pdf		CHIP,Deonstrator,Museum Tour,ISWC2007,Semantic Web,Recommentation Syste
/Users/ngh/Documents/My	DBPedia.pdf		DBPedia,Bizer,Christian,Open Data,Mike Dean,ISWC2007,Semantic Web,XML,

Fig. 10. Screenshot of Collaborative Document Finder

Figure 10 shows a screenshot of the Collaborative Document Finder. Similar to other file finders, this tool has a text field for queries from a user. A user also can choose the range of the search for the tag. The user can search the tags applied by other users from their Social Network, or their STAN, or only those they have applied themselves. In Figure 10, the Finder lists the files in the user's desktop resulting from a match between the user's input text, 'ISWC' and the combination of tags in his STAN.

7 Experimental Result

This chapter includes empirical experimental results to test the suggested framework: how collaborative annotating can help users, and how feasible it is to measure trust from this collaboration. For the experiment, eight semantic web researchers, assumed as already trusting each other, were instructed to each separately use the Collaborative Annotator. With it, they would attach annotations, semantic annotations, and tags to two PDF documents. The papers, selected from the ISWC/ASWC 2007 proceeding, were 'DBpedia: A Nucleus for a Web of Open Data,' and 'SALT: Weaving the claim web.' For this semantic annotation, the SWRC (Semantic Web for Research Communities) Ontology [16] was used. For the second part of the experiment, the acceptance process, the subjects accepted others' annotations if they were judged useful or necessary for inclusion in their file.

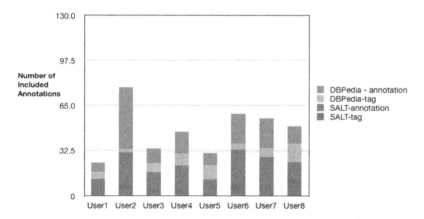

Fig. 11. Number of personally applied annotations (Before acceptance stage)

Figure 11 shows the resulting number of annotations applied personally by each experimenter. As the graph shows, on average they attached 47 annotations between the two papers. After the acceptance process, they had collected more annotations, as shown in Figure 12. On average they had 83 annotations stored in their own documents in the end. The result shows that they attach a larger variety of acceptable annotations, upon accepting suggestions from other the experimenters.

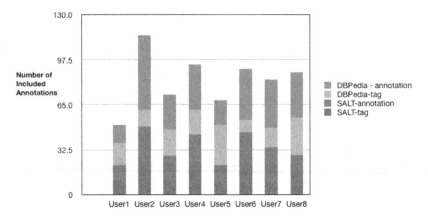

Fig. 12. Number of included annotations among users (After acceptance stage)

Also, the graphs in Figure 13 show the counts of the annotations with their usage frequencies before and after the acceptance process. As we can see, before the acceptance, 130 annotations are used only once, but some of them appear more after the acceptance. We can observe the tendency that some annotations become noticeably more common than the others. We can additionally expect if the annotation is shared in the network over a long time, more acceptable annotations in general will be more frequently used, through selection and transmission.

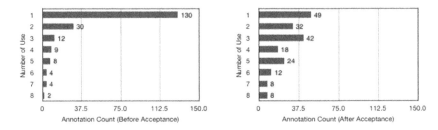

Fig. 13. Number of counted annotations by their use frequencies

Finally, Table 1 below shows the measured trust rates of experimenter 4, the experimenter who most actively accepted others' annotated data. The trust rate is calculated from the similarity frequency and the acceptance frequency divided by the total shared number, using the proposed algorithm. Because every experimenter annotated and selected at the same time, the authorship was not preserved. Instead, the acceptance increased the acceptance frequency of every experimenter who had the same annotation.

Because the number of experimenters was too small, it is difficult to determine the entire significance of the resulting trust rate. However, these trust rates correspond with a qualitative comparison of the experimenter's trust in the others, and also corresponds to other experimenters' trust rates for the same people. Therefore, we can accept that these rates signify beliefs in other person for the purposes of future annotation activity.

Table 1. Trust rates of experimenter 4

User #	Similarity Frequency	Acceptance Frequency	Trust Rate
1	9	7	0.67
2	20	21	0.38
3	6	8	0.41
5	22	6	0.90
6	24	13	0.63
7	23	15	0.68
8	9	3	0.24

8 Future Work and Conclusion

In this paper, we defined STAN (Social, Trusted Annotation Network) as a trust network derived from a social network and users' collaborative annotation behavior. With the accompanying framework enabling collaborative annotation, quantitative measures of trust can be delivered. We expect that this kind of trust network can be adapted to other areas that can benefit from measured trust in others.

Additionally, to overcome the weaknesses in this trust measurement, we need more sophisticated assessment methods and validations of our trust network in future work. Also, since the comparison of semantic annotations is an important aspect of our approach, work will be devoted to study the handling of annotations which reference multiple or different ontologies.

Acknowledgement. This work was supported in part by MKE & IITA through IT Leading R&D Support Project.

References

1. Decker, S., Frank, M.: The social semantic desktop. In: DERI Technical Report 2004-05-02 (2004)
2. Sheth, A., et al.: Semantic Content Management for Enterprises and the Web. IEEE Internet Computing, 80–87 (July/August 2002)
3. Quan, D., Huynh, D., Karger, D.R.: Haystack: A Platform for Authoring End User Semantic Web Applications. In: Proceedings of the Twelfth World Wide Web Conference (2003)
4. Groza, T., et al.: The NEPOMUK Project - On the way to the Social Semantic Desktop. In: Proceedings of I-Semantics 2007, pp. 201–211 (2007)
5. Groza, T., Möller, K., Handschuh, S., Trif, D., Decker, S.: SALT: Weaving the claim web. In: Aberer, K., Choi, K.-S., Noy, N., Allemang, D., Lee, K.-I., Nixon, L., Golbeck, J., Mika, P., Maynard, D., Mizoguchi, R., Schreiber, G., Cudré-Mauroux, P. (eds.) ASWC 2007 and ISWC 2007. LNCS, vol. 4825, pp. 197–210. Springer, Heidelberg (2007)
6. CiteULike: A free online service to organise your academic papers,
   http://www.citeulike.org/
7. Cadiz, J.J., Gupta, A., Grudin, J.: Using Web Annotations for Asynchronous Collaboration Around Documents. In: Proceedings of CSCW 2000 (2000)

8. Golbeck, J., Hendler, J.: Inferring binary trust relationships in Web-based social networks. ACM Transactions on Internet Technology 6(4), 497–529 (2006)
9. Stephen, P.M.: Formalising trust as a computational concept. PhD thesis, Department of Mathematics and Computer Science, University of Stirling (1994)
10. Golbeck, J., Hendler, J.: Computing and applying trust in web-based social networks. University of Maryland at College Park, College Park (2005)
11. Golbeck, J., Hendler, J.: FilmTrust: Movie recommendations using trust in Web-based social networks. In: Proceedings of the IEEE Consumer Communications and Networking Conference, Las Vegas, NV (2006)
12. Zaihrayeu, I., Silva, P.P., McGuinness, D.L.: IWTrust: Improving user trust in answers from the web. In: Proceedings of 3rd International Conference on Trust Management (2005)
13. Golbeck, J., Hendler, J.: Reputation Network Analysis for Email Filtering. In: Proc. of the Conference on Email and Anti-Spam (CEAS), Mountain View, CA, USA (2004)
14. Secure Hash Standard Technical Report FIPS PUB 180-1 US Department of Commerce/National Institute of Standards and Technology (1995)
15. Arms, W.Y.: Digital libraries. MIT Press, Cambridge (2000)
16. Brickley, D., Miller, L.: FOAF vocabulary specification (2005)
17. Sure, Y., et al.: The SWRC Ontology - Semantic Web for Research Communities. In: Bento, C., Cardoso, A., Dias, G. (eds.) EPIA 2005. LNCS, vol. 3808, pp. 218–231. Springer, Heidelberg (2005)

Consolidating User-Defined Concepts with StYLiD

Aman Shakya[1], Hideaki Takeda[1], and Vilas Wuwongse[2]

[1] National Institute of Informatics,
2-1-2 Hitotsubashi, Chiyoda-ku, Tokyo, Japan 101-8430
[2] Asian Institute of Technology
Klong Luang, Pathumthani, Thailand 12120
{shakya_aman,takeda}@nii.ac.jp, vw@cs.ait.ac.th

Abstract. Information sharing can be effective with structured data. However, there are several challenges for having structured data on the web. Creating structured concept definitions is difficult and multiple conceptualizations may exist due to different user requirements and preferences. We propose consolidating multiple concept definitions into a unified virtual concept and formalize our approach. We have implemented a system called StYLiD to realize this. StYLiD is a social software for sharing a wide variety of structured data. Users can freely define their own structured concepts. The system consolidates multiple definitions for the same concept by different users. Attributes of the multiple concept versions are aligned semi-automatically to provide a unified view. It provides a flexible interface for easy concept definition and data contribution. Popular concepts gradually emerge from the cloud of concepts while concepts evolve incrementally. StYLiD supports linked data by interlinking data instances including external resources like Wikipedia.

Keywords: Structured data, concept consolidation, multiple conceptualizations, social Semantic Web, linked data, information sharing.

1 Introduction

People want to share a wide variety of information on the web which is evident from the rapid rise of user generated contents on the social web. Different types of data can be modeled by structuring them. Structured data has many benefits as follows.

- It becomes easy to define the semantics of data for automated processing.
- Information sharing becomes effective with common conventions.
- Search and browsing become more effective with structured data.
- Structured data from various sources can be easily mixed and integrated.
- Interoperability between systems is possible with standard formats or mapping different formats.

However, there are several challenges for sharing structured data on the web. There is a long tail of information domains for which people have information to share [1]. There are separate solutions for dealing with few popular information types but availability of software is rare for the long tail. There are not many ontologies to cover the

J. Domingue and C. Anutariya (Eds.): ASWC 2008, LNCS 5367, pp. 287–301, 2008.
© Springer-Verlag Berlin Heidelberg 2008

wide variety of information we may want to share [2, 3]. Developing individual solutions every time is infeasible because creating new ontologies or new information systems is difficult. It is not easy to define concepts adequately. Usually, we can only have vague partial descriptions. Moreover, we conceptualize the same thing in different ways based on different contexts, requirements or preferences. Hence, multiple heterogeneous or overlapping conceptualizations always exist.

Creating ontologies should be a widely collaborative and incremental process [2]. However, to have mass participation, systems should be easy to understand and use. It is difficult for general users to understand and use ontologies. On the other hand, social software has proven to be very successful in drawing mass user participation because they are easy to understand and use. Thus, the combination of social software with Semantic Web technologies has been gaining significant attention [4, 5, 6].

This paper attempts to address the following problems for sharing a wide variety of structured data on the web

- Difficulty of creating concept definitions adequate to satisfy the evolving requirements of many people.
- Need for multiple conceptualizations for different people and contexts.

We propose the use of social software to enable general users to freely define their own conceptualizations and share structured information based on that. The original contribution of the paper is that we propose allowing multiple conceptualizations to satisfy individual requirements of different people and, at the same time, consolidating them into a single collaborative conceptualization as a virtual concept. The consolidation of multiple user-defined concepts serves as a new way to build up more complete definitions by merging separately defined concepts. We also present a formalization of the approach. We have implemented a system called StYLiD (an acronym for Structure Your own Linked Data) to integrate several aspects of social software and Semantic Web technologies into a synergetic whole. Multiple definitions are consolidated by semi-automatic alignment of concept attributes. Popular concepts can gradually emerge out and converge to stability by usage. Besides, StYLiD also supports linked data[7] by interlinking internal and external data resources.

The paper is organized as follows. We will present some motivating use case scenarios in Section 2. Section 3 contains some formalization of the consolidation of multiple conceptualizations. The details of the StYLiD platform implementing the approach are given in Section 4. Related works are discussed in Section 5. We conclude and point out some ongoing and future work in Section 6.

2   User Requirements

We discuss some use cases which identify the general requirements of users and their motivations to use the proposed system.

Sharing Structured Data with User-defined Concepts. Suppose a user wants to share some structured information, let's say details of a talk program. However, he cannot find a system for sharing such data. He may register an account on StYLiD and easily define his own "talk" concept on the fly with a list of attributes like topic, speaker, date, time, venue, etc. If a similar concept already existed in the system, he may have chosen to use that directly or to modify it to create his own version. Different users may define multiple versions of the same concept to meet their own requirements. He may update the concept anytime and add more attributes whenever needed. Once a concept has been defined, he may easily start posting data instances. He may also link attribute values to other data instances or external resources. He may easily share the post in his community using social software features. Other users may also contribute data easily using his concept. This can be used as a way of collecting data from the community. By posting data instances and having others post to the system he also maintains a useful collection of information in a structured way.

Browsing and Querying Structured Data. The user would be able to browse different types of data using the concepts defined by him and others. He can also navigate through linked data entries. Moreover, he may search data instances in a structured way. For instance, he may search all the talks by "Peter" held at "Tokyo".

Consolidated View of Multiple Concepts. The user may want to browse all instances of a concept regardless of the concept version. This would be possible with a consolidated concept which groups the multiple definitions. He may want to have all instances in a uniform table view so that he can process them uniformly. For that, the corresponding attributes of the individual concepts have to be aligned first. The system should help in this by automatically suggesting the alignment. He may verify and complete the alignment. This is also useful if he wants to search data over the multiple definitions. If he searches over a consolidated concept all aligned attributes would be searched. For instance, if he searches using the "venue" attribute of the talk, the "location" attribute aligned in a different version would also be searched.

Exploiting the Structured Data. The user may be offered useful features exploiting the structured data. For example, a "conference" concept may allow an operation to help in booking a hotel at the conference "venue" from "start date" to "end date". Moreover, developers may import the structured data into various useful applications.

3   Concept Consolidation

Different users may have multiple definitions for the same concept. As illustrated in Fig. 1, the same "Hotel" concept may be defined by 3 users in different ways. Even the same user may have multiple versions for the concept. The system groups them and consolidates them into a single virtual "Hotel" concept combining all the features of the individual definitions. In this section, we formalize our approach of consolidating multiple concept definitions. Our approach for consolidation is based on the Global-as-View (GaV) approach for a data integration system where a global schema is defined in terms of the source schemas[8]. The approach is simplified in our case because a concept schema does not have multiple relations and integrity constraints as in relational database schemas. The implication of our formalism for concept consolidation is described with the implementation in Sections 4.2 and 4.4.

Definition 1. Concept and Instances. A concept C is an entity characterized by a set of attributes given by the function $att(C) = \{a_1, a_2, \ldots a_r\}$

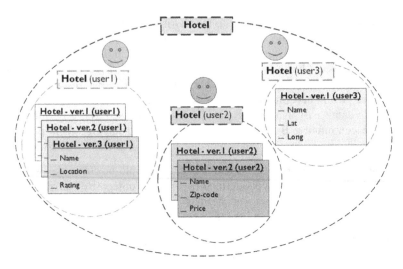

Fig. 1. Concept consolidation

The fact that x is an instance of C is denoted by the relation *instanceof*(x, C). C may have a set of instances I. The value for an attribute a of an instance k of C is given by the function $v(k, a)$.

Definition 2. Concept Consolidation. A concept consolidation C is defined as a triple $<\overline{C}, S, \mathcal{A}>$ where

- $\overline{C}$ is called the *consolidated concept*
- S is the set of *constituent concepts* $\{C_1, C_2, \ldots C_n\}$, n is the number of constituent concepts
- $\mathcal{A}$ is the *alignment* between $\overline{C}$ and S.

Let the set of attributes of $C_i \in S$ be *att*$(C_i) = \{a_i^1, a_i^2, \ldots a_i^{n_i}\}$ where n_i is the number of attributes of C_i. Let the set of attributes of $\overline{C}$ be *att*$(\overline{C}) = \{\overline{a}_1, \overline{a}_2, \ldots$ $\overline{a}_m\}$, called *consolidated attributes*, where m is the number of attributes of $\overline{C}$.

Definition 3. Alignment between Attributes. For each concept $C_i \in S$, if attribute $b_i^k \in att(C_i)$ is aligned to $\overline{d}_l \in att(\overline{C})$, we denote it as

$$aligned(\overline{d}_l, b_i^k)$$

for $l = 1, 2, \ldots r \ (r \leq m)$. All $\overline{d}_l$ are different. The mapping between $\overline{C}$ and C_i is defined as a set of ordered pairs

$$M_i = \{(\overline{d}_l, b_i^k) \mid \forall \overline{d}_l \in att(\overline{C}) \ aligned(\overline{d}_l, b_i^k) \wedge b_i^k \in att(C_i)\}$$

aligned represents a correspondence between the aligned attributes. Some relation may hold between the aligned attributes asserted by the correspondence.

Then, alignment $\mathcal{A}(\overline{C})$ between $\overline{C}$ and concepts in S is defined as the set of mappings $\{ M_1(\overline{C}), M_2(\overline{C}), \ldots\ldots M_n(\overline{C}) \}$.

Definition 4. Mapped Concepts in a Concept Consolidation. A concept $C_i \in S$ in the concept consolidation $<\overline{C}, S, \mathcal{A}>$ is said to be *mapped* if and only if

$$\exists x \in att(C_i)\ \exists y \in att(\overline{C})\ aligned(y, x)$$

i.e., at least one of its attributes is aligned to a consolidated attribute.

Definition 5. Grounded Consolidated Concept. The consolidated concept $\overline{C}$ in $<\overline{C}, S, \mathcal{A}>$ is said to be grounded if and only if

$$\forall z \in att(\overline{C})\ \exists x \in \bigcup_{i=1}^{n} att(C_i)\ aligned(z, x)$$

i.e., all the consolidated attributes are aligned to some attribute of the constituent concepts.

Definition 6. View of an attribute in a consolidated concept. The view of an attribute $b \in att(C_i)$ of concept C_i in the consolidated concept $\overline{C}$ for C is given by the following function.

$$\rho\,(b, C_i, \mathsf{C}) = \begin{vmatrix} a & \text{if } \exists a \in att(\overline{C})\ (a, b) \in M_i(\overline{C}) \in \mathcal{A}(\overline{C}) \\ \phi & \text{otherwise} \end{vmatrix}$$

Definition 7. Image of a consolidated attribute. The image of an attribute $a \in att(\overline{C})$ of the consolidated concept $\overline{C}$ for a constituent concept C_i in C is given by the following function which is the inverse function of ρ.

$$\sigma\,(a, C_i, \mathsf{C}) = \begin{vmatrix} b & \text{if } \exists b \in att(C_i)\ (a, b) \in M_i(\overline{C}) \in \mathcal{A}(\overline{C}) \\ \phi & \text{otherwise} \end{vmatrix}$$

Definition 8. Consolidated views of instances. The view of an instance k of concept C_i in the concept consolidation C is given by the following function

$$\overline{k} = w(k, C_i, \mathsf{C})$$

where *instanceof*($\overline{k}$, $\overline{C}$) and the value of each attribute $\overline{a}_j \in att(\overline{C})$ ($j = 1, 2, \ldots m$) for $\overline{k}$ is given by

$$v(\overline{k}, \overline{a}_j) = \begin{vmatrix} v(k, \sigma(\overline{a}_j, C_i, \mathsf{C})) & \text{if } \sigma(\overline{a}_j, C_i, \mathsf{C}) \neq \phi \\ \phi & \text{otherwise} \end{vmatrix}$$

The value $v(k, a)$ of each attribute a of C_i is known. The set of instances of $\overline{C}$ is exactly

$$\overline{I} = \{ \overline{k} : \overline{k} = w(k, C_i, \mathsf{C}) \wedge instanceof(k, C_i) \wedge C_i \in S \}$$

$\overline{I}$ is disjoint from the set of instances I_s of the constituent concepts in S.

Theorem 1. Translation of instances. The translation of an instance k of concept C_i to another concept C_j in the concept consolidation C, denoted by the function

$$k' = \gamma(k, C_i, C_j, \mathsf{C})$$

can be obtained as follows. If $\overline{k} = w(k, C_i, \mathsf{C})$ is the consolidated view of instance k, the value of each attribute $a_j^l \in att(C_j)$ $(l = 1, 2, \dots n_j)$ for k' is given by

$$v(k', a_j^l) = v(\overline{k}, \rho(a_j^l, C_j, \mathsf{C})) \qquad \text{(from def. 6)}$$

$$= v(k, \sigma(\rho(a_j^l, C_j, \mathsf{C}), C_i, \mathsf{C})) \quad \text{(from def. 8)}$$

Attributes of k' are exactly $att(C_j)$. However, $k' \notin I_s$.

Theorem 2. Lossless Translation. Instances of concept C_i can be translated to instances of C_j without any loss of information iff the following conditions hold.

$\forall a \in att(C_i)$
$\rho(a, C_i, \mathsf{C}) \neq \phi$ and
$\sigma(\rho(a, C_i, \mathsf{C}), C_j, \mathsf{C}) \neq \phi$

$|att(C_i)| \leq |att(C_j)|$ is a necessary condition for the lossless translation of an instance from C_i to C_j. If $k_j = \gamma(k_i, C_i, C_j, \mathsf{C})$ is lossless, $k_i = \gamma(k_j, C_j, C_i, \mathsf{C})$.

The proofs have been avoided as they are quite intuitive.

Query over a Concept. The main advantage of GaV in data integration is that queries on the global schema can simply be unfolded to the sources. The same advantage applies in our case too. Thus, we have the following theorems for unfolding and translating queries. The proofs follow from the literature for the GaV approach [8].

Theorem 3. Unfolding Queries over $\overline{C}$ in C. Any query $Q(\overline{C})$ over $\overline{C}$ can be unfolded into the union of queries $Q_1(C_1) \cup Q_2(C_2) \cup \dots \cup Q_n(C_n)$, where $C_i \in S$ ($i = 1, 2, \dots n$). Let the queries be defined over the concept attributes as follows

$$Q(\overline{C}) = Q'(\overline{a}_1, \overline{a}_2, \dots \overline{a}_r) \text{ where } \overline{a}_j \in att(\overline{C}) \ (j = 1, 2, \dots r)$$

$$Q_i(C_i) = Q_i'(a_i^1, a_i^2, \dots a_i^r) \text{ where } a_i^j \in att(C_i)$$

Each Q_i can be obtained by unfolding the attributes in Q using C

$$Q_i'(a_i^1, a_i^2, \dots a_i^r) = Q'(\sigma_i(\overline{a}_1), \sigma_i(\overline{a}_2), \dots \sigma_i(\overline{a}_r))$$

where $\sigma_i(a)$ is the short form of $\sigma(a, C_i, \mathsf{C})$.

Theorem 4. Query Translation. The query $Q_i'(a_i^1, a_i^2, \dots a_i^r)$, $a_i^k \in att(C_i)$ ($k = 1, 2, \dots r$) over C_i can be translated into a query $Q_j'(a_j^1, a_j^2, \dots a_j^r)$, $a_j^k \in att(C_j)$ over C_j in the concept consolidation C as following

$$Q_j'(a_j^1, a_j^2, \dots a_j^r) = Q_i'(\sigma_j(\rho_i(a_i^1)), \sigma_j(\rho_i(a_i^2)), \dots \sigma_j(\rho_i(a_i^r)))$$

where $\rho_i(a)$ and $\sigma_j(b)$ are short forms of $\rho(a, C_i, \mathsf{C})$ and $\sigma(b, C_j, \mathsf{C})$ respectively.

4 Implementation

StYLiD has been implemented to realize the use cases described in Section 2 using the approach of concept consolidation described in Section 3. StYLiD is available online[1] and undergoing further development.

4.1 Sharing Structured Data with User-Defined Concepts

The main interface of StYLiD is shown in Fig. 2. The users may freely define their own concepts by specifying the concept name, some description (optional) and a set of attributes. Each attribute is defined by the attribute name, description (optional) and a set of concepts as the suggested value range (optional) as shown in Fig. 3. Any user may enter instance data using system generated online forms as shown in Fig. 4.

Users do not need to define concepts from scratch. The user can modify an existing concept defined by another user to make his own version. The system creates a copy of the concept and makes modifications on it. It keeps record of the source from which the concept was derived using the *dc:source* property. Users can update their own concept definitions to add new attributes when needed. Thus, concepts can evolve incrementally along with different versions.

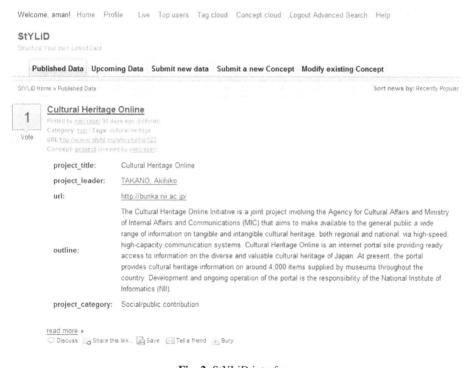

Fig. 2. StYLiD interface

---

[1] http://www.stylid.org/

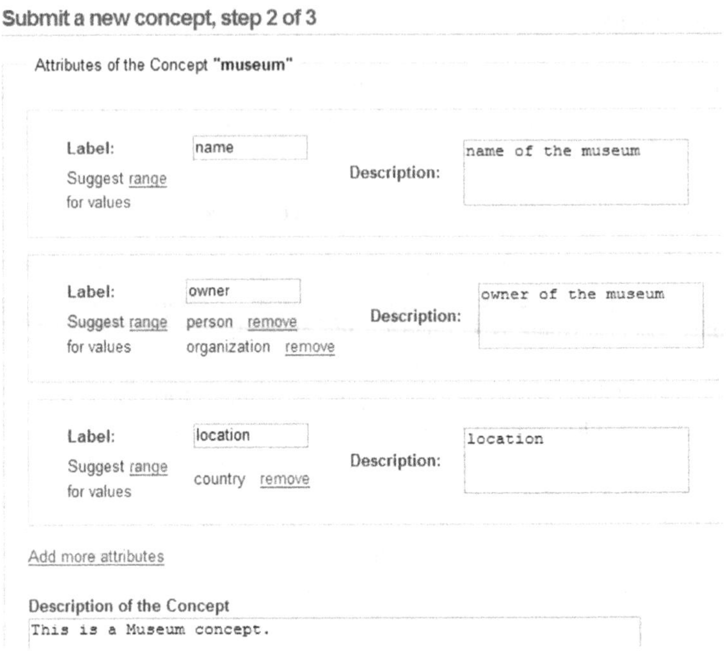

Submit a new concept, step 2 of 3

Attributes of the Concept **"museum"**

Label: `name`
Suggest range
for values

Description: `name of the museum`

Label: `owner`
Suggest range   person remove
for values   organization remove

Description: `owner of the museum`

Label: `location`
Suggest range   country remove
for values

Description: `location`

Add more attributes

Description of the Concept
`This is a Museum concept.`

Fig. 3. Interface to create a new concept

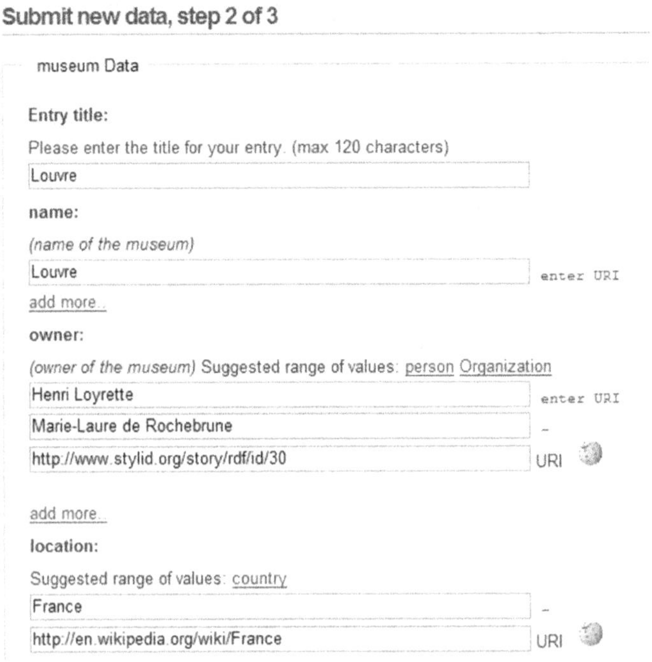

Submit new data, step 2 of 3

museum Data

Entry title:
Please enter the title for your entry. (max 120 characters)
`Louvre`

name:
(name of the museum)
`Louvre`   enter URI
add more...

owner:
(owner of the museum) Suggested range of values: person Organization
`Henri Loyrette`   enter URI
`Marie-Laure de Rochebrune`   –
`http://www.stylid.org/story/rdf/id/30`   URI

add more...

location:
Suggested range of values: country
`France`   –
`http://en.wikipedia.org/wiki/France`   URI

Fig. 4. Interface to enter instance data

Flexible Definitions and Relaxed Data Entry. It is difficult to think of all attributes and all possible value ranges. Further, while defining a concept A, if an attribute takes a resource of type B, we need to ensure that concept B has already been defined. If concept B has an attribute which takes values of type C, then concept C must be defined first, and so on. Moreover, we may not always have perfect data at the time of data entry. The system tries to avoid these difficulties by allowing flexible and relaxed definitions. The range of values defined for attributes, as seen in Fig. 3 and 4, is only suggestive and does not impose strict constraints. Rather the system assists the user to pick instances from the suggested range. The system accepts both literal values and resource URIs for any attribute. Users may input single or multiple values for any attribute. Users generally enter appropriate or sensible data as has been evidenced by systems like tagging and wiki which accumulate plenty of good data in spite of having completely relaxed interface.

The system also offers a personal structured data space. It provides a *Concept Collection* for each user, as seen in Fig. 5. Concepts created or adopted by the user are automatically added to this collection. Users can also add any other useful concepts to their collection. The concepts actually created by the user are shown in a separate tab.

4.2 Consolidation of User Defined Concepts

Concepts defined by different users with the same name are grouped together by the system forming a single virtual concept. The grouped concepts are the *constituent concepts* and the virtual concept is the *consolidated concept* as defined in Section 3 (def. 2). We are also working on consolidating similar concepts with different names. On the other hand, ambiguous concepts have to be sub-grouped by intended meaning. However, for now, we shall focus on consolidating concepts with same name only.

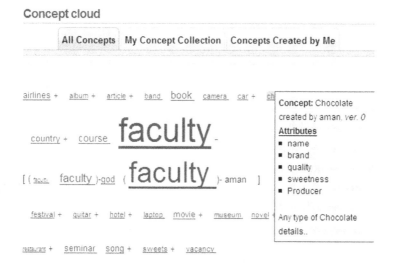

Fig. 5. Concept cloud

Consolidated Concept Cloud. All the concepts are visualized in a Concept Cloud as shown in Fig. 5. Clicking on a concept shows all its instances. Hovering on a concept shows its details. Popular concepts appear bigger in the cloud. Stable definitions gradually emerge out from the cloud as more data instances are contributed.

A consolidated concept can be expanded into a sub-cloud showing all the versions defined by different users, labeled with the creator name and version number. In the sub-cloud, multiple versions defined by the same user are subgrouped together. In Fig. 5, the "Faculty" concept has been expanded to show two versions by the user "god" and one version by "aman". The sizes of all versions in the sub-cloud add up to form the size of the consolidated concept. Clicking on the consolidated concept shows all instances of all its versions. We can also list instances of the multiple versions of a concept defined by a single user by clicking on the user name.

Semi-Automatic Concept Alignment. Different concepts in a consolidated group can be aligned to produce a uniform and integrated view. When the instances of a consolidated group of concepts are viewed as a single table, the system automatically suggests alignments between the attributes, as shown in Fig. 6. Matching attributes are automatically selected in the form-based interface. The Alignment API[2] [9] with its WordNet extension has been used for the purpose. It utilizes a WordNet based similarity measure between attribute labels to find alignments. However, more sophisticated alignment methods may be used in the future.

Fig. 6. Aligning the attributes of multiple concepts

It is not possible to make the alignment fully automatic and accurate. So it is necessary to have the user in loop to complete the process by adding or modifying mappings not correctly suggested by the system. The alignments are represented using the alignment ontology[3] and saved by the system. This forms the alignment $\mathcal{A}$ defined in Section 3 (def. 2, 3). Once a user completes the alignment others need not do it again. Thus, both machine intelligence and human intelligence are used in getting the concepts aligned. The alignment can be incrementally updated as more concepts may be added to the consolidated group of concepts.

---

[2] http://alignapi.gforge.inria.fr/
[3] http://www.atl.lmco.com/projects/ontology/

A Unified View. Each set of aligned attributes is mapped to a single consolidated attribute. This *consolidated attribute* (def. 2) is the *view* of a corresponding attribute (def. 6) from each constituent concept as defined in Section 3. The system automatically fills a name for each consolidated attribute, as shown in Fig. 6, though the user may rename it as desired. The user may even remove attributes from the unified view, if not required. Thus, the user can create a unified view, customized according to his need, and view heterogeneous data in a uniform table. This table corresponds to the *consolidated view of instances* described in Section 3 (def. 8). The table can be sorted by any field. To have all instances of all the concepts listed, all the concepts should be *mapped* (def. 4). The consolidated concept should be *grounded* (def. 5) to have no empty attributes in the unified table. The user is notified if all concepts are not mapped or the consolidated concept is not grounded.

4.3 Creating Linked Data

The system helps in creating linked data using URIs. It generates unique dereferenceable URIs for each concept, attribute and instance. Each concept is uniquely identified by the concept name, its creator and the version number. An example URI for a concept "car", version 2, defined by the user with ID 1 would be *http:// www.stylid.org/ concept_detail/rdf/concept_name/car_ver2_1#car*

An attribute is uniquely identified by the concept and the attribute name. For example, the URI for the price attribute of the car concept would be *http:// www.stylid.org/ concept_detail/rdf/concept name/car_ver2_1#price*

The hash URI for a concept retrieves the RDF document describing the concept and dereferences to the concept description. The attribute URIs are handled similarly.

An instance is uniquely identified by the system generated ID. For example, the URI for an instance with ID 623 would be *http://www.stylid.org/story/rdf/id/623*. The URI dereferences to the RDF description of the instance by an HTTP 303 redirect. For both types of URI, content negotiation is used to return the RDF description in case of 'application/rdf+xml' request and HTML otherwise [10].

Data instances can be linked to each other directly by entering resource URIs as attribute values (see Fig. 4). The system provides support for this by suggesting range of values for the attributes. The user may easily pick up instances from this range. The data appears as simple hyperlinked entries for the user (see Fig. 2). However, the linked data can be crawled by machines to enable powerful applications.

Linking to Wikipedia and External Resources. The user may directly enter any external URI as an attribute value. The system provides some support to link to Wikipedia contents. The familiar Wikipedia icon is seen next to the URI field (see Fig. 4). When the user clicks on the icon it searches for the Wikipedia page about the text attribute value typed by the user and displays it as a pop-up. The user may copy the Wikipedia page URL as the URI. Transparent to the user, the system converts it into the corresponding DBpedia URI. DBpedia[11] exposes the structured data in Wikipedia on the Semantic Web like a database. Unlike DBpedia, Wikipedia is well understood by general people and user-friendly. So the users would be motivated to link to Wikipedia pages to make their data more informative, interesting and useful.

4.4 Querying Structured Data

The system provides a structured search interface, as shown in Fig. 7, to retrieve instances of a concept by specifying attribute, value pairs as criteria. The search can be done over a consolidated concept. In that case, the query terms are unfolded to aligned attributes of all versions of the concept as described in Section 3 (theorem 3). The system also provides a SPARQL query interface for open external access.

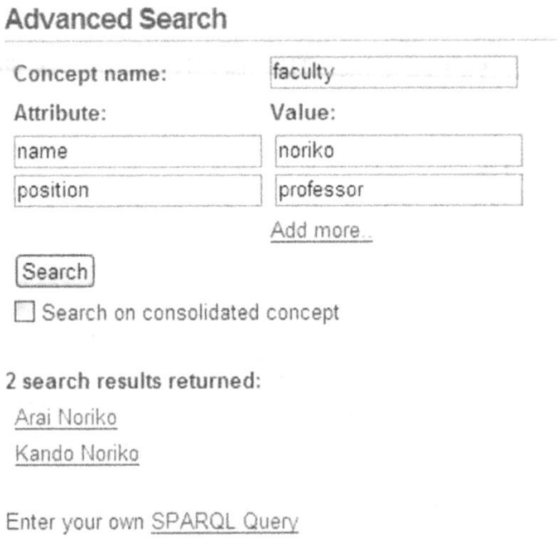

Advanced Search

Concept name: faculty

Attribute: Value:

name   noriko

position   professor

                       Add more..

[Search]

☐ Search on consolidated concept

2 search results returned:

Arai Noriko

Kando Noriko

Enter your own SPARQL Query

Fig. 7. Advanced search interface

4.5 Embedding Machine Readable Data

Besides serving RDF when URIs are dereferenced, the system also embeds machine understandable data in the HTML posts using RDFa. Many useful RDFa tools and plug-ins are available[4] and we may expect more powerful tools to be available in the future. Users with some programming knowledge may code small scripts with the Operator[5] browser extension to create useful operations for different types of data.

4.6 Technologies Used

StYLiD has been built upon Pligg[6], a popular Web 2.0 content management system. It is an open source social software with a long list of useful features and a strong community support. It uses PHP and MySQL. The structured concepts and data are stored as RDF triples in a MySQL database. We used the RDF API for PHP (RAP) as the Semantic Web framework.

---

[4] http://esw.w3.org/topic/RDFa
[5] http://www.kaply.com/weblog/operator/
[6] http://www.pligg.com/

5 Related Work

There are several related works if we consider various aspects of the proposed approach separately. However, none of the works cover all these aspects together.

There are systems that enable general users to create and share a wide variety of structured data on the web. Works like Freebase[7], Google Base[8] and Exhibit [1] allow the users to define their own schemas to model different types of data. However, the structured types defined by different users are kept separate and not consolidated or related in any way. So it is difficult to fully utilize the structured concepts defined by the mass. With Freebase, it is difficult for casual users to create their own types because of strict constraint requirements. All attributes must have strict types within the ones already defined in the system. It may also be difficult to enter instance data because of strict constraints. If an attribute takes a resource value of some type, the resource must be entered first. Moreover, it is difficult to link to external resources and other systems to link to Freebase data. Exhibit is a lightweight framework which enables ordinary users to publish web pages with structured data. However, authoring such structured data pages would be cumbersome to the users. Revyu[12] allows sharing a wide variety of data by reviewing and rating anything. Things are identified and interlinked using URIs. However, most concepts are modeled simply as things without modeling the detailed structure of the information. There had been a lot of works on semantic blogging [13, 14, 15] which exploit the easy publishing paradigm of blogs and enhance blog items with semantic structure. However, they deal with limited types of metadata and the schemas do not evolve.

Semantic wikis make the collaborative knowledge contributed by users more explicit and formal. Buffa et al. [16] have reported on the state-of-art of semantic wikis. Semantic wikis facilitate collaborative creation of resources, many of them supporting the building of ontologies. The myOntology[2] project also uses wikis for collaborative and community-driven building of horizontal lightweight ontologies by enabling general users to contribute. However, they do not consider sharing structured data in the community based on schemas. Freebase, myOntology and other semantic wikis are all based on wiki technology. With wikis, each concept or resource can only have a single prominent version which everyone is assumed to settle with. However, in practice, multiple conceptualizations may exist. Moreover, unlike a wiki, StYLiD is a dynamic information sharing platform like a community blog.

Takeda et al.[17] had discussed the significance of multiple conceptualizations and modeled heterogeneous system of ontologies using aspects. A *category aspect* is a collection of different conceptualizations and a *combination aspect* integrates various aspects. They had proposed muti-agent communication by translating messages across different aspects. This would also be possible in our case as shown in Section 3 (theorem 1, 2, 4). Some works have been done for deriving ontologies from folksonomies[3, 18]. The basic ideas include grouping similar tags, forming emergent concepts from them, making the semantics more explicit and utilizing external knowledge resources to find semantic relations. Folksonomies serve collaborative organization of objects using tags. However, the objects are still left unstructured.

---

[7] http://www.freebase.com/
[8] http://base.google.com/

There is a large body of research about schema matching [19] and ontology alignment [20] and we do not intend to develop new methods for this. Rather we propose the utilization of these techniques to align and consolidate user-defined schemas. There are also some tools for casual users like Potluck[21] which provides a user-friendly interface to align, mix and clean structured data from Exhibit-powered pages. The schema alignment is manual. We propose to have some automation in schema alignment and, moreover, saving the alignments in Semantic Web format.

6   Conclusions and Future Work

In this paper, we proposed a new approach for community-driven definition of concepts by consolidating multiple user-defined conceptualizations. Rich definitions can be formed by combining concepts created by different users facilitating collaborative knowledge formation while satisfying individual requirements. Alignments can reconcile different definitions. We proposed StYLiD, a social software for sharing any type of structured data. It combines various aspects of social semantic software into a more effective whole than the sum of the separate parts. By allowing users to define their own concepts and providing a relaxed interface, it motivates free contribution. Concepts can evolve incrementally and emerge by popularity. Ontologies can be a by-product of usual information sharing activities and even with informal social software formal linked data can be produced.

We are currently working on grouping similar concepts and computing relations between concepts. This can be done by considering the structure definitions, and utilizing lexical resources like WordNet. Ideas from works on deriving ontologies from folksonomies[3, 18] may also be adapted. Better alignment techniques and more complex alignments may be employed. Consolidation of instance data is another issue which needs to be addressed in the future. We should also reuse existing vocabularies and map concept definitions to them. Other useful features like mash-ups may be introduced to exploit the structured data. We can facilitate users to contribute plugins for handling different types of data. Scrapers may be associated to concepts for collecting data from web pages easily. We may enable users to create and share such scrapers too. Besides providing linked data and SPARQL interface, structured data may also be exposed through an API or extended RSS.

References

1. Huynh, D., Karger, D., Miller, R.: Exhibit: lightweight structured data publishing. In: Proceedings of the 16th international conference on World Wide Web, pp. 737–746. ACM Press, New York (2007)
2. Siorpaes, K., Hepp, M.: myOntology: The marriage of ontology engineering and collective intelligence. In: Bridging the Gap between Semantic Web and Web 2.0 (SemNet 2007), pp. 127–138 (2007)
3. Van Damme, C., Hepp, M., Siorpaes, K.: FolksOntology: An integrated approach for turning folksonomies into ontologies. In: Bridging the Gap between Semantic Web and Web 2.0 (SemNet 2007), pp. 57–70 (2007)

4. Ankolekar, A., Krötzsch, M., Tran, T., Vrandečić, D.: The two cultures: Mashing up web 2.0 and the Semantic Web. In: Proceedings of the 16th International World Wide Web Conference (WWW 2007), Banff, Alberta, Canada, pp. 825–834. ACM Press, New York (2007)

5. Gruber, T.: Collective knowledge systems:Where the social web meets the Semantic Web. Journal of Web Semantics 6(1), 4–13 (2008)

6. Schaffert, S.: Semantic social software: Semantically enabled social software or socially enabled Semantic Web? In: Proceedings of the SEMANTICS 2006 conference, Vienna, Austria, OCG, pp. 99–112 (2006)

7. Berners-Lee, T.: Linked data. World Wide Web design issues (July 2006)
   `http://www.w3.org/DesignIssues/LinkedData.html`

8. Lenzerini, M.: Data integration: A theoretical perspective. In: Proceedings of the twenty-first ACM SIGMOD-SIGACT-SIGART symposium on Principles of database systems, pp. 233–246 (2002)

9. Euzenat, J.: An API for ontology alignment. In: McIlraith, S.A., Plexousakis, D., van Harmelen, F. (eds.) ISWC 2004. LNCS, vol. 3298, pp. 698–712. Springer, Heidelberg (2004)

10. Bizer, C., Cyganiak, R., Heath, T.: How to Publish Linked Data on the Web (2007),
    `http://www4.wiwiss.fu-berlin.de/`
    `bizer/pub/LinkedDataTutorial/`

11. Auer, S., Bizer, C., Kobilarov, G., Lehmann, J., Cyganiak, R., Ives, Z.G.: DBpedia: A nucleus for a web of open data. In: Aberer, K., Choi, K.-S., Noy, N., Allemang, D., Lee, K.-I., Nixon, L., Golbeck, J., Mika, P., Maynard, D., Mizoguchi, R., Schreiber, G., Cudré-Mauroux, P. (eds.) ASWC 2007 and ISWC 2007. LNCS, vol. 4825, pp. 722–735. Springer, Heidelberg (2007)

12. Heath, T., Motta, E.: Revyu.com: A reviewing and rating site for the web of data. In: Aberer, K., Choi, K.-S., Noy, N.F., Allemang, D., Lee, K.I., Nixon, L.J.B., Golbeck, J., Mika, P., Maynard, D., Mizoguchi, R., Schreiber, G., Cudré-Mauroux, P. (eds.) ISWC/ASWC 2007. LNCS, vol. 4825, pp. 895–902. Springer, Heidelberg (2007)

13. Cayzer, S.: Semantic blogging and decentralized knowledge management. Communications of the ACM 47(12), 48–52 (2004)

14. Möller, K., Bojārs, U., Breslin, J.G.: Using semantics to enhance the blogging experience. In: Sure, Y., Domingue, J. (eds.) ESWC 2006. LNCS, vol. 4011, pp. 679–696. Springer, Heidelberg (2006)

15. Karger, D.R., Quan, D.: What would it mean to blog on the Semantic Web? Journal of Web Semantics 3(2), 147–157 (2005)

16. Buffa, M., Gandon, F., Ereteo, G., Sander, P., Faron, C.: A Semantic Wiki. Journal of Web Semantics 6(1), 84–97 (2008)

17. Takeda, H., Iino, K., Nishida, T.: Agent organization and communication with multiple ontologies. International Journal of Cooperative Information Systems 4(4), 321–337 (1995)

18. Specia, L., Motta, E.: Integrating folksonomies with the Semantic Web. In: Franconi, E., Kifer, M., May, W. (eds.) ESWC 2007. LNCS, vol. 4519, pp. 624–639. Springer, Heidelberg (2007)

19. Rahm, E., Bernstein, P.: A survey of approaches to automatic schema matching. The VLDB Journal The International Journal on Very Large Data Bases 10(4), 334–350 (2001)

20. Euzenat, J., Le Bach, T., Barasa, J., et al.: State of the art on ontology alignment. Knowledge Web Deliverable D2.2.3 (2004)

21. Huynh, D.F., Miller, R.C., Karger, D.R.: Potluck: Data mash-up tool for casual users. In: Aberer, K., Choi, K.-S., Noy, N., Allemang, D., Lee, K.-I., Nixon, L., Golbeck, J., Mika, P., Maynard, D., Mizoguchi, R., Schreiber, G., Cudré-Mauroux, P. (eds.) ASWC 2007 and ISWC 2007. LNCS, vol. 4825, pp. 239–252. Springer, Heidelberg (2007)

An Integrated Approach for Automatic Construction of Bilingual Chinese-English WordNet

Renjie Xu[1], Zhiqiang Gao[1], Yingji Pan[1], Yuzhong Qu[1], and Zhisheng Huang[2]

[1] School of Computer Science and Engineering, Southeast University, China
dustin.xu@gmail.com, zqgao@seu.edu.cn,
jennifer-pyj@hotmail.com, yzqu@seu.edu.cn
[2] Department of Computer Science, Vrije Universiteit Amsterdam. The Netherlands
huang@cs.vu.nl

Abstract. This paper compares various approaches for constructing Chinese-English bilingual WordNet. First, we implement three independent approaches that translate English WordNet to Chinese WordNet automatically, including Minimum Distance (MDA), Intersection (IA) and Words Co-occurrence (WCA). Minimum Distance compares the gloss of synset with the explanations of words from dictionaries. Intersection chooses the intersection part of Chinese in a synset. Words Co-occurrence counts the results of Chinese and English words from Google. Then, we integrate these three approaches into an integrated one, which is named MIWA. Experimental results show that the integrated approach MIWA has better performance: F-measure reaches 0.615, which is higher than that of each independent one.

Keywords: WordNet, Chinese WordNet, Co-occurrence.

1 Introduction

We aim to construct Chinese-English WordNet, a bilingual lexical database with wordnets for Simplified Chinese and English. Its semantic hierarchy is originated from Princeton WordNet [1-5], which consists of more than 115, 000 synsets, over 200, 000 words. We try to translate the original English WordNet into Chinese one according to automatic translation approaches. The first one is Minimum Distance Approach (MDA), which compares the gloss of synset with the explanations of words from dictionaries by Levenshtein Distance. The second is Intersection Approach (IA), which chooses the same or similar Chinese words of different English words, which are in the same synset. The third is Words Co-occurrence Approach (WCA), which counts the results of Chinese and English words from Google. Several linguistic resources have been used for building Chinese-English WordNet, such as American Heritage Dictionary (Chinese&English edition), X-Dict Dictionary (English-Chinese). According to the performance of each approach, we describe a new one (MIWA), which integrates these three approaches following some rules, to complete the automatic translation work. After this step, some experts check the results and make some proper modifications. Now, Chinese-English WordNet has more than 150, 000 Chinese words, besides the original English part.

J. Domingue and C. Anutariya (Eds.): ASWC 2008, LNCS 5367, pp. 302–314, 2008.
© Springer-Verlag Berlin Heidelberg 2008

Mapping between Chinese part and English part in Chinese-English WordNet is established on the basis of synsets. A Chinese synset is corresponding to an English synset, but the words inside these two synsets do not have any specific mapping relations. The connections among English synsets perform as bridges among Chinese ones, because Chinese synsets do not have any direct contact with each other. This is illustrated in Figure 1. {car, auto, automobile, machine, motorcar} is translated into 【汽车】, 【救护车】acts as the hyponym of 【汽车】, because its corresponding English synset {ambulance} is the hyponym of {car, auto, automobile, machine, motorcar}. 【门锁】 is a part of 【车门】 since {doorlock} is a part of {car door}.

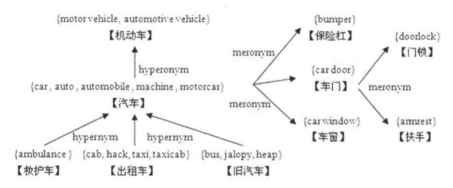

Fig. 1. Synsets related to car in its first sense in Chinese-English WordNet

We design some experiments to evaluate the performance of these approaches, it shows that IA has the highest precision, WCA has the highest recall except for MIWA, and MDA performs not bad in F-measure evaluation. From an over perspective, MIWA shows an obvious improvement than using each of the three approaches respectively, and it has been proved helpful in construction of Chinese-English WordNet.

Chinese-English WordNet is stored in a lexical database system, and each Chinese synset is linked to an English synset, and each relation is saved in an individual table. Such a bilingual database is useful for cross-language information retrieval, and it is also suitable to add more specific language wordnet in future.

This paper is structured as follows: Section 2 describes background and some related work of WordNet; Section 3 and 4 describe the approaches and their evaluation, and Section 5 presents the conclusions and future work.

2   Background

Semantic information analysis is one of the most difficult problems in Natural Language Processing (NLP). Without a large and computable semantic resource, machines cannot understand the information in natural language just as people do. As a significant part that composes semantic resources, Machine-Readable Dictionary (MRD)'s quality directly affect the development of NLP. WordNet is the representative of MRD, so its value is self-evident.

Under this background, many organizations have developed their native language wordnet, or some other electronic lexical databases for words. EuroWordNet, CoreNet and HowNet are three exotic flowers in this area.

EuroWordNet [6] is a multilingual lexical database with wordnets for several European languages, including Dutch, Italian, Spanish, English, German, French, Estonian and Czech. It is structured along the same lines as the Princeton WordNet. Mainly there are two ways to construct EuroWordNet: one is to do classification of the language internally into synsets and then link them to English as an interlingual "spine", such as Dutch part; the other is to translate the English wordnet into other languages, such as Italian part. Inter-Lingual-Index (ILI) maintains the language specific structures and to allow for the separate development of independent resources, each synset in the monolingual wordnets have at least one equivalence relation with a record in this ILI. ILI acts as the bridge to connect different synsets in specific language wordnets. The project of EuroWordNet started in March 1996, and finished in June 1999, more than ten universities and institutions have participated into this project.

CoreNet [8] is a Korean-Chinese-Japanese wordnet, it has been developed using a shared semantic hierarchy, originated in NTT Goidaikei-a Japanese Lexicon. Developers used an information retrieval technique to construct Korean wordnet. Based on the results, developers translated all of the Japanese words under NTT Goidaikei into Korean words using a Japanese-Korean electronic dictionary. Experts correct the result of automatic translation. They manually correct the erroneous assignments between two languages. Then they assign semantic categories by matching the Korean words with the translated word list under the NTT Goidaikei's semantic category. In post-processing, word sense disambiguation was done manually to assign proper semantic categories to each sense of the word and the translation errors were also removed. Two people performed independently the same post-processing. A third party examined the different parts of the results and chose the proper ones [9-10]. CoreNet has taken nearly ten years to finish it.

HowNet [11] is an electronic lexical database for words, which are mostly in Chinese. Unlike WordNet, synsets are not explicitly defined in HowNet. HowNet takes a constructive approach to build a lexical hierarchy, and the focus of it is the relations among concepts and their attributes. At the most atomic level is a set of almost 1500 basic definitions or sememes. HowNet has finished on the basis of large corpus, some programs extract language fragment between empty words [12], and then some experts make artificial amendment. It is also nearly ten years since the initial development.

From three lexical databases introduced above, we find that some automatic approaches have been used during processing, for example, CoreNet used electronic dictionary for automatic translation from Japanese to Korean. However, nearly all the mapping jobs between different language resources have been done by people. That is why they spent a lot of time and human resources on the projects.

3 Our Approaches

Because of the significant value of WordNet, we decide to construct a Chinese and English Bilingual WordNet. It is supposed to be useful for cross-language information retrieval, ontology learning and NLP, especially in Chinese. However, there is no doubt that if all construction jobs are done by people, it would take either too many people or too much time or both. Even though many language resources for translation, such as electronic dictionaries are available, there are quite a few algorithms that can finish the translation job effectively. That is why we put forward these language independent approaches as follows.

3.1 Minimum Distance Approach

Minimum Distance Approach (MDA) calculates the Levenshtein Distance [13] between gloss of synset and explanations in American Heritage Dictionary (Chinese&English edition). The explanation has the minimum LD with gloss seems to be the proper one for this synset, and then analyze the corresponding Chinese explanation to get the key words as the members of Chinese synset.

American Heritage Dictionary is one of the most authoritative dictionaries in America. The difference between AHD (Chinese&English edition) and most of other Chinese-English electronic dictionaries is that the former has not only translated the words, but also provided a small section of text to explain each meaning in two languages. It is likely that WordNet's developers have referred to the definitions when they define the glosses of synsets, because plenty of the glosses in WordNet look similar to some explanations in AHD, and even some of them are the same.

3.1.1 Description of MDA
1) Load the American Heritage Dictionary (Chinese&English edition) to an open source electronic dictionary program, searching for the English words in WordNet, and save the returned explanations, both English and Chinese, in database.
2) Set a synset as the processing unit; calculate Levenshtein Distance between gloss and each explanation corresponding to words in this synset. After some experiments, we found when the cost of MODIFY operation is 1, 0.6 is proper for that of ADD and DELETE. Because most of the glosses are a bit succinct than the explanations, if the cost of ADD and DELETE is more than 0.6, the total cost of correct explanation is more than the cost of very short but incorrect explanation; but if that is less than 0.6, many incorrect explanations will also be chosen in experiments.
3) Analyze Chinese explanation corresponding to the English one which has the minimum Levenshtein Distance with gloss; get the key words as members of Chinese synset.
4) Repeat 2-3), until all the synsets have been processed.

3.1.2 Example

"sailing" is in four synsets of WordNet, they are: (1) the work of a sailor,(2) riding in a sailboat;(3) the departure of a vessel from a port;(4) the activity of flying a glider, and it has three explanations in American Heritage Dictionary, they are illustrated in Table 1.

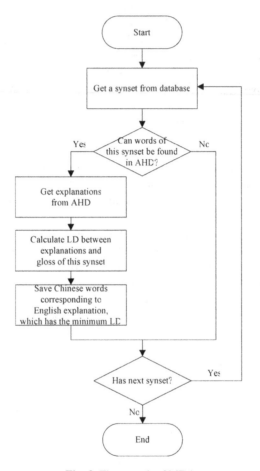

Fig. 2. Flow graph of MDA

Table 1. "sailing" in American Heritage Dictionary

English	Chinese
The skill required to operate and navigate a vessel; navigation.	航海术：驾驶和航行一条船所需的技巧；航行术
The sport of operating or riding in a sailboat.	帆船运动：驾驶或航行帆船的一项体育运动
Departure or time of departure from a port.	启航：离开港口；离开港口的时间

From Table 1 and four glosses in WordNet we can see that the third gloss in WordNet looks like the third explanation in dictionary, so using MDA, the result is showed in Table 2 that the third is the best explanation that matching gloss of this synset, then analyze the third Chinese explanation, set"启航" as the member of this Chinese synset.

Table 2. Example of MDA

explanation	Levenshtein Distance
the skill required to operate and navigate a vessel navigation	0.666
the sport of operating or riding in a sailboat	0.622
departure or time of departure from a port	0.475

3.2 Intersection Approach

Because synset is a set that words inside it have the same meaning, it is language independent, so when words in the same synset are translated into other languages, there should be an intersection among the translated results, they are the members of new synset, this is the principle of IA. Moreover, based on the construction of used electronic dictionaries, IA can also find a proper translation even synset contains only one word.

3.2.1 Description of IA
1) Load X-Dict Dictionary to an open source electronic dictionary program, searching for the English words in WordNet, and save returned translated Chinese words in database.
2) Set a synset as the processing unit:
   a) If this synset has only one word and this word has one translated Chinese word, set this Chinese word as the member of new Chinese synset.
   b) If this synset has more than one word, comparing them in two:
      i. If both of the English words have same translated Chinese words, they are set as the members of new Chinese synset.
      ii. If the English words have similar translated Chinese words, and these translations meet the Levenshtein Distance requirement set by us, they also set as the members of new Chinese synset.
   a) If none of above be met, this synset cannot translated by this approach.
3) Repeat 2), until all the synsets have been processed.

3.2.2 Example
"stop" and "halt" are in the same synset, the gloss of its is "the event of something ending; "it came to a stop at the bottom of the hill", and the pos is noun. In X-Dict Dictionary, the results of "stop" and "halt" are as follows:

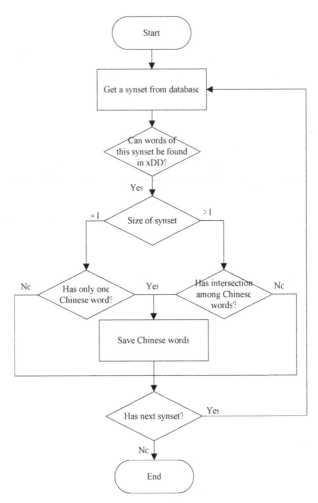

Fig. 3. Flow graph of IA

Table 4. "Stop" in X-Dict Dictionary

stop
n. 停止, 车站, 逗留, 填塞, 障碍, (风琴的)音栓
vi. 停止, 被塞住
vt. 塞住, 堵塞, 阻止, 击落, 停止, 终止, 断绝

Table 5. "Halt" in X-Dict Dictionary

halt
n. 停止, 立定, 休息
vt. 使停止, 使立定
vi. 立定, 停止, 蹒跚, 踌躇, 有缺点

Table 4, 5 shows that: as noun, both of "stop" and "halt" have the Chinese "停止", so this is the intersection of these words, it is the proper member of new Chinese synset.

3.3 Words Co-occurrence Approach

Words Co-occurrence Approach (WCA) follows the principle that the frequency of two words' co-occurrence refers to the relation between these two words, in other words, two words have more connection with each other, and they appear more at the same time. For example, "country" and "国家" are supposed to appear together more than "country" and "天空". Therefore, WCA sets an English word and a Chinese word as a group, get the result from Google, based on these results, the program automatically choose the best translations for Chinese synsets.

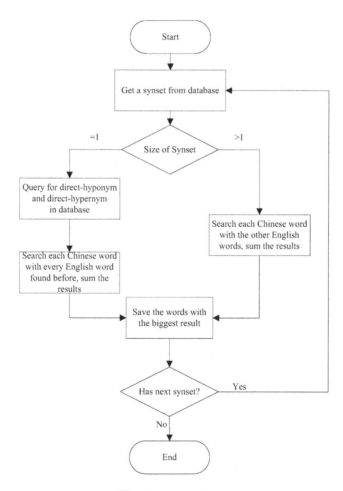

Fig. 4. Flow graph of WCA

3.3.1 Description of WCA

1) Using the backup Chinese translation saved in step 1 of IA
2) Set a synset as the processing unit:
   a) If this synset has only one word, find out all the hypernym and hyponym words if there exist, searching them with every backup Chinese word of this synset, and aggregate results for each Chinese word.
   b) If this synset has more than one words, searching each backup Chinese word with English members of this synset except for the one directly translated to it, aggregate results for each Chinese word.
   c) The Chinese word with the biggest result is set to be the member of new Chinese synset.
3) Repeat 2), until all the synsets have been processed

3.3.2 Example

For a synset has one word: {dashboard} is a synset, its gloss is {instrument panel on an automobile or airplane containing dials and controls}, X-Dict Dictionary returns three Chinese words for dashboard{挡泥板,遮水板,仪表板}, {dashboard}'s direct-hypernym synset is {control-panel, instrument-panel, control-board, board, panel}, and there is no hyponym. Set a Chinese word and an English word as a group, searching is in Google, and the results are illustrated as follows:

Table 6. Google results for {dashboard}

	仪表板	挡泥板	遮水板
control-panel	3380	631	59
instrument-panel	2590	701	745
control-board	709	127	89
board	47000	13000	15900
panel	29300	3910	4090
OVERALL	82979	18369	20824

From Table 6, we can see that total result for "仪表盘" is 82,979, higher than that of "挡泥板" and "遮水板", so "仪表盘" is the best choice for the new Chinese synset corresponding to {dashboard}.

For a synset has more than one word: {discomposure, discomfiture, disconcertion, disconcertment} is a synset, its gloss is {anxious embarrassment}, the translation for words of this synset are:

Table 7. Translations for "discomposure", "discomfiture", "disconcertion", "disconcertment"

Word	Translation in X-Dict Dictionary
discomposure	不安, 心乱, 狼狈
discomfiture	失败, 妨害计划
disconcertion	(not found in X-Dict Dictionary)
disconcertment	不平, 不满

Table 8. Google results for {discomposure, discomfiture, disconcertion, disconcertment}

	失败	妨害计划	不安	心乱	狼狈	不平	不满
discomposure	126	48	--	--	--	64	140
discomfiture	--	--	436	104	211	160	180
disconcertion	8	3	15	5	5	3	4
disconcertment	18	4	46	8	10	--	--
OVERALL	152	55	497	117	226	227	324

In Table 7, we can see that "disconcertion" is not found in X-Dict Dictionary, but it does not affect the translation work. Table 8 shows that "不安" has the highest OVERALL result, so it is the best translation for this synset.

3.4 MIW Approach

MIWA is an approach that integrates three approaches described above. Based on precision of theirs, we arrange three of them in the descending order, and use them one by one. We use the approach with the highest Precision first; the rest synsets that cannot be processed in the first step, we choose the approach with the second highest Precision to deal with; finally, the third approach is to be used to dispose the rest synsets after step one and step two. Given the precision of IA > MDA's > WCA's, according to MIWA, we will choose IA firstly to process the whole English WordNet, obviously, only a part of it can be successful translated, then MDA will be used to deal with the rest synsets of WordNet, after that, WCA will be adopted for the rest. By following this order, we can keep a high precision of translation and increase the number of synsets that can be translated.

4 Experiments and Results

4.1 Experimental Materials

Based on the organization, quantity and quality of words, we choose American Heritage Dictionary (Chinese&English edition) to implement to MDA, and X-Dict Dictionary to achieve IA and WCA.

We randomly chose 35,000 synsets as the universal set for the experiment. The manual translation finished by nearly ten experts, who are skilled in both Chinese and English, seem to be the standard results. Levenshtein Distance [13] Algorithm has been used for the automatic comparison between the results of automatic translation and manual job. Recall, Precision and F-measure are three evaluation targets: Recall represents how many synsets can be processed by using an approach; Precision means according to comparing results of experts and those of the approaches, how many synsets are correct; F-measure is the weighted harmonic mean of Precision and Recall. We can also use the following formula to calculate the Precision and Recall of MIWA according to those of the other three approaches', Group (1-3) represent three processing groups described in Section3.4.

$$\text{Recall} = \frac{\sum_{i=1}^{3} \text{Size}(\text{Group}(i))}{\sum_{i=1}^{3} \text{Size}(\text{Group}(i)) + \text{Size}(\text{Group}(N/A))} \tag{1}$$

$$\text{Precision} = \sum_{j=1}^{3} (\text{Precision}(\text{Group}(j)) \times \frac{\text{Size}(\text{Group}(j))}{\sum_{i=1}^{3} \text{Size}(\text{Group}(j))}) \tag{2}$$

4.2 Results

The results of experiments are illustrated as follows:

Table 9. Results of experiment

Approach	Recall	Precision	F-measure
MDA	0.430	0.599	0. 501
IA	0.194	0.641	0. 298
WCA	0.572	0.386	0.461
MIWA	0.703	0.547	0.615

From the results, we can see that: in the first three approaches, IA has the highest precision, but lowest recall, this is because one-synset-one-word situation takes up nearly half of WordNet, this can be seen in Figure 5; WCA covers 57.2% synsets, it has the widest processing ability, however, because results from Google contains many noises, it also has the lowest precision, and since this program has to download web pages during running, it takes much longer time than the other two programs. MDA has the median recall and median precision, it performs the best in F-measure evaluation, this result makes more sure of the fact that WordNet's developers has referred to American Heritage Dictionary while developing WordNet, and because of this, MDA can also translate some words in specific areas, that even cannot found in X-Dict Dictionary.

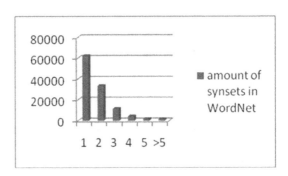

Fig. 5. The amount of different size of synsets in WordNet

The MIWA which integrates three approaches does a good job in experiment: it covers 70% of synsets. Even though precision is affected by that of WCA, its F-measure still prove that from an overall perspective, MIWA is better than using any of the approach respectively.

Table 10. Advantages and Disadvantages of MDA, IA and WCA

	MDA	IA	WCA
Advantage	1.High Precision 2.Can handle words in expertise field	Highest Precision	Highest Recall
Disadvantage	Organization of dictionary is special	Lowest Recall	1.Lowest Precision 2.It takes a long time to run the program

5 Conclusion and Future Work

We have described here a new procedure which integrated three approaches for automatic translation from English WordNet to Chinese WordNet. Compared to other similar approaches or using each approach respectively, it has the following improvements:

- Each of the approaches is greatly helpful and creative, and from an overall perspective, the procedure involved three approaches can process a great number of synsets in WordNet and has a high precision. It can save a lot of time for manual construction of WordNet.
- It is language independent; people can use it for any specified language WordNet construction, if they can find enough multilingual language resources, such as the electronic dictionaries used in this paper.

Concerning future work, we plan to use some other dictionaries for more experiments, and we also think about how to improve WCA, for example, we concern about the results of a single word in Google, it affects the results of two words. Because WCA has the highest recall but lowest precision, in other words, it is the most potential approach in these three, and if we can improve its precision, it must be very helpful to MIWA. First edition of Chinese-English WordNet has been published recently, and we plan to involve manual correct to perfect it.

Acknowledgements

This work is supported by National Science Foundation of China under Grant 60773107, and National Key Basic Research and Development Program of China under Grant 2003CB317004. We would like to thank Hui Xu for his contribution to algorithms of this work.

References

1. Miller, G.A., Beckwith, R., Fellbaum, C., Gross, D., Miller, K.J.: Introduction to WordNet: An On-line Lexical Database. International Journal of Lexicography (1990)
2. George, A.: Miller.: Nouns in WordNet: A Lexical Inheritance System. International Journal of Lexicography (1990)

3. Gross, D., Miller, K.J.: Adjectives in WordNet. International Journal of Lexicography (1990)
4. Fellbaum, C.: English Verbs as a Semantic Net. International Journal of Lexicography (1990)
5. Beckwith, R., Miller, G.A., Tengi, R.: Design and Implementation of the WordNet Lexical Database and Searching Software. International Journal of Lexicography (1990)
6. Vossen, P.: Introduction to EuroWordNet. In: Computers and the Humanities, vol. 32, pp. 73–89 (1998)
7. Vossen, P., Diez-Orzas, P., Peters, W.: Multilingual design of EuroWordNet. In: Proceedings of the IJCAI 1997 workshop Multilingual Ontologies for NLP Applications (1997)
8. Choi, K.-S., Bae, H.-S.: Procedures and Problems in Korean-Chinese-Japanese Wordnet with Shared Semantic Hierarchy. In: GWC 2004, Proceedings (2004)
9. Choi, K.-S., Bae, H.-S., Kang, W., Lee, J., Kim, E., Kim, H., Kim, D., Song, Y., Shin, H.: Korean-Chinese-Japanese Multilingual Wordnet with Shared Semantic Hierarchy. In: Proceedings of LREC 2004, Portugal (2004)
10. Choi, K.-S.: CoreNet: Chinese-Japanese-Korean wordnet with shared semantic hierarchy. In: Natural Language Processing and Knowledge Engineering (2003)
11. Zhendong, D.: Knowledge Description: What, How and Who? In: The Proceedings of the International Symposium on Electronic Dictionaries, Tokyo, Japan
12. Dorr, B.J., Levow, G.-A., Lin, D.: Construction of Chinese-English Semantic Hierarchy for Information Retrieval. In: Proceedings of the Workshop on English-Chinese Cross Language Information Retrieval (2000)
13. Levenshtein, V.: Binary codes capable of correcting deletions, insertions, and reversals: Soviet Physics Doklady (1966)
14. Chen, H.-H., Lin, C.-C., Lin, W.-C.: Construction of a Chinese-English WordNet and Its Application to CLIR. In: Proceedings of the 5th International Workshop Information Retrieval with Asian Languages
15. Maedche, A., Staab, S.: Ontology Learning for the Semantic Web. IEEE Intelligenct Systems 16(2), 72–79 (2001)
16. Alfonseca, E., Ruiz-Casado, M., Okumura, M., Castells, P.: Towards Large-scale Non-taxonomic Relation Extraction: Estimating the Precision of Rote Extractors. In: Proceedings of the 2nd Workshop on Ontology Learning and Population
17. Olsen, S.: Wordnet Wordsense Disambiguation using an Automatically Generated Ontology. In: Proceedings of the Class of, Senior Conference (2003)
18. Church, K., Gale, W., Hanks, P., Hindle, D.: Bell Laboratories and Collins.: Parsing, Word Associations and Typical Predicate-Argument Relations. In: International Workshop on Parsing Technologies, CMU (1989)
19. Liu, Y., Yu, S., Yu, J.: In: Building a Bilingual WordNet-Like Lexicon: the New Approach and Algorithms. In: Proceedings of the 19th international conference on Computational linguistics
20. Yamaguchi, T.: Constructing Domain Ontologies Based on Concept Drift Analysis. In: Proceedings of the IJCAI 1999 workshop on Ontologies and Problem-Solving Methods(KRR5) Stockholm, Sweden, August 2 (1999)

Predicting Category Additions in a Topic Hierarchy

Janez Brank, Marko Grobelnik, and Dunja Mladenić

Jožef Stefan Institute, Jamova 39, 1000 Ljubljana, Slovenia
{janez.brank,marko.grobelnik,dunja.mladenic}@ijs.si

Abstract. This paper discusses the problem of predicting the structural changes in an ontology. It addresses ontologies that contain instances in addition to concepts. The focus is on an ontology where the instances are textual documents, but the approach presented in this document is general enough to also work with other kinds of instances, as long as a similarity measure can be defined over them. We examine the changes in the Open Directory Project ontology of Web pages over a period of several years and analyze the most common types of structural changes that took place during that time. We then present an approach for predicting one of the more common types of structural changes, namely the addition of a new concept that becomes the subconcept of an existing parent concept and adopts a few instances of this existing parent concept. We describe how this task can be formulated as a machine-learning problem and present an experimental evaluation of this approach that shows promising results of the proposed approach.

Keywords: Ontologies, taxonomies, knowledge organization, semantic web, modeling human expertise, machine learning, text mining, support vector machine.

1 Introduction

Many ontologies are not static objects. If an ontology is a shared conceptualization of a domain, it is not surprising that it may have to change in response to changes in either the domain itself, or in our understanding of it, or in the purposes with which we are building a shared conceptualization of it. Thus it is natural to ask whether such changes in an ontology can be predicted automatically as an aid to the people maintaining the ontology.

In this paper we begin by discussing an example of a large real-world ontology whose evolution over the course of several years can be readily observed, namely the topic hierarchy of the Open Directory Project (ODP; see *http://www.dmoz.org/*). We identify the most common types of structural changes occurring in this ontology and analyze their frequency. Based on these observations, we decide to focus on trying to predict one specific type of structural changes: the addition of a new subconcept as a child of an existing parent concept, from which the new concept also takes a few instances. This is one of the more common types of structural changes in the ODP, and it is also amenable to an automatic prediction approach.

We then discuss how the problem of predicting this kind of subconcept additions can be formulated as a machine learning task. The main challenge here is to describe a concept by a set of features in such a way that a predictive model (obtained through

J. Domingue and C. Anutariya (Eds.): ASWC 2008, LNCS 5367, pp. 315–329, 2008.
© Springer-Verlag Berlin Heidelberg 2008

machine learning) will be able to predict, from these features, whether a new subconcept should be added below the given concept or not. Our approach is based on the assumption that the ontology contains not only concepts but also instances, and a new subconcept should be added if there exists a subgroup of closely related instances in the parent concept. We cluster the instances of the parent concept and compute several statistical properties of the resulting partition of the instances into clusters. In the case of the ODP, the instances are textual documents, so that techniques from information retrieval can be used for the needs of cluster analysis.

We also present an experimental evaluation of the proposed approach. Experiments on the ODP ontology show that this is feasible approach for predicting this type of ontology changes.

Finally we will discuss a few ideas for future work, especially with a view to predicting other types of structural changes that are not addressed by the approach presented in this report.

Related work. Maedche et al. [14] and Stojanović [12] defined three types of change discovery: structure-driven (where suggested changes are deduced from analyzing the ontology structure itself), usage-driven (changes are recommended by observing the usage patterns over time) and data-driven (which is based on changes in the underlying data that describes the domain of interest). An example of work focusing on usage-driven change discovery is [9]. [8] discussed the incorporation of data-driven change discovery into a framework for learning an ontology from a corpus of textual documents. Some authors have also discussed ontology evolution from a more formal, logical point of view, with an emphasis on the semantics of ontology evolution and reasoning in the presence of an evolving ontology [10, 11].

For a recent overview of the area of ontology change, and its relationship with ontology evolution, merging, and integration, see the survey by Flouris et al. [15].

The topic of ontology change has also been discussed recently by Maynard et al. [16], which defines a number of ontology change operations. The operations defined there are relatively low-level, whereas the changes which we attempt to predict in the work reported in this paper are somewhat higher-level, and can be seen as aggregations of several low-level operations in the sense of [16]. Our choice of the set of operations is mainly due to two reasons: (1) Higher-level operations (e.g. "move a category" instead of "insert/delete a parent-child link") correspond more closely to the way a human ODP editor (whose expertise we are in a way trying to model here) would conceptualize his/her work. (2) We require operations that can be observed in the available ODP data, and that can be modeled and predicted via a machine learning approach; thus, some operations from [16] do not apply here (e.g. creating a category without immediately attaching it to a parent).

2 Comparing Ontology Snapshots to Identify Structural Changes

2.1 The Open Directory Project Dataset

To investigate the issue of structural changes in an ontology, it is helpful to consider a real-world ontology for which it is possible to observe the changes through a period of time. Additionally, the ontology should be reasonably large, so as to provide a sufficient amount of data for the training of predictive models. We decided to use the topic ontology of the Open Directory Project (ODP, available from *www.dmoz.org/*).

In the ODP ontology, the concepts are actually topical categories; they are organized into a tree via the parent-child relationship. Each category has a name and a short description (the latter not usually shown to the user but available in the data).

In addition the ODP ontology contains instances; these are actually links to external web pages. Besides the link, each instance also contains a title and a short textual description of the page. Thus we will regard each instance as a short document as proposed in [17] for ontology population, and techniques from the area of text mining will be used in dealing with the data.

Note that our approach for predicting structural changes is dependent on the fact that the ontology is well populated with instances. It is not, however, dependent on the fact that the instances are textual documents. As we will see later, the proposed approach only assumes that the instances can be clustered; for that, the only thing one really needs is a measure of similarity (or distance) between the instances.

The ODP ontology is interesting for our purposes because snapshots of the ontology at different points in time are available. The ODP makes available approximately one snapshot per month; more than 50 such snapshots are available on the dmoz.org website, covering the period since July 2003.

One problem with the ODP dataset, from the point of view of predicting structural changes, is that any two consequent snapshots are approximately a month apart and a number of structural changes can take place during that time period. Sometimes several of these structural changes affect the same part of the ontology, and it is not possible to uniquely determine the exact sequence of structural changes that took place. We developed a set of heuristics to compare two snapshots of the ontology and output a set of operations that could change the earlier snapshot into the later one. Of course, there is no guarantee that this is exactly the same sequence of operations that was actually performed by the human editors of the ODP ontology, as the same changes in the ontology can be effected through several different sequences of operations. In addition, the sequence of operations will depend on what set of elementary transformations one is willing to employ.

2.2 Low-Level Structural Changes

Changes in the ontology may be roughly divided into those that affect the categories (i.e. concepts) and those that affect the documents (i.e. instances). The latter group consists of the inclusion of new documents (links to external web pages), and removal and rearrangement of existing ones. We will not attempt to predict these document-level operations because, first of all, most of them cannot really be understood as causing structural changes in the ontology, and secondly, because it would be difficult to predict them without additional (and often unavailable) external data (e.g. to know whether something was removed due to becoming a dead link). Similarly, predicting the inclusion of new documents would require information about which web pages were available at a certain point in the past, so that they could have been discovered by the ODP editors and considered for inclusion in the ontology. However, we consider such questions to be outside the scope of this paper.

Thus we will focus on changes involving categories instead. In principle, one snapshot of the ontology can always be transformed into another one by a sequence of two elementary operations: addition and deletion of categories. By comparing the set of categories in one snapshot with the set of categories in the previous month's snapshot, it is easy to see which categories are missing and which are new. For example, the following list shows a subset of the changes that we may notice within the *Top/Computers* subtree between April 3 and May 1, 2007. "DEL" indicates that a category was deleted (i.e. it was present on April 3 but not on May 1) and "ADD" indicates that it was added (i.e. it was present on May 1 but not on April 3).

DEL Top/Computers/Open_Source/Software/Games/FPS
DEL Top/Computers/Software/Internet/Servers/Directory/LDAP
DEL Top/Computers/Software/Internet/Servers/Directory/LDAP/Products
DEL Top/Computers/Software/Internet/Servers/Directory/LDAP/Standards_and_Organizations
DEL Top/Computers/Software/Internet/Servers/Directory/LDAP/Products/Related_Middleware
DEL Top/Computers/Software/Internet/Servers/Directory/LDAP/Products/Related_Client_Apps
ADD Top/Computers/Open_Source/Software/Games/Shooter
ADD Top/Computers/Programming/Languages/Smalltalk/Squeak/Croquet
ADD Top/Computers/Programming/Languages/Smalltalk/Squeak/Croquet/News_and_Media
ADD Top/Computers/Internet/Protocols/LDAP
ADD Top/Computers/Internet/Protocols/LDAP/Standards_and_Organizations
ADD Top/Computers/Internet/Protocols/LDAP/Software/Client
ADD Top/Computers/Internet/Protocols/LDAP/Software/Server
ADD Top/Computers/Internet/Protocols/LDAP/Software

As we can see from this list, it is unsatisfactory to describe the transformation of one snapshot to another solely through these two types of low-level operations. Although one can in principle transform the April snapshot to the May snapshot by deleting the first six categories and then adding the next eight ones, it is clear that the human editors working on the ontology must have really conceptualized their work as a sequence of more abstract operations, each of which may then be manifested in one or more low-level additions and deletions of the type from list above.

In our example, we can see that the removal of concept *.../FPS* and the addition of concept *.../Shooter* are really two related operations: in other words. "FPS" has simply been renamed "Shooter" (note that FPS is itself nothing but an acronym for "first-person shooter", a genre of computer games). Similarly, the deletions and additions related to concept LDAP show us that the whole *LDAP* subtree has been moved from *.../Internet/Servers/Directory* to *.../Internet/Protocols*. Additionally, the *Products* subtree has been renamed into *Software* and rearranged somewhat.

Finally, the *.../Squeak/Croquet*, with its *News_and_Media* child, is a genuinely new subtree; as it turns out, it also contains genuinely new documents that did not exist at all in the previous snapshot of the ontology.

There also exist other types of structural changes not illustrated by the above example. One typical ODP phenomenon, which accounts for many additions of new categories, is the creation of new subcategories in advance, without populating them immediately with documents. The human editors of the ODP sometimes create new subcategories corresponding to all the letters of the alphabet to break down categories containing long lists (e.g. of companies); similarly, they sometimes add a set of subcategories corresponding to U.S. states, or import whole subtrees of zoological or botanical taxonomy; in all these cases, many of the newly created categories are initially empty. Such changes depend too strongly on background knowledge and high-level abstract decisions by a human editor to be predictable by a computer. Therefore, our efforts to predict structural changes will focus on situations when a new category has been added and some documents from previously existing categories transferred into it; this suggests that the structural change in question was genuinely an editor's response to the available data, and it may therefore be predicted automatically given the same data. An example may be that an editor decides that a category has too many documents and is too diverse, and it may therefore be split into several subcategories, with the documents then divided among these subcategories.

2.3 Heuristics for the Identification of Higher-Level Structural Changes

As we have seen in the previous section, low-level additions and deletions of categories can be easily observed by comparing two snapshots of the ontology, but the really interesting operations are more abstract and each such operation can give rise to several low-level additions and deletions. In addition, many such operations can take place in the period of time (e.g. a whole month) between two snapshots, and where several such operations affect the same part of the ontology, it can be difficult to identify the abstract operations given the set of low-level additions and deletions that can be discerned from the data. Thus, it is helpful to develop reasonably robust heuristics that can identify at least some of these higher-level operations, with the understanding that we cannot expect them to correctly identify them in all situations.

As it turns out, the largest group of low-level additions and deletions are actually due to the renaming of categories (e.g. *FPS* to *Shooter* in the example in the previous section). If a category with many descendants is renamed, this may manifest itself as a large number of low-level additions and deletions. Although this is a common operation, these are not really structural changes and so we want to recognize them and exclude them from further consideration. For this purpose we use a heuristic based on the notions of precision and recall from information retrieval. Given a category C from the old snapshot that does not appear (under the exact same name) in the new snapshot, we consider the set S of all documents from this category and its descendants (in the old snapshot). For each category C' of the new snapshot, we can similarly form a set S' of all documents of this category and its descendants. Then the recall and precision of C' with respect to C can be defined as $|S \cap S'|/|S|$ and $|S \cap S'|/|S'|$, respectively. If C' is to be recognized as a new incarnation of C (under a new name), it should ideally have high recall and high precision as well. In information retrieval, precision and recall are traditionally combined into a value called the F_1-measure, which is simply the harmonic mean of precision and recall: $F_1 = (2 \cdot precision \cdot recall) / (precision + recall)$. As the harmonic mean, F_1 is high only if both precision and recall are high. Insisting on a high recall is obviously desirable, but a good argument can be made for requiring high precision as well.

Note that it is possible that new documents were introduced into the ontology in the time between the old and the new snapshot, and some of these documents may have ended up in C'; these would increase the size of S' but not of $S \cap S'$ (as they did not appear in the old snapshot), whereby decreasing the precision. Thus, to prevent such new documents from unfairly affecting the match between C and C', we take into S' only those documents that have already existed in the old snapshot.

Thus, for each deleted category from the old snapshot, we find its best match (i.e. the one with maximal F_1) in the new snapshot. In principle, it is possible that there is no really good match, e.g. if the category and its documents were really deleted from the ontology, rather than simply renamed. In our experience, such deletions are rare; however, since we often work with just a part of the whole ontology for reasons of faster experimentation (e.g. just the subtree rooted in *Top/Computers*, etc.), it can happen that a category is moved outside of the part of the ontology that is under consideration, which is from our viewpoint the same as if it had been deleted entirely.

For the purposes of detecting the renaming and moving of categories, we consider only matches with a recall of at least 90%. We will refer to these as "strong matches". (The purpose of using a 90% threshold is to enable us to still track the identity of a category even if a small percentage of its documents were moved elsewhere or deleted. Whether a different percentage than 90% would lead to better performance would be an interesting subject for future experimental work.) The next step is to combine the matches on the level of categories into matches on the level of entire subtrees. For example, if a deleted category *Top/A/B/C*, with children *Top/A/B/C/D1* and *Top/A/B/C/D2*, is found to match strongly with a new category *Top/E/C'*, and furthermore its two children match strongly with two new categories *Top/E/C'/D1'* and *Top/E/C'/D2'*, then it is reasonable to refer to this as a move operation on the entire subtree rooted in *Top/A/B/C*, rather than as a set of operations that happened individually and separately to *C*, *D1* and *D2*. In general, the subtree rooted in *C* may be deeper (i.e. there may be grandchildren and other descendants in addition to just children), so the heuristic we actually use is the following. We say that there is a strong match between the subtree rooted by *C* (in the old snapshot) and the one rooted by *C'* (in the new snapshot) if the following two conditions are met: (1) For each descendant *D* of *C* (in the old snapshot), there must exist an strong match $sm(D)$ (in the subtree rooted by *C'* in the new snapshot); and (2) furthermore, for each such *D* we require that $parent(sm(D)) = sm(parent(D))$. In other words, we consider a strong match between subtrees to exist in cases when a strong match exists for each category in the subtree and the matches preserve the parent-child relationships. At the same time, our definition is robust in the sense that the addition of new categories into the subtree rooted by *C'*, or the merging of several old categories into a new one, does not prevent us from recognizing the strong match between the subtrees.

The strong matches between entire subtrees, once they have been identified, are a good first step towards the identification of higher-level structural changes:

1. **Rename:** If the subtree of *C* (in the old snapshot) strongly matches the subtree of *C'* (in the new snapshot), and *C'* did not exist in the old snapshot, and *C* and *C'* have the same parent, and no other subtree of the old snapshot strongly matches that of *C'*, then we say that *C* has been renamed into *C'*.

2. **Move:** If the same conditions are true except that *C* and *C'* do not share the same parent, we say that *C* has been moved to become *C'*.

3. **Merge:** If, on the other hand, *C'* has already existed in the old snapshot or it is new but some other subtree besides that of *C* has strongly matched the subtree of *C'*, then we say that *C* has merged into *C'*. Sometimes a category may merge into its parent, for example if the editor has decided that the previous subdivision was excessively fine-grained and the topics represented by the categories were too

narrow. On the other hand, sometimes a category merges into some more distant relative rather than a parent. It can also happen that several categories merged.

As an example, the chart in Figure 1 shows the frequency of these various types of higher-level ontology changes within the entire ODP ontology, over the last three years. As described above, all the category deletions that have been observed as low-level structural changes have now been explained as either renames, moves, or merges, with merges further divided into many-to-one merges, merges into parent and merges into other (nonparent) categories. What remains are the additions of genuinely new categories, rather than categories which appear new but are included in a strong subtree match with some formerly existing category (meaning that they are really the result of a rename, move or merge). It can be seen that additions are by far the most frequent structural changes, followed by renames and moves. Merges are comparatively rare. Since it is debatable to what extent a rename can be considered a truly structural change, and since moves are already fairly rare relative to the additions, we decided to concentrate on additions from now on as the most important and most frequently occurring type of structural change in the ODP ontology.

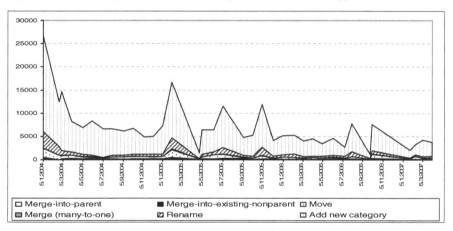

Fig. 1. Frequency of various types of ontology changes

2.4   Different Types of Category Additions

As we saw on Figure 1, the addition of new categories is the most common type of structural change, even after we exclude the categories that seem to be new but are really just old categories that have been renamed, moved or merged. In this section we will look at the additions of new categories in more detail. If we take the total over the entire three-year period covered by Figure 1, we find that there were 65105 category additions within the ODP hierarchy during this period. Figure 2 shows how these additions can be divided into several kinds.

Sometimes what is added is not just a simple leaf node of the tree but a whole sub-tree, consisting of a category and one or more children and possibly other descendants as well. Thus it turns out that approx. 34% of the newly added categories had a parent that was also newly added at the same time (or at least within the same snapshot). We will not attempt to predict the addition of such categories, as it is challenging enough to predict the addition of an individual category, much less of a whole subtree. Thus the remaining groups of additions discussed in this subsection consist of new categories added to a previously existing parent.

Approximately 11% of the new categories were empty, i.e. they contained no documents at all. As has been discussed in the previous section, these are mostly caused by systematic additions of large groups of sibling categories, e.g. corresponding to U.S. states or to letters of the alphabet. Approximately 19% of the new categories are not empty, but they contain only documents that did not exist in the ontology at the previous point in time for which a snapshot is available. This suggests that the category has been added on the basis of external web pages that were included in the ontology (e.g. the Croquet example Section 2.2).

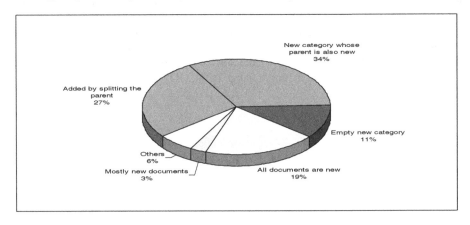

Fig. 2. Frequency of various types of category additions

Approximately 27% of the new categories could reasonably be said to have been obtained by splitting a previously existing parent category. This group of additions was defined as follows: the new category (e.g. *C*) must be the child of a previously existing parent (e.g. *P*); it must contain at least one document from the old snapshot of the ontology (although it may also contain zero or more new documents), and of these documents from the old snapshot, the majority must come from *P* or one of its descendants, rather than from some other part of the ontology that does not lie below *P*. In other words, to consider an addition to be a split of an existing category, we require that the new child adopts more documents from the parent than from other parts of the old ontology. Unfortunately most of the categories added in this way are fairly small; less than a third of them contain at least five documents from *P*.

Finally, the remaining additions result in categories that contain some mixture of old documents from *P*, old documents from other parts of the ontology, and entirely new documents. In approx. a third of these (3% of all additions), the new documents predominate; in the others most of the documents are from the old ontology, but with those from *P* outnumbered by those from outside of *P*. These new categories are thus obtained by a combination of new data (web pages newly included in the ODP directory) and of rearranging and moving of existing data (previously existing documents), and it is not clear that they can be characterized in any unified way.

2.5 Prediction of Category Additions as a Learning Problem

One can treat the problem of predicting category additions as a machine learning problem. Each example of the learning problem consists of a category and a point in time; the question to be answered is whether a new subcategory should be created below the given category at the given time. Thus, this is a binary (two-class) classification problem, with the positive class consisting of those examples where the addition of a child category is necessary, and the negative class consisting of those where it isn't.

The main open question at this point is how to describe each example by a set of features (or attributes) such that the resulting representation will be suitable as an input for a machine learning algorithm. The features should contain information that is relevant for making a decision whether a subcategory is needed or not. As discussed in Section 2.4, we ignore those additions of subcategories that are clearly based on background knowledge external to the ontology itself; the remaining additions must therefore be based at least partly on the actual contents of the ontology, i.e. the documents in the category below which a new subcategory is going to be added. Our approach is based on the idea that the human editors of the ODP probably suggest the addition of a new subcategory when they notice, within an existing category C, a few documents dealing with a reasonably well-defined narrower subtopic of the general topic of C. In this case a new subcategory would be added as a child of C, and the documents dealing with the subtopic thus identified would be moved into the new subcategory (whereas they had previously resided in C or possibly in one of its descendants). Since these documents all deal with a relatively narrow subtopic, one would hope that they are closely related to one another, use similar terminology, etc. If we represent them as points in a multidimensional space, we would expect to find them relatively closely together, closer than the average distance over all documents from C (which, covering a somewhat wider topic, would be expected to be dispersed more widely in space). To express this using data mining terminology: we would expect the documents of the new subcategory to form a cluster within the set of all documents of the parent category C. Thus, we turn to clustering as a technique that will help us assess whether such subsets of tightly related documents actually exist.

First, we represent each document by a TF-IDF vector (normalized to unit length), as is usually done in information retrieval and text mining. The cosine of the angle between two such vectors can then be used as an approximate measure of similarity between the two documents.

Now consider the set of all documents that have been assigned to a category C or any of its descendants. This is the set within which we would like to find any tightly coupled cluster; this would help us decide whether any new subcategories should be introduced below C. We will use the well-known k-means clustering algorithm [7] with $k = 2$ to split our set of documents into two clusters, then apply it recursively to each cluster. (Using $k = 2$ is convenient because it avoids the issue of choosing the number of clusters into which to split a particular cluster at a particular point.) We use the following termination criteria: we stop when there are 10 clusters, we do not try to split clusters containing less than 5 documents, nor do we split a cluster if it turns out that one of its two resulting subclusters would contain just one document.

Let A be the initial set of documents in the category C (and its descendants), and let $P = \{B_1, ..., B_k\}$ be the partition of A into k disjoint clusters obtained by the hierarchical 2-means algorithm. We will use the following features to describe this partition:

(1) One feature is the average cosine between each document and the cluster to which it belongs: $(1/|A|) \sum_{B \in P} \sum_{x \in P} \cos(\mathbf{x}, centroid(B))$. This is a measure of how tight clusters we have obtained by partitioning the initial set A into k clusters.

(2) Find the cluster with minimum variance: $B = \arg \min_{B' \in P} var(B')$, and use as features the following properties of this cluster:

- The size of this cluster. Instead of using $|B|$ directly, we use $\log |B|$, to prevent large clusters from having an excessive influence on the range of this feature.
- The relative size of this cluster, i.e. $|B| / |A|$.
- The variance of this cluster, $var(B)$.
- The variance of this cluster, relative to that of the whole set: $var(B) / var(A)$.

(3) Find the cluster with the maximum average intra-cluster similarity:
$$B = \arg \max_{B' \in P} \sum_{x, y \in B'; x \neq y} \cos(\mathbf{x}, \mathbf{y}) / (|B'| \cdot (|B'| - 1)).$$

For this cluster B, we use four features analogous to those described above in (2) for the minimum-variance cluster. The idea here is that the average intra-cluster similarity is another measure of cluster compactness, and these features may therefore help the classifier identify categories with a compact subset of documents that would be a suitable basis for creating a new subcategory.

In this way we have described the partition P by nine features. Every time that our hierarchical clustering algorithm splits a cluster, the partition changes (one of its clusters gets replaced by two smaller ones), and we add, to the feature vector for the category under observation, the nine features describing the new partition. We let the clustering continue until there are ten clusters, which means that in the end the category is described by a 90-dimensional feature vector. (If the clustering algorithm stops before ten clusters have been obtained, we repeat the features of the final partition as many times as necessary to bring the feature vector to the full 90-dimensions.) The resulting feature vectors can be used as the input into a machine learning algorithm; we used the support vector machine (SVM [4]), as it is a well-known and state-of-the-art learning method that has been found to perform well in many areas, including on tasks with a considerable number of features and training examples. Since the feature space used in our representation is relatively modest (90 features, as opposed to e.g. thousands of features as is commonly the case in text and image categorization settings), we decided to use the radial basis function (RBF) kernel rather than a simple linear kernel.

3   Experimental Evaluation

3.1   The Dataset

In this section we describe our experimental evaluation of the proposed approach for the prediction of category additions. We used the *Computers* subtree of the Open Directory Project ontology. In the period under consideration, i.e. from January 2004 through October 2006 (there being no snapshots of the ODP from November and December 2006), the Computers subtree grew from 7,732 categories to 8,309 categories, while the number of documents on average tended to decrease rather than increase, eventually shrinking from 143,760 documents in January 2004 to 133,595 documents in October 2006.

During this period, there were 964 category additions, 198 category renames, 134 category moves, and 153 merges of various types. The relative frequency of different types of structural changes was similar to that shown on Figures 1 and 2 for the entire ODP ontology. Of the category additions, 482 were such that the new category is added as the child of a previously existing parent category and more documents have been moved into the new category from the parent (or its previously existing descendants) than from other parts of the hierarchy. This, as described in Section 2.4, is the type of additions that we will be trying to predict. However, it turns out that even in these cases, the number of documents moved from the parent to the new child category is often quite small. The hypothesis underlying our approach is that the human editors of the ODP notice, in an existing category, a group of documents dealing with some narrower subtopic and then decide to create a new subcategory and move those documents into it. Thus our approach can not be reasonably expected to perform well in situations where only e.g. one or two documents have been moved from the parent to the new child, since in this case there is effectively no subgroup of closely related documents that could have been detected in the old parent category (and then be used to predict an addition). Therefore, for the purposes of defining our classification problem, we limit ourselves to the additions of categories in which at least five documents were moved from the parent category into the new child category (in addition to these, the new category may also contain documents that came from elsewhere). This leaves us with 107 category additions as the basis for our prediction task.

The question that our predictive model will attempt to answer is this: "given a category, should any new subcategories be added below it, as its children, during the next month?" Since there are two possible answers to this question, yes or no, this will be a binary (i.e. two-class) classification problem. A category at a given point in time is a positive example if some children (matching the criteria described above) have indeed been added to it between that point and the next point in time for which an ontology snapshot is available (i.e. approximately one month later). According to this definition, the above-mentioned 107 additions give rise to 98 positive examples (this is less than the number of additions because sometimes several children are added to the same parent in a certain month).

But when is a category a negative example? For example, suppose that a comparison of the snapshots for March 2005 and April 2005 shows that no suitable children have been added to category C in the intervening period, but the comparison of the snapshots for April 2005 and May 2005 shows one such addition. This suggests that the category C such as it was in April 2005 is a positive example for the purposes of our machine learning problem; but is it reasonable to say that C such as it was in March 2005 is a negative example, just because no additions were made to it between March and April? The category C has not necessarily changed much from March to April; perhaps the ODP editors would have already made the addition to C in March rather than in April, but they simply hadn't yet noticed that there exists a compact subgroup of documents that can become a new subcategory. Therefore, to avoid having an excessively narrow definition of the negative set, we declare a category to be negative at a certain point in time only if no suitable children have been added to it at that point or in the preceding or following three months. Despite this constraint, the vast majority of categories are treated as negative examples at any particular point in time, since the category additions are rare relative to the total number of categories. In total, we could obtain more than 168,000 negative examples from the *Computers* subtree in the period 2004–2006. To speed up the experiments and to prevent the positive examples from being completely overwhelmed by the negative ones during the training process, we randomly selected three times as many negative examples as there are positive examples. We then divided the resulting data into a training set (all examples from the years 2004 and 2005) and a test set (all examples from the year 2006). Thus, we end up with a training set containing 74 positive and 222 negative examples, and a test set containing 24 positive and 72 negative examples.

3.2 Experimental Setup

As has been discussed in Section 2.5, we will be using the SVM algorithm to train classifiers. We use the SVM^{light} implementation of SVM by Thorsten Joachims [5]. Training an SVM requires one to set various parameters, in particular the error cost C. For relatively unbalanced datasets, i.e. those where the positive examples are heavily outnumbered by the negative ones, it is often beneficial to treat errors on positive training examples as more problematic than those on negative training examples. Thus, one effectively uses two different error costs: the baseline cost C on the negative and its multiple $j \cdot C$ on the positive examples.

Thus, C and j are two tunable parameters which we will select via five-fold cross-validation on the training set. A third tunable parameter is γ, the width of the Gaussian functions in the RBF kernel. We tested the following parameter values: $C \in \{0.1, 1, 10, 100, 1000\}$; $j \in \{1, 2, 3, 5, 10, 20, 50, 100\}$; and $\gamma \in \{0.0001, 0.0002, 0.0005, 0.001, 0.002, 0.005, 0.01, 0.02, 0.05, 0.1, 0.2, 0.5, 1, 2, 5\}$. For most combinations of these parameter settings, training a model cost less than a second; most of the processor time was actually spent on generating the features.

To evaluate the output of the classifiers, we use well-known evaluation measures from the area of information retrieval [2]: the precision-recall breakeven point (BEP) [3] and the area under the ROC curve (AuROC) [1]. Both measures yield values in the range 0 to 1, with higher values indicating better performance. As a baseline, a model that ranked the examples in random order would achieve a breakeven point equal to the proportion of the positive examples relative to all examples (which is 0.25 for our dataset), and the area under its ROC curve would be 0.5. A perfect model would achieve a score of 1 according to both measures.

3.3 Results

We used stratified 5-fold cross-validation (CV) on the training set to investigate the influence of C, j, and γ parameters. As described in Section 3.2, we investigated 5 values of C, 8 values of j and 15 values of γ. This results in 600 combinations of parameter settings. We then select the combination that performed the best during cross-validation on the training set; using this combination of parameter settings, we train

Table 1. Performance of models selected with various model selection criteria

Model description	Performance on the validation set during 5-fold CV		Performance on the test set	
	BEP	Au-ROC	BEP	Au-ROC
Max. BEP during CV	0.5148	0.7796	0.7083	0.8893
Max. a.u.ROC during CV	0.5021	0.7850	0.6667	0.8738
Max. BEP on the test set	0.4717	0.7436	0.7500	0.9011
Max. auROC on the test set	0.4768	0.7495	0.7500	0.9155
Random ranking	0.2500	0.5000	0.2500	0.5000

the final model on the entire training set, and this model would then be evaluated on the test set. The results are summarized in the following table:

The first row, "highest BEP during CV", refers to the models having the greatest breakeven point during cross-validation. There were three models (i.e. three different combinations of parameter settings) with the maximum BEP here, so the other columns of the table show average performance over these three models. The same approach has been used in the other rows.

The rows referring to the highest BEP/AuROC on test set indicate what the best models among those tested here are capable of, with the caveat that we aren't able to identify these models without peeking at the test data. Comparing these results with the results from the first two rows tells us how much room for improvement there is if we can select our models using some better criterion than cross-validation on the training set. We can see that the difference is not really very large here, and by selecting our models through cross-validation we obtain models that also perform quite well on the test set.

For comparison, the last row of the table shows the performance of a hypothetical model that doesn't learn anything and instead just outputs random scores for all the examples.

Parameter tuning. Until now we have been looking for the best combination of parameter settings by allowing all three parameters to vary – C, j, as well as γ. But what if we hold one of these parameters fixed at some specific value and then examine only the models obtained by varying the other two parameters? Fig. 4 shows the results. For each value of each parameter, we select the other two parameters so as to maximize the AuROC measure during cross-validation. We report this AuROC value, as well as the AuROC achieved by the same combination of parameters on the test set. Regarding the j parameter, the best results were achieved with $j = 3$, which is intuitively reasonable: since there are three times as many negative examples as there are positive ones, it makes sense to treat errors on the positive examples as three times more serious than the errors on negative examples. Observations on C and γ are less useful because the optimal values of these parameters depend strongly on the properties of the dataset and the kernel settings. In general, we can say that for both C and γ there is actually a fairly broad range of parameter values where good performance can be achieved. The results for C are somewhat surprising because there, CV appears to mislead us: the larger values of C lead to poorer performance during cross-validation but actually better performance on the test set.

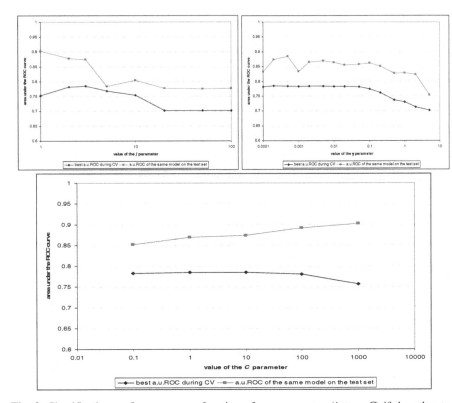

Fig. 3. Classification performance as a function of one parameter (j, γ or C) if the other two parameters are tuned using fivefold cross-validation

4   Conclusions and Future Work

In this paper we have described an approach for predicting a subset of structural changes in an ontology. Our approach aims to predict the addition of categories within a hierarchy of documents, under the assumption that the new category is a child of an existing parent category and that it contains at least a few documents that were formerly members of the parent category. We have described how this task can be formulated as a machine learning problem and presented experiments that show that the prediction of this type of changes is feasible.

There are several directions along which this work could be extended. A possible next step is to try to devise more features for our current machine-learning task. Additionally, existing descendants of a category could be taken into account: if a cluster of documents corresponds nicely to an existing descendant of a category C, it should not be taken as evidence that C needs a new child.

Another interesting extension would be to try predicting not just whether the addition of a new category is warranted, but also which documents it should include, and perhaps which keywords it should be described by. In addition to this, it would be interesting to try suggesting other types of ontology changes: additions of categories that contain mostly new documents (e.g. by trying to classify them into existing categories, and recommending a new category when this fails), and moving and merging of existing categories (e.g. by comparing the centroids of categories to see if another category is closer than the current parent).

It would be interesting to also address the operation of category renaming, since it is so frequent. One might train a model that works on individual terms and predicts how well they describe the category. If terms from the current name score poorly, recommend a rename. It might be necessary to use WordNet to find the suitably abstract terms for upper-level categories.

Acknowledgments

This work was supported by the Slovenian Research Agency and the IST Programme of the European Community under NeOn Lifecycle Support for Networked Ontologies (IST-4-027595-IP) and PASCAL Network of Excellence (IST-2002-506778). This publication only reflects the authors' views.

References

1. Provost, F., Fawcett, T.: Robust classification for imprecise environments. Machine Learning 42(3), 203–231 (2001)
2. van Rijsbergen, C.J.: Information Retrieval. Butterworth (1979)
3. Lewis, D.D.: Representation and Learning in Information Retrieval. Ph.D. Thesis, Univ. of Mass, Amherst, USA (1991)
4. Cortes, C., Vapnik, V.: Support-Vector Networks. Machine Learning 20(3), 273–297 (1995)
5. Joachims, T.: Text categorization with support vector machines: Learning with many relevant features. In: Proc. ECML 1998, Chemnitz, Germany, April 21-23, 1998, pp. 137–142 (1998)
6. Su, T., Dy, J.G.: In search of deterministic methods for initializing K-means and Gaussian mixture clusterind. Intelligent Data Analysis 11(4), 319–338 (2007)
7. Steinbach, M., Karypis, G., Kumar, V.: A comparison of document clustering techniques. In: Proc. KDD Text Mining Workshop (2000)
8. Cimiano, P., Völker, J.: Text2onto – a framework for ontology learning and data-driven change discovery. In: Proc. NLDB 2005 (2005)
9. Haase, P., Sure, Y., Völker, J.: Management of dynamic knowledge. J. of Knowledge Management 9(5), 97–107 (2005)
10. Haase, P., Stojanović, L.: Consistent evolution of OWL ontologies. In: Proc. ESWC (2002)
11. Haase, P., van Harmelen, F., Huang, Z., Stuckenschmidt, H., Sure, Y.: A framework for handling inconsistency in changing ontologies. In: Proc. ESWC (2005)
12. Stojanović, L.: Methods and Tools for Ontology Evolution. PhD thesis, University of Karlsruhe (2004)
13. Völker, J., Sure, Y.: Data-driven change discovery. Deliverable 3.3.1, SEKT Project (EU IST-2003-506826) (July 22, 2005)
14. Maedche, A., Motik, B., Stojanovic, L., Studer, R., Volz, R.: Ontologies for Enterprise Knowledge Management. IEEE Intelligent Systems (January/February 2003)
15. Flouris, G., Plexousakis, D., Anto, G.: A Classification of Ontology Change. In: Proceedings of SWAP 2006, the 3rd Italian Semantic Web Workshop, Pisa, Italy, Dec. 18-20 (2006)
16. Maynard, D., Peters, W., d'Aquin, M., Sabou, M., Aswani, N.: Dynamics of Metadata. Deliverable 1.5.1, NeOn Project (EU IST-2005-027595). March 30 (2007)
17. Grobelnik, M., Mladenic, D.: Simple classification into large topic ontology of Web documents. Journal of Computing and Inf. 13(4), 279–285 (2005)

Catriple: Extracting Triples from Wikipedia Categories

Qiaoling Liu[1], Kaifeng Xu[1], Lei Zhang[2], Haofen Wang[1], and Yong Yu[1],
and Yue Pan[2]

[1] Apex Data and Knowledge Management Lab
Shanghai Jiao Tong University, Shanghai, 200240, China
{lql,kaifengxu,whfcarter,yyu}@apex.sjtu.edu.cn
[2] IBM China Research Lab
Beijing, 100094, China
{lzhangl,panyue}@cn.ibm.com

Abstract. As an important step towards bootstrapping the Semantic Web, many efforts have been made to extract triples from Wikipedia because of its wide coverage, good organization and rich knowledge. One kind of important triples is about Wikipedia articles and their non-isa properties, e.g. (Beijing, country, China). Previous work has tried to extract such triples from Wikipedia infoboxes, article text and categories. The infobox-based and text-based extraction methods depend on the infoboxes and suffer from a low article coverage. In contrast, the category-based extraction methods exploit the widespread categories. However, they rely on predefined properties, which is too effort-consuming and explores only very limited knowledge in the categories. This paper automatically extracts properties and triples from the less explored Wikipedia categories so as to achieve a wider article coverage with less manual effort. We manage to realize this goal by utilizing the syntax and semantics brought by super-sub category pairs in Wikipedia. Our prototype implementation outputs about 10M triples with a 12-level confidence ranging from 47.0% to 96.4%, which cover 78.2% of Wikipedia articles. Among them, 1.27M triples have confidence of 96.4%. Applications can on demand use the triples with suitable confidence.

1 Introduction

Extracting as much semantic data as possible from the Web is an important step towards bootstrapping the Semantic Web. Many efforts have been made to extract triples from Wikipedia because of its wide coverage of domains and good organization of contents. More importantly, Wikipedia embraces the power of collaborative editing to harness collective intelligence, which results in rich knowledge from its articles, categories and infoboxes. Table 1 shows the volume of the rich knowledge contained in English Wikipedia[1]. One kind of important triples is about Wikipedia articles and their non-isa properties, e.g. (Beijing,

---

[1] The data used in this paper is from English Wikipedia database dump on 2008-1-3.

J. Domingue and C. Anutariya (Eds.): ASWC 2008, LNCS 5367, pp. 330–344, 2008.
© Springer-Verlag Berlin Heidelberg 2008

Table 1. Some Statistics of Wikipedia

Articles	Articles with Infobox	Articles with Category
2,390,513	1,057,563 (44.2%)	1,927,525 (80.6%)
Categories	Category-Category Pairs	Article-Category Pairs
312,422	577,579	6,136,876

country, China). As each article corresponds to an entity, such triples capture the non-isa properties of entities, thus are quite useful to the Semantic Web.

Despite the rich knowledge contained in Wikipedia, difficulties exist in extracting such triples. The key challenge is to extract properties and values for the articles. Previous work has tried to extract them from Wikipedia infoboxes, article text and categories [4,11,12,14]. The infobox-based extraction method [4] took advantage of user-edited properties and values in the Wikipedia infobox templates. As an enhancement, the text-based extraction method [14] extracted more values from the article text according to cleaned infobox properties. However, their problem is the low article coverage. As shown in Table 1, only 44.2% articles have infoboxes. In contrast, the Wikipedia category system has a much higher coverage of 80.6%. Therefore, triples extracted from the categories tend to be more diverse concerning the entities. However, previous category-based extraction methods [11,12] relied on manually specifying regular expressions corresponding to each predefined relation to match category names. This is too effort-consuming, especially considering the large number of relation types. As a result, a large portion of knowledge in the categories remains unexplored.

In this paper, we focus on *automatically* extracting properties and triples from the less explored Wikipedia categories so as to achieve a wider article coverage with less manual effort. The key challenge is then how to automatically extract for a category the property and value shared by its articles. Given a single category, e.g. "Category:Songs by Pat Ballard", it is very difficult for machines to understand that the songs in this category (e.g. "Mr. Sandman") are written by Pat Ballard, let alone to automatically extract any property. Fortunately, the hierarchical structure of Wikipedia categories provides hints for this. Given that "Category:Songs by Pat Ballard" has a supercategory "Category:Songs by songwriter", it is now much easier for machines to extract "songwriter" as property and "Pat Ballard" as value for those songs. Then, triples can be created as ("Mr. Sandman", songwriter, "Pat Ballard"). The syntax and semantics brought by super-sub category pairs makes it possible to automatically extract structured data from the unstructured category names.

Based on this idea, we accomplish an automatic extraction of triples from Wikipedia categories, which is beyond the ability of previous category-based extraction. Besides, it complements methods extracting triples from other sources such as infoboxes and text. Specifically, we make the following contributions:

- We originally observe the semantics of Wikipedia category pairs for extracting properties and values of articles. Two kinds of helpful category pairs are summarized: 1) category pairs containing both explicit property and explicit value, e.g. "Category:Songs by artist"-"Category:The Beatles songs", with "artist" as property and "The Beatles" as value; 2) category pairs containing explicit value but implicit property, e.g. "Category:Rock songs"-"Category:British rock songs", with "British" as value yet no property.

- We propose methods to automatically extract triples based on the two kinds of helpful category pairs. Natural Language Processing (NLP) technologies are employed to extract the explicit properties and values from the category names. To determine the implicit property in the latter case, we propose a voting strategy that resorts to the semantics of articles in a category. Finally, after reasoning out the complete articles belonging to a category using category hierarchies, triples are created with articles in the category, the extracted properties, and the extracted values.

- Our prototype implementation, the Catriple system, outputs about 10M triples with a 12-level confidence ranging from 47.0% to 96.4%, which cover 78.2% of Wikipedia articles. Among them, 1.27M triples have confidence of 96.4%. The extracted data can be used in many applications, e.g. realizing faceted search for Wikipedia, enriching Wikipedia infobox data, and refining the Wikipedia category system. Applications can on demand use the triples with suitable confidence.

The rest of the paper is organized as follows. Section 2 discusses the related work and Section 3 describes our extraction methods. Then, the experiment and evaluation of our methods are presented in Section 4. Finally, we discuss the pros and cons of our methods in Section 5 and conclude the paper in Section 6.

2   Related Work

There are several other systems extracting information about articles and relations from Wikipedia. Auer et al. initiated the DBpedia project that originally extracted information from the infoboxes and encapsulated it in triples for advanced queries [5]. The project successfully extracted 18M infobox triples, after being further developed [4]. However, the fact that only about 44.2% articles have infoboxes results in that only a minor portion of articles are covered by these infobox triples. Contrary to DBpedia, Catriple relies on categories that cover about 80.6% articles to extract triples. Wu et al. described the Kylin system [14] which enhanced the infobox data by extracting triples from article text. Different from Catriple in which properties are extracted from categories, Kylin depends on cleaned infobox schemata to determine the properties for triple extraction. The authors also made further efforts to create a clean infobox ontology [15] so as to better organize the extracted triples. Nguyen et al. extracted relations from articles by utilizing the syntactic and semantic information [7]. However, it only focused on a limited number of predefined relations between entities. Other studies about extracting relations from article text include [13,6].

Suchanek et al. developed YAGO [11,12], a large ontology derived from the categories, infoboxes of Wikipedia and the taxonomic relations of WordNet.

Their work is the most similar to ours in respect of extracting triples from Wikipedia categories. However, obvious distinctions exist. First, YAGO primarily focuses on *isa* relations, while Catriple targets *non-isa* relations. Second, YAGO requires much human effort to specify the relations (e.g., locatedIn) and the corresponding patterns (e.g., "Mountains|Rivers in (.*)"), while Catriple automatically extracts relations from the categories. Third, YAGO uses information in individual categories for triple extraction, while Catriple additionally utilizes the semantics of category pairs to enable an automatic triple extraction.

When property and value are extracted for a category, triples can be created for the articles in the category with the same property and value. How about the articles in the subcategories? If the category system is a taxonomy with fully-fledged subsumption hierarchy, the property and value can also be applied to the articles in the subcategories. Otherwise, there will be risk to do that. Ponzetto et al. [8] found that Wikipedia categories actually do not form a fully-fledged subsumption hierarchy, but only a thematically organized thesaurus. They proposed methods to derive a large scale taxonomy from Wikipedia by distinguishing between *isa* and *non-isa* relations in Wikipedia's category network. Based on this Wikipedia Taxonomy, the authors further proposed an automatic method to distinguish instances and classes [18]. We directly use their publicly available results to create triples about articles in the subcategories.

Catriple utilizes the good features and rich semantics of the Wikipedia category system to extract triples. There are many other interesting applications of the category system. We list some representative ones here. Yu et al. [16] argued that the category system is equivalent to a simple ontology and conducted ontology evaluation based on it. Ponzetto and Strube [9,10] computed semantic relatedness by taking the category system as a semantic network. Zesch and Gurevych [17] conducted a graph-theoretic analysis of the category system and concluded that it can be used for NLP tasks.

3   Methods

To realize an automatic extraction of triples from Wikipedia categories, the key challenge is how to automatically extract for a category the property and value shared by its articles. Then triples can be easily created with articles in the category, the extracted property, and the extracted value. Hereinafter, we call such an extracted pair (property, value) a *label* of the category.

Inspired by the fact that people would learn a new category in Wikipedia through its category system, we observed the syntax and semantics of category pairs for automatically extracting labels for categories. We have discovered two kinds of helpful category pairs: 1) category pairs containing both explicit property and explicit value, e.g. "Category:Songs by artist"-"Category:The Beatles songs". It is easy to extract "artist" as property and "The Beatles" as value for the subcategory; 2) category pairs containing explicit value but implicit property, e.g. "Category:Rock songs"-"Category:British rock songs". It is easy to extract "British" as value for the subcategory, yet without property.

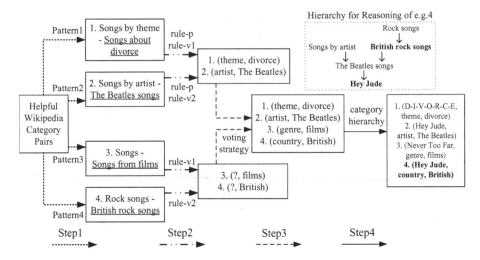

Fig. 1. Workflow of Catriple

Based on this observation, we propose methods for an automatic category-based extraction of triples. Fig. 1 shows the workflow. First, we analyze the category names using NLP technologies and summarize some helpful patterns of category pairs. Second, rules are proposed w.r.t. these name patterns to extract the explicit properties and values for categories. Third, to determine the implicit property given explicit value for a category, we propose a voting strategy that resorts to the semantics of articles in the category. Finally, when creating triples about articles given a category with the extracted property and value, we leverage category hierarchies to infer the complete articles belonging to the category. In the following, we will describe each step in a subsection in more detail.

3.1 Step1: Recognizing Useful Name Patterns

To automatically identify the helpful category pairs, we analyze the category names using NLP technologies and induce some useful name patterns.

First, we pick out property-contained category names (PCCN) and value-contained category names (VCCN). By observing the Wikipedia category system, we find that users tend to organize many category pairs using patterns "X BY Z"-"Y X" (e.g. "Category:Songs by artist"-"Category:The Beatles songs") and "X BY Z"-"X PREP Y" ("Category:Songs by theme"-"Category:Songs about divorce"). These patterns were also identified as expressing is-refined-by relations between categories and were discarded for deriving the subsumption hierarchy [8]. However, they have better structured the category system, thus are very useful in our task. The by phrase can act as an indication of PCCN (e.g. "Category:Songs by artist" and "Category:Songs by theme", with "artist" and "theme" as properties). The noun phrase or prepositional phrase can act as an indication of VCCN (e.g. "Category:The Beatles songs" and "Category:Songs about divorce", with "The Beatles" and "divorce" as values). Then, helpful category pairs can be recognized using the following name patterns:

- Pattern1: by-prep, which means the supercategory is a PCCN with by phrase and the subcategory is a VCCN with prepositional phrase, e.g. "Category: Songs by theme"-"Category:Songs about divorce";
- Pattern2: by-noun, which means the supercategory is a PCCN with by phrase and the subcategory is a VCCN with noun phrase, e.g. "Category:Songs by artist"-"Category:The Beatles songs";
- Pattern3: *-prep **except** by-prep, which means the subcategory is a VCCN with prepositional phrase and the supercategory is not a PCCN, e.g. "Category:Songs"-"Category:Songs from films";
- Pattern4: *-noun **except** by-noun, which means the subcategory is a VCCN with noun phrase and the supercategory is not a PCCN, e.g. "Category:Rock songs"-"Category:British rock songs";

Note that the usefulness of these four patterns depends on the correctness of the subsumption relationships between the category pairs. Yet, category pairs in Wikipedia do not always have subsumption relationships, e.g. "Category:Songs"-"Category:Song forms". To guarantee the correctness, we restrict to pairs of categories sharing the same lexical head. This restriction also plays an important role in the value extraction for Pattern2 and Pattern4 in Step2.

To recognize the preposition for Pattern1 and Pattern3, we use OpenNLP [1] to get the part-of-speech tags of the category names first. The tagged prepositions are then checked by a list of creditable prepositions collected from Wikipedia. We believe this would lead to a higher precision.

To parse the lexical head of category names, we use both Stanford Parser [3] and OpenNLP [1] to provide a dual fail-safe result. Only noun head is accepted. Whenever encountering a non-noun head, we remove it from the category name and repeat the head finding process. Considering that a plural noun head may be wrongly identified as a verb, we first get the stem of a category name using Porter stemmer [2], and then feed it to Stanford Parser [3].

3.2   Step2: Extracting Explicit Properties and Values

According to the name patterns of category pairs in Step1, we propose the following rules to extract the explicit properties and values for the subcategory.

- Rule-p: for Pattern1 and Pattern2, extract the by phrase in the supercategory as property, e.g. "theme" in "Category:Songs by theme" and "artist" in "Category:Songs by artist";
- Rule-v1: for Pattern1 and Pattern3, extract the prepositional phrase in the subcategory as value, e.g. "divorce" in "Category:Songs about divorce" and "films" in "Category:Songs from films";
- Rule-v2: for Pattern2 and Pattern4, extract the extra modifier of the head in the noun phrase of the subcategory w.r.t. the supercategory as value, e.g. "The Beatles" in "Category:The Beatles songs" and "British" in "Category:British rock songs";

The basic idea of the rules is simple, yet a special case needs to be considered. In Rule-v1, the prepositional phrase in the subcategory may be a by phrase. Some by phrases are useful for value extraction, e.g. "Category:Songs by songwriter"-"Category:Songs by Richard Adler", while others are not, e.g. "Category:Songs by genre"-"Category:Rock songs by subgenre". It is easy for human to recognize the two cases, by understanding that "subgenre" is a property instead of a value. Yet it is much more difficult for machines. By looking into the content of the subcategory, we find that there will be mostly subcategories instead of articles, if the by phrase is a property. Besides, we can leverage existing properties in infoboxes to do a double check. Thereby, our strategy for this case is that we judge the by phrase useful for value extraction if two conditions are satisfied: 1) the number of its articles is larger than ten times of the number of its subcategories; 2) the by phrase is not contained in the properties collected from DBpedia infobox triples. In this way, "Richard Adler" is extracted successfully as a value (0 subcategories, 7 articles, and no such property in infobox), while "subgenre" is eliminated (16 subcategories and 0 articles).

3.3   Step3: Voting Implicit Properties

Since category pairs of Pattern1-2 contain both explicit properties and values, we can already derive labels for their subcategories after Step2. However, to derive labels for subcategories in Pattern3-4, we face a problem of discovering the implicit property given the explicit value for a category.

Considering that the implicit property and explicit value are shared by all articles in the category, we try to solve the problem by gathering the semantics of those articles through existing triples about them. We propose a voting strategy, which elects the property with the maximum frequency in existing triples that contain the given value and are about articles in the given category. Currently, the DBpedia infobox triples are leveraged in this voting process, since the size is very large and also rich properties and values are contained.

The best implementation for the voting strategy is to build a Local Value Pool, which is a database where key=category+value, data=property+frequency. In each data item, the frequency denotes the number of triples about articles in the category that contain the property and the value. Then, given category c and value v, we can easily get the property with the maximum frequency satisfying key=$c + v$ in the Local Value Pool.

However, this realization is too time- and space-consuming, so we use an approximate implementation that builds a Global Value Pool and a Local Property Pool. Literally, "Local" indicates the context of a certain category, while "Global" means no restriction. The Global Value Pool is a database where key=value, data=property+frequency. In each data item, the frequency denotes the number of triples about any article that contain the property and the value. The Local Property Pool is a database where key=category, data=property. Each data item denotes that there are triples with the property and about articles in the category. Then, given category c and value v, the frequency for a property satisfying key=v in the Global Value Pool becomes the property's votes. A vote is valid if the corresponding property exists in the Local Property Pool satisfying key=c. We choose the property with the maximum valid votes as result.

For example, from "Category:Songs"-"Category:Video game songs", we can extract the explicit value "Video game" for the subcategory. Then by querying the Global Value Pool with value "Video game", we can get the properties and the corresponding frequencies as S1={industry(39), products(32), genre(16), type(14), ...}. By querying the Local Property Pool with "Category:Video game songs", we can get the properties as S2={type, genre, title, format, ...}. Finally, "genre" is chosen as the result property because it has the maximum valid votes, i.e. the maximum frequency in S1 among the properties in S2.

This can be seen as a narrow voting. We further propose a broad voting that refines it by considering more contextual information. Note that in Pattern1-2, all the subcategories of a supercategory share a property, only with different values. By inspecting category pairs of Pattern3-4, we observe similar phenomenon in most cases. Therefore, we take this phenomenon as an *assumption*. Then, to determine the implicit property shared by all the subcategories of a supercategory, we first sum up for a property the valid votes from all the subcategories, and then choose the property with the maximum valid votes. This broad voting strategy would combine the valid votes from siblings to help discover the correct property for a category. An example showing its contribution is as follows:

- "Education in China"-"Education in Macau" | country(18) | Macau;
- "Education in China"-"Education in Macau" | city(7) | Macau;
- "Education in China"-"Education in Nanjing" | city(9) | Nanjing;
- "Education in China"-"Education in Suzhou" | city(2) | Suzhou;
- "Education in China"-"Education in Hangzhou" | city(6) | Hangzhou;

By narrow voting, given category "Category:Education in Macau" and value "Macau", the wrong property "country" will be chosen since it has the maximum valid votes of 18. However, by broad voting based on the assumption, the correct property "city" will be elected since it has the maximum valid votes of 24.

3.4   Step4: Creating Triples about Articles

Until now, we have derived labels for the subcategories in all the four patterns. Based on these labels, we can generate triples about articles. Yet a basic problem is to get the articles belonging to a category. Although Wikipedia contains more than 6M basic article-category pairs, different category hierarchies would lead to different results after reasoning out the complete article-category pairs.

For this issue, we compare three different hierarchies: Null Hierarchy (NH), Wikipedia Hierarchy (WH), and Wikipedia Taxonomy (WT). The Null Hierarchy denotes no category pair at all. The Wikipedia Hierarchy denotes the complete super-sub category pairs in Wikipedia. The Wikipedia Taxonomy denotes the category pairs with subsumption relationships produced by [8]. Table 2 shows their sizes.

Based on these hierarchies, we then infer the complete articles of a category. However, it is too time-consuming to do reasoning about articles. So, we do

Table 2. Category Hierarchies: Null Hierarchy (NH), Wikipedia Hierarchy (WH), and Wikipedia Taxonomy (WT)

	NH	WH	WT
Categories	256,499	312,422	121,256
Category-Category Pairs	0	577,579	148,646

reasoning about labels instead. First, we build a Label Pool, which is a database where key=category, data=property+value. Each data item denotes a label for the category. Then, the Label Pool will be enlarged by propagating labels along the category hierarchies. Finally, for each article, we get all the labels of its categories from the enlarged Label Pool and generate triples.

For example (as shown in Fig. 1), the label (country, British) of "Category:British rock songs" will be propagated to its subcategory "Category:The Beatles songs" by WH. The article "Hey Jude" will then get this label and form a triple ("Hey Jude", country, British).

4   Experiments

4.1   Evaluation

Based on the different choices in Step1 and Step4, i.e. four patterns and three category hierarchies, Catriple has actually produced 12 methods for final triple extraction. To evaluate the precision of the triples extracted by our methods (we call them Catriple triples), we use the manual evaluation method in [12] as a reference. For each evaluation, we randomly select 500 Catriple triples and manually judge whether they are correct. Note that Null Hierarchy is included by Wikipedia Taxonomy, and Wikipedia Taxonomy is included by Wikipedia Hierarchy. To get a more accurate investigation of their specific impacts on performance, we did our evaluation on the extra labels and extra triples generated by them, namely NH, WT−NH and WH−WT.

Table 3 shows the performance of each method. We can see that some methods achieve very good results, while some get only poor results. First, when looking horizontally, we can find that for all patterns, NH achieves the best performance,

Table 3. Evaluation Results

		NH	WT−NH	WH−WT
Pattern1	precision	0.938±0.019	0.886±0.030	0.746±0.028
	size	752,558	408,759	1,502,574
Pattern2	precision	0.964±0.023	0.864±0.017	0.638±0.029
	size	1,265,782	395,381	1,057,697
Pattern3	precision	0.880±0.022	0.792±0.029	0.470±0.029
	size	504,664	167,120	1,494,621
Pattern4	precision	0.794±0.024	0.692±0.027	0.594±0.024
	size	800,328	344,821	1,652,848

followed by WT−NH and then WH−WT. According to our analysis, errors in Step4 are due to three reasons: 1) Wrong labels created in Step1-3 are propagated to subcategories, e.g. a wrong label (country, science) created in Step3 by Pattern3 for "Category:Historiography of science" is propagated to its 8 subcategories and generates about 2000 incorrect triples by WH; 2) Correct labels are propagated to wrong subcategories, e.g. a correct label (topic, science) created in Step2 by Pattern1 for "Category:History of science" are propagated to its wrong subcategory "Category:Living People" and generates more than 5000 incorrect triples by WH; 3) Incorrect triples are created by wrong article-category pairs, e.g. a wrong triple ('Willow and Wind, director, Abbas Kiarostami) is created as a result of the wrong categorization of article "Willow and Wind" into "Category:Films directed by Abbas Kiarostami" (Abbas Kiarostami is actually its writer but not its director). Since the number of errors produced by a category hierarchy depends on its quality, this result indicates that NH has the best quality, followed by WT and then WH. The fact that Wikipedia leaf categories and article-category pairs have a high quality is also observed in [12]. The fact that WH reflects merely thematic structure of articles and not suitable for inference is also discussed in [12,8]. Since WT is a subsumption hierarchy derived from WH [8], its quality lies between NH and WH.

Next, when looking at Table 3 vertically, we can see that for all category systems, Pattern1-2 achieve better performance than Pattern3-4. The difference is that in Step2 and Step3, Pattern1-2 extract the explicit property while Pattern3-4 vote the implicit property. This indicates that voting implicit properties produces more incorrect triples than extracting explicit properties. By inspecting the false positives in Pattern3-4, we found two causes for the errors: 1) the approximate implementation by the Global Value Pool and Local Property Pool instead of the Local Value Pool, e.g. given category "Category:Cities in Ontario" and value "Ontario", property "birthplace(874)" is wrongly elected since it has the maximum valid votes in the Global Value Pool, instead of "province(469)". However, in the Local Value Pool, "province" would have larger frequency than "birthplace". 2) the assumption that all the subcategories of a supercategory share a property, e.g. a wrong label (country, science) is elected based on the assumption in the following case:

- "Historiography"-"Historiography of the United States" | country(538) | the United States
- "Historiography"-"Historiography of India" | country(1793) | India
- "Historiography"-"Historiography of science" | genre(44) | science

Compared to Pattern3-4, much fewer errors are produced by Pattern1-2, thanks to the more well-formed structure owned by their category pairs. The few errors mainly come from wrong article-category pairs, and value extractions, e.g. a wrong label (year, the 1999 cricket world cup) is extracted from "Category:World Cup cricketers by year"-"Category:Cricketers at the 1999 Cricket World Cup".

Besides, we can see that *generally*, Pattern1 gets higher precision than Pattern2, while Pattern3 gets higher precision than Pattern4. The difference is that in Step2, Pattern1 and Pattern3 extract value from the subcategory's prepositional

Table 4. Other Statistics of the Methods

	Cat Pair Coverage		Labels for Categories			Article Coverage		
	Candidate	Workable	NH	WT−NH	WH−WT	NH	WT−NH	WH−WT
Pattern1	0.079	0.079	45,211	27,425	91,179	0.232	0.151	0.421
Pattern2	0.070	0.070	40,138	21,053	70,673	0.290	0.139	0.328
Pattern3	0.088	0.037	20,175	9,972	62,728	0.174	0.066	0.308
Pattern4	0.138	0.052	23,539	17,476	114,937	0.237	0.121	0.404

phrase while Pattern2 and Pattern4 from the subcategory's noun phrase. This indicates that the possibility of deriving correct triples is higher with value extracted from prepositional phrase than from noun phrase. The reason may be that Wikipedia categories with prepositional phrases are trimmed and organized better than those with noun phrases. Therefore, they better conform with the assumption that all the subcategories of a supercategory share a property.

Yet, there is a major exception that Pattern3 with WH−WT has the worst performance among all the methods. By inspecting its false positives, we found the reason is that some particular errors created in the voting step of Pattern3 are greatly enlarged by WH. From the 265 incorrect triples, we found 106 incorrect triples are created by the wrong label (country, science).

As for the size of triples generated by each method, it depends on how many category pairs are covered by the pattern, how many labels are successfully extracted, how many labels are propagated along the category hierarchy, and how many articles are covered by the categories with labels. Table 4 shows some other statistics for each method, which gives an indication of the size of triples generated. Note that it is possible that no property is elected in Pattern3-4 as a result of zero valid votes. Although Pattern3-4 cover more category pairs than Pattern1-2, their workable category pairs are less (from which labels can be successfully extracted). Therefore, they generate less triples.

4.2   Output

Triples extracted by our methods are tagged with the empirically estimated confidence values in the above evaluation. Our prototype implementation outputs about 10M triples concerning non-isa properties and covering 78.2% of Wikipedia articles with a 12-level confidence ranging from 47.0% to 96.4%. Among them, 1.27M triples have confidence of 96.4%. Applications can on demand use the triples with suitable confidence. The output triples and detailed experimental results can be downloaded from our website[2]. Since it takes less than 12 hours to complete all the methods, Catriple is efficient enough to run on a daily basis for generating triples from Wikipedia categories.

It is interesting to compare Catriple triples with existing extracted triples. When comparing with DBpedia infobox triples, two kinds of Catriple triples are interesting: the shared triples and the new triples. For this goal, we introduce *correct equal-property triples* and *article-without-infobox triples* in the following.

---

[2] http://apex.sjtu.edu.cn/apex_wiki/Demos/Catriple

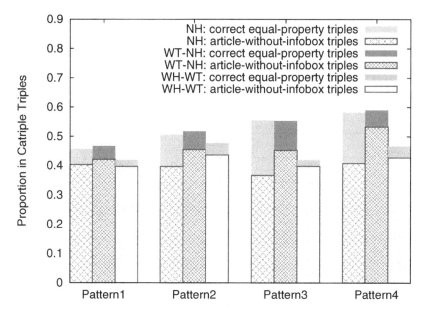

Fig. 2. Comparison with Infobox Triples

A Catriple triple (a,p,v1) is defined as an "equal-property triple" if there exists an infobox triple (a,p,v2). Note that all values are pre-normalized, e.g. we change them to lowercase and delete non-alphanumeric characters. The equal-property triple (a,p,v1) is judged correct, if match(v1,v2)=true, i.e. one condition below is satisfied: i) v1=substring(v2) or v2=substring(v1); ii) stem(v1)=stem(v2); iii) LevenshteinDistance(v1,v2)<3; iv) v1∼v2 by WordNet Synset; v) v1∼v2 by Wikipedia Redirect; vi) match(wordInPhrase(v1), wordInPhrase(v2))=true.

A Catriple triple (a, p, v) is defined as an "article-without-infobox triple" if there exists no infobox triples associated with article a.

Fig. 2 shows the proportions of these two kinds of triples in Catriple output. We can see that Catriple triples and infobox triples have little overlap. Pattern3-4 have more overlap than Pattern1-2, as they have maximally reused the properties in infobox triples during the voting process. More importantly, among all the methods, about 40% of Catriple triples concern articles that have no infobox triples. This indicates that our category-based extraction well complements the infobox-based extraction by contributing triples covering more articles.

We also compare the properties of Catriple triples with infobox triples and YAGO triples. Table 5 shows the properties with top 20 frequencies and the total number of properties in each data set. We use Catriple triples with confidence of top 9 levels to compute the statistics. It can be seen that most Catriple triples are about location, human, genre and time. They are built based on the knowledge contained in categories, thus reflect users' interested characteristics of articles and their request for navigation through Wikipedia. By contrast, the properties of infobox triples are very diverse and even chaotic, since they are collected from human edited key attributes and values in each article. Differently, YAGO

Table 5. Property Comparison

	Catriple Triples (573 properties)		Infobox Triples (28846 properties)		YAGO Triples (74 properties)	
	Property	Frequency	Property	Frequency	Property	Frequency
1	country	1,387,797	name	644,862	isCalled	1,696,493
2	nationality	483,794	relatedinstance	339,896	hasWebsite	98,262
3	topic	386,528	coordmsproperty	246,773	created	95,447
4	city	324,215	coortitledmsproperty	228,934	bornOnDate	66,880
5	genre	290,378	coortitledmproperty	223,062	hasPopulation	61,290
6	state	269,695	genre	214,384	hasArea	55,992
7	status	246,344	title	153,602	hasSuccessor	50,556
8	location	216,948	type	148,984	hasUTCOffset	43,087
9	type	131,714	subdivisionname	141,985	locatedIn	42,845
10	year	121,483	released	127,950	actedIn	42,585
11	continent	119,209	starring	125,108	hasPopulationDensity	38,787
12	artist	86,818	label	124,627	isOfGenre	38,764
13	birthplace	84,977	location	119,229	produced	35,150
14	region	84,559	producer	112,123	hasProductionLanguage	32,079
15	industry	63,967	id	107,082	bornIn	30,352
16	medium	51,149	clubs	104,086	hasImdb	28,313
17	team	49,434	votes	103,516	hasDuration	26,424
18	university or college	47,207	years	102,691	diedOnDate	23,181
19	county	46,820	party	93,345	establishedOnDate	22,446
20	occupation	45,551	artist	92,704	directed	20,913

triples enjoy clear semantics since they are extracted by human specified patterns according to each predefined relation. Yet, this results in only few properties.

5   Discussion and Future Work

The evaluation results show the strengths of our methods. We find that using super-sub category pairs for triple extraction is very important and effective, since it can largely leverage the valuable syntax and semantics within the Wikipedia category system. Based on category pairs, general patterns can be induced for automatic triple extraction, which require less manual effort and concern more categories than specific patterns. The large number of the extracted triples as well as the wide article coverage verify the effectiveness of our methods. Also, patterns exploiting better structure of category pairs lead to better performance. As Pattern1-2 (explicit properties and explicit values) leverage a more well-formed structure than Pattern3-4 (implicit properties and explicit values), they achieve better performance. Furthermore, our methods are expected to be more effective to the future Wikipedia, since an analysis of different Wikipedia versions (2007-7-16, 2007-10-23, 2008-1-3) shows that the coverage of category pairs by our methods increases with time. Thanks to the anonymous editors, the categories are being trimmed and organized better and better.

The evaluation results also reveal some deficiencies of our methods. We find that few errors occur in Step1-2 and most errors occur in Step3 (for Pattern3-4) and Step4 (for Pattern1-2). This indicates that Step3-4 become the bottleneck for the accuracy of the extracted triples. Therefore, improvements in deriving the correct implicit properties and subsumption hierarchy would be very important for improving the results. Besides, we currently summarize seven patterns and rules in Step1-2 to enable the automatic triple extraction. Although it largely increases the article coverage and reduces the manual effort w.r.t. previous work, it is still limited to explore the full knowledge in Wikipedia categories. Machine learning methods are prospective for automatic generation of patterns and rules. We will make deeper investigations and try them to go further.

5.1 Applications

The data extracted by our methods complements existing semantic web data which is the basis for advanced applications and realization of the Semantic Web. It can be used in many applications, e.g. realizing faceted search for Wikipedia, enhancing Wikipedia infobox and category system. Since not only explicit properties but also implicit properties are derived from Wikipedia categories, the data can be used to explore invisible relationships between categories as well.

First, Catriple triples can be a good data set for faceted browsing and search. This advantage comes from that they are generated from Wikipedia categories which define characteristics and help users navigate through Wikipedia. An on-line demo of faceted search for Wikipedia based on the data is available at our website (as mentioned in Section 4.2). A friendly faceted search interface is provided for users to improve their access to the large Wikipedia knowledge base. Starting with a keyword search, users can get both result articles and faceted information (e.g. the properties and values for the result articles). This enables users to stepwise refine their search results through different dimensions.

Besides, Catriple triples can also be used to enrich the infobox data. From the correct equal-property triples, we can pick a set of trusted labels for categories. Then, the trusted triples created can be directly used to enrich the infobox data (37.7% of the 4M trusted triples are new to infobox data). Furthermore, with the help of Catriple, we can possibly link categories to infobox attributes. For example, through its label (year,2008), "Category:2008 films" can be linked to infobox attribute "released" and value "2008". In this way, Catriple can be used to bridge infoboxes and categories. The integration of infobox-based properties and category-based properties will produce better structure for Wikipedia.

6   Conclusion

This paper presents Catriple, a system which automatically extracts triples about Wikipedia articles and non-isa properties from Wikipedia categories. By employing a domain independent extraction based on widespread categories, the extracted triples cover a wide range of articles. The extraction is also relation independent, thus it requires no human efforts to predefine relations. In sum, we make the following contributions: 1) We observe the semantics of Wikipedia category pairs for extracting properties and values. Helpful category pairs are discovered which contain explicit/implicit properties and explicit values. 2) We propose methods enabling an automatic extraction of triples based on the helpful category pairs. NLP technologies are employed to extract the explicit properties and values from the category names. A voting strategy resorting to the semantics of articles in a category is used to decide the implicit property. 3) We output about 10M triples with confidence of 12 levels from 47.0% to 96.4%, which cover 78.2% of Wikipedia articles. Among them, 1.27M triples have confidence of 96.4%. Applications can on demand use the triples with suitable confidence.

References

1. Opennlp, http://opennlp.sourceforge.net/
2. Porter stemmer, http://tartarus.org/martin/PorterStemmer/
3. Stanford parser, http://nlp.stanford.edu/software/lex-parser.shtml
4. Auer, S., Bizer, C., Kobilarov, G., Lehmann, J., Cyganiak, R., Ives, Z.G.: DBpedia: A nucleus for a web of open data. In: Aberer, K., Choi, K.-S., Noy, N., Allemang, D., Lee, K.-I., Nixon, L., Golbeck, J., Mika, P., Maynard, D., Mizoguchi, R., Schreiber, G., Cudré-Mauroux, P. (eds.) ASWC 2007 and ISWC 2007. LNCS, vol. 4825, pp. 722–735. Springer, Heidelberg (2007)
5. Auer, S., Lehmann, J.: What have innsbruck and leipzig in common? Extracting semantics from wiki content. In: Franconi, E., Kifer, M., May, W. (eds.) ESWC 2007. LNCS, vol. 4519, pp. 503–517. Springer, Heidelberg (2007)
6. Herbelot, A., Copestake, A.: Acquiring ontological relationships from wikipedia using RMRS. In: Proc.of the ISWC 2006 Workshop on Web Content Mining with Human Language Technologies (2006)
7. Nguyen, D.P.T., Matsuo, Y., Ishizuka, M.: Exploiting Syntactic and Semantic Information for Relation Extraction from Wikipedia. In: IJCAI Workshop on Text-Mining & Link-Analysis, TextLink 2007 (2007)
8. Ponzetto, S.P., Strube, M.: Deriving a large-scale taxonomy from wikipedia. In: AAAI 2007, pp. 1440–1445 (2007)
9. Ponzetto, S.P., Strube, M.: Knowledge derived from wikipedia for computing semantic relatedness. Journal of Artificial Intelligence Research 30, 181–212 (2007)
10. Strube, M., Ponzetto, S.P.: Wikirelate! computing semantic relatedness using wikipedia. In: AAAI (2006)
11. Suchanek, F., Kasneci, G., Weikum, G.: Yago: A large ontology from wikipedia and wordnet. Research Report MPI-I-2007-5-003, Max-Planck-Institut für Informatik, Stuhlsatzenhausweg 85, 66123 Saarbrücken, Germany (2007)
12. Suchanek, F.M., Kasneci, G., Weikum, G.: Yago: a core of semantic knowledge. In: WWW (2007)
13. Wang, G., Yu, Y., Zhu, H.: Pore: Positive-only relation extraction from wikipedia text. In: ISWC/ASWC, pp. 580–594 (2007)
14. Wu, F., Weld, D.S.: Autonomously semantifying wikipedia. In: CIKM (2007)
15. Wu, F., Weld, D.S.: Automatically refining the wikipedia infobox ontology. In: WWW, pp. 635–644 (2008)
16. Yu, J., Thom, J.A., Tam, A.M.: Ontology evaluation using wikipedia categories for browsing. In: CIKM, pp. 223–232 (2007)
17. Zesch, T., Gurevych, I.: Analysis of the wikipedia category graph for nlp applications. In: Proceedings of the TextGraphs-2 Workshop (NAACL-HLT) (2007)
18. Zirn, C., Nastase, V., Strube, M.: Distinguishing between instances and classes in the wikipedia taxonomy. In: Bechhofer, S., Hauswirth, M., Hoffmann, J., Koubarakis, M. (eds.) ESWC 2008. LNCS, vol. 5021, pp. 376–387. Springer, Heidelberg (2008)

Semantically Conceptualizing and Annotating Tables

Stephen Lynn and David W. Embley*

Brigham Young University, Provo, Utah 84602, U.S.A.

Abstract. Enabling a system to automatically conceptualize and annotate a human-readable table is one way to create interesting semantic-
web content. But exactly "how?" is not clear. With conceptualization
and annotation in mind, we investigate a semantic-enrichment procedure as a way to turn syntactically observed table layout into semantically coherent ontological concepts, relationships, and constraints. Our
semantic-enrichment procedure shows how to make use of auxiliary world
knowledge to construct rich ontological structures and to populate these
ontological structures with instance data. The system uses auxiliary
knowledge (1) to recognize concepts and which data values belong to
which concepts, (2) to discover relationships among concepts and which
data-value combinations represent relationship instances, and (3) to discover constraints over the concepts and relationships that the data values
and data-value combinations should satisfy. Experimental evaluations
indicate that the automatic conceptualization and annotation processes
perform well, yielding F-measures of 90% for concept recognition, 77%
for relationship discovery, and 90% for constraint discovery in web tables
selected from the geopolitical domain.

1 Introduction

Ontology creation is a daunting task—manual creation is tedious and time consuming, and automatic creation is often disappointingly inaccurate. But for applications such as the semantic web or making web content directly queriable,
we must facilitate ontology creation, making it reasonable to produce the vast
number and variety of ontologies required for future web applications.

In this paper we focus on one aspect of this daunting task—semantic conceptualization and annotation of tables. Because tables meant for human readers
are data-rich and semi-structured, they are a prime target for automatic conceptualization and annotation. We conceptualize a domain of interest when we
create an ontology for the domain, and we annotate documents with respect to
an ontology when we link document content with ontology components. To annotate, we identify objects within documents and link them to ontological object
sets (conceptual classes, concepts, or value sets), and we identify relationships
and link them to ontological relationship sets (conceptual properties for classes,
taxonomic structures, or associations among objects).

---

* Supported in part by the National Science Foundation under Grant #0414644.

J. Domingue and C. Anutariya (Eds.): ASWC 2008, LNCS 5367, pp. 345–359, 2008.
© Springer-Verlag Berlin Heidelberg 2008

Our particular focus in this paper is on semantic enrichment of conceptual-model instances automatically derived from given, syntactic table layout. Semantically enriching ontologies has been the focus of some recent research efforts (e.g., [3], [9]). These efforts aim mainly at enlarging the vocabulary of ontologies, and do not use tables as data and meta-data sources. Nevertheless, these efforts show that semantic enrichment is desirable and also show how to use the lexical resources available on the web to do semantic enrichment. Semantic table enrichment has also been the focus of some recent work [4,5]. In this work, researchers use established ontologies to enrich tables by adding columns and instance data (rather than the other way around—use established tables to enrich ontologies—as we do in this paper). In addition to using ontological structure in their work, these researchers also identify data instances in columns of tables based on the instance values. In our research, we follow the lead of both ontology-enrichment researchers and table-enrichment researchers, relying both on available lexical resources and on given instance-recognition semantics. These two types of semantic resources form the foundation we use for semantic enrichment.

Region and State Information

Location	Population* (2000)	Longitude[†]	Latitude[†]	Capital City
Northeast	3.120			
Maine	1.275	69° 14.0′W	45° 15.2′N	Augusta
New Hampshire	1.236	71° 34.3′W	43° 59.0′N	Concord
Vermont	0.609	72° 40.3′W	43° 55.6′N	Montpelier
Northwest	9.315			
Washington	5.894	120° 16.1′W	47° 20.0′N	Olympia
Oregon	3.421	120° 58.7′W	43° 52.1′N	Salem

*Population in Millions
[†] Geographic Center

Fig. 1. Sample Table

To be specific about what we aim to do, we provide an example. Given a table like the one in Figure 1, our semantic-enrichment algorithm generates a conceptual-model instance that accurately represents the semantics of the table. The algorithm has three main tasks: concept recognition, relationship discovery, and constraint discovery. During the steps of the semantic-enrichment process, the algorithm populates the conceptual-model instance with the data in the original table. Figure 2 shows the conceptual-model instance, the algorithm generates from the table in Figure 1. The five United States states in Figure 1 are members of the *State* object set in Figure 2. The two regions are in the *Region* object set. Together the regions and states constitute the elements of the *Location* object set. The states aggregated together constitute the different regions. The values in the population, longitude, latitude, and capital city columns of the table in Figure 1 are members of the *Population, Longitude, Latitude,* and *Capital*

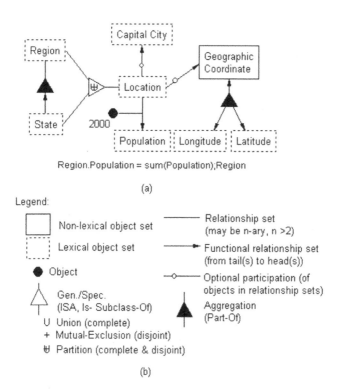

Region.Population = sum(Population);Region

(a)

Legend:

☐ Non-lexical object set	——— Relationship set (may be n-ary, n >2)
⌐¬ Lexical object set	——▶ Functional relationship set (from tail(s) to head(s))
● Object	—○— Optional participation (of objects in relationship sets)
△ Gen./Spec. (ISA, Is- Subclass-Of)	▲ Aggregation (Part-Of)
U Union (complete)	
+ Mutual-Exclusion (disjoint)	
⊎ Partition (complete & disjoint)	

(b)

Fig. 2. (a) Generated Enriched Conceptual-Model Instance for the Sample Table in Figure 1 and (b) Legend for Conceptual-Model Components

City object sets respectively. Longitude and latitude values aggregated together constitute the *Geographic Coordinate* object set. Each state has a population, a geographic coordinate, and a capital city, and each region has a population computed as the sum of the populations from the states in the region.

Automated "table understanding" has been the subject of research in the document analysis community for several decades [11,14]. Most of these efforts end, however, after only identifying table labels and table instance data. Some researchers have described a semantic-enrichment step in the table-understanding process, but as e Silva, et al. remark "no [one] has yet found a way of making [this] general" [12]. In research most similar to our own, Pivk et al. [10] have implemented a system that takes tabular data meant for human readers as input and produces F-Logic frames as output. In their work, they include a semantic-enrichment step, which has two components: (1) discovery of semantic labels and (2) the mapping of their internal model into an F-Logic frame. F-Logic [7] is a type of conceptual model, so Pivk, et al. have the same sort of output as we produce. Further, their semantic-enrichment step uses lexical resources to discover semantic labels for table data (as does ours), and their mapping step adds functional dependencies for table data (as does ours). Their semantic enrichment step, however, does not use given instance recognizers (as does ours), nor does it attempt to discover table-implied generalization/specialization, aggregation, non-table-data-specific functional dependencies, value augmentations, computed table values, or mandatory/optional participation of objects in relationship sets (as does ours).

This paper makes several contributions: (1) It describes a general, comprehensive algorithm for automatically enriching tables semantically (Section 2). (2) It shows that a prototype implementation of this algorithm works well in the geo-political domain with tables selected independently by a third-party subject (Section 3). (3) It explains, by referencing other work, how to embed the semantic-enrichment algorithm as a key component in automatically generating semantic-web content from data-rich web pages (at the end of Section 4, which also discusses points of interest and future work regarding the semantic-enrichment algorithm). The paper thus sheds light on a way to automate the generation of semantically rich web content from tables meant for human readers.

2   Semantic Enrichment Procedure

Figure 3 gives our semantic enrichment algorithm. Its input is a canonicalized table: a table whose components have been syntactically recognized based on the table's layout. The components include the title (if any), the caption (if any), the table's labels structured as dimension trees, the data values in known rows and columns, and the augmentations (footnotes and parenthetical remarks attached to any of the other components). Figure 4 shows a pictorial view of the canonicalized-table input for our running example—the table in Figure 1. The algorithm's output is a conceptual-model instance semantically enriched according to the steps in the algorithm. In what follows, we explain and illustrate each of these steps.

Before doing so, however, we describe the two semantic resources we use in our approach to semantic enrichment: a natural-language lexicon and a data-frame library. Without semantic resources no semantic enrichment can take place—semantic enrichment, by definition, consists of establishing correspondences between accepted semantic resources and the symbolic characterization being enriched. A *natural-language lexicon* should provide support for term normalization, and testing whether one word is a hypernym, hyponym, meronym, or holonym of another word. Hypernym checking, for example, allows the system to recursively check for term generalizations, which the semantic-enrichment algorithm can use to assign names to unnamed concepts or check for is-a relationships among recognized concepts. Our current prototype implementation uses Word-Net as its natural-language lexicon. Data frames in a *data-frame library* provide a mechanism for recognizing and classifying character-string representations of data values using regular-expression recognizers [1]. Suppose, for example, the string "12-08-2008" appears as a data value in a table. In looking for a concept for this data value, the semantic-enrichment algorithm can discover that the *Date* data frame recognizes dates in the form MM-DD-YYYY and can thus classify the instance value as belonging to the *Date* data frame.

1.   Input: canonicalized table
2.   Output: semantically enriched conceptual-model instance
3. $--$ recognize concepts and associate values with concepts
4.   create concept-values mappings:
5. $--$ concept-values mappings must come from the same dimension tree
6.   (column or row of table data values) instance-of (dimension leaf)
7.   (spanned table data values) instance-of (spanning dimension node)
8.   (dimension siblings/cousins) instance-of (concept)
9.   if unclassified table data values remain
10.   if no dimension tree has been established as the one with concepts,
11.   check: (data values) instance-of (title or caption concept)
12.   default: map data to (1) title, (2) caption, or (3) unnamed concept
13.   else map data to lowest unclassified nodes in established dimension tree
14.   if unclassified labels remain, classify them as non-lexical concepts
15. $--$ discover relationships, including types of relationships
16.   initialize relationships within each dimension tree
17.   refine types of relationships:
18.   (child) subclass-of (parent)
19.   (child) subpart-of (parent)
20.   (descendants in subpart-of hierarchy) subclass-of (generalization of root)
21.   molecular structure recognition
22.   value augmentations
23.   values under spanning label
24.   join dimension trees
25. $--$ adjust conceptual-model instance for discovered constraints
26.   add discovered constraints and make necessary adjustments:
27.   functional relationships
28.   is-a constraints
29.   computed values
30.   mandatory/optional participation

Fig. 3. Semantic Enrichment Algorithm

Lines 3–14: recognize concepts and associate values with concepts

As Figure 3 shows in Line 6, the first step in creating concept-values mappings is to check columns and rows of data values to determine whether they are instances of leaf concepts in dimension trees. We use both lexical services and data-frame services in our instance-of check. For our running example, the lexical service recognizes the cities in the last column in Figure 1 as *City* values and maps them to *Capital City*, which is the column header and thus also a dimension leaf node.[1] Hence, the algorithm creates a concept-values mapping between the concept *Capital City* and the set of values {*Augusta, Concord, Mont-*

---

[1] We note that both the lexical service and the data-frame service can recognize value sets even without associated names. Thus, for example, if no name had appeared as the header for the state capitals in the table in Figure 1, the algorithm would still have recognized the cities in the columns and would have given their header the name *City* (or perhaps even *State Capital* if the lexical or data-frame service contains enough specific knowledge about these cities).

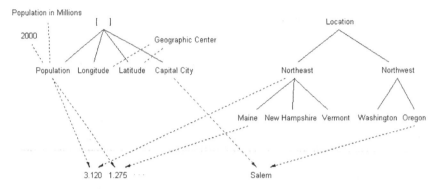

Fig. 4. Graphical View of the Sample Table in Figure 1 in Canonical Form

pelier, Olympia, Salem}. This action creates the lexical object set *Capital City* in Figure 2. In a similar way, the data-frame service recognizes the geographic coordinates as instances of the leaf nodes *Longitude* and *Latitude* in the first dimension tree in Figure 4. This action results in establishing the lexical object sets *Longitude* and *Latitude* in Figure 2. The data-frame service might also be able to recognize the *Population* values, but this recognition is more complex since the population values are in units of millions. As currently implemented, our data-frame service does not recognize the population values in Figure 1.

Observe that all three recognized concept-values mappings are for the same dimension tree. This is a property of well formed tables—the concepts for the table data values can come from at most one dimension tree—informally, the data values in rows or columns associate properly with row or column headers, but not some rows with row headers and also some columns with column headers. In our implementation, the first established mapping of a row or a column to a concept in a dimension tree determines the dimension tree for the table's concepts.

In the second step of creating concept-values mappings (Line 7), the algorithm checks multiple rows or columns of data values to see if they are instances of non-leaf nodes in dimension trees. Our sample table in Figure 1 does not have an example, but a simple variation of the table does. Consider *Population*, but instead of just one column of population values for the year 2000, imagine a table with six columns of population values, one for each year from 2000 to 2005. Further, imagine each of these columns is headed by a year label and that above the year labels, a spanning label *Population* appears. In this case the first dimension tree in Figure 4 would have a third level below *Population* with six siblings—the year labels *2000*, ..., *2005*. If we further assume that the data-frame library has a *Population* data frame that recognizes these values (perhaps, they are actual population numbers, not masked by being in units of millions), then we have an example illustrating the second step in creating concept-values mappings. A label (*Population*) spans several columns of data values, which are recognized as instances of a spanning dimension node (a non-leaf node in the dimension tree).

Having checked for mappings between the table's data values and the table's labels, the algorithm considers the possibility that some of the labels might be values for some concept associated with the table. The algorithm does not check labels already designated as concepts in established concept-values mappings, but other labels may be values. The third step in creating concept-values mappings (Line 8) uses lexical and data-frame services to check whether sibling or cousin nodes in dimension trees are values of some recognized concept. For our running example, the lexical service recognizes the states as instances of its *State* concept and recognizes *Northeast* and *Northwest* as instances of its *Region* concept. This gives rise to the object sets *State* and *Region* in Figure 2. Points of interest about checking whether labels are values include the following: (1) The data-frame service as well as the lexical service can recognize values. A data-frame for *Year*, for example, would recognize the years *2000 – 2005* as siblings under *Population* for the variation example mentioned in the previous paragraph. (2) A number of names for a concept are possible (e.g., *Area* as well as *Region* is a possible name for {*Northeast, Northwest*}). In the absence of any reason to choose one over the other the choice is arbitrary. In our implementation, if a synonym name for a concept is in the title as *State* and *Region* are in Figure 1 we prefer these names over alternative synonyms. Footnotes, captions, and other labels higher up in the dimension tree are other possible sources for selecting names from among the synonyms. (3) In our current implementation, we only consider an entire level in a dimension tree as possible value sets. Although this is typical and works in our running example for *State* and *Region* and even for our *Year* example, the label-as-value idea can be expanded to check for some, rather than all, siblings and cousins. For example, a table with population columns for years *2000 – 2005* might also have columns at the end for *Average Yearly Growth Rate* and *Five-Year Increase/Decrease*. Only the first six of these eight siblings under *Population* are year values.

After the three steps in Lines 6–8, it is possible for both data values and labels to remain unclassified. In our running example, the data values under *Population* remain unclassified, and the labels *Population, Location*, and the virtual root of the first dimension tree in Figure 4 remain unclassified. Lines 9–13 of our semantic enrichment algorithm tell how we map unclassified data values to concepts, and Line 14 tells how we classify labels. If we have already established a dimension tree as the one with concepts to which data values belong, Line 13 of the algorithm maps the set of data values indexed by a lowest level unclassified node in this dimension tree as data values for the concept named in that node. In our example the values in the column beginning with 3.120 in Figure 1 become values for the concept *Population*, yielding the lexical object set *Population* in Figure 2.

If no dimension tree has been established as the one with concepts to which data values belong, we check in Line 11 to see whether all the data values belong to a concept named in the title or caption for the table. Imagine, as an example, a table that has *only* population values for locations for several different years. Imagine further, that the labels in the year dimension consist only of these year values and that the title for the table is *Population Information*. Assuming a population data frame recognizes the data values and the keyword "Population" in the title, the algorithm would establish a mapping between the concept *Population* and all the data values in the table. If this semantic check fails, then in Line 12, the algorithm defaults to establishing a lexical object set for all the data values in the table, giving it the title as its name (if the table has a title) or the caption as its name (if the table has no title but does have a caption), and finally leaving it without a name (if the table has neither a title nor a caption).

For any unclassified labels that remain, Line 14 of the algorithm classifies them all as non-lexical concepts. In our running example, *Location* and the virtual root of the first dimension tree become non-lexical object sets. We will see later how some of these non-lexical concepts can be semantically resolved into something better. In the absence of additional semantic information to resolve these non-lexical object sets into something better, keeping them as non-lexical concepts turns out to make sense. In our running example, if the semantic "instance-of" check in Line 8 had not succeeded, each of the labels in the second dimension tree in Figure 4 would have become non-lexical object sets. We would then, for example, have a *Maine* object set, which would have a single object identifier in it denoting Maine, and a *Northwest* object set whose single object identifier would denote the concept Northwest.

Lines 15–24: discover relationships, including types of relationships

As the first step in discovering relationships (Line 16), the algorithm initializes the conceptual-model instance with relationship sets that correspond to the dimension trees. Figure 5 shows the result for our running example. If siblings/cousins have been recognized in the creation of concept-values mappings in Line 8, some edges in the dimension trees will coalesce. In the second dimension tree in Figure 4, for example, all the edges at each level will coalesce resulting in the second tree in Figure 5.

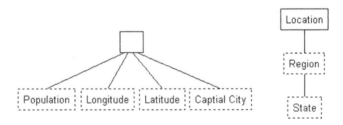

Fig. 5. Relationship Sets from Dimension Trees

Following the relationship-set-initialization step, the algorithm checks for the possibility of making several refinements (Lines 17–23). Specifically, the algorithm checks for the possibility that initialized relationship sets might represent *generalization/specialization hierarchies* (is-a relationships), *aggregation hierarchies* (part-of relationships), *molecular structures* (known structures over concept groups), and *n*-ary relationship sets ($n > 2$).

The refinement in Line 18 checks for the possibility that the concept name of a child node is a hyponym of the concept name of its parent node, or, equivalently, a hypernym from parent to child. If so, the algorithm replaces the relationship set with a generalization/specialization constraint. In our running example, our lexical service recognizes that "Region" is a hyponym of "Location" (a region is a location), and thus the algorithm makes *Location* a generalization of *Region*. Since is-a relationships require that the object sets in generalizations and specializations correspond lexically or non-lexically, if ever a mismatch occurs, non-lexical object sets become lexical. In our example, the introduction of the is-a relationship between *Location* and *Region* causes *Location* to become lexical.

Checking further, the refinement in Line 18 fails to recognize "State" as being a hyponym of "Region." Instead (since a state can be part of a region), the meronym/holonym check in Line 19 succeeds. "State" is a meronym of "Region," or, equivalently, "Region" is a holonym of "State." Thus, the algorithm introduces an aggregation relationship between *Region* and *State*. Then, in Line 20, since we have a subpart-of hierarchy from *State* to *Region* and an is-a relationship between *Region* and *Location*, the algorithm checks the descendant *State* in the part-of hierarchy to see if it is also a hyponym of *Location*, the generalization of *Region*, which is the root of the subpart-of hierarchy. In our running example, *State* is a hyponym of *Location* (a state is a location), and thus the *State* object set becomes a specialization of the *Location* object set. Figure 2 shows the result of the steps in Lines 18–20 as the aggregation between *State* and *Region* and the generalization/specialization with *Location* as the generalization and *Region* and *State* as its specializations.

The algorithm introduces a molecular structure (Line 21) whenever it finds the constituent components of the molecular structure appropriately configured in the conceptual-model instance. In our running example, *Longitude* and *Latitude* are both associated with some (as yet unnamed) concept. In our implementation, the *Geographic Coordinate* data frame in our data-frame library recognizes this *Longitude/Latitude* configuration as being a geographic coordinate. The algorithm thus introduces the non-lexical object set *Geographic Coordinate* as Figure 2 shows.

Value augmentations and values under spanning labels both indicate the presence of *n*-ary relationship sets. The value augmentation *2000* in Figure 1 indicates the presence of a ternary relationship set among the locations, the population values, and the value object *2000*. Our implementation of value augmentations (Line 22) turns this pattern into a ternary relationship set, which (with the addition of some downstream operations) eventually becomes the ternary relationship among the object *2000* and the lexical object sets *Location* and *Population* in Figure 2. The *values-under-spanning-label* step (Line 23) applies, when, for example, we have the year values *2000*, ..., *2005* discussed earlier. When the algorithm recognizes these labels as year values under the spanning label *Population*, the step in Line 23 creates a ternary relationship among *Year*, *Population*, and the unnamed object set which eventually becomes *Location*.

The algorithm's final step in discovering relationships (Line 24) joins the conceptual-modeling fragments for each of the table's n dimensions into a single conceptual-model instance. Temporarily, until the algorithm does constraint analysis in its final phase, the algorithm simply creates an n-ary relationship set among the root object sets of the n dimension trees. Further, in the case when the algorithm had established no dimension tree as the one with concepts to which the data values belong (Line 10), but rather added a lexical object set for all the data values in the table (Lines 11–12), the algorithm in Line 24 adds this lexical object set to the n-ary relationship set among the n root object set making the relationship set an $(n + 1)$-ary relationship set.

Lines 25–31: adjust conceptual-model instance for discovered constraints

Functional constraints, which we consider in Line 27 of Figure 3, arise in three ways: (1) molecular structures that include functional relationship sets, (2) relationship sets established in Lines 16–23 whose instance values indicate that the relationship set should be functional, and (3) table-implied functional dependencies —the data values of a table depend functionally on their indexing labels.

1. Molecular structures bring all their constraints with them. The bijection between *Geographic Coordinate* and the aggregate pair (*Longitude*, *Latitude*) in Figure 2 comes from the given molecular structure.
2. Relationship sets established within dimension trees may be functional. In our implementation, we check for this possibility by checking for 1-1 and many-1 relationships among instance values. The functional dependency, *State* → *Region*, in Figure 2 arises because of the many-1 relationship between the state instances and the region instances in Figure 1.
3. Fundamentally, each data value in a table depends functionally on its indexes (usually its row and column headers). Variations in how the data values and indexes become part of the conceptual model dictate where these functional dependencies appear in the evolving conceptual-model instance. Two basic variations depend on whether the table does (a) or does not (b) have a dimension tree with multiple concepts for the data values in the table.

   (a) *One dimension tree with multiple concepts for table values.* Our running example in Figure 1 illustrates one of the most common cases. One dimension tree has concept nodes for the data values in the table, and a second dimension tree has a root node whose conceptual object set is lexical and logically contains all the instance values of the dimension tree. In our example, the concept nodes in one dimension tree are *Population*, *Longitude*, *Latitude*, and *Capital City*, and the root node in the other dimension tree is *Location*, which contains all the *State* and *Region* values. In this case, the algorithm adjusts the relationship set created in Line 24 that joins the dimension trees. Specifically, it first removes the root object set of the dimension tree that contains the concepts for the table's data values and also all its connecting relationship sets. It then adds in their place functional relationship sets from the root in which the values appear to the object sets representing the concepts. Figure 2 shows the result for our running example: *Location* → *CapitalCity*, *Location* → *GeographicCoordinate*, and *Location 2000* → *Population*. In the example discussed earlier in which *Population* is a non-leaf node with children *2000*, ..., *2005*, the functional dependency would be *Location Year* → *Population*.

All other variations involve (i) non-root-contained values or (ii) more than two dimension trees. (i) When the root does not contain all the values for a dimension tree, the algorithm uses the highest level nodes that together contain all the values. In the worst case, the values are all in the leaves. In a variation of our running example in which none of the hypernym/hyponym and holonym/meronym relationships are recognized, the algorithm would yield many functional relationship sets: *Northeast 2000* → *Population*, ..., *Vermont* → *GeographicCoordinate*, ..., *Oregon* → *CapitalCity*. (ii) When a table has n dimension trees ($n > 2$), each dimension tree except the dimension tree with multiple concepts for table values provides domains for the functional dependencies (the codomains are always in the concept-providing dimension tree). The functional relationship sets are thus always n-dimensional.

(b) *No dimension tree with multiple concepts for table values.* This variation has two cases: (i) one dimension tree has a conceptual root node representing all the data values in the table and (ii) no dimension tree has a conceptual root node for the table's data values. In both cases, the algorithm makes the n-ary relationship set created in Line 24 functional. For case (i), the domains for the functional relationship set come from all the dimension trees, except the dimension tree that has the conceptual root node, and the codomain is the object set for this conceptual root node. For case (ii), all the dimension trees contribute domains for the functional relationship set, and the codomain is the lexical object set established either in Line 11 or Line 12. For both cases, the object set(s) in a dimension tree that become domain object set(s) are the highest level node(s) that together contain all the values for a dimension tree. As an example, consider a table like the table in Figure 1 but with just population values for the years 2000–2005. If the root of the dimension tree for years is *Population*, the linking relationship set between *Population* and *Location* would become functional from *Location* to *Population*.

For is-a constraints (Line 28), the algorithm considers generalization/specialization relationships identified in Lines 18 and 20. It constrains the is-a to have a union constraint if all values in the generalization object set are also in at least one of the specialization object sets, to have a mutual-exclusion constraint if there is no overlap in the values in the specialization object sets, and to have a partition constraint if the values satisfy both union and mutual-exclusion requirements. As a result of these checks for our running example, the ⊎ in Figure 2 appears: every value in the *Location* object set is also in either the *Region* object set or the *State* object set, and no value is in both.

Tables often include columns or rows that contain summations, averages, or other value aggregates. Because checking all possible combinations for all possible aggregate operators is prohibitive, the algorithm (Line 29) should only check probable combinations with likely operators. Our current implementation checks only for summations and averages for data cells associated with non-leaf nodes in dimension trees. Thus, our algorithm examines values such as *3.120*, which is indexed by the non-leaf node *Northeast*, computes aggregates of values from related object sets, and compares them. The algorithm captures constraints that hold and adds them to the conceptual-model instance. In our running example, these checks add the constraint *Region.Population = sum(Population); Region*, which means that a region's population is the sum of the population values grouped by *Region*.

In Line 30, the algorithm determines whether objects in an object set participate mandatorily or optionally in associated relationship sets. The algorithm identifies object sets whose objects have optional participation in relationship sets by considering empty value cells in the table. As Figure 2 shows, the step in Line 30 discovers that *Location* optionally participates with *Geographic Coordinate* and also with *Capital City* because some locations, namely *Northeast* and *Northwest*, have no associated longitude, latitude, and city values.

3   Experimental Evaluation

We evaluated our implemented version of the semantic-enrichment algorithm in Figure 3 using a test set of tables found by a third-party participant. We asked the participant for twenty different web pages that contain HTML tables—stipulating that the test tables should come from at least three distinct sites, should contain a mix of simple and complex tables, and should all be from the geopolitical domain. To canonicalize the tables, we used a tool [6] that makes it easy to designate a table's labels, data values, and augmentations. Our algorithm processed each of the twenty canonicalized tables and saved the resulting conceptual-model instances for manual evaluation with respect to its ability to do concept/value recognition, relationship discovery, and constraint discovery.[2]

We use precision and recall to evaluate how well our implementation of the semantic-enrichment algorithm performs. We observe how many concept-value mappings, relationships, and constraints the algorithm correctly identifies C, how many it identifies incorrectly I, and how many it misses M. We then compute precision by $C/(C + I)$ and recall by $C/(C + M)$. For the experimental test, our implemented prototype achieved 87% precision and 94% recall for the concept/value-recognition task, 73% precision and 81% recall for the relationship-discovery task, and 89% precision and 91% recall for the constraint-discovery task. As a combined measure of precision and recall, we also com-

---

[2] When building semantically enriched conceptual-model instances, there is often no "right" answer. Many tables correspond to multiple valid instances. Our evaluation permitted only valid conceptualizations, but did allow for reasonable alternatives.

puted F-measures. Concept recognition and constraint discovery both have an F-measure of 90% while relationship discovery has an F-measure of 77%.

4   Discussion Points and Future Work

As a result of empirically investigating our prototype implementation we identified several potential enhancements. We should:

- check for totals and other aggregates in all columns or rows of numeric data values, not just in data cells for non-leaf nodes in dimension trees;
- check for lists of values rather than a single value in data cells;
- check label instances in flat dimensions tree for generalization/specialization and aggregation relationships (the canonicalization step may not be able to syntactically discern nestings that indicate these possibilities);
- combine multiple columns (or rows) that corresponded to the same concept— for example, when a table about mountain peaks contains two columns labeled *Height*, one in meters and in the other in feet; and
- discard columns that merely provide rank sortings based on some other column—sort order is always recoverable.

An in-depth discussion of canonicalization issues, an assumed preprocessing step to our semantic-enrichment algorithm, is beyond the scope of this paper. We mention, however, that the motivation is to split the work of canonicalization (based on observations of syntactic layout) and semantic enrichment (based on observations with respect to semantic resources such as WordNet and a data-frame library). We and many others, especially in the document-analysis community, are investigating the problem of table canonicalization [11]. Some good results have been found, and better results are likely forthcoming. As a direction for further work, it appears possible to synergistically exploit syntactic/semantic interplay. For example, syntactic discovery of table orientation may suggest semantic label/value-set associations when semantic resources fail to discover them, and semantic label analysis may suggest dimension-tree nesting even when it is not obvious from syntactic layout.

Our semantic-enrichment algorithm assumes the existence of a good lexicon and data-frame library, both rich in the domain knowledge about a table's content. But what if these resources are unavailable or insufficiently provide semantic information for a domain of interest? We offer two answers:

1. The semantic-enrichment algorithm (Figure 3) degrades gracefully. When little or no semantic knowledge applies, the algorithm still successfully creates a semantic-model instance from a canonicalized table. Although additional syntactic clues enable the algorithm to perform better, it can succeed based only on a proper division between values and labels, split into n dimensions for an n-dimensional table.
2. The semantic resources—the data-frame library in particular—can improve itself with use through self-adaptation. Whenever the algorithm establishes a

concept-values mapping, the system can update its data-frame recognizers by adding any values not currently recognized—for example, the system could add cities in Figure 2 not already recognized by the *City* data frame. The system could also establish a new data frame for the library and initialize its recognizers with information in the table. If, for example, a data frame for *Population* did not already exist, the system could establish one based on the information in the table in Figure 2. The system could also update keywords and units for data frames—adding, for instance, from the table in Figure 2 that populations can be expressed in units of millions and that "Geographic Center" is a phrase connected with geographic coordinates.

The algorithm for semantic-enrichment, presented here, does not stand alone. We envision it as part of a much larger system that automatically, or at least semi-automatically, generates interesting semantic-web content from data-rich web pages [2,13].

5   Concluding Remarks

We have described an an algorithm that automates the generation of semantically rich conceptual-model instances from canonicalized tables. The algorithm uses a novel approach to semantic enrichment based on semantic knowledge contained in lexicons and a data-frame library. Experimental results show that the algorithm is able to automatically identify the concepts, relationships, and constraints for data in a table with a relatively high level of accuracy—with F-measures of 90%, 77%, and 90% respectively in web tables selected from the geopolitical domain. These results are encouraging in our effort to automate the conceptualization and annotation of semi-structured data and make the data available on the semantic web.

Acknowledgements

We wish to thank the anonymous referees and also George Nagy for their insightful comments on our paper. Although in agreement with their suggestions for additional explanation, commentary, and illustrations, space constraints prevent us from complying with their requests. We offer, however, on-line access to [8],[3] which includes the requested figures illustrating an evolving conceptual model and also the tables used in our experiments.

References

1. Embley, D.W., Campbell, D.M., Jiang, Y.S., Liddle, S.W., Lonsdale, D.W., Ng, Y.-K., Smith, R.D.: Conceptual-model-based data extraction from multiple-record web pages. Data & Knowledge Engineering 31(3), 227–251 (1999)

---

[3] www.deg.byu.edu/papers

2. Embley, D.W., Liddle, S.W., Lonsdale, E., Nagy, G., Tijerino, Y., Clawson, R., Crabtree, J., Ding, Y., Jha, P., Lian, Z., Lynn, S., Padmanabhan, R.K., Peters, J., Tao, C., Watts, R., Woodbury, C., Zitzelberger, A.: A comceptual-model-based computational alembic for a web of knowledge. In: Proceedings of the 27th International Conference on Conceptual Modeling, Barcelona, Spain (October 2008)
3. Faatz, A., Steinmetz, R.: Ontology enrichment with texts from the www. In: Proceedings of ECML—Semantic Web Mining, Helsinki, Finland (August 2002)
4. Gagliardi, H., Haemmerlé, O., Pernelle, N., Saïs, F.: An automatic ontology-based approach to enrich tables semantically. In: Proceedings of The First International Workshop on Context and Ontologies: Theory, Practice and Applications, Pittsburgh, Pennsylvania, July 2005, pp. 64–71 (2005)
5. Hignette, G., Buche, P., Dibie-Barthélemy, J., Haemmerlé, O.: Semantic annotation of data tables using a domain ontology. In: Proceedings of the 10th International Conference on Discovery Science (DS 2005), Sendai, Japan, pp. 253–258 (October 2007)
6. Jha, P., Nagy, G.: Wang notation tool: Layout independent representation of tables. In: Proceedings of the 19th International Conference on Pattern Recognition ICPR 2008, Tampa, Florida (December 2008) (in press)
7. Kifer, M., Lausen, G., Wu, J.: Logical foundations of object-oriented and frame-based languages. Journal of the Association for Computing Machinery 42(4), 741–843 (1995)
8. Lynn, S.: Automating mini-ontology generation from canonical tables. Master's thesis, Department of Computer Science, Brigham Young University (2008)
9. Pazienza, M.T., Stellato, A.: An open and scalable framework for enriching ontologies with natural language content. In: Ali, M., Dapoigny, R. (eds.) IEA/AIE 2006. LNCS, vol. 4031, pp. 990–999. Springer, Heidelberg (2006)
10. Pivk, A., Sure, Y., Cimiano, P., Gams, M., Rajkovič, V., Studer, R.: Transforming arbitrary tables into logical form with TARTAR. Data & Knowledge Engineering 60, 567–595 (2007)
11. Rahman, F., Klein, B.: Special issue on detection and understanding of tables and forms for document processing applications. International Journal of Document Analysis 8(2), 65 (2006)
12. Silva, A.C.e., Jorge, A.M., Torgo, L.: Design of an end-to-end method to extract information from tables. International Journal of Document Analysis and Recognition 8(2), 144–171 (2006)
13. Tijerino, Y.A., Embley, D.W., Lonsdale, D.W., Ding, Y., Nagy, G.: Toward ontology generation from tables. World Wide Web: Internet and Web Information Systems 8(3), 261–285 (2005)
14. Zanibbi, R., Blostein, D., Cordy, J.R.: A survey of table recognition: Models, observations, transformations, and inferences. International Journal of Document Analysis and Recognition 7(1), 1–16 (2004)

Semantic Assistants – User-Centric Natural Language Processing Services for Desktop Clients

René Witte[1] and Thomas Gitzinger[2]

[1] Department of Computer Science and Software Engineering
Concordia University, Montréal, Canada
[2] Institute for Program Structures and Data Organization (IPD)
University of Karlsruhe, Germany

Abstract. Today's knowledge workers have to spend a large amount of time and manual effort on creating, analyzing, and modifying textual content. While more advanced semantically-oriented analysis techniques have been developed in recent years, they have not yet found their way into commonly used desktop clients, be they generic (e.g., word processors, email clients) or domain-specific (e.g., software IDEs, biological tools). Instead of forcing the user to leave his current context and use an external application, we propose a "Semantic Assistants" approach, where semantic analysis services relevant for the user's current task are offered directly within a desktop application. Our approach relies on an OWL ontology model for context and service information and integrates external natural language processing (NLP) pipelines through W3C Web services.

1 Introduction

Consider the following scenarios: (1) A scientific journalist, while writing an article on the global climate change, needs to find information on the role of DMSP[1] in the Atlantic marine biology. A *Google* search finds thousands of hits on this topic, forcing our user to interrupt his writing in order to manually evaluate the results. (2) A software developer, while editing code in his IDE, needs to trace a method back to the requirements document in order to understand why a certain feature was implemented. But requirements are not directly linked at the source code level, forcing our developer to interrupt her code analysis and switch to document retrieval and editing tools.

Both scenarios highlight two particularities of today's desktop environments: First, whenever dealing with textual information, users do not get any semantic analysis support besides full-text information retrieval (and document search). Although research in natural language processing (NLP) and text mining has developed a large number of tools and applications within the last decade, like

---

[1] DMSP stands for "Dimethylsulfoniopropionate" and is a component of the organic sulfur cycle.

J. Domingue and C. Anutariya (Eds.): ASWC 2008, LNCS 5367, pp. 360–374, 2008.
© Springer-Verlag Berlin Heidelberg 2008

machine translation, question-answering, summarization, topic detection, cluster analysis, and information extraction [1], none of these newly developed technologies have materialized in the standard desktop tools commonly used by today's knowledge workers—such as email clients, software development environments (IDEs), or word processors. This directly leads to the second observation: The vast majority of users still relies on manual retrieval of relevant information through an information retrieval tool or website and subsequent manual processing of the (often millions of) results—forcing the user to interrupt his workflow by leaving his current client and performing all the "natural language processing" himself, before returning to his actual task.

The core idea of our *Semantic Assistants* approach is to take the existing NLP frameworks, wrap concrete analysis pipelines in an OWL-based semantic description that can be brokered through a service-oriented architecture (SOA), and allow desktop clients to connect to this architecture with a plug-in interface using Web services. The semantic analysis services, like question-answering or summarization, then become available directly in the desktop tool used for manipulating content. The following figure illustrates this idea:

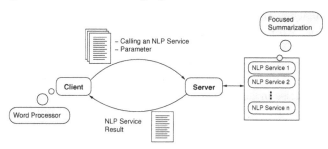

The user is not concerned with the implementation or integration of these services, from his point of view he only sees context-sensitive Semantic Assistants relevant for his task at hand.

2 Requirements Analysis

In this section, we define the requirements for our *Semantic Assistants* more precisely. As stated before, the central goal is to bring semantic text analysis support to the end user, by integrating NLP systems with desktop clients:

REQUIREMENT #0: SEMANTIC ASSISTANTS ARCHITECTURE. *Provide the infrastructure for bringing semantic services for content analysis and development offered by NLP systems directly to end-user clients in a context-sensitive fashion.*

An immediate observation is that a number of different user groups will be involved in the creation of these assistants: language engineers, software engineers, and end users. Hence, we adopt a *separation of concerns:* First, there is the role of the *end user.* Considering his perspective, we think about a user working with a client program enriched through semantic services, and not knowing or caring much about the underlying technology. Secondly, we adopt the perspective of the *system integrator*. His role consists of "plugging" new clients or NLP services into our architecture. System integrators form an important user group, because with every client and every NLP service, they add value to the overall system. Also, this group's requirements can be quite different from those of the end user group. Additionally, we define the role of the *language engineer*, who is actually creating and developing the semantic services, like question-answering, summarization, or information extraction. However, within this paper we do not discuss the engineering of specific language services—a topic already widely discussed in the literature[1]—since for the integration capabilities of our approach it should be of no concern *how* a language service was created. Likewise, the language engineer should not need to care about whether his NLP service will be integrated or not. Therefore, within this paper, we simply assume that NLP services exist, and do not inquire where they came from. Finally, we assume the *system* perspective, where we take care of properties desirable for the system as a whole.

2.1   End User Requirements

"End users" is a very broad term, and necessarily so: we want to facilitate the work of knowledge workers who use any kind of desktop client while working with textual content. Word processors, email clients, PIMs and IDEs are used by a large variety of users for numerous tasks. There should be no *a priori* assumptions; End user characters can range from secretaries to researchers to school kids to housewives. In particular, this means that we cannot expect these users to have any expertise in semantic technologies, language engineering, or software engineering. The most important and obvious consequence from this observation is the requirement to design a client-independent system architecture:

REQUIREMENT #1: CLIENT INDEPENDENCE.   *The architecture must be open and flexible with respect to the type of clients integrated with it.*

Thus, the architecture must not be limited to any particular user group or software type. It should even allow the connection of future clients that do not currently exist.

REQUIREMENT #2: CONTEXT SENSITIVITY.   *Not every analysis method is suitable for every situation, user, or document, but can rather depend on language capabilities (of both the users and analysis services), data formats, and the user's current goal and task.*

In other words, Semantic Assistants must be equipped with some kind of context model that captures the user's context and matches it with applicable services in order to be able to recommend helpful assistants.

2.2   System Integrator Requirements

The second group of users are *system integrators:* developers who integrate either some client software or a new semantic service into our architecture. We want to make their job as easy as possible by taking care of some important issues:

REQUIREMENT #3: FACILITATE CLIENT INTEGRATION.   *Clients in the form of end-user applications or user agents partly acting on their own are the user's "entry door"* to our architecture. *Without such clients, the architecture is rather useless, so Requirement #3 is a very fundamental one: Allow for the integration of any client, and enable developers to do this in an effective and efficient manner.*

After all, the main reason current desktop clients do not offer sophisticated semantic NLP services is that their integration involves taking care of many cumbersome details, like connection settings and remote procedure calls. Further refining this requirement, we have to take care of two details: *(1) Server Abstraction:* the integrator must be shielded from low-level communication details between the client and the server. *(2) Implementation of Common Client Functionality:* Many clients will most likely have to fulfill some of the same or similar tasks, like finding available NLP services, passing text from the desktop client, or retrieving results from the service. To avoid duplication of these common functions, a reusable abstraction layer should be provided to system integrators.

REQUIREMENT #4: FACILITATE NLP SERVICE INTEGRATION. *NLP services must not be hardcoded but rather discovered on-the-fly by the architecture.*

This requirement allows for new services to be plugged into the architecture, thereby becoming immediately available to any connected end-user client. In order words, this requires the development of service metadata, based on a formal specification.

2.3   System Requirements

Finally, we address some additional core architectural requirements that stem from the point of view of the "system role." First, it is quite plausible that clients need to have a way to provide input to the semantic services they wish to use—otherwise, these services would have nothing to work on:

REQUIREMENT #5: INTUITIVE INPUT PASSING FOR CLIENTS. *Allow for an easy transfer of unstructured data from the client to the analysis service.*

Language services might require individual parameters, such as the length of a summary to be generated. Some of these parameters can be automatically determined by the architecture (such as input/output connections), but some should be configurable by the end user. Therefore, we have:

REQUIREMENT #6: CONTROL OF LANGUAGE SERVICES. *Provide a way to transmit parameter values from the client to the actual language processing component(s).*

In particular, the architecture must enable parameter detection, client notification of required parameters, and handle correct value assignment of individual parameters to language services (e.g., the length of a summary to be generated).

REQUIREMENT #7: FLEXIBLE RESULT HANDLING. *The final step in the process of invoking a language service includes taking the result or results of the service, wrapping it up in a response message and sending that message back to the client.*

Here, the architecture must be capable of detecting the output(s) provided by
the language services: these can be new documents, stored in a file or database,
or annotations attached to an existing or newly retrieved document. Our architecture must therefore allow for a description of these outputs along with the
description of a language service. This description has to be detailed enough so
that the architecture knows how to capture and handle the produced output.
Furthermore, to allow for a simple client-side integration, we postulate a uniform
response format that must be expressive enough to capture all the various result
forms.

2.4   Related Work

Some previous work exists in building personalized information retrieval agents,
e.g., for the Emacs text editor [2] or Microsoft Word [3]. These approaches are
typically focused on a particular kind of application (e.g., emails or word processing), whereas our approach is general enough to define NLP services independent
from the end-user application through an open, client/server infrastructure.

The most widely found approach for bringing NLP to an end user is the
development of a new interface (be it Web-based or a "fat client"). These applications, in turn, embed NLP frameworks for their analysis tasks, which can be
achieved through the APIs offered by frameworks such as GATE [4] or UIMA
[5]. The BioRAT system [6] targeted at biologists is an example for such a tool,
embedding GATE to offer advanced literature retrieval and analysis services.
In contrast with these approaches, we provide a service-oriented architecture to
broker any kind of language analysis service in a network-transparent way. Our
architecture can just as well be employed on a local PC as it can deliver focused
analysis tools from a service provider. For example, a commercial scientific publisher might want to offer a "related work finder" analysis service, similar to the
one presented by [7], to scientists writing research papers or proposals.

Recent work has been done in defining Web services for integrating NLP
components. In [8], a service-oriented architecture geared towards terminology
acquisition is presented. It wraps NLP components as Web services with clearly
specified interface definitions and thus allows language engineers to easily create
and alter concatenations of such components, also called processing configurations. Their work is complimentary to our approach, since it addresses the
composition of NLP components into pipelines, whereas we are concerned with
semantic descriptions of existing pipelines and their integration with desktop
clients.

Close in spirit to our approach is the work performed by the *Semantic Desktop*
community [9]. The architectures developed in this area, like IRIS [10],[2] aim
at deriving, maintaining, and exchanging semantic metadata between desktop
applications to provide better semantic support for end users. This work can
again be seen as complimentary to our approach, as the type of services offered
and their data granularity differs greatly.

---

[2] The same holds for similar approaches, like Nepomuk, Gnowsis, or Haystack.

3   Semantic Assistants Design

We now describe the design of our Semantic Assistants, in particular the overall system architecture (Section 3.1) and our ontology-based NLP service and context model (Section 3.2).

3.1   System Architecture

An overview of our system architecture developed for the stated requirements is shown in Fig. 1. It is based on a typical multi-tier information system design.

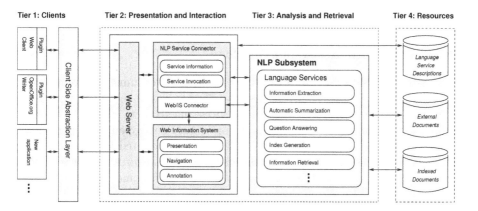

Fig. 1. Architecture for integrating text analysis services and end-user clients

Tier 1: Clients. This tier has the main purpose of providing access to the system. Typically, this will be an existing client (like a word processor or email client), extended to connect with our architecture through a plug-in interface. Besides facilitating access to the whole system, clients are also in part responsible for presentation, e.g., of language service results. In addition to the actual client applications, an abstraction layer is part of Tier 1. It shields the clients from the server and provides common functionality for NLP services, as stipulated by Requirement #3.

Tier 2: Presentation and Interaction. Tier 2 consists of a standard Web server and a module labeled "NLP Service Connector" in the diagram. One responsibility of this module is interaction, in that it handles the communication flows between the NLP framework and the Web server. Moreover, it prepares language service responses, by collecting results from the NLP services and transforming them into a format suitable for transmission to the client (Requirement #7). Finally, the NLP Service Connector reads the descriptions of the language services, and therefore "knows" the vocabulary used to write these descriptions, as discussed in Requirement #4. In addition, our architecture can also provide services to other (Web-based) information system, like a Wiki [11].

Tier 3: Analysis and Retrieval. Tiers 1 and 2 are the ones the user has direct contact with. Tier 3 is only directly accessed by the NLP Service Connector. It contains the core functionality that we want, through Tiers 1 and 2, to bring to the end user. Here, the semantic services reside, and the NLP subsystem in whose environment they run (such as GATE or UIMA). Language services can be added and removed from here as required.

Tier 4: Resources. The final tier "Resources" contains the descriptions of the language services. These descriptions are read by the NLP Service Connector so that it can satisfy its clients' information needs. Whenever a new language service is added to the architecture, its description must be added here, too (Requirement #4). The vocabulary of these descriptions is based on an OWL ontology using description logics (OWL-DL). Furthermore, indexed documents as well as external documents (e.g., on the Web) count as resources.

3.2   The Semantic Assistants Ontology

As discussed in Requirement #2, in order to be able to recommend semantic services relevant for the user's current task, we need to model the current context, which includes the user's task, available services, the artifacts involved (services, documents, etc.) and their languages. We previously developed an upper ontology for supporting software processes [12], which we adapted for the Semantic Assistants setting. It includes five essential concepts that form the basis of this upper ontology, namely *Artifact*, *Format*, *User*, *Language*, and *Task*:

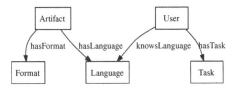

Artifacts include both content (documents, Web pages, etc.) and tools (such as an NLP tool). Formats (e.g., OWL, XML, or PDF) have been modeled separately from languages (both natural languages and artificial languages) as they are largely orthogonal. Users are modeled with their language capabilites and the tasks they need to perform.

Modeling Artifacts.   We can now start to extend the *Artifact* concept of the upper ontology with concepts required for Semantic Assistants (Fig. 2), as stipulated by Requirement #4. We have just mentioned *Tool* as a sub-concept of *Artifact*, a tool possibly processing artifacts as input and producing artifacts as output (*consumesInput* and *producesOutput* relations). Additionally, a tool may require parameters (*hasParameter* relation). For Semantic Assistants, *documents* are a focal point. NLP services are often language-specific, so to be able to only offer assistants relevant for the language of a document a user is working on, and the language(s) he understands, we make use of the *hasLanguage* relation.

In order to work with documents, they often must somehow be identified and retrieved. In particular, the server must be able to pull documents from a networked source, including the Internet. Thus, when we model such an input

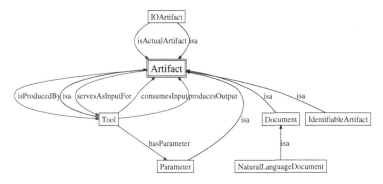

Fig. 2. Modeling Artifacts in the Upper Ontology

document in our ontology, we must have a way to specify its URI (Uniform Resource Identifier) by which we can address it. As not only documents, but also other artifacts like Web services have to be uniquely identified and addressed, we introduce *hasIdentifier* as an optional property for artifacts. We define *IdentifiableArtifact* as a class whose members are artifacts and have this property. In practice, the identifier can, and often will, be a URI, but it does not have to be. For example, if we have a set of elements with unique names, a simple string can be enough. With *hasIdentifier*, we provide an important property on the highest abstraction level, while leaving the exact semantics to the concrete ontologies and the applications that use them.

The same tool can often produce different output (types), depending on both *how* it is invoked (parameters) and the type of the input artifacts (if any). To model this fact, we introduced a concept called *IOArtifact*, where information on input and output relationships can be stored. We will show a more concrete use of this concept in the following section, when we concretize the upper ontology. By means of an *isActualArtifact* relation, we have *IOArtifact* individuals "point" to *Artifact* individuals. They can be seen as proxies for artifacts.

Specializing the Upper Ontology. The upper ontology that we have just introduced provides us with several concepts we need: artifacts, users, parameters, tools, etc. However, to integrate NLP analysis tools on a semantic level, we have to refine this abstract ontology for language services. While the overall design is largely independent from a concrete NLP system, at this point we also introduce concepts specific to our environment, which is based on GATE [4]. However, other NLP subsystems (like UIMA [5]) can be integrated in a similar fashion.

To be able to offer semantic services to end users, we need to model existing NLP analysis services, like summarization (Requirement #4). Towards this end, we introduce new child concepts to the *Tool* concept, classifying language services into two categories: *IRTool* and *NLPTool*. The semantics that we want to convey by this separation is that an information retrieval tool *(IRTool)* finds documents, but leaves them untouched, while an NLP tool processes documents and typically generates some new artifact(s) from them. For NLP tools, input and output natural languages can be specified, if they are language-specific. A *GATEPipeline* is an analysis service with a certain format that can perform either type (or both) of document processing. Instances of *GATEPipeline* are the language services we offer the user through our architecture.

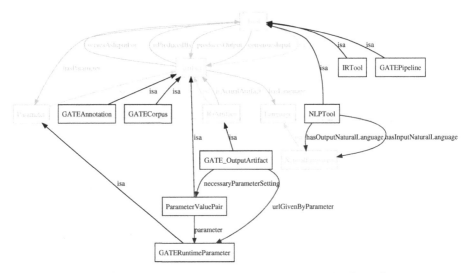

Fig. 3. Additional classes introduced in the specialized ontology

To be able to invoke concrete services, we also have to extend the *Artifact* concept. A *GATECorpus* represents a collection of documents, which typically serves as input for a language service. *GATEAnnotation* instances describe the information that language services add to a document during processing, including their result output. *GATERuntimeParameter* models parameters that control certain aspects of a GATE pipeline, and is introduced as a sub-concept to the already existing *Parameter* concept. Along with these artifact types, we have to introduce corresponding new formats (remember that artifacts are required to have a format). These are defined under the common parent concept of *GATEFormat*, which is a child of the *Format* concept.

We mentioned earlier that output formats can change depending on both parameters passed to a tool (like a GATE pipeline) and its input document, which led to the introduction of the *IOArtifact* concept. To allow our architecture to automatically retrieve the obtained result and deliver it to the end user, we introduce the *GATE_OutputArtifact*. Instances of *GATE_OutputArtifact* have a property *necessaryParameterSetting*, thus connecting an output artifact to a certain parameter value. This parameter value is represented through an instance of a newly introduced concept called *ParameterValuePair*.

4   Implementation

We now discuss selected aspects of the current implementation of our architecture and briefly describe the process of integrating new (desktop) clients in Section 4.5.

4.1 Language Service Description and Management

We start discussing the implementation from the status quo, a common component-based NLP framework—in our case, GATE [4]. The NLP subsystem allows us to load existing language services, provide them with input documents and parameters, run them, and access their results. The GATE framework's API also permits to access all the artifacts involved in this process (documents, grammars, lexicons, ontologies, etc.). Language services take the form of (persistent) *pipelines* or *applications* in GATE, which are composed of several sequential *processing resources* or *PRs*. As mentioned in Section 2, creating a language service (e.g., for summarization or question-answering) is the responsibility of a language engineer, and need not further concern us here.

Our ontology described in the previous section has been implemented using OWL-DL.[3] For each deployed language service, the corresponding entries in the Semantic Assistants ontology need to be created and stored on the server-side (Fig. 1, Tier 4). This permits us to dynamically find, load, parametrize, and execute available language services, based on the user's current task and language capabilities (this is further discussed in Section 4.4). Additionally, the ontology contains all information needed to to locate and retrieve the result(s) delivered (Requirement #7).

4.2 Web Services

Thus far, we can search, load, parametrize, and execute language services. However, all input/output channels are still local to the context of the NLP framework's process. To make NLP services available in a distributed environment, we have to add network capabilities, which we achieve using *Web services*, a standard defined by the W3C:[4] *"A Web service is a software system designed to support interoperable machine-to-machine interaction over a network. It has an interface described in a machine-processable format (specifically WSDL[5]). Other systems interact with the Web service in a manner prescribed by its description using SOAP[6] messages, typically conveyed using HTTP with an XML serialization in conjunction with other Web-related standards."* In essence, a requester agent has to know the description of a Web service to know how to communicate with it, or, more accurately, with the provider agent implementing this Web Service. It can then start to exchange SOAP messages with it in order to make use of the functionality offered by the service. Provider agents are also referred to as Web service *endpoints*. Endpoints are referenceable resources to which Web service messages can be sent. Within our architecture (Fig. 1), the central piece delivering functionality from the NLP framework as a Web service endpoint is the NLP Service Connector. Our implementation makes use of the Web service code generation tools that are part of the Java 6 SDK and the Java API for XML-Based Web services (JAX-WS).[7]

---

[3] OWL Web Ontology Language Guide, http://www.w3.org/TR/owl-guide/

[4] Web Services Architecture, see http://www.w3.org/TR/ws-arch/

[5] Web Services Description Language (WSDL), see http://www.w3.org/TR/wsdl

[6] Simple Object Access Protocol, see http://www.w3.org/TR/soap/

[7] Java API for XML-Based Web Services (JAX-WS), see https://jax-ws.dev.java.net/

With all necessary artifacts in place, we can now generate and publish the Web service. The JAX-WS API provides convenient functions for this, so that, with two lines of source code (comments not counted), we can start a Web server integrated with the Java environment, and publish the Web service at an address of our choice:

```
// Create SSB instance
SemanticServiceBroker agent = new SemanticServiceBroker();
// Publish SSB instance as Web service endpoint
Endpoint endpoint = Endpoint.publish("http://localhost/...", agent);
```

4.3   The Client-Side Abstraction Layer (CSAL)

We have just published a Web service endpoint, which means that the server of our architecture is in place. On the client side, our *client-side abstraction layer* (CSAL) is responsible for the communication with the server. This CSAL offers the necessary functionality for clients to detect and invoke brokered language services. The implementation essentially provides a proxy object (of class SemanticServiceBroker), through which a client can transparently call Web services. A code example, where an application obtains such a proxy object and invokes the getAvailableServices method on it to find available language analysis services, is shown below:

```
// Create a factory object
SemanticServiceBrokerService service = new SemanticServiceBrokerService();
// Get a proxy object, which locally represents the service endpoint (= port)
SemanticServiceBroker broker = service.getSemanticServiceBrokerPort();
// Proxy object is ready to use. Get a list of available language services.
ServiceInfoForClientArray sia = broker.getAvailableServices();
```

4.4   Dynamic Assistant Generation

The ontology described in Section 3.2 contains the information needed to dynamically find, load, parametrize, and execute available language services, based on user's current task and language capabilities. In our implementation, it is queried using Jena's SPARQL[8] interface, using the context information delivered by the client plug-in, in order to recommend applicable Semantic Assistants.

For example, when a recommendation request is received, with a context object saying the user knows English and German, the generated SPARQL query should restrict the available services to those that deliver English or German as output language. A simplified version of such a generated query is shown below:

```
SELECT ?x ?name
WHERE { ?x sa:hasGATEName ?name .
       {?x cu:hasFormat sa:GATECorpusPipeline_Format} . {
          {?x sa:hasOutputNaturalLanguage cu:en} UNION
             {?x sa:hasOutputNaturalLanguage cu:de}}
}
```

Once the SPARQL query has been generated, it is passed to the OntModel instance containing the language service descriptions. The results are then retrieved from this object, converted into the corresponding client-side versions, and returned to the client.

---

[8] SPARQL Query Language for RDF, see http://www.w3.org/TR/rdf-sparql-query/

4.5   Client Integration

After describing the individual parts of our architecture's implementation, we now show how they interact from the point of view of a system integrator adding Semantic Assistants to a client application. The technical details depend on the client's implementation: If it is implemented in Java (or offers a Java plug-in framework), it can be connected to our architecture simply by importing the CSAL archive, creating a `SemanticServiceBrokerService` factory, and calling Web services through a generated proxy object. After these steps, a Java-enabled client application can ask for a list of available language services, as well as invoke a selected service. The code examples shown above demonstrate that a developer can quite easily integrate his application with our architecture, without having to worry about performing remote procedure calls or writing network code.

A client application developer who cannot use the CSAL Java archive still has access to the WSDL description of our Web service. If there are automatic client code generation tools available for the programming language of his choice, the developer can use these to create CSAL-like code, which can then be integrated into or imported by his application.

5   Application

In this section, we present a real-world application scenario for our architecture: the integration of Semantic Assistants into a word processor.

5.1   The OpenOffice.org Writer Plug-In

Word processor applications are one of the primary tools of choice for many users when it comes to creating or editing content. Thus, they are an obvious candidate for our approach of bringing advanced NLP support directly to end users in form of Semantic Assistants. We selected the open source OpenOffice.org[9] application *Writer* for integration. With its plug-in framework, extensions can easily be added to any OpenOffice.org application.

Following the steps described in Section 4.5, we developed a Java plug-in for *Writer* that offers the functionality to connect with our architecture, inquire about available language services, offers additional dialogs for selecting services and setting required parameters, and handles passing of input documents and NLP results. Depending on the selected language service, either the full document a user is working on can be sent to the language service, or only a highlighted text segment. On the back-end, we integrated some of the language services we developed for other projects, which include index generation, automatic summarization, and question-answering (focused summarization).

Our plug-in creates a new menu entry "Semantic Assistants," as shown in Fig. 4. In this menu, the user can inquire about available services, which are selected based on the client (here *Writer*) and the available languages, as described in Section 4.4. The dynamically generated list of available services is then

---

[9] Open source office suite OpenOffice.org, see http://www.openoffice.org/

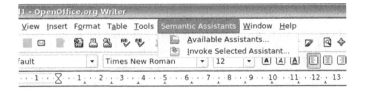

Fig. 4. "Semantic Assistants" menu entry in OpenOffice.org Writer

presented to the user, together with a brief description, in a separate window. The user can then select an assistant and execute it. In case the service requires additional parameters, such as the length of a summary to be generated, they are detected by our architecture through the OWL-based service description and requested from the user through an additional dialog window.

Once invoked, the language service is executed asynchronously by our architecture, allowing the user to continue his work (he can even execute additional services). Note that all low-level details of handling language services, such as metadata lookup, parametrization, and result handling, are hidden from the client plug-in through our client-side abstraction layer.

5.2 Example Use Case

One direct use case of our Semantic Assistants is to satisfy information needs of a knowledge worker. As motivated in Section 1, language services can deliver focused analysis results directly within the client—here a word processor—needed to perform a task, rather than interrupting the user's workflow by forcing him to perform an external (Web) search.

Let us go back to the example scenario stated in the introduction: a scientific journalist who is writing a report on the global climate change and needs information on the role of *"DMSP in the Atlantic marine biology."* Using our Semantic Assistants, he can simply highlight this phrase in the editor and select the "Web Retrieval Summarizer." This is a compound assistant that performs

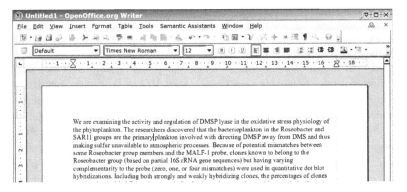

Fig. 5. The result of the "Web Retrieval Summarizer" Semantic Assistant, answering the user's question, is presented as a new *Writer* document

two tasks: In a first step, a selected number of hits from a Yahoo! search using the highlighted phrase is retrieved to build a corpus on-the-fly. This corpus is then fed into the multi-document summarizer ERSS [13] to produce a summary answering the question(s) (a so-called *focused summary*). All these actions are performed in the background, allowing the user to continue with other parts of his report. When the summary is ready, the architecture notifies the plug-in, which then presents the generated summary in a new window (Fig. 5). The user can now inspect the result, refine it, and continue with other parts of the report.

6 Conclusions and Future Work

In this paper, we presented the idea of "Semantic Assistants" that aim to help users dealing with the proverbial information overload. In particular, we address end-users that need to find, analyze, or write any kind of textual content. The central idea of our work is that such support should be offered directly integrated into the clients users are accustomed to when working on natural language data: their email clients, Web browsers, word processors or editors. Thus, instead of offering analysis services through a Web interface or custom-build applications, we propose to integrate them directly into end-user clients by means of a service-oriented architecture, based on W3C Web services. An important design goal of our architecture is that it should be as easy as possible to integrate existing clients through plug-in frameworks on the user side, and new semantically-oriented NLP services on the server side.

Our work is the first that aims to bring existing NLP analysis services directly to end users. We believe this is an important goal as there have been numerous advances in the areas of NLP and text mining over the last decade—but none of the newly developed tools have yet found their way into today's desktop environments. Rather, in order to find and process content, users still have to leave their application of choice and perform an external (desktop or Internet) search, forcing a mentally expensive context-switch. In our paradigm, desktop applications can directly react to the user's need for retrieving and analyzing content by offering semantically-oriented services, such as question-answering or summarization, within the same interface. We achieve this by adding a layer of "semantic glue" using an OWL-DL context and service ontology that permits us to connect the existing, but so far separated, worlds of desktop applications and NLP frameworks in a way that brings added value to end users.

Our implementation shows that these ideas can be implemented with current, off-the-shelf tools and open standards. A first practical evaluation of our approach, by integrating a major open source application, the OpenOffice.org *Writer* program, proves that the Semantic Assistants concept can be deployed on a contemporary desktop environment. Future user studies will need to be performed to evaluate the impact of such services on the completion of prede-fined tasks; however, this will obviously highly depend on the selected type of client, the tasks, and the deployed language services and must therefore not be confused with the evaluation of our architecture and ontology model as such.

Obviously, many extensions are still possible throughout the architecture, but the one most beneficial to end users will be the development of new client plug-ins, bringing further semantic support to, e.g., email clients (searching for relevant information and providing answers to questions), software development environments (linking code to its documentation and offering support when modifying either side), or domain-specific tools (like for biologists or architects). We believe that bringing the existing, hard-won advances in natural language processing, like question-answering, summarization, or opinion mining, to a larger user base will have significant impact on the fields of NLP, semantic desktop research, and software engineering work. Users can directly benefit from a large number of developed technologies that so far have been limited to expert users and proprietary commercial applications. The developed architecture will be made available under an open source license, which we hope will foster a vibrant ecosphere of client plug-ins.

References

1. Feldman, R., Sanger, J.: The Text Mining Handbook: Advanced Approaches in Analyzing Unstructured Data. Cambridge University Press, Cambridge (2006)
2. Rhodes, B.J., Maes, P.: Just-in-time Information Retrieval Agents. IBM Syst. J. 39(3-4), 685–704 (2000)
3. Colbath, S., Kubala, F.: TAP-XL: An Automated Analyst's Assistant. In: Proc. NAACL 2003, ACL, pp. 7–8 (2003)
4. Cunningham, H., Maynard, D., Bontcheva, K., Tablan, V.: GATE: A Framework and Graphical Development Environment for Robust NLP Tools and Applications. In: Proc. of the 40th Anniversary Meeting of the ACL (2002), http://gate.ac.uk
5. Ferrucci, D., Lally, A.: UIMA: An Architectural Approach to Unstructured Information Processing in the Corporate Research Environment. Natural Language Engineering 10(3-4), 327–348 (2004)
6. Corney, D.P., Buxton, B.F., Langdon, W.B., Jones, D.T.: BioRAT: Extracting Biological Information from Full-Length Papers. Bioinformatics 20(17), 3206–3213 (November 2004)
7. Zeni, N., Kiyavitskaya, N., Mich, L., Mylopoulos, J., Cordy, J.R.: A lightweight approach to semantic annotation of research papers. In: Kedad, Z., Lammari, N., Métais, E., Meziane, F., Rezgui, Y. (eds.) NLDB 2007. LNCS, vol. 4592, pp. 61–72. Springer, Heidelberg (2007)
8. Cerbah, F., Daille, B.: A service oriented architecture for adaptable terminology acquisition. In: Kedad, Z., Lammari, N., Métais, E., Meziane, F., Rezgui, Y. (eds.) NLDB 2007. LNCS, vol. 4592, pp. 420–426. Springer, Heidelberg (2007)
9. Decker, S., Park, J., Quan, D., Sauermann, L. (eds.): Proc. of the 1st Workshop on The Semantic Desktop, Galway, Ireland, CEUR Workshop Proceedings, vol. 175 (November 6, 2005), CEUR-WS.org
10. Cheyer, A., Park, J., Guili, R.: IRIS. Integrate. Relate. Infer. Share. In: [9] (2005)
11. Witte, R., Gitzinger, T.: Connecting Wikis and Natural Language Processing Systems. In: Proc.of the 2007 Intl. Symp. on Wikis, WikiSym 2007 (2007)
12. Rilling, J., Meng, W.J., Witte, R., Charland, P.: A Story Driven Approach to Software Evolution. IET Software (2008)
13. Witte, R., Bergler, S.: Fuzzy clustering for topic analysis and summarization of document collections. In: Kobti, Z., Wu, D. (eds.) Canadian AI 2007. LNCS, vol. 4509, pp. 476–488. Springer, Heidelberg (2007)

Exploiting Gene Ontology to Conceptualize Biomedical Document Collections

Hai-Tao Zheng, Charles Borchert, and Hong-Gee Kim*

Biomedical Knowledge Engineering Laboratory, Seoul National University
28 Yeongeon-dong, Jongro-gu, Seoul, Korea
hgkim@snu.ac.kr

Abstract. As biomedical science progresses, ontologies play an increasingly important role in easing the understanding of biomedical information. Although much research, such as Gene Ontology annotation, has been proposed to utilize ontologies to help users understand biomedical information easily, most of the research does not focus on capturing gene-related terms and their relationships within biomedical document collections. Understanding key gene-related terms as well as their semantic relationships is essential for comprehending the conceptual structure of biomedical document collections and avoiding information overload for users. To address this issue, we propose a novel approach called 'GOClonto' to automatically generate ontologies for conceptualization of biomedical document collections. Based on GO (Gene Ontology), GOClonto extracts gene-related terms from biomedical text, applies latent semantic analysis to identify key gene-related terms, allocates documents based on the key gene-related terms, and utilizes GO to automatically generate a corpus-related gene ontology. The experimental results show that GOClonto is able to identify key gene-related terms. For a test biomedical document collection, GOClonto shows better performance than other clustering algorithms in terms of F-measure. Moreover, the ontology generated by GOClonto shows a significant informative conceptual structure.

1 Introduction

In the biomedical domain, ontologies have been widely used to represent sets of concepts and the relationships between those concepts. With increasing biomedical information availability, much research, such as Gene Ontology annotation, has been proposed to utilize ontologies to help users understand the information easily. However, most of the existing methods do not use ontologies to help users directly capture key gene-related terms and their relationships within biomedical document collections. In this paper, key gene-related terms are considered as the most important gene-related terms to which a biomedical document collection are related. Understanding key gene-related terms and their semantic relationships is

---

* Corresponding author.

J. Domingue and C. Anutariya (Eds.): ASWC 2008, LNCS 5367, pp. 375–389, 2008.
© Springer-Verlag Berlin Heidelberg 2008

essential for comprehending the conceptual structure of biomedical document collections and avoiding information overload for users. Since GO (Gene Ontology) [1] provides a controlled vocabulary to describe gene and gene product attributes in any organism, it can be used to represent knowledge related to biomedical document collections on a conceptual level. Using a corpus-related gene ontology, which is a subset of Gene Ontology, users can easily visualize not only to which key gene-related terms the documents are related, but also the semantic relationships between groupings of the documents, via these gene-related terms.

In this study, we propose a novel approach called 'GOClonto' to identify the key gene-related terms and automatically generate corpus-related gene ontologies based on these key gene-related terms, for conceptualization of biomedical document collections. GOClonto has been developed from the Clonto method, which focuses on using WordNet to conceptualize general document corpora [2]. Conceptualization of biomedical document collections here means representing document collections with a set of key gene-related terms and their semantic relationships, which can help users more easily understand biomedical document contents. First, GOClonto extracts gene-related terms that are contained in GO, which we call GO-terms, from a biomedical document collection. Then, GOClonto applies LSA (latent semantic analysis) to identify key GO-terms, allocates documents based on these key GO-terms, and uses GO to automatically generate a corpus-related gene ontology. Finally, the biomedical documents are linked to the ontology through key GO-terms. The main contribution of this paper is proposing a novel method that exploits GO to automatically generate ontologies for conceptualizing biomedical document collections.

The rest of the paper is organized as follows: Section 2 discusses the related work. Section 3 elaborates the GOClonto method to show the process of key GO-term identification and ontology generation. Section 4 presents our experimental results. We give our conclusion and future work in section 5.

2   Related Work

To help users better understand the structure of document collections, several clustering algorithms that extract meaningful labels for documents have been proposed. Zamir et al [3,4] proposed a phrase-based document clustering approach based on suffix tree clustering, which uses shared suffixes to identify and label base clusters of documents and combines them into final clusters. Schockaert et al [5] developed a clustering method using Fuzzy Ants, which uses ant colony optimization principles to find good partitions of the data. Lang et al [6] presented an algorithm for web search results clustering based on Tolerance Rough Set (TRS), which is able to deal with vagueness and fuzziness and is used to model relations between terms and documents. Osinski et al [7] proposed a concept-driven algorithm for clustering search results, the Lingo algorithm, which uses LSA (Latent Semantic Indexing) techniques to separate search results into meaningful labeled groups.

To better visualize document collections, a number of exploratory visualization tools are described in [8]. Olsen [9] developed a document visualization system called VIBE. Grobelnik et al [10] presented a system for visualization of large amounts of new stories. Fortuna et al [11] used LSI techniques for visualization of text document collections. Zhu [12] introduced the design and application of an integrated exploratory visualization system called Storylines. Shaw et al [13] describes a three-dimensional volumetric interactive information visualization system for management and analysis of document collections. In the Semantic Web community, Fluit et al [14] described several applications for ontology-based information visualization. For example, an ontology-based visualization tool, AutoFocus, was proposed to display search results for documents on desktops as clusters of populated concepts. Thai et al [15] presented IVEA (Information Visualization for Exploratory Document Collection Analysis), a visualization tool which employs the PIMO (Personal Information Model) ontology to provide knowledge workers with an interactive interface allowing them to browse for information in a personalized manner.

However, the above methods are only proposed for general purposes of document conceptualization, and they do not identify GO-terms found in biomedical text. To deal with this problem, much research has been proposed related to GO-term annotation[16,17,18,19,20]. Hill et al [19] addressed the question of what GO annotations signify and of how they are created by working with biologists. Camon et al [17] presented the GOA (Gene Ontology Annotation) database, which aims to provide high-quality electronic and manual annotations to the UniProt Knowledgebase (Swiss-Prot, TrEMBL and PIR-PSD) using the standardized vocabulary of the Gene Ontology. Bada et al [16] presented GOAT, which aims to aid the user in the annotation of gene products with GO terms by displaying those field values that are most likely to be appropriate based on previously entered terms. Seki et al [20] described an application of IR (Information Retrieval) and text categorization methods to a highly practical problem in biomedicine, specifically, Gene Ontology annotation. In addition, many GO based annotation tools are listed in [18]. Differing from the GO based annotation methods' focus on identifying GO-terms in biomedical text, GOClonto is focused on applying LSA to identify the key GO-terms and exploit their semantic relationships within biomedical document collections. The ontology generated by GOClonto is specifically used to help users understand the conceptual structure of a biomedical document collection. To the best of our knowledge, the idea of GOClonto has not been researched in detail until now.

3 The GOClonto Method

Figure 1 shows the overview of the GOClonto method. First, a biomedical document collection is preprocessed into GO-term frequency files, in which each document is represented as a list of its GO-term frequencies. Second, the inverted document frequency of each term is calculated and each term weight is computed by multiplying the term frequency and inverted document frequency. Inverted term-document files are generated for each GO-term and the term-document matrix is constructed based on term weights. Third, based on the term-document

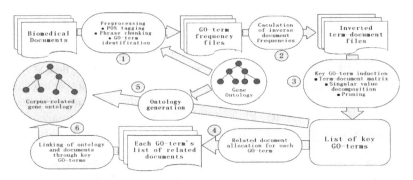

Fig. 1. An overview of the GOClonto method

matrix, we conduct key GO-term induction using LSA techniques. Fourth, with the list of key GO-terms, we allocate the related documents for each of the key GO-terms. Fifth, using GO, each key GO-term's superclass GO-terms are detected and used to construct a corpus-related gene ontology, which is a subset of GO. Sixth, documents are linked to the ontology through the identified key GO-terms.

To identify the key GO-terms for a biomedical document collection, extracting all of the literally contained GO-terms is the first step. Since GO-terms can be single terms or phrases, GOClonto employs CRFTagger [21] and CRFChunker [22] to perform POS (Part-of-Speech) tagging and phrase chunking respectively. Based on the tagging and chunking results, GOClonto utilizes GO to determine whether or not the terms in the result are GO-terms. With the extracted GO-terms, GOClonto manages documents based on the vector space model (VSM). VSM is a method of information retrieval that uses linear-algebra operations to compare textual data. VSM associates a single multidimensional vector with each document in a collection, and each component of that vector reflects a particular keyword or term related to the document. Based on VSM, GOClonto represents a set of documents by arranging their vectors in a term-document matrix.

Next, GOClonto applies LSA to analyze the constructed term-document matrix. Unlike VSM, LSA aims to represent the input collection using abstract terms found in the documents rather than the literal terms appearing in them. To do this, LSA approximates the original term-document matrix using a limited number of orthogonal factors. These factors represent a set of abstract terms, each conveying some idea common to a subset of the input collection. In the GOClonto method, these terms are used as key GO-terms for representing the biomedical document collections. Since subclasses of GO-terms are useful to capture more specific meanings of the GO-terms, the subclasses of the key GO-terms are identified among the frequent GO-terms. Then, documents are clustered and assigned to groups based on each GO-term along with its subclass GO-terms. Based on GO, superclasses of these GO-terms are detected and used to generate a corpus-related gene ontology automatically. Documents are linked to the ontology through the key GO-terms we have identified.

3.1 Preprocessing and Term-Document Matrix Construction

At the preprocessing stage, we first conduct the tokenization to split a biomedical document into sentences. Second, using CRFtagger [21], which is a Java-based conditional random fields POS Tagger for English, we perform the POS tagging. Third, to identify the noun phrases in a document, CRFChunker [22], a Java-based conditional random fields phrase chunker, is employed. With the identified nouns and noun phrases, GOClonto determines whether or not the nouns or noun phrases are GO-terms by referencing GO. To illustrate the ideas of GOClonto, we use a simple example collection of $d = 8$ documents (Fig. 2(a)), in which $t = 5$ GO-terms (Fig. 2(b)) appear more than once and thus are treated as frequent. We can see that GOClonto not only extracts single-word GO-terms, but also multi-word GO-terms.

```
(a)
D1: All of the contents of a cell excluding the plasma membrane and nucleus.
D2: The lipid bilayer surrounding an organelle.
D3: A septum which spans a cell and does not allow exchange of organelles or cytoplasm between compartments.
D4: Caveolaes may be pinched off to form free vesicles within the cytoplasm.
D5: A cell junction at which the cytoplasmic face of the plasma membrane is attached to actin filaments.
D6: The process by which cells digest parts of their own cytoplasm.
D7: A cellular organelle, found close to the nucleus in many eukaryotic cells.
D8: The change in shape of the spermatid nucleus from a spherical structure to an elongated organelle.

(b)
GO-T1: Cell
GO-T2: Nucleus
GO-T3: Organelle
GO-T4: Plasma membrane
GO-T5: Cytoplasm
```

Fig. 2. A biomedical document collection example

To construct the term-document matrix, the $tfidf$ (term frequency-inverted document frequency) is applied to calculate the weights of terms. In the vector space model, a document d is represented as a feature vector $\boldsymbol{d} = (tf_{t_1}, ..., tf_{t_i})$, where tf_t returns the absolute frequency of term $t \in \mathcal{T}$ in document $d \in \mathcal{D}$, where $\mathcal{D}$ is the document collection and $\mathcal{T} = \{t_1, t_2, ..., t_i\}$ is the set of all different terms occurring in $\mathcal{D}$. To weigh the frequency of a term in a document with a factor that discounts its importance when it appears in almost all of the documents, the idf (inverted document frequency) of term t in document d is proposed by Salton et al [23] as follows:

$$idf_t = log_2 n - log_2 df_t + 1 \qquad (1)$$

where df_t is the document frequency of term t that counts how many documents in which term t appears. Consequently, the $tfidf$ measure is calculated as the weight w_t of term t:

$$w_t = tf_t \times idf_t \qquad (2)$$

With the weight w_t of term t, the inverted file of term t is constructed, which contains the related documents of term t. For the example we used (Fig. 2), after calculating the term weights, each document is represented as a feature vector, which is used to compose the term-document matrix $\mathbf{A}$ as shown in Figure

3(a). In this matrix, each column vector represents each document, and each row vector denotes each term extracted to represent the documents' features. In our example, the first row represents the term GO-T1 'cell', the second row represents the term GO-T2 'nucleus' and so on through the terms listed in Figure 2(b). Similarly, the columns denote the documents listed in Figure 2(a). The first column represents document D1, the second column represents document D2, and so on.

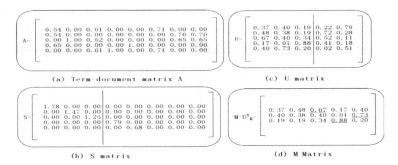

Fig. 3. Matrices used for the key GO-term induction

3.2 Key GO-Term Induction

To conduct the key GO-term induction, we apply LSA to process the term-document matrix by performing the singular value decomposition (SVD) of matrix $\mathbf{A}$ (Fig. 3(a)), which breaks it into three matrices ($\mathbf{U}$, $\mathbf{S}$, and $\mathbf{V}$) such that $\mathbf{A} = \mathbf{U}\mathbf{S}\mathbf{V}^T$. LSA finds a low-rank approximation to the term-document matrix. It turns out that when we select the k largest singular values from $\mathbf{S}$ (Fig. 3(b)), and their corresponding singular vectors from $\mathbf{U}$ (Fig. 3(c)) and $\mathbf{V}$, we get the rank k approximation to $\mathbf{A}$ with the smallest error. In addition, SVD translates the term and document vectors into a concept space. The first r columns of $\mathbf{U}$ (where r is $\mathbf{A}$'s rank) form an orthogonal basis for the term-document matrix's term space. Therefore, basis vectors, which are the column vectors in $\mathbf{U}$, are vector representations of the documents' abstract terms.

In practice, if we take all r basis vectors as abstract terms it would result in an unmanageable number of GO-terms selected as key. The singular values of the $\mathbf{A}$ matrix (lying on the $\mathbf{S}$ matrix shown in Figure 3(b)) are used to determine how many columns of $\mathbf{U}$ (Fig. 3(c)) should actually proceed to the next stage of the algorithm. In our example, based on the threshold we set as 1.2 empirically, the number of singular values that is higher than the threshold is $k = 3$. Consequently, the matrix $\mathbf{M}$, which represents the importance of abstract terms, is obtained by setting $\mathbf{M} = \mathbf{U}_k^T$ (Fig. 3(d)). The importance of a GO-term means the extent to which a given GO-term relates to the biomedical document collection. Based on $\mathbf{M}$, we extract the highest value of each row because each column of $\mathbf{M}$ denotes the importance of corresponding GO-terms. Finally, the corresponding GO-terms with the highest value in each row are selected as key GO-terms. In our example, in the first row, the third column, which corresponds to the GO-term GO-T3 'organelle', has the highest absolute value, 0.67. Therefore, GO-T3 'organelle' is selected as the first GO-term to represent the document collection. Similarly, in the second row, the fifth column has the highest absolute value, 0.73, and its corresponding term GO-T5 'cytoplasm' is selected. In the third row, the fourth column has the highest absolute value, 0.88, and its corresponding term GO-T4 'plasma membrane' is selected.

3.3 Document Allocation and Ontology Generation

We allocate the biomedical documents by matching them to related key GO-terms. This process is similar to extracting related documents for a query in information retrieval models. GOClonto uses the key GO-terms as queries. To improve the recall of information retrieval, query expansion is a widely used method. Similarly, we employ GO to find subclass GO-terms of the key GO-terms among the frequent GO-terms in the biomedical document collections. Documents are allocated to each key GO-term based on the cosine similarity between each document and the set including key GO-terms and their subclass GO-terms. For each key GO-term, if the cosine similarity between a document and the key GO-term exceeds a predefined threshold, the document is allocated to the corresponding group represented by the key GO-term. This assignment method naturally creates overlapping groups and well handles cross-topic documents. In our example, the key GO-terms are GO-T3 'organelle', GO-T5 'cytoplasm', and GO-T4 'plasma membrane'. The allocation results of our example are shown in Figure 4.

To generate an ontology based on a set of key GO-terms, we develop an algorithm called corpus-related gene ontology generation algorithm (Algorithm 1). This algorithm uses a set of key GO-terms and their subclass GO-terms, which are identified among the frequent GO-terms in a document collection, as input. A tree structure is used to store GO-terms and their subclass GO-terms, each tree node representing a GO-term, and its subclass GO-terms stored as subnodes of this tree node. α is a list of tree nodes storing GO-terms, initially containing the original input. For each iteration, a tree node t_j whose GO-term is not the root in GO is selected from α. We obtain p, the direct superclass

```
(a) Organelle
D2: The lipid bilayer surrounding an organelle.
D7: A cellular organelle, found close to the nucleus in many eukaryotic cells.
D8: The change in shape of the spermatid nucleus from a spherical structure to an elongated organelle.
D3: A septum which spans a cell and does not allow exchange of organelles or cytoplasm between compartments.

(b) Cytoplasm
D4: Caveolaes may be pinched off to form free vesicles within the cytoplasm.
D3: A septum which spans a cell and does not allow exchange of organelles or cytoplasm between compartments.
D6: The process by which cells digest parts of their own cytoplasm.

(c) Plasma membrane
D5: A cell junction at which the cytoplasmic face of the plasma membrane is attached to actin filaments.
D1: All of the contents of a cell excluding the plasma membrane and nucleus.
```

Fig. 4. Related document allocation results for each key GO-term

---

Algorithm 1. Corpus-related gene ontology generation algorithm

---

Input: $\eta \leftarrow$ a set of key GO-terms and their subclass GO-terms

Output: a corpus-related gene ontology $\mathcal{O}$ in OWL format

$\alpha \leftarrow$ Empty list {α is a list of tree nodes storing GO-terms used to construct the ontology}

for each GO-term g_i in η **do**

    Create tree node t_i that represents g_i

    Add t_i to α

end for

while α has more than one tree nodes **do**

    Get a tree node t_j from α whose GO-term is not the root in GO

    Get the direct superclass GO-term p of t_j's GO-term from GO

    Create tree node pr that represents p

    if pr is not found in α **then**

        Add pr to α

    end if

    Set t_j as subnode of pr

    Remove t_j from α

end while

Output the last tree node in α as a corpus-related gene ontology $\mathcal{O}$ in OWL format

---

GO-term of t_j's GO-term from GO, and create tree node pr corresponding to p. pr is checked for presence in α by recursively looking up all the tree nodes and their subnodes. If pr is not contained in α, pr is added to α. Next, the tree node t_j is added as pr's subnode. t_j is removed from α because t_j has been added as a subnode of pr. Finally, when the common superclass GO-term of all the input GO-terms is found, the tree node having this common superclass GO-term as its root, the last item in α, represents the generated ontology. GOClonto recursively stores the whole tree into an OWL file [24].

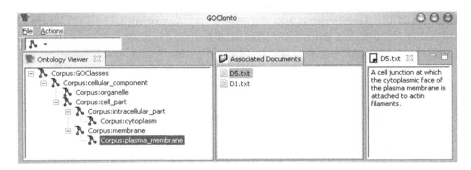

Fig. 5. Generated ontology of the document collection example in GOClonto

Figure 5 shows the generated ontology of our example in the user interface of the tool 'GOClonto', which helps users conceptualize a biomedical document collection by automatically generating a corpus-related gene ontology. The documents, having been allocated to their related key GO-terms, are then linked to the ontology through these GO-terms. When users select GO-terms in the ontology, their corresponding documents automatically display in the right panel. Note that all of the documents allocated to a GO-term's subclass GO-terms are also allocated to that GO-term. A conceptual structure of the biomedical document collection in the example (Fig. 2) is easily visualized in the generated ontology (Fig. 5). The documents allocated to GO-T4 'plasma membrane' are also related to the documents allocated to GO-T5 'cytoplasm', because the two GO-terms have the same superclass GO-term 'cell part'. GO-term 'cell part' incorporates all documents allocated to GO-T4 'plasma membrane' and GO-T5 'cytoplasm'. In addition, since all the GO-terms have the same superclass 'cellular component', users can see that the whole document collection is related to this more general GO-term. Therefore, with this ontology, users not only see the potential GO-terms related to the document collection, but can also more easily understand the semantic relationships between groupings of the documents.

4   Experimental Results

4.1   Experimental Setup

To examine the effectiveness of GOClonto, we conducted a series of experiments. First, to evaluate the results of key GO-term identification, we combined documents that belong to pre-defined categories and examined whether or not GOClonto can identify the category topics as key GO-terms. Second, to evaluate the effectiveness of the related document allocation of each key GO-term, we performed a clustering evaluation by comparing GOClonto with the STC (Suffix Tree Clustering) algorithm [3], the Lingo algorithm [7], the Fuzzy Ants clustering algorithm [5], and clustering based on Tolerance Rough Set (TRS) [6]. All the above algorithms were tested using the $Carrot^2$ software [25]. Finally, to evaluate informativeness of the generated ontology, we compared the ontology generated by GOClonto with the hierarchical tree generated by the Fuzzy Ants clustering algorithm [5]. The experiments were performed on J2SE 5.0, Windows XP, Pentium 4, 3.0GHz with 2GB RAM.

We collected document sets related to various GO-terms from PubMed [26]. We used the 'MajorTopic' tag along with the GO-terms as queries to PubMed. Since the retrieved documents are tagged manually with GO-terms as a result of

Table 1. Experimental Biomedical Document Collection

Category name	Number of Documents	Description
Chromosome	20	Abstracts of biomedical literature related to chromosome
Membrane	20	Abstracts of biomedical literature related to membrane
Cilium	10	Abstracts of biomedical literature related to cilium
Axoneme	7	Abstracts of biomedical literature related to axoneme
Centrosome	10	Abstracts of biomedical literature related to centrosome

common sense agreement of many users, we use them as the answer set for experiments. The GO-terms used for the queries were also used as category names. For each category, documents were assembled from the titles and abstracts retrieved from PubMed. Five categories were constructed: chromosome, membrane, cilium, axoneme, and centrosome (Table 1). Next, we combined the documents obtained from different categories into a biomedical document collection with 67 documents.

To evaluate the quality of the clustering results, we adopted a quality measure, F-measure, which is widely used in the text mining literature for the purpose of document clustering [27]. F-measure combines the precision and recall ideas found in the information retrieval literature. Each cluster is treated as if it were the result of a query and each class is treated as if it were the desired set of documents for a query. The precision and recall of a cluster j with respect to a class i are defined as:

$$\mathcal{P} = Precision(i, j) = \frac{n_{ij}}{n_i} \tag{3}$$

$$\mathcal{R} = Recall(i, j) = \frac{n_{ij}}{n_j} \tag{4}$$

where n_{ij} is the number of members of class i in cluster j, n_j is the number of members of cluster j and n_i is the number of members of class i. The F-measure of cluster j and class i is then given by $\mathcal{F}(i, j) = 2\mathcal{P}\mathcal{R}/(\mathcal{P} + \mathcal{R})$. The overall F-measure is computed by taking the weighted average of all values for the F-measure as given by the following:

$$\mathcal{F} = \sum_i \frac{n_i}{n} \max\{\mathcal{F}(i, j)\} \tag{5}$$

where $\mathcal{F}(i, j)$ is the highest F-measure to the cluster j that maps to class i, n is the number of documents.

4.2   Results and Discussion

Based on the biomedical document collection, the key GO-terms extracted by GOClonto are: centrosome, microtubule, centriole, membrane, flagellum, spindle, cilium, growth, chromatin. Among the nine key GO-terms, we found that category names 'centrosome', 'membrane', and 'cilium' are correctly selected. Although category name 'chromosome' is not highlighted, the selected GO-term 'chromatin' is a subclass GO-term 'chromosomal part', which is closely related to GO-term 'chromosome'. Since the size of category 'axoneme' is smaller than the other categories, the GO-term 'axoneme' is less important than the other identified key GO-terms for the whole collection. We can see above that GOClonto is able to recognize key GO-terms from the document collection.

According to different threshold values, different numbers of documents are allocated based on their related GO-terms. We tested various thresholds and set it to 1.8, allocating the documents to their related key GO-terms with similarities higher than that threshold. The results were compared with other clustering

Table 2. Comparison of GOClonto and other clustering algorithms

	GOClonto	STC	Lingo	Fuzzy Ants	Clustering based on TRS
F-measure	**0.6356**	0.4859	0.3718	0.4888	0.1598

algorithms. The desired clusters of other algorithms were set from three to seven. We tested each algorithm with various parameters and chose the best clustering results. The F-measure values of all the methods are listed in Table 2, which shows that GOClonto has the highest F-measure value 0.6356. The F-measure of GOClonto is 0.1497 higher than STC, 0.2638 higher than Lingo, 0.1468 higher than Fuzzy Ants, and 0.4758 higher than clustering based on TRS. GOClonto is focused on extracting key GO-terms and allocating documents based the key GO-terms, while the other clustering algorithms focus on general meaningful groupings, but do not specialize in the biomedical domain. Therefore, we can see that GOClonto outperforms other clustering algorithms on biomedical documents, allocating them to their related GO-terms with a relatively high precision.

Figure 6 shows a corpus-related gene ontology generated by the GOClonto method. Figure 7 shows a hierarchical tree created by the Fuzzy Ants clustering algorithm. We found that the ontology is much more informative than the hierarchical tree. GO is structured as directed acyclic graphs and many GO-terms are inherited from different superclasses in the generated ontology. For example, GO-term 'centrosome' has superclass GO-term 'intracellular non-membrane-bounded organelle' and superclass GO-term 'microtubule organizing center'. In the generated ontology, although GO-term 'heterochromatin' is not identified as a key GO-term, it is a frequently found GO-term in the biomedical collection and found to be a subclass GO-term of key GO-term 'chromatin' by GOClonto. Therefore, GO-term 'heterochromatin' is also used in the generated ontology, giving a more specific meaning than the key GO-term 'chromatin'. This shows how the generated ontology includes not simply the key GO-terms used to distinguish document groups, but also uses other important GO-terms that may be helpful to navigate the whole document collection.

From the generated ontology, we can easily observe the conceptual structure of the biomedical document collection. For instance, the documents allocated to GO-term 'membrane' are related to the documents allocated to GO-term 'cilium' because they share the same superclass GO-term 'cell part'. Specifically, the documents allocated to GO-term 'cell part' also incorporates all documents allocated to its child GO-terms, including 'cilium' and 'membrane'. The generated ontology guarantees an 'is-a' relationship between GO-terms. The documents are thus sorted and categorized in an intuitive and semantically sound way. However, the hierarchical tree generated by the Fuzzy Ants algorithm does not maintain relationship meaning. For the document collection we used, the created hierarchical tree is meaningless for the purposes of conceptualization. We attribute this to the fact that GOClonto specifically aims at conceptualizing biomedical document collections, while the Fuzzy Ants clustering algorithm does not.

To conclude, the GOClonto method is able to identify the key GO-terms and generate corpus-related gene ontologies to represent the biomedical document

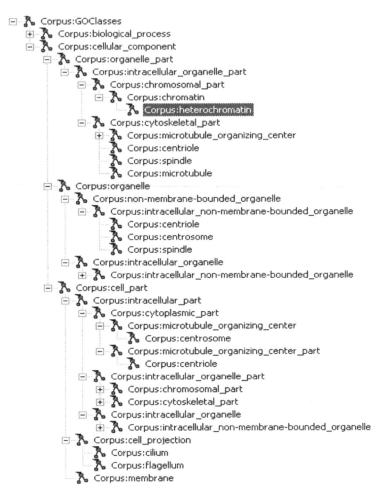

Fig. 6. A corpus-related gene ontology generated by GOClonto

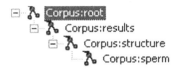

Fig. 7. A hierarchical tree created by the Fuzzy Ants clustering algorithm

collection. The ontology generated by GOClonto is more informative than the hierarchical tree created by Fuzzy Ants clustering algorithm. This ontology can help users easily visualize the conceptual structure of the biomedical document collection and intuitively navigate its document groupings.

5   Conclusion and Future Work

In this paper, we proposed a novel method, GOClonto, which exploits GO to automatically generate corpus-related gene ontologies for users. The generated ontologies can help users conceptualize biomedical document collections. Based on the vector space model, LSA techniques are used to identify the meaningful key GO-terms. The documents are allocated to these GO-terms using cosine similarities. By determining the superclass GO-terms of these GO-terms, ontologies are automatically generated and documents are linked to the generated ontologies through the GO-terms. The experimental results show that GOClonto is able to identify key GO-terms from document corpora. The generated ontologies are more informative than the hierarchical tree created by Fuzzy Ants clustering algorithm. We believe that GOClonto will play an important role helping users visualize and conceptualize biomedical document collections.

One limitation of the GOClonto method is that its performance depends on the precision of GO-term extraction. Since many GO-terms are implicitly represented in biomedical text, more sophisticated NLP (Natural Language Processing) techniques are necessary to discover potential GO-terms.

We will conduct further research to improve our work in the following ways. First, we will study more NLP methods to extract potential GO-terms from biomedical text. Also, addition of other visualization techniques alongside GOClonto can further aid user navigation of biomedical document collections. Furthermore, we can consult with biomedical researchers and other professionals in order to gauge how best GOClonto can be used to support their work. Finally, other biomedical-related ontologies can be used to generate the ontologies. Good examples are FMA (the Foundational Model of Anatomy) [28], which is known to be ontologically well designed, and SNOMED CT (Systematized Nomenclature of Medicine - Clinical Terms) [29], which is a practical clinical ontology used by many hospitals.

Acknowledgements

This work was supported in part by MKE & IITA through IT Leading R&D Support Project.

References

1. Ashburner, M., Ball, C.A., Blake, J.A., Botstein, D., Butler, H., Cherry, J.M., Davis, A.P., Dolinski, K., Dwight, S.S., Eppig, J.T., Harris, M.A., Hill, D.P., Issel-Tarver, L., Kasarskis, A., Lewis, S., Matese, J.C., Richardson, J.E., Ringwald, M., Rubin, G.M., Sherlock, G.: Gene ontology: tool for the unification of biology. The Gene Ontology Consortium. Nat Genet 25(1), 25–29 (2000)
2. Zheng, H.T., Borchert, C., Kim, H.G.: A concept-driven automatic ontology generation approach for conceptualization of document corpora (unpublished manuscript, 2008)

3. Zamir, O., Etzioni, O.: Web document clustering: a feasibility demonstration. In: SIGIR 1998: Proceedings of the 21st annual international ACM SIGIR conference on Research and development in information retrieval, pp. 46–54. ACM, New York (1998)
4. Zamir, O., Etzioni, O.: Grouper: a dynamic clustering interface to web search results. Comput. Netw. 31(11-16), 1361–1374 (1999)
5. Schockaert, S.: Het clusteren van zoekresultaten met behulp van vaagmieren (clustering of search results using fuzzy ants). Master thesis, University of Ghent (2004)
6. Lang, N.C.: A tolerance rough set approach to clustering web search results. Master thesis, Warsaw University (2004)
7. Osinski, S., Weiss, D.: A concept-driven algorithm for clustering search results. IEEE Intelligent Systems 20(3), 48–54 (2005)
8. Plaisant, C., Fekete, J.D., Grinstein, G.: Promoting insight-based evaluation of visualizations: From contest to benchmark repository. IEEE Transactions on Visualization and Computer Graphics 14(1), 120–134 (2008)
9. Olsen, K.A., Korfhage, R.R., Sochats, K.M., Spring, M.B., Williams, J.G.: Visualization of a document collection: the vibe system. Inf. Process. Manage. 29(1), 69–81 (1993)
10. Grobelnik, M., Maldenic, D.: Visualization of news articles. Informatica 28, 32–35 (2004)
11. Fortuna, B., Grobelnik, M., Mladenic, D.: Visualization of text document corpus. Informatica 29, 497–504 (2005)
12. Zhu, W., Chen, C.: Storylines: Visual exploration and analysis in latent semantic spaces. Computers & Graphics 31(3), 338–349 (2007)
13. Shaw, C.D., Kukla, J.M., Soboroff, I., Ebert, D.S., Nicholas, C.K., Zwa, A., Miller, E.L., Roberts, D.A.: Interactive volumetric information visualization for document corpus management. Int. J. on Digital Libraries 2(2-3), 144–156 (1999)
14. Fluit, C., Sabou, M., van Harmelen, F.: Ontology-based information visualisation: Towards semantic web applications. In: Visualising the Semantic Web, 2nd edn. (2005)
15. Thai, V., Handschuh, S., Decker, S.: IVEA: An information visualization tool for personalized exploratory document collection analysis. In: Bechhofer, S., Hauswirth, M., Hoffmann, J., Koubarakis, M. (eds.) ESWC 2008. LNCS, vol. 5021, pp. 139–153. Springer, Heidelberg (2008)
16. Bada, M., Turi, D., McEntire, R., Stevens, R.: Using reasoning to guide annotation with gene ontology terms in goat. SIGMOD Rec. 33(2), 27–32 (2004)
17. Camon, E., Magrane, M., Barrell, D., Lee, V., Dimmer, E., Maslen, J., Binns, D., Harte, N., Lopez, R., Apweiler, R.: The gene ontology annotation (goa) database: sharing knowledge in uniprot with gene ontology. Nucleic Acids Res. 32 (database issue) (2004)
18. Gene_Ontology_Annotation_Tool,
    http://www.geneontology.org/go.tools.annotation.shtml
19. Hill, D.P., Smith, B., McAndrews-Hill, M.S., Blake, J.A.: Gene ontology annotations: what they mean and where they come from. BMC bioinformatics 9 (suppl. 5) (2008)
20. Seki, K., Mostafa, J.: An application of text categorization methods to gene ontology annotation. In: SIGIR 2005: Proceedings of the 28th annual international ACM SIGIR conference on Research and development in information retrieval, pp. 138–145. ACM, New York (2005)
21. Phan, X.H.: Crftagger: Crf english pos tagger (2006),
    http://crftagger.sourceforge.net/

22. Phan, X.H.: Crfchunker: Crf english phrase chunker (2006),
    http://crfchunker.sourceforge.net/
23. Salton, G., Wong, A., Yang, C.S.: A vector space model for automatic indexing.
    Commun. ACM 18(11), 613–620 (1975)
24. OWL_Web_Ontology_Language, http://www.w3.org/tr/owl-ref/
25. Carrot[2], http://project.carrot2.org/
26. PubMed, http://www.ncbi.nlm.nih.gov/sites/entrez/
27. Steinbach, M., Karypis, G., Kumar, V.: A comparison of document clustering tech-
    niques. In: KDD Workshop on Text Mining (2000)
28. Rosse, C., Mejino, J.L.V.: A reference ontology for biomedical informatics: the
    foundational model of anatomy. J. of Biomedical Informatics 36(6), 478–500 (2003)
29. Stearns, M., Price, C., Spackman, K., Wang, A.: Snomed clinical terms: overview
    of the development process and project status. In: Proc. AMIA Symp., pp. 662–666
    (2001)

Extracting Semantic Frames from Thai Medical-Symptom Phrases with Unknown Boundaries

Peerasak Intarapaiboon, Ekawit Nantajeewarawat,
and Thanaruk Theeramunkong

School of Information and Computer Technology
Sirindhorn International Institute of Technology, Thammasat University
Pathumthani, Thailand
{ipeerasak,ekawit,thanaruk}@siit.tu.ac.th

Abstract. Due to the limitations of language-processing tools for the Thai language, pattern-based information extraction from Thai documents requires supplementary techniques. Based on sliding-window rule application and extraction filtering, we present a framework for extracting semantic information from medical-symptom phrases with unknown boundaries in Thai free-text information entries. A supervised rule learning algorithm is employed for automatic construction of information extraction rules from hand-tagged training symptom phrases. Two filtering components are introduced: one uses a classification model for predicting rule application across a symptom-phrase boundary, the other uses extraction distances observed during rule learning for resolving conflicts arising from overlapping-frame extractions. In our experimental study, we focus our attention on two basic types of symptom phrasal descriptions: one is concerned with abnormal characteristics of some observable entities and the other with human-body locations at which symptoms appear. The experimental results show that the filtering components improve precision while preserving recall satisfactorily.

1 Introduction

Standard formalisms for knowledge representation such as RDF and OWL have been recently developed by the semantic web community and are now in place. A crucial question still remains: how will we feed machines with the relevant knowledge in an application domain? In this paper, we present an information extraction (IE) framework towards bridging the gap between the world of symbols in Thai text in the domain of medical symptoms, i.e., words used in medical symptom descriptions in Thai, and the world of concepts, which represent abstractions of human thought. IE techniques usually involve linguistic patterns and domain-specific lexicons, coupled with a conceptual description of an application domain, i.e., a domain ontology. While an ideal domain ontology is possibly language-independent, linguistic patterns and lexicons rely heavily on the language in which the source textual information appears. Due to

J. Domingue and C. Anutariya (Eds.): ASWC 2008, LNCS 5367, pp. 390–404, 2008.
© Springer-Verlag Berlin Heidelberg 2008

language-structure differences, some basic language-processing tools available in one language may be unavailable in another language. When an IE framework that works well in one language is applied in a different language, the framework often needs modification and supplementary components are often necessary.

As part of a larger project on constructing a large-scale medical-related knowledge base in Thailand from information sources available on the Internet, we aim to develop a system for extracting semantic information from symptom descriptions in the Thai language and representing the extracted results in a form of machine-processable frames. IE based on linguistic analysis of Thai text is not currently feasible due to the lack of basic supporting language-processing tools. Not to mention a full parser for Thai sentences, neither a shallow-parsing (chunking) tool nor a fairly accurate part-of-speech analyzer is currently available, much owing to the high ambiguity of the structure of written Thai.

However, by incorporation of ontology-based semantic annotations and appropriate extraction filtering techniques, it is expected that IE based on patterns of triggering class tags and triggering plain words can be realized for Thai documents without text chunking and part-of-speech tagging. We focus on two types of symptom descriptions: one is concerned with abnormal characteristics of some observable entities and the other with human-body locations at which primitive symptoms appear. A well-known supervised rule learning algorithm, called WHISK [6], is used as the core algorithm for constructing IE rules automatically from a set of hand-tagged training symptom phrases. The technical challenges we address in this paper are twofold:

1. *IE from free text with unknown phrase boundary:* From Thai symptom textual phrases tagged with desired extraction outputs in a training corpus, WHISK generates a set of IE rules, each of which yields an extracted frame when its pattern matches a newly incoming symptom phrase. However, the information sources of our target IE task are collections of free-text information entries describing diseases, rather than collections of text portions identified beforehand as potential symptom phrases. Each such information entry is a "paragraph-like" textual description, typically containing several symptom phrases along with other text portions. Locating potential symptom phrases in an information entry requires a chunk parser and is thus not currently achievable for Thai text. A method, called *rule application using sliding windows (RAW)*, is introduced for applying IE rules to free-text information entries without predetermining symptom-phrase boundaries.

2. *Extraction filtering techniques:* Using sliding windows, IE rules are applied to text portions regardless of symptom-phrase boundaries and, therefore, tend to make many false-positive extractions. Two extraction filtering modules, called *wildcard-instantiation filtering (WIF)* and *overlapping-frame filtering (OFF)*, are proposed for removal of incorrect extractions. The first module uses a binary classifier for prediction of rule application across a symptom-phrase boundary; the second one uses extraction distances observed during rule learning to resolve extraction conflicts arising from overlapping

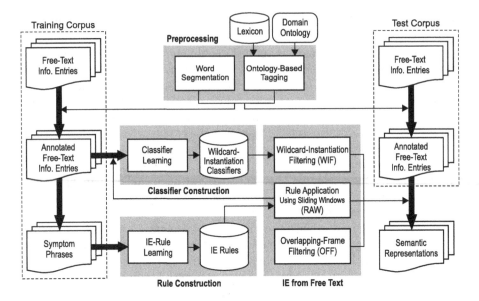

Fig. 1. An overview of the presented framework

extracted frames. It is desirable that these modules improve extraction precision while minimally sacrificing recall.

To begin with, Section 2 briefly describes Thai language-processing barriers and then gives an overview of our framework. After introducing a phrase-boundary extension technique for preparation of training instances, Section 3 explains the RAW method. Section 4 describes WIF and OFF. Section 5 presents experimental results and Section 6 discusses some related works.

2   Framework Overview

Like many other South Asian and South East Asian languages, Thai is an analytic language—its syntax and meaning are shaped by the use of particles and word order rather than by word inflection. Working with Thai text introduces many problems, primarily founded on the ambiguity of the language structure. In the Thai writing system, words are consecutively written without delimiters and a sentence comprises a series of words without an explicit sentence boundary. Spaces occur only occasionally between words or phrases within sentences—there is no standard rule for using spaces. A paragraph in written Thai often contains chunks of phrases that do not together constitute sentences grammatically. Sometimes, the main subject, verb, or object can be omitted from a sentence and it is still considered valid. The high language-structure ambiguity seriously hinders the development of basic language-processing tools. Among very few such tools available, only a word segmentation program (a word boundary detector) is commonly used.

Preprocessing. Considering the langauge-processing limitations, the IE framework outlined in Fig. 1 is proposed. Paragraph-like Thai free-text descriptions, referred to as *information entries*, are taken as input documents for our target IE task. Word segmentation is applied to all information entries as part of a preprocessing step. A domain-specific ontology, along with a lexicon for it, is then employed to partially annotate word-segmented phrases with tags denoting the semantic classes of occurring words with respect to the lexicon. Fig. 2 illustrates an obtained word-segmented and partially annotated information entry, where '|' indicates a word boundary, '~' signifies a space, and the tags "sec," "col," "ptime," "sym," and "org" denote the semantic classes "Secretion," "Color," "Time period," "Symptom," and "Organ," respectively, in the domain ontology. This information entry contains four symptom phrases of our target types. They are underlined in the figure.

เป็น|โรค|ที่|พบ|บ่อย|หลัง|จาก|เป็น|ไข้หวัด|~|ผู้ป่วย|จะ|มี|[sec เสมหะ]|เป็น|[col สีเขียว]|เป็น|ระยะ|

เวลา|~|[ptime 4-10 วัน]|~|และ|อาจ|มี|อาการ|อื่น|ด้วย|~|ที่|พบ|บ่อย|ได้แก่|~|[sym เบื่ออาหาร]|~|มี|

[sym อาการเจ็บ]|ที่|[org หน้าอก]|อยู่|นาน|~|[ptime 6-12 วัน]|~|มี|[sym อาการเจ็บ]|[org คอ]|~|และ|มี|

[sym อาการไอ]|จน|เกิด|[org คอ]|[col แดง]|นาน|~|[ptime 3-4 วัน]|~|โดย|ความรุนแรง|ขึ้น|อยู่|กับ|ชนิด|

ของ|เชื้อโรค|ที่|ได้|รับ|~|เป็น|โรค|ที่|พบ|ใน|ผู้ใหญ่|มาก|กว่า|ใน|เด็ก|~|ผู้ป่วย|อาจ|มี|สุขภาพ|ทั่วไป|แข็งแรง

Fig. 2. An information entry describing acute bronchitis

Extracted Frames and IE Rules. Fig. 3 and Fig. 4 show the frames required to be extracted from the first and the second underlined symptom phrases, respectively, in Fig. 2. Each of them contains three slots, i.e., OBS, ATTR, and PER in Fig. 3 and SYM, LOC, and PER in Fig. 4, where OBS, ATTR, PER, SYM, and LOC stand for "observed entity," "attribute," "period," "symptom," and "location," respectively. Fig. 5 depicts the semantic representations of the two extracted frames. Fig. 6 gives a typical example of an IE rule. Its pattern part contains three triggering class tags, one triggering plain word, and four instantiation wildcards. The three triggering class tags also serve as *slot markers*—the terms into which they are instantiated are taken as fillers of their respective slots in the resulting extracted frame. When instantiated into the symptom phrase in Fig. 4, this rule yields the extracted frame shown in the same figure.

Rule Construction and Classifier Construction. From manually identified symptom phrases in a training corpus, the WHISK algorithm is used for automatically constructing a set of IE rules. During rule learning, symptom-phrase lengths and extraction distances are observed when a rule makes correct extractions on the training instances: the former kind of observed information is used for determining an appropriate window size for RAW while the latter one for removal of overlapping extracted frames in the OFF module. In order to prepare training data for construction of classification models used for predicting

Symptom phrase:　|มี|[sec เสมหะ]|เป็น|[col สีเขียว]|เป็น|ระยะ|เวลา|~|[ptime 4-10 วัน]|

Extracted frame:　{OBS [sec เสมหะ]}{ATTR [col สีเขียว]}{PER [ptime 4-10 วัน]}

Fig. 3. A symptom phrase and an extracted frame

Symptom phrase:　|มี|[sym อาการเจ็บ]|ที่|[org หน้าอก]|อยู่|นาน|~|[ptime 6-12 วัน]|

Extracted frame:　{SYM [sym อาการเจ็บ]}{LOC [org หน้าอก]}{PER [ptime 6-12 วัน]}

Fig. 4. A symptom phrase and an extracted frame

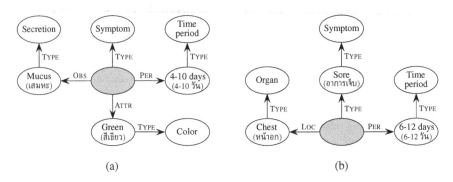

(a)   (b)

Fig. 5. Semantic representations of the extracted frames in Fig. 3 and Fig. 4

Pattern:　*(sym)*(org)*นาน*(ptime)

Output template:　{SYM $1}{LOC $2}{PER $3}

Fig. 6. An IE rule example

rule instantiation across symptom-phrase boundaries in the WIF module, the IE rules obtained from WHISK are applied to the partially annotated free-text information entries in the training corpus using RAW (cf. Fig. 1).

3   Rule Learning and Rule Application

3.1   Rule Learning from Phrases with Extended Boundaries

WHISK [6] uses a covering algorithm to construct a set of multi-slot extraction rules. It takes a corpus of training instances that are hand-tagged with desired extraction outputs to guide rule creation. The algorithm induces rules top-down, starting from the most general rule that covers all training instances, and then specializing the initial rule by adding triggering terms one at a time in order to prevent rule application with incorrect extractions. Reasons for selecting WHISK include not only its previous success in a wide variety of English-text IE tasks but also its capability to generate multi-slot extraction rules, which enable extracted pieces of symptom-related information to be semantically connected, e.g., an observed entity and its abnormal characteristic. Other IE-rule learning algorithms with performance comparable to WHISK, e.g., RAPIER [1] and SRV [3], can generate only single-slot extraction rules (i.e., individual-field extraction rules) and, hence, do not suit our application requirements.

For rule learning, symptom phrases are manually collected from a training corpus of information entries. Our rule application targets are, by contrast, information entries containing symptom phrases whose boundaries are not identified beforehand. It is thus desirable to construct rules with the ability to make extractions from text portions in which symptom phrases appear alongside some other words. To prepare training instances for construction of such rules, each collected symptom phrase is extended with a prefix and a suffix so as to make its boundary unknown. Since each rule produces only one (multi-slot) output frame when it is applied, symptom-phrase extension is subject to the constraint that no extra frame can be extracted from an extended phrase. The notion of an n-word-bound extension is introduced for this purpose: the *n-word-bound extended version* of a symptom phrase S with respect to an information entry E is the longest text portion T comprising m_1 words preceding S in E (called m_1-*word prefix*), S itself, and m_2 words following S in E (called m_2-*word suffix*) such that $m_1, m_2 \leq n$ and only one extracted frame can be obtained from T.

Prefix: |ที่|พบ|บ่อย|ได้แก่|~|[sym เบื่ออาหาร]|~|

Symptom phrase: |มี|[sym อาการเจ็บ]|ที่|[org หน้าอก]|อยู่|นาน|~|[ptime 6-12 วัน]|

Suffix: |~|มี|[sym อาการเจ็บ]|

Hand-tagged frame: {SYM [sym อาการเจ็บ]}{LOC [org หน้าอก]}{PER [ptime 6-12วัน]}

Fig. 7. A hand-tagged symptom phrase extended with a prefix and a suffix

Example 1. Fig. 7 illustrates a hand-tagged training instance prepared by extending the symptom phrase in Fig. 4 using 5-word-bound extension with respect to the information entry in Fig. 2. (This symptom phrase is the second underlined phrase in Fig. 2.) The prefix part of this extension contains five words (four plain words and one annotated word). The suffix part, however, contains only two words (one plain word and one annotated word) by the constraint that no extra frame is obtained from the extended phrase—adding one more word to the suffix part causes the resulting phrase to contain the third underlined symptom phrase in Fig. 2 entirely and an extra frame is obtained.   □

3.2   Rule Application Using Sliding Windows (RAW)

A WHISK rule does not have ability to automatically segment a free-text information entry so that the rule can be applied to the relevant portion of text. When working with free text, a WHISK-based IE system normally uses a part-of-speech tagger and a shallow parser to group together words into larger syntactic chunks, from which the boundaries of relevant text portions are determined. In the absence of such supporting tools, a sliding window technique is introduced

Table 1. Frames extracted from the text portions in Fig. 8 by the rule in Fig. 6

Portion	Extracted frame	Correctness
[35, 44]	{SYM [sym เบื่ออาหาร]}{LOC [org หน้าอก]}{PER [ptime 6-12 วัน]}	Incorrect
[36, 45]	{SYM [sym อาการเจ็บ]}{LOC [org หน้าอก]}{PER [ptime 6-12 วัน]}	Correct
[50, 59]	{SYM [sym อาการไอ]}{LOC [org คอ]}{PER [ptime 3-4 วัน]}	Incorrect

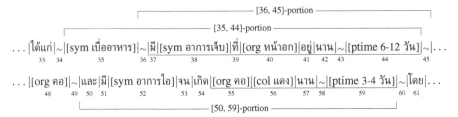

Fig. 8. Text portions from which extractions are made when the rule in Fig. 6 is applied to the information entry in Fig. 2 using a 10-word sliding window

to locate text portions for rule application, and filtering methods are used to remove potentially incorrect extractions afterwards. *Rule application using sliding windows (RAW)* is explained below along with an example.

Using a k-word sliding window, a rule r is applied to each k-word portion of an information entry one-by-one sequentially. More precisely, assume that an information entry E consisting of n words is given and for any l, m such that $1 \leq l \leq m \leq n$, the $[l, m]$-*portion* of E is the portion beginning at the lth word position and ending at the mth word position of E. Then r is applied to the $[i, i + k - 1]$-portion of E for each i such that $1 \leq i \leq n - k + 1$. An application that results in a duplicated frame is discarded.

Example 2. As shown in Fig. 8, when the rule in Fig. 6 is applied to the information entry in Fig. 2 using a 10-word sliding window, it makes extractions from the $[35, 44]$-portion, the $[36, 45]$-portion, and the $[50, 59]$-portion of the entry. Table 1 shows the resulting extracted frames. Only the extraction made from the $[36, 45]$-portion is correct. When the rule is applied to the $[35, 44]$-portion, the slot filler taken through the first slot marker of the rule, i.e., "sym," does not belong to the symptom phrase containing the filler taken through the second slot marker of it, i.e., "org," whence an incorrect extraction occurs. The same situation arises when the rule is applied to the $[50, 59]$-portion. As will be described in the next section, WIF is designed to filter out incorrect extractions of this kind.   □

The possibility that a symptom phrase fits into a window increases as the window size increases; but the risk of rule application across a symptom-phrase boundary is also higher. Appropriate window size for different rules may be different. In the evaluation presented in Section 5, the base window size for an individual rule is set to the length of the longest training symptom phrase observed when the rule makes correct extractions in the rule generation process.

4 Extraction Filtering

Two modules are proposed for filtering out incorrect extractions, i.e., false positives, resulting from RAW. The first module, called *wildcard-instantiation filtering (WIF)*, employs a classification model to predict incorrect extractions based on instantiation features of rule internal wildcards. The second one, called *overlapping-frame filtering (OFF)*, uses extraction distances and rule Laplacian errors observed during rule learning to resolve conflicts caused by overlapping frames.

4.1 Wildcard-Instantiation Filtering (WIF)

WHISK learns an extraction pattern in terms of triggering terms for making an extraction from a "single" symptom phrase. Using RAW, however, an obtained IE rule may be instantiated across a symptom-phrase boundary (cf. Example 2), yielding an extracted frame containing unrelated slot fillers, which is definitely a false positive. Instantiations of wildcards occurring between the first and the last slot markers of a rule, called *internal wildcards*, provide a clue to detect such an undesirable extraction—if an internal wildcard of a rule is instantiated across a symptom-phrase boundary, then unrelated slot fillers are extracted by the resulting rule instantiation. A wildcard that is not internal does not give the same clue since it is never instantiated into a word string enclosed by two slot fillers of the same extracted frame.

   Predicting whether an internal wildcard is instantiated across a symptom-phrase boundary can be regarded as a binary classification problem. A classifier is constructed for each rule from observations obtained from applying the rule to free-text information entries in the training corpus by assuming that rules obtained from WHISK are error-free with respect to their training symptom phrases,[1] i.e., if each internal wildcard of a rule is instantiated into a portion of a single symptom phrase, then the rule yields a correct extraction. Under this assumption, an incorrect extraction implies an existence of an internal-wildcard instantiation across a symptom-phrase boundary. We construct a feature vector characterizing a rule wildcard instantiation as follows: Let T be a text portion and r a rule containing m internal wildcards $w_1, w_2, \ldots, w_m$. The *instantiation feature vector* for w_i observed at T is a vector $\boldsymbol{f} = [f_a, f_p, f_s]$, where f_a, f_p and f_s are the number of annotated words, the number of plain words and the number of spaces, respectively, in the text portion of T into which w_i is instantiated when r is applied to T. The *wildcard-instantiation feature vector* for r observed at T is then defined as a vector $\boldsymbol{x} = \boldsymbol{f}_1 \parallel \boldsymbol{f}_2 \parallel \cdots \parallel \boldsymbol{f}_m$, where $\boldsymbol{f}_i$ is the instantiation feature vector for w_i observed at T and '$\parallel$' denotes vector

---

[1] WHISK uses an error tolerance threshold to accept rules with incorrect extractions. However, this threshold is normally small.

Table 2. Instantiation features obtained from rule application using RAW in Fig. 8

Portion	Internal-wildcard instantiation			Feature vector	Label
	1st wildcard	2nd wildcard	3rd wildcard		
[35, 44]	|∼|มี|[sym อาการเจ็บ]|ที่|	|อยู่|	|∼|	[1, 2, 1, 0, 1, 0, 0, 0, 1]	-1
[36, 45]	|ที่|	|อยู่|	|∼|	[0, 0, 0, 0, 1, 0, 0, 0, 1]	1
[50, 59]	|จน|เกิด|	|[col แดง]|	|∼|	[0, 2, 0, 1, 0, 0, 0, 0, 1]	-1

concatenation. Fisher's linear discriminant analysis [2], a well-known baseline method for supervised linear classification, is used for classifier construction in our experiments.

Example 3. The rule in Fig. 6 has three internal wildcards, i.e., those occurring between "sym" and "ptime" in its pattern. Referring to Example 2, Table 2 shows the instantiations of these wildcards and the obtained wildcard-instantiation feature vectors when the rule is applied to the three text portions in Fig. 8. For example, when the rule is applied to the [35, 44]-portion, the first internal wildcard is instantiated into a string consisting of one annotated word, two plain words, and one space, yielding [1, 2, 1] as the resulting instantiation feature vector. Likewise, the instantiation feature vectors for the second and the third internal wildcards observed at the same portion are [0, 1, 0] and [0, 0, 1], respectively. The three vectors constitute the wildcard-instantiation feature vector for the rule observed at the [35, 44]-portion. □

4.2   Overlapping-Frame Filtering (OFF)

WHISK directly supports multi-slot extraction using a single rule application, allowing one to design a multi-slot template containing all required slot fillers from an individual symptom phrase. In our IE application, one symptom phrase is independent of another symptom phrase. Accordingly, when two distinct extracted frames overlap, i.e., when they contain a slot filler extracted from the same text position, one of them is necessarily a false positive. Overlapping frames are resolved based on two kinds of information collected during rule learning, i.e., the most frequently observed extraction distances and the Laplacian expected errors; the former takes priority over over the latter. OFF resolves overlapping frames using the procedure described below.

An extracted frame F is identified with a pair $\langle T, r \rangle$, where T is the text portion from which F is obtained and r is the rule that extracts F from T. The *extraction distance* of $\langle T, r \rangle$ is the distance in T between the word positions from which the first and the last slot fillers of F is obtained when r is applied to T. Let $\delta(T, r)$ be the difference between the extraction distance of $\langle T, r \rangle$ and the extraction distance most frequently observed when r makes correct extractions on the training set. Let $L(r)$ denote the Laplacian expected error of r, i.e., $L(r) = (e + 1)/(n + 1)$, where n is the number of extractions made by r on the training set and e is the number of errors among those extractions.

Now suppose that $\mathcal{F} = \{\langle T_1, r_1 \rangle, \langle T_2, r_2 \rangle, \ldots, \langle T_n, r_n \rangle\}$ is an initially given set of extracted frames. Overlapping frames in $\mathcal{F}$ are filtered out by repeatedly performing the following steps until no overlapping frame remains in $\mathcal{F}$:

1. Determine the set $\mathcal{O}$ of all overlapping frames currently belonging to $\mathcal{F}$.
2. Determine the subsets Max_δ and Max_L of $\mathcal{O}$ by:
   (a) $Max_\delta = \{\langle T, r \rangle \in \mathcal{O} \mid \forall \langle T', r' \rangle \in \mathcal{O} : \delta(T, r) \geq \delta(T', r')\}$,
   (b) $Max_L = \{\langle T, r \rangle \in Max_\delta \mid \forall \langle T', r' \rangle \in Max_\delta : L(r) \geq L(r')\}$.
3. Select an arbitrary frame in Max_L and remove it from $\mathcal{F}$.

5   Experimental Results and Discussion

5.1   Data Sets and Output Templates

Information Entries. This work is part of a project supported by the Thailand Research Fund (TRF) and the National Electronics and Computer Technology Center (NECTEC), aiming at development of a framework for constructing a large-scale medical-related knowledge base in Thailand from various information sources available on the Internet. The overall project framework includes data acquisition, keyword extraction, link construction, and ontology-based knowledge representation. A set of supporting tools for gathering medical text data from Thai web pages were developed and a number of medicinal and pharmaceutical web sites (2759 URLs) were selected as seeds. The obtained data covers 474 diseases and 770 medicinal chemical substances, with approximately 6600 and 3350 information entries, respectively. Disease information entries were organized into disease characteristics, symptoms, cause, treatment, etc.

From textual data gathered in this knowledge-base construction project, free-text symptom information entries are collected and divided into 3 data sets, i.e., D1, D2, and D3, based on their disease groups. D1 comprises distinct information entries obtained from 5 disease groups, i.e., the circulatory system, the urology system, the reproductive system, the eye system, and the ear system; D2 from 6 groups, i.e., the skin/dermal system, the skeletal system, the endocrine system, the nervous system, parasitic diseases, and venereal diseases; D3 from 4 groups, i.e., the respiratory system, the gastrointestinal tract system, infectious diseases, and accidental diseases. The collected information entries are preprocessed using a word segmentation program, called CTTEX, developed by NECTEC, and are

Table 3. Data set characteristics

Data set	No. of info. entries	No. of words per info. entry			No. of distinct symptom phrases		No. of symptom phrase occurrences	
		Max.	Avg.	Min.	Type-A	Type-B	Type-A	Type-B
D1	59	130	44	9	179	77	213	84
D2	56	146	45	7	136	66	160	69
D3	58	140	55	8	161	65	210	73

Table 4. Top five rules for each template type obtained from WHISK

Type	Pattern	Output template	Lap. error
A	*(org)*(col)	{Obs $1}{Attr $2}	0.148
A	*(org)*(szq)	{Obs $1}{Attr $2}	0.148
A	*(sec)*(col)	{Obs $1}{Attr $2}	0.200
A	*(ch)*ที่*(org)	{Obs $2}{Attr $1}	0.250
A	*(ch)*ตาม*(org)	{Obs $2}{Attr $1}	0.250
B	*(sym)*(org)	{Sym $1}{Loc $2}	0.029
B	*(sym)*บริเวณ*(org)	{Sym $1}{Loc $2}	0.125
B	*(sym)*ใน*(org)	{Sym $1}{Loc $2}	0.166
B	*(sym)*(org)*นาน*(ptime)	{Sym $1}{Loc $2}{Per $3}	0.166
B	*(sym)*(org)*เป็น*(ptime)	{Sym $1}{Loc $2}{Per $3}	0.166

then partially annotated with semantic class tags using a predefined ontology lexicon. The second column of Table 3 shows the number of information entries in each data set. It is followed by a column group showing the maximum number, the average number, and the minimum number of words per information entry in each data set. The last two column groups of this table characterize the three data sets in terms of the number of symptom phrases and their occurrences. They are detailed below.

Symptom Phrases and Output Templates. A collected information entry typically contains several symptom phrases, which provide several kinds of symptom-related information. Two basic types of symptom phrases, referred to as *Type-A* and *Type-B*, are considered in our experiments. A symptom phrase of Type-A describes a symptom in terms of abnormal characteristics of some observable entity. The output template for Type-A takes the form:

$$\{\text{Obs } X\}\{\text{Attr } Y\}\{\text{Per } Z\},$$

where X is an observed entity, Y its abnormal characteristic, and Z the time period in which the abnormality occurs. The semantic representation graphically shown in Fig. 5a provides an example of a result generated by the Type-A template. A symptom phrase of Type-B is concerned with a symptom that is described in terms of a primitive named symptom in the predefined domain ontology. The output template for Type-B takes the form:

$$\{\text{Sym } X\}\{\text{Loc } Y\}\{\text{Per } Z\},$$

where X specifies a primitive named symptom, Y the location at which the primitive symptom occurs, and Z the time period. The representation in Fig. 5b illustrates an instance of the Type-B template. The slot Per in the Type-A template is optional. One of the slots Loc and Per, but not both, may be omitted in the Type-B template. One symptom phrase may occur in more than one information entry and may therefore have multiple occurrences in one data set. The number of all distinct symptom phrases of the two types and the number of their occurrences are given in the last two column groups of Table 3.

5.2   Experimental Results

Training Process. D1 is used as the training corpus. All Type-A and Type-B symptom phrases occurring in D1 are manually tagged with desired output frames and their 5-word-bound extended versions are used as training instances for rule learning. Using our implementation of WHISK, 27 rules for the Type-A template and 11 rules for the Type-B template are generated. Tables 4 shows the top five rules with the lowest Laplacian errors from the obtained rule set for each template. The length of the longest symptom phrase observed when a rule yields correct extractions on the training set is taken as the window size for the rule. By applying the obtained rules to the information entries in D1 using RAW, wildcard-instantiation feature vectors are constructed and then used as training data for constructing WIF classifiers. Fisher's linear discriminant analysis [2] is employed for classifier learning.

Evaluation. The proposed framework is evaluated using D2 and D3 as test sets. Recall and precision are used as performance measures, where the former is the proportion of correct extractions to relevant symptom phrases and the latter is the proportion of correct extractions to all obtained extractions. Table 5 shows the evaluation results obtained from using RAW without any extraction filtering, RAW with WIF (RAW+WIF), RAW with OFF (RAW+OFF), and RAW with both WIF and OFF (RAW+WIF+OFF),[2] where 'R' and 'P' stand for recall and precision, which are given in percentage. Compared to the results obtained using RAW alone, each of RAW+WIF, RAW+OFF, and RAW+WIF+OFF improves precision while satisfactorily preserving recall in every row of this table. For Type-A, where RAW has high precision (85.4 for D2 and 89.4 for D3), the three combinations yield similar precision improvement. However, for Type-B, where RAW has low precision (37.5 for D2 and 31.9 for D3), RAW+WIF and RAW+WIF+OFF significantly outperform RAW+OFF in terms of precision gain.

Table 6 compares WIF, OFF, and the combination of them (WIF+OFF) in terms of removal correctness. The third row, for example, shows that when applied to D2, RAW generates 115 incorrect extractions for Type-B, 111 of which are removed by WIF+OFF. The same row also indicates that WIF+OFF removes two correct extractions, while WIF and OFF remove one each—on closer examination, the frame incorrectly removed by WIF in this case differs from that removed by OFF. Altogether, the results in this table suggest that WIF and OFF have different but complementary abilities.

Recall Improvement. As seen in Table 5, recall for Type-A appears to be relative lower than that for Type-B. A detailed examination of Type-A symptom phrases in D2 and D3 reveals that there are 79 target symptom phrases from which RAW fails to make any extraction, i.e., 79 false negatives. These false

---

[2] In RAW+WIF+OFF, OFF is applied to the set of frames obtained from RAW+WIF.

Table 5. Evaluation results

Type	Data set	RAW		RAW+WIF		RAW+OFF		RAW+WIF+OFF	
		R	P	R	P	R	P	R	P
A	D2	76.9	85.4	76.9	94.6	76.3	94.6	76.9	96.1
A	D3	80.0	89.4	80.0	96.0	78.6	96.5	79.0	97.1
B	D2	100.0	37.5	98.6	86.1	98.6	66.0	97.1	94.4
B	D3	98.6	31.9	98.6	92.3	94.5	67.6	97.3	93.4

Table 6. Removal correctness evaluation

Type	Data set	RAW		WIF		OFF		WIF+OFF	
		$\#E_{cor}$	$\#E_{inc}$	$\#R_{cor}$	$\#R_{inc}$	$\#R_{cor}$	$\#R_{inc}$	$\#R_{cor}$	$\#R_{inc}$
A	D2	123	21	14	0	14	1	16	0
A	D3	168	20	13	0	14	3	15	2
B	D2	69	115	104	1	80	1	111	2
B	D3	72	154	148	0	121	3	149	1

$\#E_{cor}$: No. of correct extractions $\#R_{cor}$: No. of correct removals
$\#E_{inc}$: No. of incorrect extractions $\#R_{inc}$: No. of incorrect removals

Table 7. Recall improvement by doubling the window size and rule generalization

Data set	Improved by	RAW		RAW+WIF		RAW+OFF		RAW+WIF+OFF	
		R	P	R	P	R	P	R	P
D2	2W	81.3	73.4	81.3	94.9	80.0	97.0	80.0	98.5
D2	RG	85.0	76.4	85.0	94.4	84.4	93.1	84.4	95.7
D2	2W+RG	88.1	60.3	88.1	93.4	86.3	93.9	86.9	97.9
D3	2W	83.3	71.4	83.3	96.2	82.4	97.2	82.4	97.2
D3	RG	89.0	77.0	88.1	95.9	89.0	96.4	88.1	98.9
D3	2W+RG	89.0	55.3	89.0	93.0	88.6	94.9	88.6	96.9

negatives are divisible into two disjoint groups: symptom phrases that match the pattern part of some existing rule, and those that do not. 23 of the 79 false negatives belong to the first group, and the rest of them belong to the second group. By increasing the size of a window in use, extractions can be made from false negatives in the first group but not from those in the second group.

Further analysis of the obtained Type-A rules shows that most of them have low coverage, i.e., they tend to be overfitting. Each of them contains a tag denoting a subclass of "Gradable quantity" in the domain ontology, e.g., the tags "col" (Color), "szq" (Size), and "ch" (Characteristic) in the five Type-A rules in Table 4. Moreover, some rule differs from another only at tags denoting subclasses of "Gradable quantity" and can be merged together by generalizing such tags into "gq" (Gradable quantity). For example, consider the first two Type-A rules in Table 4. The first rule differs from the second one only at their second tags, i.e., "col" and "szq." Merging them by generalizing "col" and "szq" upwards into "gq" yields a new rule with higher coverage.

Two approaches are taken so as to reduce the number of false negatives: first, double the window size previously used for each rule, and, secondly, merge rules that can be made identical by merely generalizing tags denoting subclasses of "Gradable quantity" upwards into "gq." Referring to the first approach as 2W and the second one as RG, Table 7 presents the obtained evaluation results, where 2W+RG denotes the combination of the two approaches. The table shows that RG and 2W+RG yield higher recall gain than 2W in both D2 and D3. Compared to the results obtained from using RAW alone for Type-A in Table 5, RG and 2W+RG increase recall from 76.9 to 85.0 and 88.1, respectively, in D2, and from 80.0 to 89.0 in D3. It is noteworthy that, like in Table 5, each of RAW+WIF, RAW+OFF, and RAW+WIF+OFF improves precision while satisfactorily preserving recall in every row of Table 7.

6   Related Works

Application of WHISK to IE tasks in medical-related domains was reported in [4] and [5]—the former is concerned with information on biomedical events and the latter with drug treatment information. Both of them take English documents as information sources and use linguistic tools, such as part-of-speech taggers and chunk parsers, along with ontology-based semantic tagging for text preprocessing. While a preliminary investigation was given in [5] without introducing any supplementary technique, an extraction verification module was proposed in [4] for removing incorrectly extracted biomedical events based on a maximum entropy (ME) classification method. The verification module in [4] uses a learned ME classifier to predict a class of the an extracted slot filler, and removes an extracted frame whose components contradict the class assigned by the classifier. By contrast, the WIF module in our framework uses a classifier to predict rule application across symptom-phrase boundary, compensating for the unavailability of a phrase boundary analyzer, which is a fundamental problem for text processing in Thai. Very few works on IE from Thai text were reported in the literature. Although some IE techniques were applied in [7], its target application was word boundary identification in Thai text rather extraction of semantically related slot fillers. An approach to Thai-text IE using triggering terms and corpus-based syntactic surface analysis was discussed in [8]. However, only hand-crafted IE rules were demonstrated; rule learning was not considered.

7   Conclusions

From a set of manually collected symptom phrases, IE rules are created using our implementation of WHISK. To apply the obtained rules to free-text information entries without predetermining symptom-phrase boundaries, rule application using sliding windows is introduced. Filtering techniques are proposed for removal of false positives resulting from rule application across symptom-phrase boundaries and those resulting from overlapping-frame extractions. The experimental results show that these techniques improve extraction precision while satisfactorily preserving recall. Further works include extension of the types of target phrases, empirical investigation of framework application in different data domains, and in-depth analysis of how the ontology-based semantic frames extracted from symptom phrases facilitate logic-based medical diagnosis reasoning.

Acknowledgement. This work has been supported by the Thailand Research Fund (TRF), under Grant No. BRG50800013 and under TRF Royal Golden Jubilee Ph.D. Program Grant No. PHD/0056/2550, and was supported by the National Electronics and Computer Technology Center (NECTEC), under Grant No. NT-B-22-I4-38-49-05.

References

1. Califf, M.E., Mooney, R.J.: Bottom-up Relational Learning of Pattern Matching Rules for Information Extraction. Journal of Machine Learning Research 4, 177–210 (2003)
2. Duda, R.O., Hart, P.E., Stork, D.G.: Pattern Classification, 2nd edn. Wiley Interscience, Hoboken (2000)
3. Freitag, D.: Machine Learning for Information Extraction in Informal Domains. Machine Learning 39(2–3), 169–202 (2000)
4. Kim, E., Song, Y., Lee, C., Kim, K., Lee, G., Yi, B.-K.: Two-Phase Learning for Biological Event Extraction and Verification. ACM Transactions on Asian Language Information Processing 5(1), 61–73 (2006)
5. Lee, C.-H., Na, J.-C., Khoo, C.S.G.: Towards ontology enrichment with treatment relations extracted from medical abstracts. In: Sugimoto, S., Hunter, J., Rauber, A., Morishima, A. (eds.) ICADL 2006. LNCS, vol. 4312, pp. 419–428. Springer, Heidelberg (2006)
6. Soderland, S.: Learning Information Extraction Rules for Semi-Structured and Free Text. Machine Learning 34(1–3), 233–272 (1999)
7. Sornlertlamvanich, V., Potipiti, T., Charoenporn, T.: Automatic Corpus-based Thai Word Extraction with the C4.5 Learning Algorithm. In: Proc. 18th International Conference on Computational Linguistics, Saarbrucken, Germany, pp. 802–807 (2000)
8. Sukhahuta, R., Smith, D.: Information Extraction Strategies for Thai Documents. International Journal of Computer Processing of Oriental Languages 14(2), 153–172 (2001)

Refining Instance Coreferencing Results Using Belief Propagation

Andriy Nikolov, Victoria Uren, Enrico Motta, and Anne de Roeck

Knowledge Media Institute, The Open University, Milton Keynes, UK
{a.nikolov,v.s.uren,e.motta,a.deroeck}@open.ac.uk

Abstract. The problem of coreference resolution (finding individuals, which describe the same entity but have different URIs) is crucial when dealing with semantic data coming from different sources. Specific features of Semantic Web data (ontological constraints, data sparseness, varying quality of sources) are all significant for coreference resolution and must be exploited. In this paper we present a framework, which uses Dempster-Shafer belief propagation to capture these features and refine coreference resolution results produced by simpler string similarity techniques.

1 Introduction

A major problem, which needs to be solved during information integration, is coreference resolution: finding data instances, which refer to the same real-world entity. This is a non-trivial problem due to many factors: different naming conventions used by the authors of different sources, usage of abbreviations, ambiguous names, data variations over time. This problem for a long time has been studied in the domains of database research and machine learning and multiple solutions have been developed. Although in the Semantic Web community information integration has always been considered as one of the most important research directions, so far the research has been primarily concentrated on resolving schema-level issues. However, semantic data represented in RDF and formatted according to OWL ontologies, has its specific features: instances often have only a few properties, relevant information is distributed between interlinked instances of different classes, an OWL ontology allows expressing a wider range of data restrictions than a standard database schema, different sources may significantly differ in quality. Some of these features make it hard to directly reuse the algorithms developed in the database domain, while others may provide valuable clues, which should be exploited.

The main motivation for our work comes from the enterprise-level knowledge management use case. In this scenario a shared corporate ontology is populated automatically with information extracted from multiple sources: text documents, images, database tables. Although there is no schema alignment required in this scenario, the data-level integration problems listed above are present. In addition to the usual issues related to heterogeneity, the data may also contain noise

J. Domingue and C. Anutariya (Eds.): ASWC 2008, LNCS 5367, pp. 405–419, 2008.
© Springer-Verlag Berlin Heidelberg 2008

caused by incorrect extraction results. Data sparseness often prevents the use of sophisticated machine-learning algorithms and requires simple techniques such as string similarity metrics applied to instance labels. The output of these techniques is not completely reliable. In order to improve coreferencing results we have to utilize the links between data instances, to take into account uncertainty of sources and coreferencing algorithms and to consider logical restrictions defined in the domain ontology. In this paper we describe an approach, which uses the Dempster-Shafer belief propagation in order to achieve this goal.

The rest of the paper is organized as follows: in the section 2 we briefly discuss the most relevant existing approaches. Section 3 provides a short description of the approach and its place in the overall integration architecture. Section 4 summarizes the theoretical background of our belief propagation algorithm. Section 5 describes in detail the usage of belief networks and provides examples. In the section 6 we present the results of our experiments performed with test datasets. Finally, section 7 summarizes our contribution and outlines directions for future work.

2   Related Work

The problem of coreference resolution during data integration has been studied for a long time [1]. In different communities it has been referred to as record linkage [1], object identification [2] and reference reconciliation [3]. A large number of approaches (see [4] for a survey) are based on a vector similarity model initially proposed in [1]: similarity scores are calculated for each pair of instances' attributes and their aggregation is used to make a decision about whether two instances are the same. This procedure is performed for instances of each single class in isolation. Different string similarity techniques have been proposed to measure the similarity between attribute values (e.g., edit distance, Jaro, Jaro-Winkler, Monge-Elkan [5]) and different machine learning algorithms to adjust the parameters of decision models have been developed (e.g., [6], [2]).

Such approaches assume that all attributes, which are relevant for determining the equivalence of two instances, are contained in the attribute vector. This assumption does not hold for scenarios where relevant data is distributed between different instances, which are related to each other. Thus, approaches, which analyze relations between data instances of different classes, have received significant attention in recent years (e.g., [3], [7], [8], [9]). One algorithm focusing on exploiting links between data objects for personal information management was proposed in [3], where the similarities between interlinked entities are propagated using dependency graphs. RelDC [7] proposes an approach based on analyzing entity-relationship graphs to choose the best pair of coreferent entities in case when several options are possible. The authors of these algorithms reported good performance on evaluation datasets and, in particular, significant increase in performance achieved by relation analysis. These algorithms, however, assume data representation similar to relational databases. The OWL language used for formatting Semantic Web data allows more advanced restrictions over data to be defined (e.g., class disjointness, cardinality restrictions, etc.), which are relevant for the validation of coreference mappings. Given the variable quality of semantic annotations, information about provenance of the data is also valuable: if a mapping between two individuals violates an ontological restriction, it is possible that some piece of data is wrong, rather than a mapping. These factors require development of specific solutions adjusted to the needs of the Semantic Web domain.

The problem of data integration in the Semantic Web context also requires dealing with data sparseness and the distribution of data between several linked individuals. In the Semantic Web community so far the research effort has been primarily concentrated on the schema-level ontology matching problem [10]. Some of the schema-matching systems utilize links between concepts and properties to update initial mappings created using other techniques. One such technique is similarity flooding [11], which uses links in the graph to propagate similarity estimations. It is, however, more suitable to schema matching rather than data integration: it relies, for example, on the assumption that the graph is complete. Ontological restrictions and uncertainty of mappings between concepts are analyzed in [12]. Now, with a constantly increasing amount of RDF data being published and the emergence of the Linked Data initiative, the problem of instance-level integration is also gaining importance. The issue of recognizing coreferent individuals coming from different sources and having different URIs has been raised by different research groups and several architectural solutions were developed, such as OKKAM [13], Sindice [14], RKBExplorer [15]. Sindice [14] relies on inverse functional properties explicitly defined in corresponding ontologies. The authors of OKKAM entity name service [13] have employed Monge-Elkan string similarity for their prototype implementation. Data aggregation for RKBExplorer [15], to our knowledge, was performed using techniques specially developed for the scientific publication domain (e.g., analyzing co-authorship, etc.). The L2R/N2R algorithm recently proposed in [16] and [17] focuses on employing ontological restrictions (in particular, functionality and disjointness) in order to make coreferencing decisions. Their approach is probably the most similar to ours, but emerged as a purely logical inference-based algorithm and treats some aspects in a different way. In particular, data uncertainty is not considered (data statements are treated as correct) and similarity between individuals is aggregated using maximum function, which does not allow capturing cumulative evidence.

In our view, there is still a need for data integration methods adjusted to the needs of the Semantic Web domain. First, as was said, the algorithms developed in the database community do not take into account the specific properties of semantic data. Ontology matching techniques, on the other hand, focus primarily on the schema-matching issues. Our approach tries to analyze together relations between individuals of multiple classes, logical restrictions imposed by ontologies and data uncertainty in order to improve the quality of instance coreferencing.

3   Overview

The algorithm described in the paper represents a module of the knowledge fusion architecture KnoFuss initially developed to integrate semantic annotations produced from different sources using automatic information extraction algorithms. The architecture receives as its input a source knowledge base (KB) containing a set of RDF assertions extracted from a particular source. The system processes this source KB and integrates it into the target KB. KnoFuss aims to solve two main problems: find and merge coreferent individuals and ensure consistency of the integrated KB. The structure of the KnoFuss system and the initial stage of its workflow is described in [18]. This stage involves producing mappings between individuals (interpreted as *owl:sameAs* relations) using a library of coreferencing algorithms. In this paper we focus on the second stage of the fusion workflow where these initially produced mappings are refined using additional factors, which are not considered by attribute-based similarity algorithms but can serve as evidence for revising and refining the results of coreferencing stage. We consider three kinds of such factors:

- *Ontological schema restrictions.* Constraints and restrictions defined by the schema (e.g., functionality relations) may provide both positive and negative evidence. For instance, having two individuals as objects of a functional property with the same subject should reinforce a mapping between these individuals. The reverse also applies: the fact that two potentially identical individuals belong to two disjoint classes should be considered negative evidence.
- *Coreference mappings between other entities.* Even if there is no explicit functionality restriction defined for an ontological property, related individuals still may reduce the ambiguity: the fact that two similar individuals are both related to a third one may reinforce the confidence of the mapping.
- *Provenance data.* Knowledge about the quality of data may be used to assign the confidence to class and property assertions. This is important when we need to judge whether a mapping, which violates the domain ontology, is wrong or the conflict is caused by a wrong data statement. Knowledge about the "cleanness" of a source (e.g., whether duplicates occur in a given source) provides additional evidence about potential mappings.

Most information, which we have to deal with in the fusion scenario, is uncertain. Mappings are created by attribute-based matching algorithms, which do not provide 100% precision. Class and property assertions may come from unreliable sources or be extracted incorrectly. Various ontological relations provide different impact as evidence for mappings: if two similar *foaf:Person* individuals are both connected to a *sweto:Publication* individual via a *sweto:author* relation, it is a much stronger evidence for identity mapping than if they were related to a *tap:Country* individual *#USA* via a *#citizenOf* relation. In order to manage uncertainty adequately, the framework needs to have well-defined rules for reasoning about the confidence of both data statements and coreference mappings, combining multiple uncertain pieces of evidence and propagating beliefs. This can be achieved by employing an uncertainty representation formalism. Our architecture utilizes the Dempster-Shafer theory of evidence [19], which generalizes the Bayesian probability theory. We proposed the initial version of the algorithm as a means to resolve ABox inconsistencies in knowledge bases [20]. The next section briefly summarizes our previous work.

4   Dempster-Shafer Belief Propagation

Our algorithm uses the Dempster-Shafer theory of evidence as a theoretical basis for uncertainty representation. The reason for this choice (in comparison with the more commonly used Bayesian probability) is its ability to represent a degree of ignorance in addition to the positive and negative belief [20]. This feature is valuable when we deal with the output of coreferencing algorithms. By default, these algorithms can only produce positive evidence: a positive result produced by a low-quality algorithm (e.g., with a precision 0.2) can only be considered as insufficient evidence rather than negative evidence. The uncertainty of a statement is described by belief masses, which can be assigned to sets of possible values. In our case each statement is described by three mass assignments: (i) belief that the statement is true $m(1)$, (ii) belief that the statement is false $m(0)$ and (iii) unassigned belief $m(\{0;1\})$, specifying the degree of our ignorance about the truth of the statement. Given that $\sum_i m_i = 1$, these assignments are usually represented using two values: *belief* (or *support*) ($m(1)$) and *plausibility* ($m(1) + m(\{0;1\})$). Bayesian probability is a special case, which postulates that no ignorance is allowed and $m(1) + m(0) = 1$. Our workflow for processing a conflict involves three steps:

- *Constructing a belief propagation network.* At this stage an OWL subontology is translated into a belief network.
- *Assigning mass distributions.* At this stage the belief mass distribution functions are assigned to nodes.
- *Belief propagation.* At this stage the uncertainties are propagated through the network and the confidence degrees of statements are updated.

As the theoretical base for belief propagation we used valuation networks as described in [21]. Valuation networks contain two kinds of nodes: *variable nodes*, which represent the uncertain assertions, and *valuation nodes*, which represent the belief propagation rules (converted from TBox axioms). We use a set of rules to convert an OWL subontology into a corresponding valuation network (this procedure is described in more detail in [20]). Then, initial beliefs are propagated through the network and updated values are produced according to the standard axioms for valuation networks formulated in [21]. The basic operators for belief potentials are marginalization $\downarrow$ and combination $\otimes$. Marginalization takes a mass distribution function m on domain D and produces a new mass distribution on domain $C \subseteq D$. It extracts the belief distribution for a single variable or subset of variables from a complete distribution over a larger set.

$$m^{\downarrow C}(X) = \sum_{Y^{\downarrow C} = X} m(Y)$$

For instance, if we have the function m defined on the domain $\{x, y\}$ as $m(\{0; 0\})$ $= 0.2$, $m(\{0; 1\}) = 0.35$, $m(\{1; 0\}) = 0.3$, $m(\{1; 1\}) = 0.15$ and we want to find a marginalization on the domain $\{y\}$, we will get $m(0) = 0.2 + 0.3 = 0.5$ and $m(1) = 0.35 + 0.15 = 0.5$. Combination calculates an aggregated belief distribution based on several pieces of evidence. The combination operator is represented by Dempster's rule of combination [19]:

$$m_1 \otimes m_2(X) = \frac{\sum_{X_1 \cap X_2 = X} m_1(X_1) m_2(X_2)}{1 - \sum_{X_1 \cap X_2 = \emptyset} m_1(X_1) m_2(X_2)}$$

Belief propagation through the network is performed by passing messages between nodes according to the following rules:

1. Each node sends a message to its inward neighbour (towards the arbitrary selected root of the tree). If $\mu^{A \to B}$ is a message from a node A to a node B, $N(A)$ is a set of neigbours of A and the potential of A is m_A, then the message is specified as a combination of messages from all neighbours except B and the potential of A: $\mu^{A \to B} = (\otimes \{\mu^{X \to A} | X \in (N(A) - \{B\}) \otimes m_A\})^{\downarrow A \cap B}$
2. After a node A has received a message from all its neighbors, it combines all messages with its own potential and reports the result as its marginal.

Loops must be eliminated by replacing all nodes in a loop with a single node combining their belief functions. The initial version of the algorithm deals with inconsistency resolution and does not consider coreference mappings and identity uncertainty. In the following section we describe how we further develop the same theoretical approach in order to reason about coreference mappings.

5   Refining Coreference Mappings

The algorithm receives as its input a set of candidate mappings between individuals of source and target KBs. In order to perform belief propagation, these mappings along with relevant parts from both knowledge bases must be translated into valuation networks. Building a large network from complete knowledge bases is both computationally expensive and unnecessary, as not all triples are valuable for analysis. We select only relevant triples, which include (i) values of object properties, which can be used to propagate belief between two *owl:sameAs* mappings (functional, inverse functional and "influential" as described in 5.2) and (ii) class and property assertions, which produce conflicts. Conflicts are detected by selecting all statements in the neighborhood of potentially mapped individuals and checking their consistency with respect to the domain ontology (we use the Pellet OWL reasoner with the explanation service). If the reasoner found an inconsistency, all statements which contribute to it are considered relevant. Then, belief networks are constructed by applying the rules defined in ([20] and the extended set described in subsection 5.1) and initial beliefs are assigned to variable nodes. For each *owl:sameAs* variable node the belief is determined according to the precision of the corresponding coreferencing algorithm, which produced it. Each algorithm could produce two kinds of mappings: "probably correct" exceeding the optimal similarity threshold for the algorithm (the one, which maximized the algorithm's F-measure performance), and "possibly correct" with similarities below the optimal threshold, but achieving at least 0.1 precision. Each variable node representing a class or property assertion receives its initial belief based on its attached provenance data: the reliability of its source and/or its extraction algorithm. After that the beliefs are updated using belief propagation and for each mapping the decision about its acceptance is taken.

The most significant part of the algorithm is network construction. At this stage we exploit the factors listed in the section 3. In the following subsections we describe how it is done in more detail.

5.1 Exploiting Ontological Schema

Logical axioms defined by the schema may have both positive and negative influence on mappings. First, some OWL axioms impose restrictions on the data. If creating an *owl:sameAs* relation between two individuals violates a restriction, the confidence of the mapping should be reduced. Second, object properties defined as *owl:FunctionalProperty* and *owl:InverseFunctionalProperty* allow us to infer equivalence between individuals. The initial set of rules and possible network nodes we proposed in [20] does not capture instance equivalence and thus is insufficient for reasoning about coreference relations. Therefore, in this section we present a novel set of additional rules (Table 1), which allow us to reason about coreference mappings. Table 2 lists the additional belief assignment functions for corresponding valuation nodes.

Table 1. Belief network construction rules

N	Axiom	Pre-conditions	Nodes to create	Links to create
1	*sameAs*	$I_1 = I_2$	$N_1 : I_1 = I_2$ (variable)	
2	*differentFrom*	$I_1 \neq I_2$	$N_1 : I_1 \neq I_2$ (variable)	
3	*sameAs*	$N_1 : I_1 = I_2$ (variable), $N_2 : R(I_2, I_3)$	$N_3 : I_1 = I_2$ (valuation), $N_4 : R(I_1, I_3)$	$(N_1, N_3), (N_2, N_3),$ (N_3, N_4)
4	*differentFrom*	$N_1 : I_1 = I_2$ (variable), $N_2 : I_1 \neq I_2$ (variable)	$N_3 : I_1 \neq I_2$ (valuation)	$(N_1, N_3), (N_2, N_3)$
5	*Functional Property*	$\top \sqsubseteq\, \leq 1R$, $N_1 : R(I_3, I_1)$, $N_2 : R(I_3, I_2)$, $N_3 : I_1 = I_2$	$N_4 : \top \sqsubseteq\, \leq 1R$	$(N_1, N_4), (N_2, N_4),$ (N_3, N_4)
6	*InverseFunctional Property*	$\top \sqsubseteq\, \leq 1R^-$, $N_1 : R(I_1, I_3)$, $N_2 : R(I_2, I_3)$, $N_3 : I_1 = I_2$	$N_4 : \top \sqsubseteq\, \leq 1R^-$	$(N_1, N_4), (N_2, N_4),$ (N_3, N_4)

The axioms *owl:sameAs* and *owl:differentFrom* (Table 1, rows 1-4) lead to the creation of both variable and valuation nodes. This is because each one represents both a schema-level rule, which allows new statements to be inferred, and a data-level assertion, which has its own confidence (e.g., produced by a matching algorithm). *owl:FunctionalProperty* and *owl:InverseFunctionalProperty* (rows 5-6) can only be linked to already existing *owl:sameAs* nodes, so that they can only increase similarity between individuals, which were already considered potentially equal. Otherwise the functionality node is treated as in [20]: as a strict constraint violated by two property assertion statements. This is done to prevent the propagation of incorrect mappings.

To illustrate the work of the algorithm we will use an example from our experiments with datasets from the citations domain (see Section 6). One such dataset (DBLP) contains an individual *Ind1* describing the following paper:

> D. Corsar, D. H. Sleeman. Reusing JessTab Rules in Protege. Knowledge-Based Systems 19(5). (2006) 291-297.

Another one (EPrints) also contained a paper *Ind2* with the same title:

> Corsar, Mr. David and Sleeman, Prof. Derek. Reusing JessTab Rules in Protege. In Proceedings The Twenty-fifth SGAI International Conference on Innovative Techniques and Applications of Artificial Intelligence (2005), pages pp. 7-20, Cambridge, UK.

This illustrates a common case when the same group of researchers first publishes their research results at a conference and then submits the extended and revised paper to a journal. An attribute-based coreferencing algorithm (Jaro-Winkler similarity applied to the title), which had a good overall performance (precision about 0.92 and F-measure about 0.94), incorrectly considered these two papers identical. However, a mapping between these individuals violated two restrictions: the individual belonged to two disjoint classes simultaneously and had two different values for the functional property *year*. The inconsistencies were detected by the algorithm, which produced two sets of relevant statements: {*owl:sameAs(Ind1, Ind2)*; *Article(Ind1)*; *Article_in_Proceedings(Ind2)*; *owl:disjointWith(Article, Article_in_Proceedings)*} and {*owl:sameAs(Ind1, Ind2)*; *year(Ind1, 2006)*; *year(Ind2, 2005)*; *owl: Functional Property(year)*}. Since these sets share a common statement (*sameAs* link), they are translated into a single valuation network (Fig. 1). Although in our example the initial support of the mapping was higher than the support of both statements related to Ind2 (*Article_in_Proceedings(Ind2)* and *year(Ind2, 2005)*), after belief propagation the incorrect *owl:sameAs* mapping was properly recognized and received the lowest plausibility (0.21 - obtained as $m(1) + m(0; 1) = 0.20 + 0.01$).

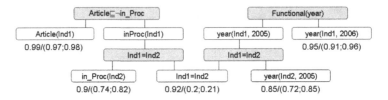

Fig. 1. Example of a belief network constructed during the experimental testing. The numbers show the support before propagation and support and plausibility after propagation for variable nodes (white). Leaf variable nodes are given in the KB while non-leaf ones are inferred using axioms corresponding to valuation nodes (blue).

5.2 Influence of Context Mappings

Belief propagation for properties explicitly defined as functional is a trivial case. However, properties which allow multiple values are also valuable as a means to narrow the context of matched individuals and increase similarity between them. We have to estimate the impact of the relation and model this in the network. As shown in Table 2 (row 1), by default the valuation node for the $owl{:}sameAs$ relation is defined in such a way that the belief in $I_1 = I_2$ is completely independent from a strong belief for both $R(I_3, I_1)$ and $R(I_3, I_2)$. The functionality axiom represents an opposite scenario: having a belief 1.0 for both $R(I_3, I_1)$ and $R(I_3, I_2)$ implies the belief 1.0 for $I_1 = I_2$. The actual strength of influence for a property may lay between these extreme cases. In order to utilize such links the network construction algorithm receives for each relevant property a vector $< n_1, n_2 >$, where n_1, n_2 determine the impact of the link in direct (subject to object) and reverse (object to subject) directions. The impact in two directions may be different: having two people as first authors of the same paper strongly implies people's equivalence, while having the same person as the first author of two papers with the similar title does not increase the probability of two papers being the same. The $owl : sameAs$ valuation node, combining variables $I_1 = I_2$, $R(I_2, I_3)$, $R(I_1, I_3)$ will receive two belief assignments instead of one: m({0;0;0}, {0;0;1}, {0;1;0}, {1;0;0}, {1;1;1})=n_1 and m({0;0;0}, {0;0;1}, {0;1;0}, {0;1;1}, {1;0;0}, {1;1;1})=$1 - n_1$. One possible way to determine coefficients $< n_1, n_2 >$ is to learn them from training data, as we did in our experiments, or to assign them based on expert estimations or the number of statements per individual as in [11].

Table 2. Belief distribution functions for valuation nodes

N	Axiom	Node type	Variables	Mass distribution
1	$sameAs$	$I_1 = I_2$	$I_1 = I_2$, $R(I_1, I_3)$, $R(I_2, I_3)$	m({0;0;0}, {0;0;1}, {0;1;0}, {0;1;1}, {1;0;0}, {1;1;1})=1
2	$differentFrom$	$I_1 \neq I_2$	$I_1 = I_2$, $I_1 \neq I_2$	m({0;1},{1;0})=1
3	Functional Property	$\top \sqsubseteq\, \leq 1R$	$R(I_3, I_1)$, $R(I_3, I_2)$, $I_1 = I_2$	m({0;0;0}, {0;0;1}, {0;1;0}, {1;0;0}, {1;1;1})=1
4	Inverse Functional Property	$\top \sqsubseteq\, \leq 1R^-$	$R(I_1, I_3)$, $R(I_2, I_3)$, $I_1 = I_2$	m({0;0;0}, {0;0;1}, {0;1;0}, {1;0;0}, {1;1;1})=1

Also some relevant relations may be implicit and not defined in the ontology. For instance, the same group of people may be involved in different projects. If the link between a project and a person is specified using a property $akt{:}has{-}project{-}member$, when two knowledge bases describing two non-overlapping sets of projects are combined, the relations between people cannot be utilized. In order to capture these implicit relations we can add artificial properties, which connect individuals belonging to the same sets, into the ontology. Co-authorship analysis, commonly used in the citation matching domain, is a special case of this scenario (Fig. 2a).

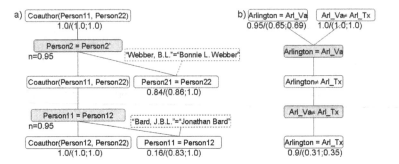

Fig. 2. Examples of belief networks illustrating (a) the usage of artificial set membership relations and (b) processing competing mappings knowing that a source does not contain duplicates. The numbers show the belief before propagation and belief and plausibility after propagation.

5.3   Provenance Data

The estimated reliability of a source is directly used at the starting stage when initial beliefs are assigned to variable nodes representing class and property assertions. Thus, if a violation of a functional restriction is caused by a property assertion with a low belief, its impact will be insufficient to break the *owl:sameAs* link. Another important factor is the knowledge about duplicate individuals in a knowledge base. For instance, one knowledge base (AGROVOC) contains an individual "fao:arlington". If we match this against the UTexas geographical ontology, which contains two individuals "arlingtonVa" and "arlingtonTx", then although the similarity of one pair is slightly greater than another one, both values are above the threshold and both these individuals can potentially be matched to the first individual. However, knowing that this particular knowledge base does not contain duplicates, allows us to add a corresponding *owl:differentFrom* variable node into the network (Fig.2b). Updating beliefs allows us to reject one of the two competing options.

6   Evaluation

In order to test the system we used the following datasets from the domain of scientific publications:

1. AKT EPrints archive[1]. This dataset contains information about papers produced within the AKT research project.

---

[1] http://eprints.aktors.org/

2. Rexa dataset[2]. The dataset extracted from the Rexa search server, which was constructed in the University of Massachusetts using automatic IE algorithms.
3. SWETO DBLP dataset[3]. This is a publicly available dataset listing publications from the computer science domain.
4. Cora(I) dataset[4]. A citation dataset used for machine learning tests.
5. Cora(II) dataset. Another version of the Cora dataset used in [3].

AKT, Rexa and SWETO-DBLP datasets were previously used by the authors in [18]. The SWETO-DBLP dataset was originally represented in RDF. AKT and Rexa datasets were extracted from the HTML sources using specially constructed wrappers and structured according to the SWETO-DBLP ontology (Fig. 3). The Cora(I) dataset was created in the University of Massachusetts for the purpose of testing machine-learning clustering algorithms. It contains 1295 references and is intentionally made noisy: e.g., the gold standard contains some obviously wrong mappings[5]. We translated this dataset into RDF using the SWETO-DBLP ontology. The authors of Cora(II)[3] translated the data from Cora(I) into RDF according to their own ontology and cleaned the gold standard by removing some spurious mappings, so the results achieved on Cora(I) and Cora(II) are not comparable. Data and gold standards mappings in Cora(II) are significantly cleaner than in Cora(I). Also in Cora(II) all *Person* individuals were initially considered different while in Cora(I) individuals with exactly the same name were assigned the same URI, which led to a significant difference in the number of individuals (305 vs 3521) and, consequently, in performance measurements. In our tests we tried to merge each pair of datasets 1-3 and to find duplicates in the Cora datasets.

To the SWETO ontology we added the restrictions specifying that (i) classes *Article* and *Article_in_Proceedings* are disjoint, (ii) datatype property *year* describing the publication year is functional and (iii) object property *author* connecting a publication with a set of authors is functional. Given that both Cora datasets did not distinguish between journal and conference articles, instead we used venues as individuals and added functionality relations for them. Also the Cora(II) ontology described pages as two integer properties *pageFrom* and *pageTo*, which allowed us to add a functionality restriction on them as well.

For attribute-based coreferencing we used string similarity metrics applied to a paper title or person's name. In particular, we used Jaro-Winkler and Monge-Elkan metrics applied to the whole strings or tokenized strings (L2 Jaro-Winkler). L2 Jaro-Winkler is a mixture of string similarity and set similarity

---

[2] http://www.rexa.info/
[3] http://lsdis.cs.uga.edu/projects/semdis/swetodblp/august2007/
opus_august2007.rdf
[4] http://www.cs.utexas.edu/users/ml/riddle/data/cora.tar.gz
[5] For instance, two papers by N. Cesa-Bianchi et al. "How to use expert advice. 25th ACM Symposium on the theory of computing (1993) 382-391" and "On-line prediction and conversion strategies. Eurocolt'93 (1993) 205-216" were considered the same in Cora(I).

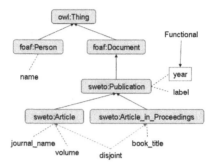

Fig. 3. Class hierarchy in the SWETO-DBLP ontology

measures: it tokenizes both compared values, then each pair of tokens is compared using the standard Jaro-Winkler algorithm and the maximal total score is selected. Initial belief mass distribution for each *owl:sameAs* relation was assigned according to the precision of the algorithm, which produced it. Initial belief assignments for the class and property assertions are shown in the Table 3. We assigned the values based on our knowledge about how each dataset

Table 3. Initial belief mass assignment

Dataset	Class assertions	Datatype assertions
DBLP	0.99	0.95
Rexa	0.95	0.81 (<2 citations)
		0.855 (>2 citations)
EPrints	0.9	0.85
Cora(I & II)	N/A	0.6

was produced and manual reviewing of the datasets. We did not further classify publications in Cora datasets into journal and conference articles, so class assertions were not relevant. Knowing that the data in Cora datasets was noisy, we assigned beliefs in such a way that disagreement on a single property value was not sufficient to break the mapping.

We measured the quality of coreference before and after belief propagation. The results of the tests are shown in the Table 4. As expected, in almost all cases the refinement procedure led to an improvement in overall performance expressed by the F1-measure. For *sweto:Publication* instances (rows 1, 2, 4, 6, 7, 8) the recall has decreased: the algorithm incorrectly resolved some inconsistencies, which in fact occurred due to wrong data statements. The decrease was slight for AKT/Rexa/DBLP datasets and more significant for Cora where the degree of noise was higher. However, in all cases this decrease was more than compensated by the increase in precision. For *foaf:Person* individuals the effect of belief propagation primarily influenced recall: links between instances reinforced the potential mappings, which would otherwise be rejected. Because Cora(II) was better formatted than Cora(I) there were very few "dubious" mappings produced during initial coreferencing and belief propagation was not able to catch them.

Table 4. Test results

Dataset	No	Matching algorithm	sweto:Publication					
			Before			After		
			Precision	Recall	F1	Precision	Recall	F1
EPrints/Rexa	1	Jaro-Winkler	0.950	0.833	0.887	0.969	0.832	0.895
	2	L2 Jaro-Winkler	0.879	0.956	0.916	0.923	0.956	0.939
EPrints/DBLP	3	Jaro-Winkler	0.922	0.952	0.937	0.992	0.952	0.971
	4	L2 Jaro-Winkler	0.389	0.984	0.558	0.838	0.983	0.905
Rexa/DBLP	5	Jaro-Winkler	0.899	0.933	0.916	0.944	0.932	0.938
	6	L2 Jaro-Winkler	0.546	0.982	0.702	0.823	0.981	0.895
Cora(I)	7	Monge-Elkan	0.735	0.931	0.821	0.939	0.836	0.884
Cora(II)	8	Monge-Elkan	0.698	0.986	0.817	0.958	0.956	0.957
			foaf:Person					
EPrints/Rexa	9	L2 Jaro-Winkler	0.738	0.888	0.806	0.788	0.935	0.855
EPrints/DBLP	10	L2 Jaro-Winkler	0.532	0.746	0.621	0.583	0.921	0.714
Rexa/DBLP	11	Jaro-Winkler	0.965	0.755	0.846	0.968	0.876	0.920
Cora(I)	12	L2 Jaro-Winkler	0.983	0.879	0.928	0.981	0.895	0.936
Cora(II)	13	L2 Jaro-Winkler	0.999	0.994	0.997	0.999	0.994	0.997

Considering the F1 measure obtained for Cora(I) publication (row 7) in comparison with the state-of-the art algorithms from the database and machine learning communities, we found that it is higher than those reported in [22] (0.867), [9] (0.87), but lower than in [2] (0.93)[6]. As was said before, in order to minimize the number of attributes processed by basic coreferencing methods, in our tests we only used the title comparison for determining candidate individuals. This was the main factor, which reduced the performance: e.g., the algorithm used in [2] achieved similar F-measure (0.88) on the test set when trained only on the *title*, *year* and *venue* attributes. For Cora(II) the F-measure was similar to that reported for [3]: slightly higher for publications (0.957 vs 0.954) while slightly lower for people (0.997 vs 0.999). The difference is due to the fact that the authors of [3] used better similarity measures (reported F-measure for publications 0.948 without exploiting links) while exploiting data uncertainty by our approach increased recall (e.g., having different years for papers was not enough to break the mapping if there was an agreement for the venue name and pages).

7   Conclusion and Future Work

In the paper we have presented an approach which uses Dempster-Shafer belief propagation in order to improve the quality of data integration, in particular coreferencing of individuals. We consider this extension and application of the

---

[6] Note that the authors of [8] and [7], and [16] used different versions of the Cora dataset where, in particular, more mappings were removed from the gold standard so that the dataset contained 132 clusters [8] rather than 125 in Cora(II), and papers with the same title and year were considered identical [16]. This does not allow direct comparison of reported performance with our algorithms.

Dempster-Shafer belief propagation mechanism as the main contribution of this paper. Our initial experiments performed with test datasets have shown an improvement in the output quality of basic string similarity algorithms. However, there are still issues which have to be resolved in the future work.

First, the Dempster-Shafer belief propagation mechanism is sensitive to the initial belief distribution, which may be an issue if initial belief does not adequately reflect the actual data, e.g., if the estimated precision of a coreferencing algorithm was measured using a test set with a different distribution of data. Second, at the moment the algorithm assumes that the data to be merged is formatted according to the same ontology. In order to be employed on a Web scale, the ability to work in a multi-ontology environment is necessary. In particular, the output of ontology matching algorithms must be considered. Another important feature would be automatic discovery of ontological restrictions by retrieving other ontologies covering the same domain (e.g., using Watson[7] or Swoogle[8] engines) and analyzing them.

Acknowledgements

This work was funded by the X-Media project (www.x-media-project.org) sponsored by the European Commission as part of the Information Society Technologies (IST) programme under EC grant number IST-FP6-026978. The authors would like to thank Steffen Rendle and Karen Tso for providing their object identification tool [2], Luna Dong for providing the Cora(II) dataset and Fatiha Saïs for providing materials about L2R/N2R algorithm [17].

References

1. Fellegi, I.P., Sunter, A.B.: A theory for record linkage. Journal of American Statistical Association 64(328), 1183–1210 (1969)
2. Rendle, S., Schmidt-Thieme, L.: Object identification with constraints. In: 6th IEEE International Conference on Data Mining, ICDM (2006)
3. Dong, X., Halevy, A., Madhavan, J.: Reference reconciliation in complex information spaces. In: SIGMOD 2005: Proceedings of the 2005 ACM SIGMOD international conference on Management of data, pp. 85–96. ACM, New York (2005)
4. Elmagarmid, A.K., Ipeirotis, P.G., Verykios, V.S.: Duplicate record detection: A survey. IEEE Transactions on Knowledge and Data Engineering 19(1), 1–16 (2007)
5. Winkler, W.E.: Overview of record linkage and current research directions. Technical Report RRS2006/02, US Bureau of the Census, Washington, DC 20233 (2006)
6. Sarawagi, S., Bhamidipaty, A.: Interactive deduplication using active learning. In: 8th ACM SIGKDD International Conference on Knowledge Discovery and Data Mining (KDD-2002), Edmonton, Alberta, Canada. ACM Press, New York (2002)
7. Chen, Z., Kalashnikov, D.V., Mehrotra, S.: Adaptive graphical approach to entity resolution. In: ACM IEEE Joint Conference on Digital Libraries 2007 (ACM IEEE JCDL 2007), Vancouver, British Columbia, Canada, pp. 204–213 (2007)

---

[7] http://watson.kmi.open.ac.uk/WatsonWUI/
[8] http://swoogle.umbc.edu/

8. Singla, P., Domingos, P.: Object identification with attribute-mediated dependences. In: 9th European Conference on Principles and Practice of Knowledge Discovery in Databases (PAKDD-2005), Porto, Portugal, pp. 297–308 (2005)
9. Parag, Domingos, P.: Multi-relational record linkage. In: KDD Workshop on Multi-Relational Data Mining, Seattle, CA, USA, pp. 31–48. ACM Press, New York (2004)
10. Euzenat, J., Shvaiko, P.: Ontology matching. Springer-Verlag, Heidelberg (2007)
11. Melnik, S., Garcia-Molina, H., Rahm, E.: Similarity flooding: A versatile graph matching algorithm. In: 18th International Conference on Data Engineering (ICDE), San Jose (CA US), pp. 117–128 (2002)
12. Castano, S., Ferrara, A., Lorusso, D., Näth, T.H., Möller, R.: Mapping validation by probabilistic reasoning. In: Bechhofer, S., Hauswirth, M., Hoffmann, J., Koubarakis, M. (eds.) ESWC 2008. LNCS, vol. 5021, pp. 170–184. Springer, Heidelberg (2008)
13. Bouquet, P., Stoermer, H., Bazzanella, B.: An Entity Name System (ENS) for the Semantic Web. In: Bechhofer, S., Hauswirth, M., Hoffmann, J., Koubarakis, M. (eds.) ESWC 2008. LNCS, vol. 5021, pp. 258–272. Springer, Heidelberg (2008)
14. Tummarello, G., Delbru, R., Oren, E.: Sindice.com: Weaving the open linked data. In: Aberer, K., Choi, K.-S., Noy, N., Allemang, D., Lee, K.-I., Nixon, L., Golbeck, J., Mika, P., Maynard, D., Mizoguchi, R., Schreiber, G., Cudré-Mauroux, P. (eds.) ASWC 2007 and ISWC 2007. LNCS, vol. 4825, pp. 552–565. Springer, Heidelberg (2007)
15. Glaser, H., Millard, I.C., Jaffri, A.: RKBExplorer.com: A knowledge driven infrastructure for linked data providers. In: Bechhofer, S., Hauswirth, M., Hoffmann, J., Koubarakis, M. (eds.) ESWC 2008. LNCS, vol. 5021, pp. 797–801. Springer, Heidelberg (2008)
16. Saïs, F., Pernelle, N., Rousset, M.C.: L2R: a logical method for reference reconciliation. In: 22nd AAAI Conference on Artificial Intelligence (AAAI 2007), Vancouver, BC, Canada, pp. 329–334. AAAI Press, Menlo Park (2007)
17. Saïs, F., Pernelle, N., Rousset, M.C.: Combining a logical and a numerical method for data reconciliation. Journal of Data Semantics 12 (2008)
18. Nikolov, A., Uren, V., Motta, E., de Roeck, A.: Handling instance coreferencing in the KnoFuss architecture. In: Workshop on Identity and Reference on the Semantic Web, ESWC 2008, Tenerife, Spain (2008)
19. Shafer, G.: A mathematical theory of evidence. Princeton University Press, Princeton (1976)
20. Nikolov, A., Uren, V., Motta, E., de Roeck, A.: Using the Dempster-Shafer theory of evidence to resolve ABox inconsistencies. In: Aberer, K., Choi, K.-S., Noy, N., Allemang, D., Lee, K.-I., Nixon, L., Golbeck, J., Mika, P., Maynard, D., Mizoguchi, R., Schreiber, G., Cudré-Mauroux, P. (eds.) ASWC 2007 and ISWC 2007. LNCS, vol. 4825. Springer, Heidelberg (2007)
21. Shenoy, P.P.: Valuation-based systems: a framework for managing uncertainty in expert systems. In: Fuzzy logic for the management of uncertainty, pp. 83–104. John Wiley & Sons, Inc., New York (1992)
22. Bilenko, M., Mooney, R.J.: Adaptive duplicate detection using learnable string similarity measures. In: 9th ACM SIGKDD International Conference on Knowledge Discovery and Data Mining (KDD-2003), Washington DC, pp. 39–48 (2003)

Named Entity Disambiguation: A Hybrid Statistical and Rule-Based Incremental Approach

Hien T. Nguyen[1] and Tru H. Cao[2]

[1] Ton Duc Thang University, Vietnam
hien@tut.edu.vn
[2] Ho Chi Minh City University of Technology, Vietnam
tru@cse.hcmut.edu.vn

Abstract. The rapidly increasing use of large-scale data on the Web makes named entity disambiguation become one of the main challenges to research in Information Extraction and development of Semantic Web. This paper presents a novel method for detecting proper names in a text and linking them to the right entities in Wikipedia. The method is hybrid, containing two phases of which the first one utilizes some heuristics and patterns to narrow down the candidates, and the second one employs the vector space model to rank the ambiguous cases to choose the right candidate. The novelty is that the disambiguation process is incremental and includes several rounds that filter the candidates, by exploiting previously identified entities and extending the text by those entity attributes every time they are successfully resolved in a round. We test the performance of the proposed method in disambiguation of names of people, locations and organizations in texts of the news domain. The experiment results show that our approach achieves high accuracy and can be used to construct a robust named entity disambiguation system.

1 Introduction

In Information Extraction and Natural Language Processing areas, named entities (NE) are people, organizations, locations, and others that are referred to by names. A wider interpretation of the term includes any token referring to something specific in reality, such as numbers, addresses, amounts of money, dates, etc. ([7]). Having been raised from research in those areas, named entities have also become a key issue in development of Semantic Web ([20]). According to the vision of the Semantic Web, metadata about named entities would be widely available with high quality for easily sharing, integrating and processing by software agents. In that spirit, extracting named entities in texts and linking them to some ontology or knowledge base (KB) such as KIM[1], OpenCyc[2], Wikipedia[3], etc. have been increasingly attracting researchers' attention.

---

[1] http://www.ontotext.com/kim/
[2] http://www.opencyc.org/
[3] http://www.wikipedia.org/

J. Domingue and C. Anutariya (Eds.): ASWC 2008, LNCS 5367, pp. 420–433, 2008.
© Springer-Verlag Berlin Heidelberg 2008

One great challenge in dealing with named entities is that one name may refer to different entities in different occurrences and one entity may have different names which may be written in different ways and with spelling errors. For example, the name *"John McCarthy"* in different occurrences may refer to different named entities such as a computer scientist from Stanford University, a linguist from University of Massachusetts Amherst, a British journalist who was kidnapped by Iranian terrorists in Lebanon in April 1986, an Australian ambassador, etc. Such ambiguity makes identification of NEs more difficult and raises NE disambiguation problem as one of the main challenges to research not only in the Semantic Web but also in areas of natural language processing in general.

Our work aims at detecting named entities in a text, disambiguating and linking them to the right ones in Wikipedia. The proposed method utilizes NEs and related terms co-occurring with the target entity in a text and Wikipedia for disambiguation because the intuition is that these respectively convey its relationship and attributes. For example, suppose that in a KB there are two entities named *"Jim Clark"*, one of which has a relation with the *Formula One* car racing championship and the other with *Netscape*. Then, if in a text where the name appears there are occurrences of *Netscape* or web-related referents and terms, then it is more likely that the name refers to the one with *Netscape* in the KB. We exploit Wikipedia as a source of NE annotations due to its size, variation, accuracy and quantity of hyper-links ([23]) to construct annotated corpus for disambiguation.

The contribution of this paper is three-fold as follows:

- First, we propose a hybrid method that combines heuristics and a learning model for disambiguation and identification of NEs in a text with respect to Wikipedia.
- Second, the proposed disambiguation process is incremental and includes several rounds that filter the candidates, by exploiting previously identified entities and extending the text by those entity attributes every time they are successfully resolved in a round. Importantly, we explore context in several levels, from local to the whole text, where diverse clues are extracted for disambiguation at high accuracy.
- Finally, our method utilizes disambiguation texts in titles of articles in Wikipedia as an important feature not only to choose the right entity for an ambiguous referent by search for their occurrences in local context, but also to disambiguate other ambiguous referents in a text.

The rest of the paper is organized as follows. Section 2 states the problem as well as describes its scope. Section 3 describes resources of information in Wikipedia that are essential for our method. Section 4 describes extraction of named entities in Wikipedia to create a disambiguation dictionary. Section 5 presents in details the disambiguation method. Section 6 describes evaluation of our method. Section 7 presents related works. Finally, we draw a conclusion in Section 8.

2 Background

The problem of disambiguation is to determine whether two named entities refer to the same entity. For instance, do *"J. Smith"* and *"John Smith"* refer to the same person?

Do different occurrences of *"John Smith"* refer to the same person? This paper addresses the problem that aims at mapping referents that are not resolved yet in a text to the right referents in a predefined list of resolved referents. For instance, for the text *"the computer scientist John McCarthy coined the term artificial intelligence in the late 1950's,"* our method detects whether `John McCarthy` and the resolved referent `John McCarthy (computer scientist)` in Wikipedia refer to the same entity and then link the referent `John McCarthy` to `John McCarthy (computer scientist)` in Wikipedia.

In [15], the authors identify several levels of named entity ambiguity. The first level of ambiguity is structured ambiguity, where the structures of names are ambiguous. For instance, in the name *"Victoria and Albert Museum"* the word *and* is a part of this name, whereas, in the case of *"IBM and Bell Laboratories"*, *and* is a conjunction joining two computer company names. The second level is semantic ambiguity, where entity-type is ambiguous. For instance, *"John Smith"* may refer to a company or a person. Referent ambiguity is the next level, when one name may be used to refer to different entities, e.g. *"Paris"* may refer to Paris, France, or a small town in Texas, the United States. Our work performs both the semantic ambiguity and the referent ambiguity resolution with assumption that the structured ambiguity is resolved in pre-processing steps.

The task of NE disambiguation bears a resemblance to Word Sense Disambiguation (WSD) in that it comprises two key steps of which a look-up step retrieves candidate referents in a KB of discourse (*sense inventory* in WSD) and the second one chooses the most likely candidate. However, this task is different from WSD in that NEs, roughly speaking, represent specific individuals in the world of discourse, while words denote generic concepts such as types, properties, and relations. Reasoning with words thus requires only lexical semantics and common senses, while reasoning with NEs requires specific knowledge about the world of discourse.

This problem has attracted much research effort, with various methods introduced for different domains, scopes, and purposes. Most of those methods fit into categories described below:

- Rule-based methods, the methods that use some heuristics to disambiguate NEs of the Location ([17], [18]), the Person ([13]) or arbitrary types from a given ontology ([16]).
- Machine learning methods, the methods that extract information from Wikipedia to form language models ([2], [4]) or co-occurrence model ([21]), and then use those models to disambiguating named entities.
- Data-driven methods, the methods of disambiguation apply semi-supervised techniques, combined with additional un-annotated corpus, to learn contextual information for disambiguation ([22]).

The method of disambiguation presented here combines heuristics and a learning model. To maximize accuracy of mapping a NE referred to in a text to the right one in Wikipedia poses a significant question that how contexts in which referents occur are exploited and how corresponding NEs can be represented. In our case, we represent NEs by their attributes and relations. The attributes are birthday, career, occupation, alias, first name, last name, and so on. The relations of an entity represent relations to others such as *part-of*, *located-in*, etc. The way we exploit the contexts is based on

Harris' Distributional Hypothesis [14] stating that words occurring in similar contexts tend to have similar senses. We adapt the hypothesis to NE instead of word sense disambiguation.

3 Wikipedia

Wikipedia is a free encyclopedia written by a collaborative effort of a large number of volunteer contributors. It is a multilingual resource and growing exponentially so that it has become the largest organized knowledge repository on the Web ([19]). We utilize Wikipedia data because of its size, quality, growth speed, as well as a source of information about synonyms, spelling variations, and abbreviations of NEs. Also, it is a fertile source for exploiting related terms and co-occurring NEs; those actually provide explicit information about the important features of the corresponding entity, such as location and industry of a company, etc. In addition, many-to-many correspondence between names and entities can be captured from Wikipedia by utilizing *redirect pages* and *disambiguation pages*.

Pages

A basic entry in Wikipedia is a *page* (or *article*) that defines and describes a single entity or concept. It is uniquely identified by its title. Typically, the title is the canonical name for the entity described in the page. When the name is ambiguous, the title contains further information that we call *disambiguation text* to distinguish the entity described from others. The disambiguation text separates from the name by parentheses, or a comma, e.g. `John McCarthy (computer scientist)`, `Columbia, South Carolina`. Each title is an identifier (ID) of a specific named entity in Wikipedia, because it identifies a unique entity in Wikipedia entity space.

Links

Each page consists of many links whose role is not only to point from the page to others but also to guide readers to pages that provide additional information about the entities mentioned. Each link is associated with an anchor text that represents the surface form of the corresponding entity. Note that if the anchor text denotes an ambiguous name or is an alternative name instead of canonical name, a piped link is used instead in wiki source code. For instances, a typical link might look like `[[Midland, Texas | Midland]]`, where `Midland, Texas` is the link target and `Midland` is its surface form.

Categories

The Wikipedia category system is also a source of meaningful information. The Wikipedia category tree is an example of a folksonomy, a kind of collaborative tagging system that enables the users to categorize the content of the encyclopedic entries. Thus, the taxonomy of Wikipedia can express not only hyponymic relations but also meronymic relations as well. As an example, the Wikipedia page for the *George Bush* belongs not only to the categories *Presidents of the United States* and *Texas Republicans* (is-a) but also to the *1946 births* (has-property). In Wikipedia, every entity page is associated with one or more categories, each of which can have subcategories expressing meronymic or hyponymic relations. Note that we extract not only direct category information of an entity but also all its parent and ancestors.

Redirect pages

A redirecting page typically contains only a reference to an entity or a concept page. Title of redirecting page is an alternative name of that entity or concept. For example, from redirect pages of United States, we extract alternative names of the *United States* entity such as *"US"*, *"USA"*, *"United States of America"*, etc.

Disambiguation pages

A disambiguation page is created for an ambiguous name which denotes two or more entities in Wikipedia. It consists of links to pages that define the different entities with the same name. For instance, the disambiguation page for *"John McCarthy"* lists pages discussing John McCarthy (referee), John McCarthy (journalist), John McCarthy (journalist), John McCarthy (computer scientist), etc. From the disambiguation pages we detect all entities that have the same name in Wikipedia for creating a disambiguation dictionary.

4   Creating the Disambiguation Dictionary

Based on the resources of information aforementioned, we follow the method presented in [2] to create a disambiguation dictionary. Since our work focuses on NEs, we first consider which pages in Wikipedia define NEs. In [2], the authors consider a page describing a NE if it satisfies one of the following heuristics:

1. If its title is a multiword title, check the capitalization of all content words in the title, i.e. words other than prepositions, determiners, conjunctions, relative pronouns or negations. Consider the page describing a NE if and only if all the content words are capitalized.
2. If its title is a one-word title that contains at least two capital letters, then the page describes a NE. Otherwise, go to step 3.
3. Count how many times its title occurs in the text of the page, in positions other than at the beginning of sentences. If at least 75% of these occurrences are capitalized, then the page describes a NE.

Following this way, a dictionary is constructed so that the set of entries in the dictionary consists of all strings that may denote a named entity. In particular, if *e* is a named entity, its title name, its redirect names, and disambiguation names are all added as entries in the dictionary. Then each entry string in the dictionary is mapped to a set of entities that the string may denote in Wikipedia. As a result, a named entity *e* is included in the set if and only if the string is one of title name, redirect names, or disambiguation names of *e*.

Note that although we utilize information from Wikipedia to create the disambiguation dictionary, our method can be adapted for an ontology or knowledge base in general. In particular, one can generate a profile for each of KB entities by making use of ontology concepts and properties of the entities. In other words, one can take advantage of hierarchy of classes for feature by extracting direct class and parent classes of the entities. Also, value of properties of entities was exploited. For attributes, their values were directly extracted. For relation properties, one can utilize names and ID of the corresponding entities. All the extracted feature values of an entity will be concatenated into a text snippet, which can be considered a profile of the entity, for further processing.

5 Proposed Method

In a news article, co-occurring entities are usually related to each other. Furthermore, the identity of a named entity is inferable from nearby and previously identified NEs in the text. For example, when the name *"Georgia"* occurs with *"Atlanta"* in a text and if *Atlanta* is already recognized as a city in the *United States*, then it is more likely that *"Georgia"* refers to a state of the *U.S.* rather than the country *Georgia*, and vice versa when it occurs with *Tbilisi* as a country capital in another text. Furthermore, the words surrounding ambiguous names may denote attributes of NEs referred to. If those words are automatically extracted, the ambiguous names may be disambiguated. For example, in the text *"Michael Collins, assistant professor"*, the word *"professor"* helps to discriminate *"Michael Collins"* who works at MIT from *"Michael Collins"* who flew on the *Apollo 11* and others with the same name. From those observations, we propose a method with the following essential points:

- It is a hybrid method containing two phases the first one of which is a rule-based phase that filters candidates and disambiguates named entities if possible and the second one employs a statistical learning model to rank the candidates.
- It is an iterative and incremental process in which a resolved referent at each iteration step is intermediately utilized to disambiguate others in the next step.
- It exploits both entity IDs and keywords as means of named entity disambiguation in two phases. In particular, in the first phase, based on NE identifiers of previously identified NEs in the local context, it searches for occurrences of candidates' disambiguation texts not only to filter the candidates but also to disambiguate ambiguous referents; then in the second phase, it utilizes words surrounding ambiguous referents in consideration and entity IDs in the whole text for ranking the candidates.

The disambiguation process comprises the following steps in each iteration:

1. Looking up candidates for referents in text using the disambiguation dictionary as a gazetteer.
2. Narrowing down the candidates for ambiguous referents in text using textual information, IDs in local context and disambiguation texts of candidates. After this step, the text will be extended by disambiguation text of the chosen candidate.
3. Ranking candidates using features extracted from the extended text and Wikipedia to disambiguate the referents that have not been resolved yet.

5.1 Looking Up Candidates

Prior to looking up candidates in the disambiguation dictionary, we perform preprocessing steps. In particular, we perform NE recognition and NE coreference resolution using natural language processing resources of Information Extraction engine based on GATE ([6]), a general architecture for developing natural language processing applications. The NE recognition applies pattern-matching rules written in JAPE's grammar of GATE, in order to identify the class of an entity in the text. After detecting all mentions of entities occurring in the text, we run NE co-reference resolution ([3]) module in the GATE system to resolve the different mentions of an NE into one group that uniquely represents the NE. After pre-processing steps, for each entity name in the text, we send it as a query to the dictionary to retrieval candidates. If there is only one candidate in the result, the corresponding referent is resolved.

It can also be observed in practice, in particular news articles, that the use of short names in place of full names is very common. For example, the names *"Bush"*, *"George Bush"* may be used to refer to the current president of the United States stand for *"George W. Bush"* in a news article, while the name *"Bush"*, if taken out of a particular context, can refer to `Laura Bush`, or `Samuel D. Bush`, for instance. If *"George W. Bush"* and *"Bush"* are found to be coreferent in a text, then it is likely that they refer to the `president of the United States`. Our work is based on the assumption that all various representations of a name in a text mention the same entity. Therefore, we propagate resolved referents to others in their coreference chains. For example, assumption that in a text, there are occurrences of *"Denny Hillis"* and *"Hillis"* (*"Hillis"* may refer to `Ali Hillis`, American actress, `Horace Hillis`, American politician, `W. Daniel Hillis`, American inventor, entrepreneur, and author, etc.), if *"Denny Hillis"* is recognized that it refers to the `W. Daniel Hillis` and *"Hillis"* also refers to `W. Daniel Hillis`.

5.2 Narrowing Down Candidates

In this step we exploit the local context to narrow down candidates and disambiguate ambiguous referent if possible. The local context of a location referent is its neighbor referents (i.e. the previous and successive ones) in the text. For example, if *"Paris"* is a location mention and followed by *"France"*, then *"France"* is in local context of *"Paris"*. Local context of a person or an organization referent are words and referents occurring in a limit length window, whose size is set to 10 tokens, centered on the entity mention.

In particular, we utilize disambiguation texts of candidates to choose the correct one for each referent. For a location referent, the right one is the candidate either whose disambiguation text is identical to the successive entity name or whose name is identical to disambiguation text of the previous resolved referent. For example, in the text *"Columbia, South Carolina"*, the candidate `Columbia, South Carolina`, the largest city of South Carolina, in Wikipedia is chosen because its disambiguation text is *"South Carolina"*, or in the text *"Atlanta, Georgia"*, the candidate, a major city of state Georgia of United States, with the name *"Atlanta"* and disambiguation text *"Georgia"* is chosen, and `Georgia` is also resolved as a `U.S. state` because previous resolved referent with identifier `Atlanta, Georgia` has disambiguation text identical to *"Georgia"*.

For a person or an organization referent, chosen candidates are ones that have disambiguation text occurring in its local context. After this step, if there is only one candidate in the result, the referent is considered being resolved. For example, in the text *"Veteran referee (Big) John McCarthy, one of the most recognizable faces of mixed martial arts"*, the word *"referee"* helps choose the candidate `John McCarthy (referee)` as the right one instead of `John McCarthy (computer scientist)`, `John McCarthy (linguist)`, etc. in Wikipedia.

After that, we extend the text by disambiguation texts of the resolved referents. Those will be exploited to resolve the remaining ambiguous referents in the next step. For example, for the text *"Atlanta, Georgia"*, after `Atlanta` was recognized as a city of state Georgia of the United Sates and `Georgia` was recognized as a state of the United States, the extended text is *"Atlanta, Georgia, Georgia (U.S. sate)"* in which `Atlanta, Georgia` is the identifier of the city Atlanta, and `Georgia (U.S. state)` is the identifier of the state Georgia in Wikipedia named entity space.

5.3 Ranking Candidates

After extracting all information about NEs in Wikipedia based on features that are titles of entity pages, titles of redirect pages, categories, hyperlinks, we represent those NEs by their information. For each of the remaining ambiguous referents in the extended text, we extract its features' values as follows:

- All words occurring in a window context centered on the entity name in the extended text. The window size is set to 55, which is the value that was observed to give optimum performance in the related task of cross-document coreference ([12]).
- All NE identifiers of identified named entities in the extended text.

After that we concatenate the entire feature' values of the referents in the extended text and NEs in Wikipedia into text snippets and represent them in form of token-based feature vectors. Then we need a similarity metric to calculate the similarity between the vectors. Cohen *et al.* ([5]) presents various string similarity schemes for the task of matching English entity names and they report TFIDF (or *cosine similarity*), which is widely used in the information retrieval community, as the best among the token-based distance metrics. Given a pair of feature vectors $S = (s_1, s_2,..., s_n)$, and $T = (t_1, t_2, ..., t_m)$, in which s_i $(i=1,2, ..., n)$ and t_j $(j=1,2, ..., m)$ are words (or tokens). Then the TFIDF is defined as

$$TFIDF = \sum_{w \in S \cap T} V(w,S) \times V(w,T)$$

$$V'(w, S) = \log(TF_{w,S} + 1) . \log(IDF_w), \text{ and } V(w, S) = \frac{V'(w, S)}{\sqrt{\sum_{w \in S} V'(w, S)^2}}$$

where $TF_{w,S}$ is the frequency of word w in S, IDF_w is the inverse of the fraction of snippets in a snippet collection that contains w.

Let *CE* be a set of entities in Wikipedia that have the same name with the target entity, *e*, in consideration in the extended text. We cast the named entity disambiguation problem as a ranking problem with assumption that there is an appropriate scoring function to calculate semantic similarity between feature vectors of an entity $ce \in CE$ and the entity *e*. We build a ranking function that takes input as a set of feature vectors of entities in *CE* and the feature vector of the entity *e*, then based on the scoring function to return the entity *ce* ∈ *CE* with the highest score. Fig.1 presents an algorithm using TFIDF as the scoring function. At the line 3 in Fig.1, the score function takes input as a token-based feature vector of a candidate and a token-based feature vector of an ambiguous referent, and then it ranks and returns the candidate with the highest score.

```
Algorithm RankingCandidates (a set of ambiguous referents R)
  1: for each referent r ∈ R do
  2:      let C a set of candidates of r
  3:      c* = arg max score(Vector(c_i), Vector(r))
              c_i∈C
  4:      assign c* to r
  5:      extend the text by c* disambiguation text
  6: end for
```

Fig. 1. An algorithm ranking candidates using TFIDF

5.4 Algorithm

Fig.2 presents our disambiguating process. First we resolve some trivial ambiguous cases and retrieve candidates for ambiguous referents (line 1). Line 2 performs coreference resolution. From line 3 to line 11, we use some heuristics to disambiguate. Line 6 searches for disambiguation text of a candidate in a window containing the ambiguous referent in consideration, its previous and successive ones in the ambiguous case of location referents and in a window containing 10 tokens centered at ambiguous referent in the ambiguous cases of person and organization referents. If only one candidate for which we found it disambiguated, the corresponding referent is resolved. Line 12 performs extending the text by disambiguation text of the resolved

```
Algorithm: Disambiguation
 1: resolve trivial (unambiguous) referents
 2: resolve coreference of referents
 3: for each ambiguous referent r do
 4:    let C a set of candidate referents of r
 5:    forall candidate c ∈ C do
 6:      search for c's disambiguation text in a window context
 7:      if found for only one candidate c* then
 8:         propagate c* to all referents in the coref-chain of r
 9:      end if
10:    end forall
11: end for
12: extend the text
13: RakningCandidates(a set of remaining ambiguous referents)
```

Fig. 2. Disambiguating process

referents. Finally, we use TFIDF to rank candidates for each of the remaining ambiguous referents and choose the candidate with the highest score. Note that after each iteration step, the algorithm extends the text by disambiguation text of the chosen candidate. Therefore, in the next iteration, the algorithms reform the feature vector of the ambiguous referent in consideration.

6   Evaluation

We downloaded top two stories in the five CNN news categories (Travel, Entertainment, World, World Business, and Americas) on July 22, 2008 to build a dataset for evaluation. For later testing, all the NEs referred to in this dataset are manually disambiguated with respect to Wikipedia, by two persons for the quality of the dataset. The Wikipedia version we used that is of Zesch[4] ([24]). Note that, due to the incompleteness of Wikipedia, an ambiguous name may be used to refer to some NEs not in Wikipedia, which are out of our work's target. Also note that, we evaluate our method on named entities of three types – Person, Location, and Organization.

Table 1. Statistic about named entities in the manually disambiguated dataset

Category	# of referents	# of found entities in Wikis	# of ambiguous referents
Person	261	213	123
Location	168	159	94
Organization	89	84	45
Total	518	454	262

There are 518 proper names occurring in the dataset 454 names of which refer to NEs in Wikipedia and 262 names refer to two or more than different NEs in Wikipedia. We evaluate our method in two scenarios. In the first scenario, we use GATE to detect and tag boundaries of names occurring in the dataset and then categorize corresponding referents as Person, Location and Organization. After that, we gain *D1* dataset. We found that GATE fails to detect boundaries of some names (12 names). For example, *"Omar al-Bashir"* is recognized as separate names *"Omar"* and *"al-Bashir"*, *"Sony Ericsson"* is recognized as separate names *"Sony"* and *"Ericsson"*, and *"African National Congress"* recognized as *"African National"*, etc. Also there are many names (77 names) that GATE does not recognize as entity names. For example, *"Darfur"*, *"Qunu"*, *"Soweto"*, *"Interfax"*, *"Rosoboronexport"*, and so on are not recognized as entity names. Then we manually fix all errors in the dataset *D1* by adjusting wrong boundaries, added tagging, and re-categorizing all the wrong cases and gain dataset *D2* with no error. Table 1 presents the number of referents, the

---

[4] http://www.ukp.tu-darmstadt.de/software/jwpl/

number of ambiguous referents, and the number of entities in Wikipedia referred to in the manually disambiguated dataset.

After that we run our method on *D1* and *D2*, respectively. The results are matched against the manually disambiguated dataset. We apply the way that Fernandez *et al.* ([10]) measure the effect of their method to ours. In particular, we measure accuracy as the total number of right assignments *NE* (in text)/*Wiki NE* divided by the total number of assignments.

Table 2 presents statistic information about named entities in dataset *D1*. The data on the Table show that the number of names detected less than it would be, which is because GATE fails to detect many proper names. Table 2 shows that there are 482 referents in dataset *D1* which is less than one presented in Table 1 because GATE does not recognize some referents as the case of *Darfur* and fails to detect boudaries of proper names as the case of *"Omar al-Bashir"*. Table 3 presents accuracy results when we run our method on this dataset.

Table 2. Statistic about named entities in the *dataset D1*

Category	# of referents	# of found entities in Wikis	# of ambiguous referents
Person	245	195	118
Location	148	140	88
Organization	89	76	41
Total	482	411	247

Table 3. Accurracy results on the dataset *D1*

Category	Person	Location	Organization	All
Correct disambiguation	174	120	62	356
Accuracy	89.23%	85.70%	81.60%	86.61%

Table 4 presents accuracy results when we run our method on the dataset *D2*. The results show that our method achieves high accuracy. The accuracy results in Table 3 when we run our methods on this dataset D1 decrease comparable to those results achieved when we run the method on the dataset *D2*, which is because of noise accumulated from the pre-processing steps. There are three reasons as follows:

- GATE fails to dectect booundaries of names, e.g *"Christopher Nolan"* detected as *"Nolan"*, *"Luis Moreno-Ocampo"* detected as *"Luis Moreno-"*, etc.
- GATE fails to categorize named entites, e.g *Robben Island Prison* is recognized as a person.
- GATE does not detect some proper names that could be clues providing meaningful information to disambiguate other entities.

Table 4. Accurracy results on dataset *D2*

Category	Person	Location	Organization	All
Correct disambiguation	207	149	73	428
Accuracy	97.18%	93.70%	86.90%	94.27%

7 Related Works

Some works resolve semantic ambiguity by performing named entity tagging which is usually seen as task of identifying text span and classifying it into a broad category such as Person, Organization, Location, etc. ([6], [9]), or into a more fine-grained category that is specified by a given ontology ([8], [11]). However, those works are not disambiguation and identification of NEs in texts.

Some other works use heuristic rules and focus on only one type of NE such as location ([18]), or person ([13]). The proposed method in [13] relies on affiliation, text proximity, areas of interest, and co-author relationship as clues for disambiguating person names in calls for papers only. Meanwhile, the domain Raphael ([18]) is that of geographical names in texts. In [18], the authors use some patterns to narrow down the candidates of ambiguous geographical names. For instance, *"Paris, France"* more likely refers to the capital of France than a small town in Texas. Then, it ranks the remaining candidate entities based on the weights that are attached to classes of the constructed Geoname ontology. The shortcoming of those methods is that it omits relationships between named entities with different classes, such as between person and organization, or organization and location, etc. The statistical method in [10], although it leverages the co-occurrence relation between NEs with different classes, is semi-automatic and uses user feedback to disambiguated results for updating heuristics and rules considered as a training dataset.

Closely related works to ours are presented in [2], [4], [16]. Bunescu *et al.* [2] and Cucerzan [4] exploited several of the disambiguation resources such as Wikipedia entity pages, redirection pages, categories, and hyperlinks, whereas in [16], the authors proposed a rule-based method that utilizes KB-based relationship and named entity co-occurring with ambiguous ones to disambiguate. Bunescu *et al.* and Cucerzan extracted that information in Wikipedia to form language models and then used those models to disambiguate named entities. Those language models are used as means for capturing different contexts in which different names referring to the same entity occur.

In this paper, we propose a hybrid statistical and rule-based incremental method that combines heuristics and a learning model. We first use heuristics and pattern matching for entity disambiguation and then extract that information from Wikipedia to form a language model to disambiguate named entities. The proposed method is incremental process which includes several rounds that filter the candidates by exploiting previously identified entities and extending the text by those entity attributes every time they are successfully resolved in a round. Importantly, our method utilizes entity IDs instead of entity names in literature to identify the right entity for ambiguous referents. Furthermore, we explore context in several levels, from local to the whole text, where diverse clues are extracted for disambiguation at high accuracy.

8   Conclusion

We have proposed an original approach to named entity disambiguation. It is a hybrid and incremental process that utilizes previously identified NEs and related terms co-occurring with ambiguous names in a text for entity disambiguation. Firstly, it is quite natural and similar to the way humans do, relying on co-occurring entities and terms to resolve other ambiguous referents in a given context. Secondly, it is robust to free texts without well-defined structures or templates. Next, currently Wikipedia editions are available for approximately 253 languages, which mean that our method can be used to build named entity disambiguation systems for a large number of languages. Finally, despite the exploitation of Wikipedia as a means of named entity disambiguation, our method can be adapted for any ontology and KB in general. The experiment results have shown that our method achieves high accuracy.

References

[1]   Baeza-Yates, R., Ribeiro-Neto, B.: Modern Information Retrieval. Addison-Wesley, Reading (1999)

[2]   Bunescu, R., Paşca, M.: Using encyclopedic knowledge for named entity disambiguation. In: Proc. of the 11th Conference of EACL, pp. 9–16 (2006)

[3]   Bontcheva, K., Dimitrov, M., Maynard, D., Tablan, V., Cunningham, H.: Shallow Methods for Named Entity Coreference Resolution. In: Proc. of TALN 2002 Workshop, Nancy, France (2002)

[4]   Cucerzan, S.: Large-Scale Named Entity Disambiguation Based on Wikipedia data. In: Proc. of EMNLP-CoNLL Joint Conference (2007)

[5]   Cohen, W., Ravikumar, P., Fienberg, S.: A Comparison of String Metrics for Name-Matching Tasks. In: IJCAI-03 II-Web Workshop (2003)

[6]   Cunningham, H., et al.: GATE: A Framework and Graphical Development Environment for Robust NLP Tools and Applications. In: Proc. of the 40th ACL (2002)

[7]   Chinchor, N., Robinson, P.: MUC-7 Named Entity Task Definition. In: Proc. of MUC-7 (1998)

[8]   Cimiano, P., Völker, J.: Towards large-scale, open-domain and ontology-based named entity classification. In: Proc. of RANLP 2005, pp. 166–172 (2005)

[9]   Tjong Kim Sang, E.F., De Meulder, F.: Introduction to the CoNLL-2003 Shared Task: Language Independent Named Entity Recognition. In: Proc. of CoNLL 2003, pp. 142–147 (2003)

[10]   Fernandez, N., et al.: IdentityRank: Named entity disambiguation in the context of the NEWS project. In: Franconi, E., Kifer, M., May, W. (eds.) ESWC 2007. LNCS, vol. 4519. Springer, Heidelberg (2007)

[11]   Fleischman, M., Hovy, E.: Fine grained classification of named entities. In: Proc. of Conference on Computational Linguistics (2002)

[12]   Gooi, C.H., Allan, J.: Cross-document coreference on a large-scale corpus. In: Proc. of HLT-NAACL for Computational Linguistics Annual Meeting, Boston, MA (2004)

[13]   Hassell, J., Aleman-Meza, B., Arpinar, I.B.: Ontology-driven automatic entity disambiguation in unstructured text. In: Cruz, I., Decker, S., Allemang, D., Preist, C., Schwabe, D., Mika, P., Uschold, M., Aroyo, L.M. (eds.) ISWC 2006. LNCS, vol. 4273, pp. 44–57. Springer, Heidelberg (2006)

[14]   Harris, Z.: Distributional structure. Word 10(23), 146–162 (1954)

[15]   Wacholder, N., Ravin, Y., Choi, M.: Disambiguation of proper names in text. In: Proc. of ANLP, pp. 202–208 (1997)

[16] Nguyen, H.T., Cao, T.H.: A knowledge-based approach to named entity disambiguation in news articles. In: Orgun, M.A., Thornton, J. (eds.) AI 2007. LNCS (LNAI), vol. 4830, pp. 619–624. Springer, Heidelberg (2007)

[17] Peng, Y., He, D., Mao, M.: Geographic Named Entity Disambiguation with Automatic Profile Generation. In: Proc. of WI 2006 (2006)

[18] Raphael, V., Joachim, K., Wolfgang, M.: Towards Ontology-based Disambiguation of Geographical Identifiers. In: Proc. of the 16th WWW Workshop on I3: Identity, Identifiers, Identifications (2007)

[19] Remy, M.: Wikipedia: The free encyclopedia. Information Review 26(6), 434 (2002)

[20] Shadbolt, N., Hall, W., Berners-Lee, T.: The Semantic Web Revisited. IEEE Intelligent Systems 21(3), 96–101 (2006)

[21] Overell, S., Rüger, S.: Geographic Co-occurrence as a Tool for GIR. In: Proc. of CIKM Workshop on Geographic Information Retrieval, Lisbon, Portugal, pp. 71–76 (2007)

[22] Smith, D., Mann, G.: Bootstrapping toponym classifiers. In: HLT-NAACL Workshop on Analysis of Geographic References, pp. 45–49 (2003)

[23] Weaver, G., Strickland, B., Crane, G.: Quantifying the accuracy of relational statements in Wikipedia: a methodology. In: Proc. of JCDL, pp. 358–358 (2006)

[24] Zesch, T., Gurevych, I., Mühlhäuser, M.: Analyzing and Accessing Wikipedia as a Lexical Semantic Resource. In: Rehm, G., Witt, A., Lemnitzer, L. (eds.) Data Structures for Linguistic Resources and Applications, pp. 197–205 (2007)

Exposing Heterogeneous Data Sources as SPARQL Endpoints through an Object-Oriented Abstraction

Walter Corno, Francesco Corcoglioniti, Irene Celino, and Emanuele Della Valle

CEFRIEL - Politecnico di Milano,
Via Fucini 2, 20133 Milano, Italy
walter.corno@students.cefriel.it, {name.surname}@cefriel.it
http://swa.cefriel.it

Abstract. The Web of Data vision raises the problem of how to expose existing data sources on the Web without requiring heavy manual work. In this paper, we present our approach to facilitate SPARQL queries over heterogeneous data sources.

We propose the use of an object-oriented abstraction which can be automatically mapped and translated into an ontological one; this approach, on the one hand, helps data managers to disclose their sources without the need of a deep understanding of Semantic Web technologies and standards and, on the other hand, takes advantage of object-relational mapping (ORM) technologies and tools to deal with different types of data sources (relational DBs, but also XML sources, object-oriented DBs, LDAP, etc.).

We introduce both the theoretical foundations of our solution, with the analysis of the relation and mapping between SPARQL algebra and *monoid comprehension calculus* (the formalism behind object queries), and the implementation we are using to prove the feasibility and the benefits of our approach and to compare it with alternative methods.

1 Introduction

The Semantic Web has as ultimate goal the construction of a Web of Data, i.e. a Web of interlinked information expressed and published in a machine-readable format which enables automatic processing and advanced manipulation of the data itself. In this scenario, initiatives like the Linking Open Data community project[1], guidelines and tutorials on how to publish data on the (Semantic) Web [1,2] and standards for querying this Web of Data like SPARQL [3,4] play a central role in the realization of the Semantic Web vision.

To achieve this aim, however, it is necessary to find an easy and automated – as much as possible – way to expose existing data sources on the Web. To this end, two big classes of approaches are being studied to help data managers to prepare their data sources for the Semantic Web: conversion and wrapping.

---

[1] http://esw.w3.org/topic/SweoIG/TaskForces/CommunityProjects/
LinkingOpenData

J. Domingue and C. Anutariya (Eds.): ASWC 2008, LNCS 5367, pp. 434–448, 2008.
© Springer-Verlag Berlin Heidelberg 2008

With *conversion* we mean all the techniques to effect a complete translation of the data source from its native format to a Semantic Web model (pure RDF or RDF triples described by some kind of ontologies in RDFS/OWL). This approach assures a complete replication and porting of the data but raises several concerns like frequency of re-conversion, synchronization, conversion processing and space occupation.

With *wrapping*, on the other hand, we mean all the techniques aimed at building an abstraction layer over the original data source which hides the underlying format and structure and exposes a SPARQL endpoint to be queried. This approach requires the run-time translation of the SPARQL query request to the source's specific query language and the run-time conversion of the query results back to the requester. Several proposals, that are gaining ground within the Semantic Web community, belong to this wrapping approach through the declaration of a mapping between the original data format and the respective RDF data model. Most of those approaches, however, deal with the wrapping of relational databases (see Section 2) and do not consider other kinds of data sources. Moreover, this direct mapping requires the developer to have quite a deep understanding of Semantic Web languages, technologies and formats to express the declaration of correspondences.

In this paper, we present our proposal for a new approach that tries to overcome the aforementioned problems. Our approach belongs to the wrapping category, but tries to give a solution to the heterogeneity of data sources as well as to the problem of adoption by the larger community of developers. For the former issue, we provide a solution based on the availability of approaches and tools to wrap different data sources with an object-oriented abstraction; for the latter problem, we propose an automatic mapping between the object-oriented model (in ODL) and the correspondent one at the ontological level (in OWL-DL, see Section 3). Our approach is both theoretically sound – because of the affinity between object-orientation and ontology modelling and because of the accordance of the respective formalisms (SPARQL algebra [3] and monoid comprehension calculus [5], see also Section 2) – and technologically and technically practicable – because we realized SPOON, the reference implementation of our approach.

The remainder of the paper is structured as follows: Section 2 introduces related approaches and theoretical foundations; Section 3 explains the automatic mapping between object-oriented models and ontologies (with its constraints); we illustrate our approach to query translation in Section 4 and our implementation and evaluation in Section 5; concluding remarks and next steps are finally presented in Section 6.

2   Related Work

In order to enable a faster expansion of the Web of Data, in the last few years some efforts arose with the aim to expose existing datasources, especially relational databases (RDBMs), on the Web and to query them through SPARQL [3]. For instance, Cyganiak and Bizer studied similarities between SPARQL algebra and relational algebra [6] and developed D2RQ [7] and D2R Server, to build

SPARQL endpoints over RDBMs. SquirrelRDF[2] exposes both relational and LDAP sources, but it is still incomplete; SPASQL [8] is a MySQL module that adds native SPARQL support to the database; Relational.OWL [9], Virtuoso Universal Server [10], R2O [11] and DB2OWL [12] are other projects that aim to expose relational data on the Web.

While relational databases are the most widespread, the most common programming paradigm is object-oriented (OO) programming. Since the OODBMS Manifesto [13] many projects developed proprietary object datasources (e.g. O2, Versant, EyeDB, and so on[3]), but none of them strictly follows the ODMG Standard [14], so these technologies failed in being either widely used or interoperable. As a consequence, new technologies were born from the cited ones: the Object-Relational Mappings (ORMs), that allow to use relational sources as if they were object datasources and to query them in an object-oriented way. Well-known ORMs are Hibernate, JPOX, iBatis SQL Maps and Kodo[4]. Among these, JPOX and Kodo implement the JDO specification [15], a standard Java-based model of persistence, that allows to use not only RDBMs but also many other types of source (OODBMS, XML, flat files and so on)[5]. Even if different in syntax and characteristics, all the object query languages developed so far are based on the Object Query Language (OQL) developed by the ODMG consortium [14]. The *monoid comprehension calculus* [5] is a framework for query processing and optimization supporting the full expressiveness of object queries; it can be considered as a common formalism and theoretical foundation for all OQL-like languages. This formalism has been used in [16] to translate queries in description logics to object-oriented ones.

3   Schema and Data Mapping

The first step of our proposed approach is to help the data manager to expose his data-source schema (already wrapped by an ORM) as an ontological model. To this end, we propose to adopt a specific mapping strategy to make this step completely automatic (albeit some restrictions/constraints on the OO model).

Object-oriented model is much more similar to ontological model than relational one. In particular, these models share a common set of primitives (e.g. *classes, properties, inheritance,...*), and can describe relationships between classes directly, whereas the relational model may require complex expedients such as the use of *join tables*.

OO and ontological models are not fully equivalent, as shown in [17,18] (e.g. single vs. multiple inheritance and local vs. global properties); however the issues highlighted in these works are only relevant for the problem of describing an existing ontology as an object model (due to some limitations of the OO model),

---

[2] http://jena.sourceforge.net/SquirrelRDF/

[3] Versant Object DB: http://www.versant.com/, EyeDB http://www.eyedb.org/

[4] Hibernate: http://www.hibernate.org/, JPOX: http://www.jpox.org/, Apache iBatis: http://ibatis.apache.org/, BEA Kodo: http://bea.com/kodo/

[5] For these reasons in our implementation we chose JPOX as ORM tool.

while in our approach we deal with the opposite problem (i.e., to expose an OO model as an ontology).

In our approach we propose a one-to-one mapping as simple as possible (similar to the one shown in [19]), because we aim to automatize it, simplifying the development process. We use ODL [14] as OO formalism and a subset of OWL-DL [20] (represented by the constructs in Table 1 and disjointness, as explained below) as the ontology language. The schema mapping is described in Table 1.

Table 1. Schema mapping

Concept	ODL	OWL-DL
Class	class	owl:Class
Subclass	class A extends B	rdfs:subClassOf
Property	attribute/relationship	owl:DatatypeProperty/ObjectProperty
Inverse relationship	inverse	owl:inverseOf
Property domain	*implicit*	rdfs:domain
Property range	*property type*	rdfs:range
Primitive types	int, double,…	*XSD datatypes*
Functional property	*non-collection types*	owl:FunctionalProperty
Non-functional prop.	set<T>[6]	*implicit*

In addition to these correspondences, we add *disjoint* constraints to the ontology because objects can belong only to a single OO class (with its parents):

$$\forall \ class \ C_1, C_2 : \nexists \ class \ C \ subClassOf \ C_1, C_2,$$
$$generate \ \langle C_1 \ \text{owl:disjointWith} \ C_2 \rangle$$

Moving from schema to instance mapping, primitive instances are mapped to RDF literals, while to map objects we need a way to create URIs for them (because they become RDF resources). The simplest approach we adopt is to combine a fixed *namespace* with a variable *local name*, formed by the values of a particular *ID* property; we prefer not to use the OIDs commonly employed in OODBMS, due to their limited support among ORMs. Object attributes and relationships are then translated into RDF triples, using the corresponding predicates as defined by the schema mapping.

To keep the mapping simple and ease its automatization, we introduce some constraints on the OO model:

- all classes have to contain an alphanumeric property *ID*, with a unique value (in class hierarchies it can be inherited from a parent class).
- OO properties having the same name in unrelated classes can only be mapped to different ontological predicates, thus having distinct semantics.
- *collection* properties are limited to the *set* type (no *bag*, *list* or *map*).
- interfaces are not supported (and thus multiple inheritance).

---

[6] *Set* is the collection type and T is the type of the elements contained in the collection.

– all classes have an *extent* in order to be directly used in OO queries (see [14] for *extent* definition).

Figure 1 shows the translation to an OWL ontology of a simple ODL schema, which will be used as a running example throughout the paper.

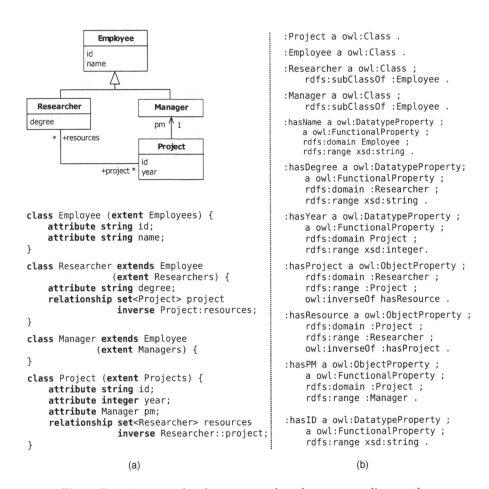

(a)   (b)

Fig. 1. Running example schema mapped to the corresponding ontology

4   Query Translation

In this section we present our framework to translate a SPARQL query into one or a few object queries. The general process is shown in Figure 2, sketched hereafter and explained in details throughout the whole section.

When a new SPARQL query is sent to our system, first we perform an *analysis* process both at the *syntactic* and *semantic* levels. During *syntactic analysis*

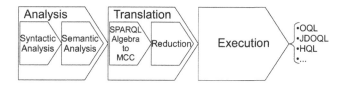

Fig. 2. The general query processing framework

the query is parsed, checked for syntactic errors and then translated into its equivalent SPARQL algebraic form [3], which is then *normalized*. In this format, a query is represented as a tree composed of *basic graph patterns* (BGP) as leaves and of the algebraic operators *filter* (σ_{pred}), *union* ($\cup$), *join* ($\bowtie$), *left join* ($\bowtie_{pred}$) and *diff* ($-_{pred}$)[7] as internal nodes. Then we perform *semantic analysis*: we apply some checks and rewriting rules to ensure that the query can be processed by the next phases. Variables on predicates are resolved and variables on subjects/objects are assigned to the corresponding OO classes.

The second step is the *core* one of our framework: the *query translation*. In this step the SPARQL algebraic form of the query is translated in a *monoid comprehension calculus* [5] expression, so the initial SPARQL query is now expressed as a query on the OO model. The translation starts processing basic graph patterns (BGPs) and then translating each SPARQL algebraic operator we meet when traversing the SPARQL algebraic form of the query in a *bottom-up* approach. When this translation is completed, we apply the *normalization rules* demonstrated in [5] to the global expression (*reduction* phase), so that we get a simpler expression (as we will see, a *union* of *monoid comprehensions*).

The last step is the *query execution*. In this step the obtained *union* of *monoid comprehensions* is translated into queries of the particular OO query language used for the implementation of the framework and then executed. Eventually the final result-set is translated into the one compatible with the original SPARQL query (*select, construct, describe, ask*).

In the remaining of this section we explain these three steps in detail, continuing the running example introduced in Section 3 with the following SPARQL query, whose effect is to return the URI, the names and (optionally) the degree of all the employee related to projects of 2006 and later.

```
SELECT ?e ?n ?d
WHERE {
    ?p hasYear ?y ;
       ?r ?e .
    ?e hasName ?n .
    OPTIONAL { ?e hasDegree ?d }
    FILTER ( ?y >= 2006 )
}
```

---

[7] To ease the notation, we borrow the symbols of *relational algebra*.

4.1   Analysis

The analysis phase takes care of parsing, checking and transforming the SPARQL query in order to prepare it for the subsequent translation phase. Query analysis is performed both at the *syntactic* and *semantic* levels.

Syntactic analysis. The first step is to parse the input query string, check its syntax and produce as output its equivalent representation in SPARQL algebra [3], as shown in Figure 4 (a) for the query of the running example. The parsed algebraic representation is then normalized, in order to "collapse" as far as possible the BGPs of the query and to reach a form easier to analyse and translate. The *normalization procedure* consists of three steps:

1. *Left joins replacement*, with a combination of *union, join, filter* and *diff* operations, according to the rule [3]:

$$A \ltimes_{pred} B \Rightarrow \sigma_{pred}(A \bowtie B) \cup (A -_{pred} B) \quad (1)$$

2. *Variable substitution*; for each *diff* node, change the names of the variables which appear in the right-hand operand (the "subtrahend") but not in the left-hand (the "minuend") with new, globally unique names.

3. *Transformation*; the algebraic structure of the query is transformed, by exploiting the commutativity of $\bowtie$ and $\cup$, the distributivity of $\bowtie$ and the left distributivity of $-_{pred}$ with respect to $\cup$ and the rules listed below, until no more transformations are possible[8]:

$$\sigma_{pred}(A \cup B) \Rightarrow \sigma_{pred}(A) \cup \sigma_{pred}(B) \quad (2)$$
$$A -_{pred} (B \cup C) \Rightarrow (A -_{pred} C) -_{pred} B \quad (3)$$
$$\sigma_{pred}(A) \bowtie B \Rightarrow \sigma_{pred}(A \bowtie B) \quad (4)$$
$$\sigma_{pred_1}(A) -_{pred_2} B \Rightarrow \sigma_{pred_1}(A -_{pred_2} B) \quad (5)$$
$$(A -_{pred} B) \bowtie C \Rightarrow (A \bowtie C) -_{pred} B \quad (6)$$
$$BGP_1 \bowtie BGP_2 \Rightarrow \text{merge of } BGP_1 \text{ and } BGP_2 \quad (7)$$

The effect of these rules is to rearrange the operators to obtain the following order (from the top) $\cup, \sigma_{pred}, -_{pred}$; note that *join* operators are all removed by rule 7. As shown in Figure 3, a normalized query consists of an (optional) union of *basic queries* each one consisting of a BGP whose results can be filtered by one or more *diff* operations. Roughly, each basic query will originate a SELECT ... FROM ... WHERE ... object query with nested sub-queries for *diff* operators; the final result-set will be obtained by executing these queries and merging their results. Figure 4 (b) shows the normalized algebra for the example query.

Semantic analysis. This step aims at *transforming* the normalized query so that (1) constraints on URIs are restated in terms of constraints on the ID attribute (2) variables on triple predicates are removed (by enumerating the possible cases) and (3) each URI or non-literal variable is associated to a single OO class. Each of these goals is addressed in a different analysis step:

---

[8] Rule 6 is only valid thanks to the variable substitution performed in the previous step, which avoids variable names clashes when moving up the *diff* node.

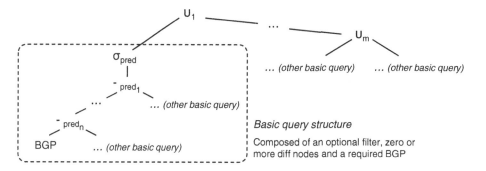

Fig. 3. Normalized query structure

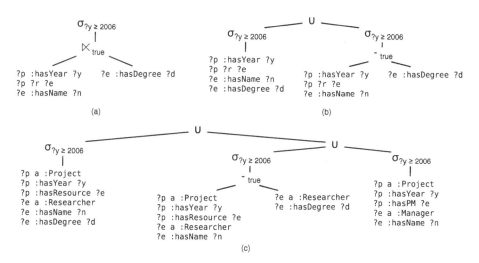

Fig. 4. Analysis of the example query: (a) parsed query, (b) normalized query (c) resulting query

1. *Rewriting of URIs*; for each URI $\langle x \rangle$ in the query, all of its occurrences are replaced with a new ♦ariable ?x, while a new $\langle$?x :hasID ID$\rangle$ triple is added to each BGP of the query which uses the variable. Note that the ID can be extracted from the URI textual representation (see Section 3).

2. *Rewriting of variables on predicates*; each BGP containing such variables is replaced with a *union* of multiple BGPs, each one corresponding to an acceptable combination of predicate assignments to these variables. A reasoner can be used to identify the alternatives, by classifying nodes in the BGP and exploiting the domain and range constraints of predicates[9]. The assignment of predicates to variables is recorded in an auxiliary data structure for each

---

[9] The reasoner can be used as explained in step *BGP validation and class assignment*; note, however, that the choice of predicates is not critical, because even if invalid predicates are considered, the next validation step will remove them.

basic query, in order to return them together with the results in case variables on predicates are included in the SELECT clause (or CONSTRUCT template) of the query. Finally, since the algebraic structure is modified, at the end of this step the query may need to be re-normalized again.

3. *BGP validation and class assignment.* A check is done that each variable is used only as a literal or URI, but not both. Then, a graph is built for each BGP by removing all the triples containing literal values, and a blank node is introduced for each other variable. An OWL DL reasoner is used to (1) check if this graph is consistent with the ontology and (2) infer new *rdf:type* triples for resources and variables (the blank nodes), which allow to associate an OO class to each node. If any of the checks fails, the BGP is discarded and the algebraic structure is adjusted accordingly (e.g. by removing parent *diff* or *filter* nodes too).

The query resulting from semantic analysis is ready to be translated. Figure 4 (c) shows the result of the semantic analysis for the query of the running example.

4.2   Translation

This phase is divided in two steps: translation in *monoid comprehension calculus* and normalization of the resulting expression. The first step starts translating the BGPs and then each SPARQL algebraic operator, using a *bottom-up* approach; the second step aims at reaching a normalized form of the expression, through a set of normalization rules defined in [5].

The *monoid comprehension calculus* is a framework for object query processing and optimization. We now give a brief overview of this *calculus*, readers are referred to [5] for more detailed information.

Object query languages deal with collections of homogeneous (i.e. of the same type) objects and primitive values such as *sets*, *bags* and *lists*, whose semantics can be captured by the notion of a *monoid*. A monoid is an algebraic structure consisting in a set of elements and a binary operation defined on them having particular algebraic properties. Collections of objects and operations on them (such as set and bag union and list concatenation, but also aggregate operations like max and count) can be represented as *collection monoids*; similarly, operations like conjunctions and disjunctions on booleans and integer addition over collections can also be expressed in terms of so-called *primitive monoids*.

The basic structure of the calculus is the *monoid comprehension*, that can describe a query or a part of it. This structure takes the form $\oplus\{e \mid \bar{q}\}$, where:

- $\oplus$ is a function called *accumulator*, that identifies the type of monoid by specifying how to compose (i.e. which operation should be used) the elements obtained by the evaluation of the comprehension;
- e is called *head* and it is the expression that defines the result;
- $\bar{q}$ is a sequence of *qualifiers*; these can be *generators* of the form $v \leftarrow e'$, where v is a variable ranging over the collection produced by the expression e' (which can be a monoid comprehension too), or *filters* of the form *pred*, which express constraints over the variable bindings produced by the generators.

For instance, this monoid comprehension: $\uplus\{v_1, v_2 | v_1 \leftarrow X, v_2 \leftarrow X.y, v_2 > n\}$ can be read as: "for all v_1 in X and for all v_2 in $X.y$ such that $v_2 > n$ consider the pairs v_1, v_2 and merge them (by applying the $\uplus$ accumulator) to obtain a *bag*". The accumulator functions in our translation are only $\uplus$ and $\vee$: the former defines a bag of solutions, while the latter is used to define the existential quantification.

BGP translation. A generic BGP contains a set of *triple patterns*. At the beginning of this step we reorder these triples. A set of triple patterns can be viewed as a directed graph, with vertices corresponding to subjects and objects and edges between them corresponding to triples and labelled with their predicates; if the graph contains some cycles, we break them by duplicating a vertex, thus obtaining a directed acyclic graph (DAG). To order the triples we perform a depth-first visit, starting from the root nodes of the DAG. Triples with *rdf:type* as predicate are not considered during the reordering process: they are removed and used later to resolve the assignment of variables to OO classes (as described below). Figure 5 shows the reordering process for a BGP of the running example (the leftmost in Figure 4 (c)).

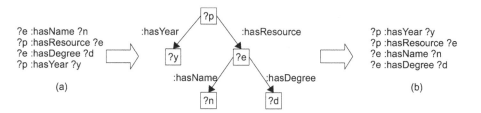

Fig. 5. Triples reordering

Now the BGP is translated in the corresponding monoid comprehension following these criteria:

1. the *accumulator* function is always $\uplus$, because a BGP returns a bag of solutions;
2. the set containing all the variables contained in the triples forms the *head* of the monoid comprehension;
3. the *qualifiers* in the *body* of the comprehension are generated by iterating over the ordered triples $\langle var_{sub}$ pred $obj \rangle$ and applying the following rules to each one:
   - **if** var_{sub} occurs for the first time, a new *generator* $var_{sub} \leftarrow Class$ (where *Class* is the OO class assigned to the variable) is added;
   - **if** obj is a variable var_{obj} occurring for the first time, a *generator* of the form $var_{obj} \leftarrow var_{sub}.pred$ (the symbol $\leftarrow$ is changed with $\equiv$ when *pred* is a functional property) is created. If *pred* is a functional property and

var_{obj} does not appear as the *subject* of other triples, a *filter* of the form $var_{obj} \neq null$ is added too[10];
- if *obj* is a literal or a variable already encountered, a new *filter* is created:
  - if *pred* is a functional property, the *filter* takes the following form: $var_{sub}.pred = obj$;
  - else the *filter* takes the form: $var' \leftarrow var_{sub}.pred, var' = obj$ (where var' is a new globally unique variable).

Equation 8 shows the resulting comprehension for the BGP of Figure 5.

$$\uplus \{p, y, e, n, d \mid p \leftarrow Project, y \equiv p.year, y \neq null, e \leftarrow p.resources,$$
$$n \equiv e.name, n \neq null, d \equiv e.degree, d \neq null\} \quad (8)$$

Compound constructs translation. Each SPARQL algebraic operator can be translated to a corresponding monoid comprehension expression. Using P to describe a generic *pattern* (BGPs or group-graph-patterns, i.e. BGPs composed with algebraic operators), we indicate with $\tau(P)$ the translation of P.

In Table 2 are shown the translation rules. These rules are applied using a *bottom-up* approach, starting from the leaves of the tree and moving up towards the root (see Figure 4(c)). We do not define rules for *join* ($\bowtie$) and *left join* ($\bowtie_{pred}$) because these operators are eliminated in the *analysis* step (see Section 4.1).

Table 2. Translation of SPARQL Algebra constructs

Rule	SPARQL algebra	Monoid Comprehension
T1	P	$\tau(P)$
T2	$\sigma_{pred}(P)$	$\uplus\{x\|x \leftarrow \tau(P), pred\}$
T3	$\cup(P_A, P_B)$	$\tau(P_A) \uplus \tau(P_B)$
T4	$-_{pred}(P_A, P_B)$	$\uplus\{x\|x \leftarrow \tau(P_A), \neg \vee \{pred \mid y \leftarrow \tau(P_B)\}\}$

Simplification rules. At the end of the translation step, we obtain a composition of nested monoid comprehensions. In their work [5], Fegaras and Maier suggest a set of *meaning-preserving* normalization rules, to unnest many kinds of nested monoid comprehension. The relevant rules for our approach are shown in Table 3.

Table 3. Relevant normalization rules

Rule	Before	After
N1	$\oplus\{e \mid \bar{q}, v \leftarrow (e_1 \otimes e_2), \bar{s}\}$	$(\oplus\{e \mid \bar{q}, v \leftarrow e_1, \bar{s}\}) \oplus (\oplus\{e \mid \bar{q}, v \leftarrow e_2, \bar{s}\})$ for commutative $\oplus$ or empty $\bar{q}$
N2	$\oplus\{e \mid \bar{q}, v \leftarrow \otimes\{e' \mid \bar{r}\}, \bar{s}\}$	$\oplus\{e \mid \bar{q}, \bar{r}, v \equiv e', \bar{s}\}$

The monoid comprehension expression resulting from the example query (Figure 4 (c)) is the following:

---

[10] Not null constraints are required because all variables must be bound to a value in solutions of a BGP pattern.

$$(\uplus\{p, y, e, n, d \mid p \leftarrow Project, y \equiv p.year, y \neq null, e \leftarrow p.resources,$$
$$n \equiv e.name, n \neq null, d \equiv e.degree, d \neq null, y \geq \text{``2006''}\})$$

$$\uplus$$

$$(\uplus\{p, y, e, n \mid p \leftarrow Project, y \equiv p.year, y \neq null, e \leftarrow p.resources,$$
$$n \equiv e.name, n \neq null, y \geq \text{``2006''}, \neg \vee \{true \mid d \equiv e.degree, d \neq null\}\})$$

$$\uplus$$

$$(\uplus\{p, y, e, n \mid p \leftarrow Project, y \equiv p.year, y \neq null, e \equiv p.pm, n \equiv e.name,$$
$$n \neq null, y \geq \text{``2006''}\}) \tag{9}$$

The expression obtained at the end of these steps can be already translated into object queries. However, exploiting the comprehension calculus it can be further optimized, e.g., simplifying some variables or collapsing some monoid comprehensions. We do not describe this process here due to limited space and because we are still working to identify a set of general simplification rules. To give an idea of the possible improvements, however, we show in Equation 10 the optimized expression for the example query.

$$(\uplus\{p, e, n, d \mid p \leftarrow Project, e \leftarrow p.resources, e.name \neq null,$$
$$p.year \geq \text{``2006''}, n \equiv e.name, d \equiv e.degree\})$$

$$\uplus$$

$$(\uplus\{p, e, n \mid p \leftarrow Project, e \equiv p.pm, e.name \neq null,$$
$$p.year \geq \text{``2006''}, n \equiv e.name\}) \tag{10}$$

4.3 Execution

In this last step we translate the normalized monoid comprehension expression into object queries, execute them on the datasource and convert the results in the format expected by the original SPARQL query. In this section we describe the translation to OQL; note, however, that the translation to other OQL dialects (such as JDOQL used by SPOON) is similar.

The normalized expression produced by the translation phase is a *union* of monoid comprehensions. Each of these monoid comprehensions is translated to a separate object query in a straightforward manner: all the expressions for the variables in the *head* are returned in the SELECT clause (for object variables we extract the object IDs, not the full objects), *generators* become the collections on which variables iterate in the FROM clause and *filters* become a conjunction of constraints in the WHERE clause. The monoid comprehension of the form: "$\neg \vee \{\ldots\}$" (that appears in rule T4 of Table 2) becomes a *subquery* of the form: "NOT EXISTS (SELECT...)", also belonging to the WHERE clause.

The OQL translation of the running example query is reported below. We show the translation of the simplified comprehensions of Equation 10; however, translation to object queries is applicable starting from the comprehensions of Equation 9 (but the resulting queries would be not so compact.).

```
SELECT p.id, e.id, e.name, e.degree
FROM Projects p, p.resources e
WHERE e.name != null AND
        p.year >= 2006
```

```
SELECT p.id, p.pm.id, p.pm.name
FROM Projects p
WHERE p.pm.name != null AND
        p.year >= 2006
```

The queries obtained so far are executed one by one, then the result-sets are merged together and SPARQL *solution sequence modifiers* [3] (*order by*, *distinct*, *reduced*, *offset* and *limit*) are applied to the whole result-set. The last thing to do is the conversion of the obtained result-set in the format expected by the SPARQL query:

- for SELECT queries, we select from the result-set only the requested variables and return a *table*-form result-set;
- for ASK queries, we return *true* if the result-set is not empty, *false* otherwise;
- for CONSTRUCT queries, we create an RDF graph with the data from the result-set;
- DESCRIBE queries are currently not directly supported by our approach, however a DESCRIBE query can always be translated to a CONSTRUCT query that asks for all the triples with the desired resource as subject or object, and this kind of query is supported by our approach.

5   Implementation and Evaluation

With regards to the comprehensive framework we presented in Section 4, to prove the feasibility of our approach, we implemented SPOON – SParql to Object Oriented eNgine – a tool based on Jena and JPOX which helps the automatic mapping between an OO model and the respective ontological abstraction and translates SPARQL queries in JDOQL [15] queries. The first implementation of SPOON is focused on the main constructs, namely BGP and FILTER, and it does not yet support variables on predicates.

In order to compare our approach with existing and competing systems, we chose to set up an evaluation framework, by applying different approaches to the same data source. We chose Gene Ontology data source (GO), which is available in different formats among which a SQL dump and a RDF format[11].

Our evaluation, therefore, is conducted as follows: given a SPARQL query, (1) it is translated by SPOON into JDOQL and executed by JPOX over the relational source of GO, (2) it is mapped to the GO relational database through D2R and (3) it is executed directly to the RDF version of GO loaded in a Sesame Native store (we also used the respective SQL query run on MySQL as

---

[11] We modified a bit the RDF version of GO available at http://www.geneontology.org/, because it contains some errors that make it not well-formed RDF.

a baseline reference). We stressed the system with three different queries with increasing complexity; the comparison of results is offered in Table 4, while in Table 5 we distinguish SPOON performances in *translation time* (from SPARQL to JDOQL) and *execution time* (by JPOX).

Table 4. Response time of the evaluated systems with the test queries

Query	SPOON	D2R	Sesame	MySQL
Query nr.1	291ms	695ms	280ms	95ms
Query nr.2	313ms	774ms	281ms	70ms
Query nr.3	540ms	3808ms	63620ms	179ms

Table 5. SPOON response time divided in translation (τ) and execution (χ) time

Query	τ	χ
Query nr.1	14ms	277ms
Query nr.2	14ms	299ms
Query nr.3	177ms	363ms

The recorded performances, although preliminary and partial, show an evident advantage in using our approach. A detailed report with more discussion about SPOON and its evaluation (queries, testing environment, configurations, etc.) is available at `http://swa.cefriel.it/SPOON`.

6   Conclusions

In this paper we presented our approach to the wrapping of heterogeneous data sources to expose them as SPARQL endpoints; we employ an object-oriented paradigm to abstract from the specific source format, as in ORM solutions, and we base the run-time translation of SPARQL queries into an OO query language on the correspondence between SPARQL algebra and monoid comprehension calculus. Finally, we realized a proof of concept with SPOON, which implements (a part of) our proposed framework, to evaluate it against competing approaches and we proved the effectiveness and the potentials of our approach.

Our future work will be devoted to the extension of SPOON implementation to cover other SPARQL options (like OPTIONAL and UNION); we also plan to extend the evaluation of our approach, from the point of view of the expressivity and variance of the automatic mapping between the models.

Acknowledgments

The work described in this paper is the main topic of Walter Corno's Master Thesis in Computer Engineering at Politecnico of Milano. The research has been partially supported by the NeP4B project, co-funded by the Italian Ministry of University and Research (MIUR project, FIRB-2005). We would also like to thank professor Stefano Ceri for his guidance and our colleagues at CEFRIEL for their support.

References

1. Bizer, C., Cyganiak, R., Heath, T.: How to Publish Linked Data on the Web (2007)
2. Berrueta, D., Phipps, J.: Best Practice Recipes for Publishing RDF Vocabularies – W3C Working Draft (2008)
3. Seaborne, A., Prud'hommeaux, E.: SPARQL Query Language for RDF – W3C Recommendation (2008)
4. Torres, E., Feigenbaum, L., Clark, K.G.: SPARQL Protocol for RDF – W3C Recommendation (2008)
5. Fegaras, L., Maier, D.: Optimizing object queries using an effective calculus. ACM Trans. Database Syst. 25(4), 457–516 (2000)
6. Cyganiak, R.: A relational algebra for SPARQL. Technical report, HP Labs (2005)
7. D2RQ: The D2RQ Platform - Treating Non-RDF Relational Databases as Virtual RDF Graphs
8. Prud'hommeaux, E.: Adding SPARQL Support to MySQL (2006)
9. de Laborda, C.P., Conrad, S.: Relational.OWL - A Data and Schema Representation Format Based on OWL. In: Proceedings of the Second Asia-Pacific Conference on Conceptual Modelling, APCCM 2005 (2005)
10. Blakeley, C.: Virtuoso RDF Views. OpenLink Software (2007)
11. Barrasa, J., Corcho, O., Gómez-Pérez, A.: R₂O, an Extensible and Semantically Based Database-to-ontology Mapping Language. In: Proceeding of the Second International Workshop on Semantic Web and Databases (2004)
12. Cullot, N., Ghawi, R., Yétongnon, K.: DB2OWL: A Tool for Automatic Database-to-Ontology Mapping. Université de Bourgogne (2007)
13. Atkinson, M., et al.: The Object-Oriented Database Manifesto. In: Proceedings of the First Intl. Conference on Deductive and Object-Oriented Databases (1989)
14. Cattell, R., Barry, D.K., Berler, M., Eastman, J., Jordan, D., Russell, C., Schadow, O., Stanienda, T., Velez, F. (eds.): The Object Data Standard: ODMG 3.0. Morgan Kaufmann, San Francisco (1999)
15. Russell, C.: Java Data Objects 2.0 JSR243. Sun Microsystems Inc. (2006)
16. Peim, M., Franconi, E., Paton, N.W., Goble, C.A.: Querying Objects with Description Logics
17. Oren, E., Delbru, R., Gerke, S., Haller, A., Decker, S.: ActiveRDF: Object-Oriented Semantic Web Programming. In: Proceedings of the Sixteenth International World Wide Web Conference (2007)
18. Kalyanpur, A., Pastor, D.J., Battle, S., Padget, J.: Automatic Mapping of OWL Ontologies into Java. In: Proceedings of the International Conference of Software Engineering and Knowledge Engineering (2004)
19. Athanasiadis, I.N., Villa, F., Rizzoli, A.E.: Enabling knowledge-based software engineering through semantic-object-relational mappings. In: Proceedings of the 3rd International Workshop on Semantic Web Enabled Software Engineering (2007)
20. McGuinness, D.L., van Harmelen, F.: OWL Web Ontology Language (2004)

Integrating Lightweight Reasoning into Class-Based Query Refinement for Object Search

Gong Cheng and Yuzhong Qu

Institute of Web Science, School of Computer Science and Engineering
Southeast University, Nanjing 210096, P.R. China
{gcheng,yzqu}@seu.edu.cn

Abstract. More and more RDF data have been published online to be consumed. Ordinary Web users also expect to experience more intelligent services promised by the Semantic Web, such as object search based on structured data. We implemented the Falcons search engine to meet the challenge. To enable keyword search, for each object, we construct and index a virtual document that includes textual descriptions of its neighboring resources. Typing information is used to serve class-based query refinement, and class-inclusion reasoning is performed to discover implicit types of objects. A method of recommending subclasses is implemented to enable navigating class hierarchies for incremental query refinement. We also report on lessons learned from Web-scale experiments.

1 Introduction

Recently, a large amount of RDF data have become available online, which is a significant step towards facilitating the development of the Semantic Web. For example, the Linking Open Data project[1] has published and interlinked RDF data sets consisting of over two billion RDF triples till October 2007. To exploit their untapped potential and commercial value, a naturally emerging problem is how to find needed data. In particular, we focus on how to efficiently find a *Semantic Web object* (*SW object* in short), i.e., a URI that identifies an object described in RDF, and we call it *object search* on the Semantic Web.

Object search is a fundamental service on the Semantic Web. It serves both data producers and consumers. When data producers prepare for publishing RDF data, they need to know existing URIs of referred objects. They can also create their own URIs as identifiers to denote the objects they want to refer to. However, it is likely to lead to the "information islands" and then isolate applications, which weakens RDF's power of exchanging data semantics and benefiting data integration. It should be a best practice to reuse existing URIs as far as possible. Even if data producers want to create some URIs, they are also encouraged to connect them with existing URIs into a Web of data.[2] All these activities

---

[1] http://esw.w3.org/topic/SweoIG/TaskForces/CommunityProjects/breakLinkingOpenData

[2] http://www.w3.org/DesignIssues/LinkedData.html

J. Domingue and C. Anutariya (Eds.): ASWC 2008, LNCS 5367, pp. 449–463, 2008.
© Springer-Verlag Berlin Heidelberg 2008

need object search. On the consumption side, ordinary users may expect that Semantic Web technology can improve search engines' ability to understand and answer queries. For example, by submitting a keyword query "ESWC 2008 session", can users immediately obtain a list of "sessions" at ESWC2008, rather than a list of hyperlinks to webpages? It is actually another application of object search. Users may also want straightforward ways to refine queries with more precise semantics, e.g., tell the search engine what type of objects they are seeking for.

To meet these challenges, several Semantic Web search engines [9,4] have been developed to serve object search in various ways. Basically, these systems index local names, labels, and maybe other associated literals of SW objects to enable keyword search. However, mainly due to the computational complexity, none of these systems considers reasoning, an attractive feature closely associated with the birth and growth of the Semantic Web [2,1]. We will later show that integrating even lightweight reasoning can improve search engines' ability to answer queries.

In this paper, we detail the object search service provided by the Falcons search engine.[3] It accepts keyword queries and serves a ranked list of SW objects. Different from other Semantic Web search engines, to enable keyword search, Falcons indexes a virtual document for each SW object, which includes not only its local name, labels, and other associated literals, but also the textual descriptions of all other neighboring resources. Classes of SW objects are indexed, and users are served with user-friendly navigation of class hierarchies to incrementally refine queries and filter results. In particular, class-inclusion reasoning over thousands of vocabularies is performed. So for SW objects, not only their explicitly specified classes but also inferred ones are indexed. We also report on experimental results on Web-scale reasoning. All the developed techniques have been proved to be effective based on a large enough data set collected from the real Semantic Web and a period of successful running of the system beginning from February 2008.

The remainder of this paper is structured as follows. We start in Sect. 2 with a demonstration of provided functions, spotted challenges, and an overview of the proposed approach. Section 3 elaborates on constructing virtual documents for SW objects. Section 4 presents the implementation of class-based query refinement and class-inclusion reasoning over thousands of vocabularies. Section 5 details how to recommend subclasses to enable navigating class hierarchies for incremental query refinement. Section 6 reports on Web-scale experiments. Related work is discussed in Sect. 7. Finally, we conclude the paper with a summary and suggestions for future work in Sect. 8.

2   Overview

This section firstly demonstrates the functions of the system and spots challenges. Then we give an overview of the proposed approach.

---

[3] http://iws.seu.edu.cn/services/falcons/

2.1   Demonstration of Functions

After a user submits a keyword query "ESWC 2008" to the system, based on an inverted index, the system serves a ranked list of SW objects of which the textual descriptions contain all the terms in the keyword query. Meanwhile, based on the results, the system collects and sorts their classes, and recommends several ones of which the instances are most probably those the user is seeking for, such as "Event". Then the user selects "Event", which submits a refined query to the system. The results page is updated to the one shown in Fig. 1. The bottom part is a filtered list of SW objects of which the textual descriptions contain all the terms in the query, and all these SW objects are instances of some "Event" class. At the top of the results page, several classes (different from the previous ones) are recommended, such as "Conference Event" and "Session Event", all of

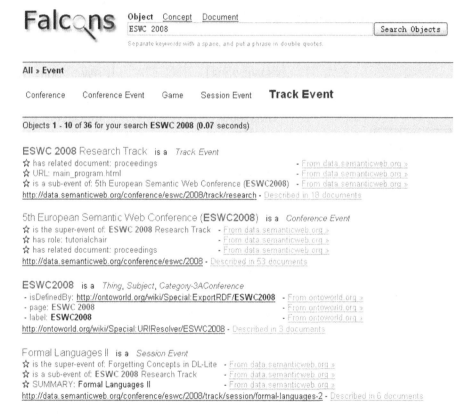

Fig. 1. A screenshot of Falcons Object Search. A user has submitted a query "ESWC 2008" and then has specified that he/she is seeking for an "Event". The system serves a list of SW objects of which the virtual documents contain all the terms in the query and are instances of some "Event" class. Besides, several subclasses of "Event", such as "Conference Event", are recommended to the user for further query refinement.

which are subclasses of "Event". The user can click on these classes to further refine queries and filter results, or trace back to relax restrictions.

2.2   Challenges

To implement the proposed functions, several challenges are to be met. The first challenge is: which descriptions of SW objects should be indexed. A commonly adopted method is to, for each SW object, index the terms in its local name and some/all of its associated literals. It is based on the assumption that terms in a keyword query indicate some properties of a SW object. However, in RDF, there are various ways to describe a property of a SW object. As shown in Fig. 2, to describe that a session is a subevent of ESWC2008, we can associate it with a literal, connect it to a URI that denotes the conference, or connect it to a blank node. In the latter two cases, the terms "ESWC 2008" will not be indexed to the session by the commonly adopted methods, so the session will not be returned to answer keyword queries like "ESWC 2008 session".

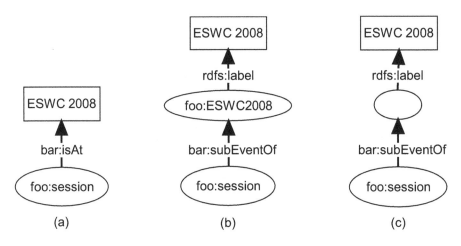

Fig. 2. Three ways to describe that a session is a subevent of ESWC2008: (a) by associating a literal, (b) by connecting to a URI, or (c) by connecting to a blank node

Secondly, for the proposed class-based query refinement, SW objects may be filtered out undesirably. For example, maybe a session is explicitly specified as an instance of some "SessionEvent" class, but is not explicitly specified as an instance of some "Event" class. When "Event" is selected to refine the query, such a session will be filtered out even though it is known that "SessionEvent" is a subclass of "Event". Therefore, reasoning techniques are necessary, although it seems difficult to be performed on the Web scale.

Thirdly, the system should also determine that, out of possibly millions of classes, which ones will be recommended to users as candidate restrictions. A similar problem is that, after a class has been selected by a user, which of its subclasses will be recommended. Recommending too many classes is boring, so the system should only recommend a limited number of ones that can best capture users' intentions.

2.3   Overview of the Approach

In this subsection, we present an overview of the proposed approach to meet the above challenges, as shown in Fig. 3. Details will be separately described in the next sections.

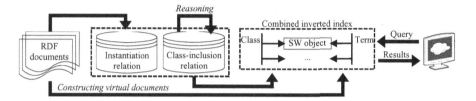

Fig. 3. Overview of the approach

For each SW object, a virtual document is constructed, which includes not only its local name and associated literals, but also the textual descriptions of all other neighboring resources in all the discovered RDF documents. As an example, in all the three cases shown in Fig. 2, the terms "ESWC 2008" will be indexed to `foo:session`. In other words, we expand textual descriptions of SW objects. Terms are weighted based on their provenance, e.g., terms in labels are assigned a higher weighting coefficient than those in other literals.

We also extract class inclusion and class equivalence axioms from schema-level documents (ontologies), write them to database, and perform class-inclusion reasoning (lightweight transitive reasoning) to obtain, for each class, all its superclasses including explicitly specified and inferred ones. For each SW object, we extract its explicitly specified classes, and look up database to find all their superclasses.

With the above SW-object-to-term and SW-object-to-class mappings, we build a combined inverted index, including an inverted index from terms to SW objects and an inverted index from classes to SW objects.

A query can be viewed as a set of terms and a set of classes (as restrictions). A SW object satisfies a query iff its virtual document contains all the terms in the query and it is an instance of at least one class in the query. Each term or class in a query can be mapped to a set of SW objects based on the combined inverted index, and the final results can be obtained by computing a conjunction of these sets of SW objects.

Besides, to recommend classes as further candidate restrictions, we iterate over the resulting SW objects to collect their classes, and filter them by preserving only those as subclasses of the current selected classes. The preserved classes are ranked according to their coverage of the resulting SW objects, and a limited number of the highest ranked ones will be recommended.

3 Constructing Virtual Documents for SW Objects

IR technology has been adapted for enabling object search by Semantic Web search engines [9,4], such as inverted index from terms to SW objects. A SW object is just a URI, with no persistent-state text except for the URI itself. To build the inverted index, search engines often use local names and associated literals of SW objects to construct their virtual documents [15]. It is based on the assumption that terms in a keyword query indicate some properties of a SW object, so indexing the terms in their properties can answer keyword queries. However, in RDF, there are various ways to characterize a property of a SW object, e.g., by associating a literal or by connecting it to a URI or a blank node. Therefore, as described in Sect. 2.3, we expand the virtual document of a SW object to include the textual descriptions of its neighboring resources.

3.1 Neighbors in RDF Graph

In graph theory, the neighbors of a vertex is defined to be the vertices adjacent to it (excluding itself), i.e., an edge exists between a vertex and its neighbor. We consider that such definition is not suitable for RDF graph due to the use of blank nodes. In RDF, blank nodes indicate the existence of things without being assigned global identifiers, and evidently no terms can be collected from blank nodes. Actually, blank nodes are created mainly for connecting other resources. So in the graph view, we collect neighbors for a SW object by starting from it, traversing the graph, and stopping until reaching URIs or literals (but not blank nodes).

To formalize, a notion of *RDF sentence* [17] is used here. In brief, an RDF sentence is a set of RDF triples that contain common blank nodes, as illustrated by Fig. 2(c). An RDF sentence is still an RDF graph, and is also called a *minimum self-contained graph* [13]. An RDF graph g can be decomposed into a unique set of RDF sentences [13], denoted by $\mathrm{Sent}(g)$. Then, let $\mathrm{N}(o, g)$ be the neighbors of a SW object o in g. A resource $r \in \mathrm{N}(o, g)$ iff $\exists \tilde{s} \in \mathrm{Sent}(g)$, $o \in \mathrm{Subj}(\tilde{s})$ and $r \in \mathrm{Obj}(\tilde{s})$, or $o \in \mathrm{Obj}(\tilde{s})$ and $r \in \mathrm{Subj}(\tilde{s})$, where $\mathrm{Subj}(\tilde{s}) = \{s | \exists \langle s, p, o \rangle \in \tilde{s}\}$ and $\mathrm{Obj}(\tilde{s}) = \{o | \exists \langle s, p, o \rangle \in \tilde{s}\}$.

3.2 Construction of Virtual Documents

We use the well-known vector space model to represent virtual documents as well as keyword queries, i.e., the virtual document of a SW object or a keyword query is represented as a term vector in the term vector space.

Let U, B, and L be the sets of all URIs, all blank nodes, and all literals, respectively. Let g be the RDF graph serialized by an RDF document. $\forall r \in U$, let $\mathrm{LN}(r)$ be the term vector representing its local name. $\forall r \in U \cup B$, let $\mathrm{Lbl}(r, g)$ be the term vector representing its labels (values of `rdfs:label`) in g. $\forall r \in L$, let Lex(r) be the term vector representing its lexical form. Then, $\forall r \in U \cup B \cup L$, define Name($r, g$) as the term vector representing its *name* as follows:

$$\text{Name}(r, g) = \begin{cases} \text{LN}(r) + \text{Lbl}(r, g) & r \in U \\ \text{Lbl}(r, g) & r \in B \\ \text{Lex}(r) & r \in L. \end{cases} \tag{1}$$

For a SW object o, its virtual document constructed from an RDF graph g is defined as follows:

$$\text{VDoc}(o, g) = \alpha \cdot \text{LN}(o) + \beta \cdot \text{Lbl}(o, g) + \gamma \cdot \sum_{r \in N(o,g)} \text{Name}(r, g), \tag{2}$$

where α, β, and γ are the weighting coefficients to be tuned. Evidently, compared to the commonly adopted methods, more SW objects are indexed from each term due to the introduction of terms from neighboring resources, which will generally improve the recall but reduce the precision of the system. If we assume that terms in queries are biased in indicating names of SW objects, we can set α and β to higher values than γ, so that those SW objects whose local names and/or labels contain the terms in a query are ranked higher. Currently, we set α, β, and γ to 10, 5, and 1, respectively.

A SW object may be used by many RDF documents. Different data producers may describe different properties of a SW object in various aspects. To enable cross-document search, for a SW object o, its virtual document on the Semantic Web is constructed by aggregating its virtual documents constructed from all the RDF documents (discovered by the system):

$$\text{VDoc}(o) = \sum_g \text{VDoc}(o, g). \tag{3}$$

Finally, an inverted index is built from terms in virtual documents to SW objects. In particular, for virtual documents (vectors) of SW objects, their components are revised by inverse document frequency.

In the system, SW objects in query results are ranked based on a combination of their relevance to the query and their popularity. The relevance score is calculated based on the cosine similarity measure. The popularity score is evaluated according to the number of RDF documents that SW objects are used by.

4   Refining Keyword Queries with Class Restrictions

Compared to the hypertext Web, the Semantic Web brings more structured data with rich semantics, which cannot be satisfactorily utilized by a purely IR-based search engine to serve object search. Recently, cognitive science shows that people are predisposed to use typing information rather than other property information to perform human reasoning [16], while typing information (rdf:type) is also widely used by data producers. Therefore, we exploit typing information in the system to improve object search. To be precise, after submitting a keyword query, the user can specify the type of objects they are seeking for to filter results.

4.1   Class-Based Refinement

In the system, a query q is formulated as $\langle T_q, C_q \rangle$, where T_q is a set of query terms and C_q is a set of classes as restrictions. A SW object o is an *answer* to q iff two conditions are satisfied: the virtual document of o contains all the terms in T_q; and o is an instance of at least one class in C_q. A set of classes, rather than just one class, are allowed to be specified because, as shown in Fig. 1, users are served with tags rather than URIs of classes, and one tag may stand for more than one classes.

As described in Sect. 2.3, we extract typing information of SW objects and build an inverted index from classes to SW objects. Afterwards, by combining it with the inverted index from terms to SW objects, the system can enhance keyword queries with the ability of specifying the class of target SW objects.

Merely using explicitly specified typing information is insufficient. As is often the case, data producers may only specify that a SW object is an instance of some class, but search engine users may specify its superclass when submitting queries. To serve this, class-inclusion reasoning is performed. Afterwards, for each SW object, we index not only its explicitly specified classes but also their superclasses inferred by reasoning. Nevertheless, reasoning on the Web scale faces more difficulties, such as trust. The following presents techniques to address these problems.

4.2   Filtering Axioms

On the Semantic Web, named classes and properties are identified by URIs, and they are organized into vocabularies [3]. A *vocabulary* on the Semantic Web, such as FOAF, is a non-empty set of URIs with a common URI namespace that denote classes or properties. A vocabulary v is formulated as $\langle \mathrm{id}(v), \mathrm{C}(v), \mathrm{P}(v) \rangle$, where $\mathrm{id}(v)$ is the URI namespace that identifies v, $\mathrm{C}(v)$ is the set of classes in v, and $\mathrm{P}(v)$ is the set of properties in v. A URI $u \in \mathrm{C}(v)$ (or $u \in \mathrm{P}(v)$) iff two conditions are satisfied: (a) the URI namespace of u is $\mathrm{id}(v)$; (b) the RDF graph merged from those decoded from the RDF documents obtained by dereferencing u and $\mathrm{id}(v)$ entail the RDF triple $\langle u, \texttt{rdf:type}, \texttt{rdfs:Class} \rangle$ (or $\langle u, \texttt{rdf:type}, \texttt{rdf:Property} \rangle$).

In this paper, we focus on named classes and class-inclusion reasoning. For each vocabulary v, let $\mathrm{CIR}(v)$ be the explicitly specified class-inclusion relation decoded from the RDF documents obtained by dereferencing $\mathrm{id}(v)$ and every $c \in \mathrm{C}(v)$. Each pair $\langle c_1, c_2 \rangle \in \mathrm{CIR}(v)$ should satisfy that $c_1 \sqsubseteq c_2$ and both c_1 and c_2 are named classes. Besides, for each class equivalence axiom $c_1 \equiv c_2$, we also include $\langle c_1, c_2 \rangle$ and $\langle c_2, c_1 \rangle$ in $\mathrm{CIR}(v)$. However, not all the pairs in $\mathrm{CIR}(v)$ will be accepted by the following reasoning engine, considering that anyone can say anything on the real Semantic Web. To ensure the rationality of axioms,

CIR(v) is filtered to its subset F-CIR(v) subject to that $\langle c_1, c_2 \rangle \in$ F-CIR(v) iff $\langle c_1, c_2 \rangle \in$ CIR(v) and $c_1 \in C(v)$. This is inspired by [7] that a vocabulary is allowed to reuse classes in other vocabularies but can only constrain the meaning of its own classes. If no such filtering is performed, one can easily mess up the system (after reasoning) by, e.g., encoding `rdfs:Resource` $\sqsubseteq$ `foaf:Person` in his/her own RDF document.

4.3   Transitive Reasoning

Let F-CIR $= \bigcup_v$ F-CIR(v), v for every vocabulary discovered by the system. Then the last task is to compute its transitive closure F-CIR$^+$. Initially we store F-CIR in a two-column table of a relational database. During reasoning, inferred axioms are also stored in the table, so after reasoning, the table is stored with F-CIR$^+$.

Computing all the superclasses of a class is to, in the digraph view, find all its reachable vertices. Based on this, we implement a parallel program to compute F-CIR$^+$. Each thread starts with a class c, and recursively looks up the table to obtain all its superclasses Sup(c), and supplements the table with $\langle c, c' \rangle$ for all $c' \in$ Sup(c). In practice, it is not necessary to apply this computation to all the classes. Clearly, for the system, if a class has never been instantiated, we do not need to compute all its superclasses since the computational results will not be used in practice.

With F-CIR$^+$, it is easily to look up superclasses when building the inverted index from classes to SW objects.

5   Recommending Subclasses for Incremental Query Refinement

As shown in Sect. 2.1, users can refine queries by navigating class hierarchies. Class hierarchies on the Semantic Web are different from category hierarchies used by many E-Commerce sites. On those sites, a category hierarchy is single, carefully designed, and relatively small-scale, while the whole class hierarchy on the Semantic Web comprises a large number of class hierarchies from different vocabularies of different qualities, so techniques will be different. This section discusses how to implement the proposed mode of user interaction that allows users to specify class restrictions.

Initially, for a query $\langle T_q, C_q \rangle$, no classes have been specified, and C_q is simply set to { `rdfs:Resource` }, which is considered as a superclass of all other classes. Navigation of the class hierarchy is viewed as submitting a sequence of queries with the same set of query terms but different sets of class restrictions. In particular, moving down (or up) the hierarchy is to replace the class restrictions with more specific (or general) ones.

To determine which subclasses should be provided to users as candidate restrictions, we devise a method composed of the following steps:

1. Find out all the answers to a given query $\langle T_q, C_q \rangle$;
2. Collect the classes of the answers and rank them;
3. Select the top K ranked classes that satisfy the following conditions: (a) each selected class must be a strict subclass of some class in C_q, and (b) the class inclusion relation does not hold between any pair of the selected classes;
4. Map the selected classes to tags and present them to users.

Step 1 is implemented based on the combined inverted index.

In Step 2, for each SW object, all its classes (including explicitly specified and inferred ones) are physically stored in the index, so we simply iterate over the answers to collect their classes. However, for some queries, there are a large amount of answers, and iterating over all of them online takes too much time. To make a trade-off between coverage and efficiency, the classes of at most the first 1,000 answers are considered in the system, and let C be the set of classes collected from the first 1,000 answers. The ranking of these classes is based on the consideration that: if a class covers more answers, it will be ranked higher. For example, out of the first 1,000 answers, if 600 ones are instances of c_1 but only 300 ones are instances of c_2, c_1 is ranked higher than c_2.

In Step 3, suppose that all the classes in C are sorted according to their ranking scores in descending order. Let C_K be the set of selected classes, which is initially empty. We scan the sorted list and stop if K classes have been selected based on the following rules or we reach the end of the list. To be selected, a class c must satisfy two conditions. Firstly, $\exists c' \in C_q, c \sqsubseteq c'$, i.e., c is a strict subclass of some class specified by the user. In Sect. 4, we have computed F-CIR$^+$, and here we can use it to check whether $c \sqsubseteq c'$ is satisfied: $c \sqsubseteq c'$ iff $\langle c, c' \rangle \in$ F-CIR$^+$ and $\langle c', c \rangle \notin$ F-CIR$^+$. Secondly, $\nexists c' \in C_K, c \sqsubseteq c'$, i.e., the class inclusion relation never holds between any pair of selected classes. This is for increasing the variety of selected classes, whereas their total number is limited.

In Step 4, selected classes are mapped to user-friendly tags, and we simply use their normalized local names as tags. Tags are sorted in lexicographic order, and tags with ranking scores higher than a threshold are highlighted, as illustrated by Fig. 1.

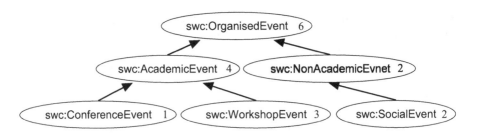

Fig. 4. A class hierarchy sampled from the SWC vocabulary. Each class is associated with the number of its instances as answers to a query.

We illustrate the whole process by an example. Figure. 4 depicts a class hierarchy sampled from the SWC vocabulary.[4] Each class is associated with the number of its instances as answers to a query. Suppose that the user has specified `swc:OrganisedEvent`, i.e., C_q = { `swc:OrganisedEvent`}. The sorted list C is: `swc:OrganisedEvent`, `swc:AcademicEvent`, `swc:WorkshopEvent`, `swc:NonAcademicEvent`, `swc:SocialEvent`, and `swc:ConferenceEvent`. Here the positions of `swc:NonAcademicEvent` and `swc:SocialEvent` are interchangeable since they are with the same ranking score. Let K = 2. Firstly, `swc:OrganisedEvent` will not be selected again because it is not a strict subclass of itself. Next, its strict subclass `swc:AcademicEvent` is selected, but then `swc:WorkshopEvent` will not be selected because it is a subclass of `swc:AcademicEvent`, which has been selected. Similarly, `swc:NonAcademicEvent` should be selected but `swc:SocialEvent` should not. However, we apply another heuristic here: if two classes are with equal ranking scores and the class inclusion relation holds between them, the more specific one of them will be selected. So our algorithm will select `swc:SocialEvent` instead of `swc:NonAcademicEvent`. Finally, two classes, `swc:AcademicEvent` and `swc:SocialEvent`, have been selected, and the recommended tags are "Academic Event" and "Social Event".

6   Experiments

We performed experiments to evaluate the feasibility and performance of the proposed approach. All the experiments were implemented in Java and performed on a 4-Core Xeon 2.50GHz server with 16GB of main memory. A MySQL database stored on a RAID-5 disk system was used.

At the time of writing, the crawler of the system has discovered 9.8 million RDF documents. A total of 1, 159, 425 classes in 3, 039 vocabularies have been recognized. Figure 5 shows such distribution, which approximates a power law. About half of the vocabularies (47.58%) contain no more than 10 classes, whereas the largest vocabulary contains 196, 591 classes.

However, only 37, 208 classes (3.2%) in 1, 174 vocabularies (38.6 %) were discovered to have been explicitly instantiated. To obtain better performance, following computations only considered these vocabularies as well as those referred by them (both directly and indirectly).

It took a 10-thread program 44 minutes to parse documents by using Jena,[5] compute F-CIR, and write F-CIR to database. Most of the time was spent in parsing documents. In particular, several vocabularies are defined by a large amount of small documents rather than a single document, resulting in extra I/O cost. We also observed several large vocabularies, each with tens of thousands of classes, which are rarely instantiated, but it took a large proportion of time to process them. So they can be ignored in practice if necessary. Finally, it took a 10-thread program 3 minutes to compute F-CIR$^+$.

---

[4] http://data.semanticweb.org/ns/swc/ontology#
[5] http://jena.sourceforge.net/

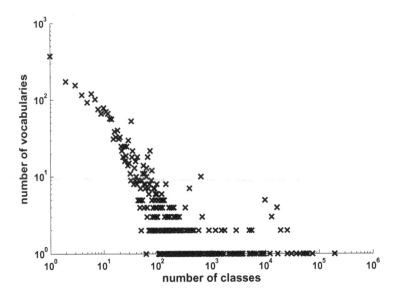

Fig. 5. Distribution of the number of vocabularies versus the number of classes in a vocabulary

Table 1. Statistics before and after reasoning

	#class inclusion relationships	Avg. #classes per SW object	Avg. #intances per class
Before reasoning	302,679	1.74	41.69
After reasoning	2,834,615	4.13	98.96

A comparison of statistical data before and after reasoning is shown in Table 1. Before reasoning, the cardinality of F-CIR is $302,679$, out of which $10,394$ pairs (3.4%) are with classes in different vocabularies. After reasoning, the cardinality of the resulting F-CIR$^+$ has increased to $2,834,615$.

The crawler has discovered $27,828,528$ SW objects that are with explicitly specified typing information. The average number of explicitly specified classes of a SW object is 1.74. After reasoning, the average number of classes of a SW object has increased to 4.13, which is 2.37 times the number before reasoning. Meanwhile, the average number of instances of a class has increased from 41.69 to 98.96. Such increases promise the usefulness of class-inclusion reasoning in the case of refining queries with classes, despite the lack of gold standard data for estimation. And note that it costs only $44+3$ minutes of offline computation.

7   Related Work

TAP [8] is one of the earliest keyword-based Semantic Web search systems. It maps keywords to labels of SW objects and serves with a SW object (and its surrounding subgraph) based on popularity, user profile, and search context. Swoogle [5], one of the most popular Semantic Web search engines, serves class/property search and ontology search, based on a PageRank-like ranking algorithm. SWSE [9], another keyword-based search engine, enables users to filter resulting SW objects by specifying a class. Semantic Web Search[6] focuses on specific types of SW objects, such as FOAF Person and RSS Item. It organizes search results by documents. Similarly, WATSON [4] also organizes results by documents and highlights mapped SW objects. It enables users to specify the scope that query terms should be mapped to, such as local name, label, or any literal. Sindice [12] allows property-value pair look up to find documents knowing a property of a SW object, and also allows keyword-based RDF document and Microformats search.

Different from these search engines, the proposed approach expands textual descriptions of SW objects in order to improve the recall of the system, while it also uses a weighting scheme to preserve the precision. Both SWSE and our Falcons system allow users to specify class restrictions. Falcons further integrates reasoning techniques to capture implicit instantiation relation, and enables navigating class hierarchies for incremental query refinement.

Semantic search [14,11,18] promises to provide more accurate results than IR-based search, by translating keyword queries to formal queries, such as SPARQL queries. However, existing approaches are more suitable for single vocabularies, and applying semantic search to the Web scale is still a great challenge because it is difficult to generate a limited number of formal queries that can cover thousands of vocabularies on overlapped topics. Besides, the performance of formal queries on the Web scale is another big challenge.

On reasoning, instance Store [10] is an approach to a restricted form of ABox reasoning that combines a DL reasoner with a database. It can only deal with a role-free ABox, i.e., an ABox that does not contain any axioms asserting role relationships between pairs of individuals. The reasoning technique implemented in this paper can be viewed as a simplified version of instance Store since we only deal with named classes. However, we further perform axiom filtering, considering the Web's characteristics.

8   Conclusion

We have detailed how our Falcons system serves object search. It is designed as a domain-independent search engine that works on the real Semantic Web. The main technical contributions of this paper include: (a) constructing virtual documents for SW objects with textual descriptions of neighboring resources, (b) supplementing IR-based object search engines with class-based query refinement, (c) integrating class-inclusion reasoning techniques to uncover implicit instantiation, and (d) recommending subclasses to enable navigating class hierarchies for incremental query refinement.

---

[6] http://www.semanticwebsearch.com/

The current research can be extended in several directions. Firstly, ontology matching techniques [6] can be integrated to offer additional class inclusion/equivalence axioms, to improve the integration of different vocabularies. Nevertheless, it requires that generated axioms should be of high precision to avoid undesirable results after reasoning. Secondly, besides typing information, it is possible to allow query refinement based on arbitrary properties. One challenge is how to extend existing faceted search techniques to cover multiple vocabularies with overlapped topics. It is also interesting to investigate other possibilities of combining IR-based search and formal query, to be adapted to the Web scale.

Acknowledgments. The work is supported in part by the NSFC under Grant 60773106, and in part by the 973 Program of China under Grant 2003CB317004. We would like to thank Weiyi Ge for his effort in implementing the system. We would also like to thank Wei Hu for his comments on the manuscript.

References

1. Berners-Lee, T., Hall, W., Hendler, J.A., O'Hara, K., Shadbolt, N., Weitzner, D.J.: A Framework for Web Science. Foundations and Trends in Web Science 1(1), 1–130 (2006)
2. Berners-Lee, T., Hendler, J., Lassila, O.: The Semantic Web. Sci. Am. 284(5), 34–43 (2001)
3. Berrueta, D., Phipps, J.: Best Practice Recipes for Publishing RDF Vocabularies. W3C Working Draft (2008)
4. d'Aquin, M., Sabou, M., Dzbor, M., Baldassarre, C., Gridinoc, L., Angeletou, S., Motta, E.: WATSON: A Gateway for the Semantic Web. In: Posters of the 6th International Semantic Web Conference (2007)
5. Ding, L., Pan, R., Finin, T.W., Joshi, A., Peng, Y., Kolari, P.: Finding and ranking knowledge on the semantic web. In: Gil, Y., Motta, E., Benjamins, V.R., Musen, M.A. (eds.) ISWC 2005. LNCS, vol. 3729, pp. 156–170. Springer, Heidelberg (2005)
6. Euzenat, J., Shvaiko, P.: Ontology Matching. Springer, Heidelberg (2007)
7. Grau, B.C., Horrocks, I., Kazakov, Y., Sattler, U.: A Logical Framework for Modularity of Ontologies. In: 20th International Joint Conference on Artificial Intelligence, pp. 298–303 (2007)
8. Guha, R., McCool, R., Miller, E.: Semantic Search. In: 12th International World Wide Web Conference, pp. 700–709. ACM Press, New York (2003)
9. Harth, A., Hogan, A., Delbru, R., Umbrich, J., O'Riain, S., Decker, S.: SWSE: Answers Before Links! In: Semantic Web Chanllenge (2007)
10. Horrocks, I., Li, L., Turi, D., Bechhofer, S.: The Instance Store: DL Reasoning with Large Numbers of Individuals. In: 2004 International Workshop on Description Logics, pp. 31–40 (2004)
11. Tran, T., Cimiano, P., Rudolph, S., Studer, R.: Ontology-based interpretation of keywords for semantic search. In: Aberer, K., Choi, K.-S., Noy, N., Allemang, D., Lee, K.-I., Nixon, L., Golbeck, J., Mika, P., Maynard, D., Mizoguchi, R., Schreiber, G., Cudré-Mauroux, P. (eds.) ASWC 2007 and ISWC 2007. LNCS, vol. 4825, pp. 523–536. Springer, Heidelberg (2007)
12. Tummarello, G., Delbru, R., Oren, E.: Sindice.com: Weaving the open linked data. In: Aberer, K., Choi, K.-S., Noy, N., Allemang, D., Lee, K.-I., Nixon, L., Golbeck,

J., Mika, P., Maynard, D., Mizoguchi, R., Schreiber, G., Cudré-Mauroux, P. (eds.) ASWC 2007 and ISWC 2007. LNCS, vol. 4825, pp. 552–565. Springer, Heidelberg (2007)

13. Tummarello, G., Morbidoni, C., Bachmann-Gmür, R., Erling, O.: RDFSync: Efficient remote synchronization of RDF models. In: Aberer, K., Choi, K.-S., Noy, N., Allemang, D., Lee, K.-I., Nixon, L., Golbeck, J., Mika, P., Maynard, D., Mizoguchi, R., Schreiber, G., Cudré-Mauroux, P. (eds.) ASWC 2007 and ISWC 2007. LNCS, vol. 4825, pp. 537–551. Springer, Heidelberg (2007)

14. Wang, H., Zhang, K., Liu, Q., Tran, T., Yu, Y.: Q2Semantic: A lightweight keyword interface to semantic search. In: Bechhofer, S., Hauswirth, M., Hoffmann, J., Koubarakis, M. (eds.) ESWC 2008. LNCS, vol. 5021, pp. 584–598. Springer, Heidelberg (2008)

15. Watters, C.: Information Retrieval and the Virtual Document. J. Am. Soc. Inf. Sci. 50(11), 1028–1029 (1999)

16. Yamauchi, T.: The semantic web and human inference: A lesson from cognitive science. In: Aberer, K., Choi, K.-S., Noy, N., Allemang, D., Lee, K.-I., Nixon, L., Golbeck, J., Mika, P., Maynard, D., Mizoguchi, R., Schreiber, G., Cudré-Mauroux, P. (eds.) ASWC 2007 and ISWC 2007. LNCS, vol. 4825, pp. 609–622. Springer, Heidelberg (2007)

17. Zhang, X., Cheng, G., Qu, Y.: Ontology Summarization Based on RDF Sentence Graph. In: 16th International World Wide Web Conference, pp. 707–716. ACM Press, New York (2007)

18. Zhou, Q., Wang, C., Xiong, M., Wang, H., Yu, Y.: SPARK: Adapting keyword query to semantic search. In: Aberer, K., Choi, K.-S., Noy, N., Allemang, D., Lee, K.-I., Nixon, L., Golbeck, J., Mika, P., Maynard, D., Mizoguchi, R., Schreiber, G., Cudré-Mauroux, P. (eds.) ASWC 2007 and ISWC 2007. LNCS, vol. 4825, pp. 694–707. Springer, Heidelberg (2007)

A Segmentation-Based Approach for Approximate Query over Distributed Ontologies

Yimin Wang[1], Guilin Qi[2], and Min Chen[3]

[1] Lilly Singapore Centre for Drug Discovery, #02-04 Biomedical Grove, Singapore
wangyimin@lilly.com
[2] Institute AIFB, University of Karlsruhe (TH), Karlsruhe, Germany
gqi@aifb.uni-karlsruhe.de
[3] Boston University, One Silber Way, Boston, MA 02215, USA
anthem16@bu.edu

Abstract. With the popularity of semantic information systems distributed on the Web, there is an arising challenge to provide efficient query answering support for these systems. However, common approaches for distributed query answering either exhibit performance disadvantages or loss of completeness in an unbalanced way. In this paper, we introduce a novel approach for *segment-based conjunctive query answering* over distributed ontologies. Our approach balances the trade-off between performance and completeness by introducing segmentation-based distributed ontology integration. We define the notions of segment and approximate conjunctive query answering. Corresponding algorithms are designed, implemented and evaluated. The evaluation results show that our approach is very promising in processing ontologies in modern semantic information systems.

1 Introduction

Today, many semantic information systems on the Web are going beyond a centralized setting and working in a distributed scenario. In those systems, ontologies are increasingly applied as the data schemata, sources, mediators, etc. [4,6,9], thus, querying the distributed ontologies is one major task in distributed semantic information systems.

Let's first look at a common scenario as a motivated example: Large organizations, such as international enterprises and universities, often have many departments, maintaining distributed data that are reasonably interconnected among departments (e.g., the shared information about people and projects across departments) on their web information systems. Query answering over distributed ontologies is a major task in these web information systems where data schemata and sources are more and more represented as ontologies. More importantly, people in these organization may like to have part of query results in an efficient manner, which raises up a new challenge.

There are two extreme situations for query answering over distributed ontologies: (1) We can integrate the distributed ontologies into a single local node and perform query answering in a centralized way (e.g. [6,9]). This approach apparently lacks the optimization for query answering in the distributed scenario, because queries are not executed in a distributed way. (2) On the other hand, we can also query over distributed ontologies without integration. In this case, the ontologies are queried in a pure distributed way

J. Domingue and C. Anutariya (Eds.): ASWC 2008, LNCS 5367, pp. 464–478, 2008.
© Springer-Verlag Berlin Heidelberg 2008

on an individual ontology in parallel, so overall execution time is reduced. However, by using this approach, we may lose significant information that is inferred by considering the interrelationships (e.g. in the form of mappings) between the ontologies. The first extreme keeps completeness by losing performance, while the second one pursues performance but loses completeness. Our aim is to find a reasonable balance between the two extremes: We argue that performance is a critical issue on today's Web, while there is often a tradeoff between the completeness and the performance when querying distributed ontologies.

Because it's difficult to achieve complete answers and performance advances at the same time, approximate query answering based on a *proper* integration of distributed ontologies but provides quick response times is an important issue to be addressed. This issue has been widely recognized and discussed in the traditional database community in order to improve query answering performance [1]. However, there is little existing work on approximate query answering over distributed ontologies in this field. The popular description logics (DL) reasoners, such as FaCT++[1], Pellet[2] and KAON2[3], only support exact query answering over ontologies and are not particularly optimized for a distributed scenario.

In this paper, we introduce a novel approach for approximate conjunctive query answering over distributed ontologies in semantic information systems on the Web. We focus on improving the overall performance by sacrificing part of the completeness. We integrate ontologies in a segmentation-based approach, where distributed ontologies are divided into several groups that are connected via mappings. We also find an appropriate way to identify the segments of the integrated global ontology. As central contribution of our approach, we have designed three algorithms for segmentation, query distribution, and termination and results collection, respectively. They have been implemented and evaluated within real-life application. The evaluation results show our approach well meets our target mentioned above.

In the following, we first discuss the related work (Section 2) and introduce some required preliminaries (Section 3) for understanding the rest of this paper. We then present the foundations of our approximate conjunctive query answering approach (Section 4) and describe the algorithms (Section 5). Afterwards we present and discuss the evaluation of our approach (Section 6). Finally, we conclude the major contributions of this paper and discuss possible future extensions (Section 7).

2   Related Work

The problem of query answering in distributed systems has been discussed in traditional databases and the Semantic Web. The distributed query answering in traditional databases simply distributes the queries and aggregates the results [13]. However, if it is directly applied to process distributed ontologies, then it is very possible to lose results without a proper integration. In [5], the authors present a logical foundation for peer-to-peer data integration and query answering using mappings but DL ontologies

---

[1] http://owl.man.ac.uk/factplusplus/

[2] http://www.mindswap.org/2003/pellet/

[3] http://kaon2.semanticweb.org

are not considered. Cai and colleagues introduces an approach to managing RDF data in a scalable manner [3], while still, DL ontologies are not in their scope.

An ontology-based approach for querying distributed data is introduced in [9], where classic view-based mappings are adopted, but more expressive DL mappings with proper ontology integration are not introduced. Several work on reasoning with distributed ontologies has been discussed, such as distributed description logics (DDL) [21], $\mathcal{E}$-Connections approach [10] and package-based description logics (PDL) [2]. However, DL ontologies can not be directly supported without adding extra syntax and semantics, such as bridge rules in DDL [21]. The actual implementation of KAONp2p [12] on query answering over distributed ontologies establishes the distributed processing of ontologies but again integrates schemata information (i.e., TBoxes of ontologies) upfront. Our approach is different from the approach given in [12] in that we first segment the distributed ontologies then integrate ontologies in each segment.

Approximate query answering is often discussed in the centralized case. Approximate query answering is introduced to improve the performance [1] for traditional centralized databases. Approximate reasoning over common OWL DL ontologies is discussed in many papers, such as in [14] and [19], but they doesn't concern conjunctive query answering over distributed ontologies. Although our approach approximates the conjunctive query answering in distributed systems, the meaning of approximate query answering in our paper is fundamentally different from that in centralized case. That is, we do not change the semantics of the local ontologies in the distributed system. Instead, our approach optimizes the query answering procedure for distributed ontologies by segmenting the integrated ontologies into several subsets that can be distributively processed.

3   Preliminaries

We first introduce some prerequisite knowledge about conjunctive query answering [16] problem over description logics $\mathcal{SHIQ}$ KB that is proved to be decidable in [8]. We also introduce an ontology mapping system by following the definition of conjunctive query.

Let N_R be a set of *role names* with both transitive and normal role names $N_{tR} \cup N_{nR} = N_R$, where $N_{tR} \cap N_{nR} = \varnothing$. A $\mathcal{SHIQ}$-role is either some $R \in N_R$ or an *inverse role* R^- for $R \in N_R$. Trans(R) and $R \sqsubseteq S$ represent the transitive and inclusion role axioms, respectively, where R and S are roles. $\mathcal{R}$ as a finite set of transitive and inclusion role axioms. A simple role is a $\mathcal{SHIQ}$-role that neither its sub-roles nor itself is transitive. Given a set of *concept names* N_C, the set of $\mathcal{SHIQ}$-concepts is the smallest set such that: (1) Every concept name is a concept; (2) if C and D are concepts, R is a role, S is a simple role, and n is a positive integer, then the following expressions are also concepts: $(\top), (\bot), (\neg C), (C \sqcap D), (C \sqcup D), (\exists R.C), (\forall R.C), (\leqslant nSC)$ and $(\geqslant nSC)$. A TBox $\mathcal{T}$ is a finite set of axioms with the form $C \sqsubseteq D$ where C and D are $\mathcal{SHIQ}$-concepts, and an ABox $\mathcal{A}$ is a finite set of axioms with the form $C(x), R(x, y)$, and $x \approx y$ ($x \napprox y$). A $\mathcal{SHIQ}$ knowledge base (KB) is a triple $(\mathcal{R}, \mathcal{T}, \mathcal{A})$ which is also considered as ontology in the semantic information systems here.

The semantics of $\mathcal{SHIQ}$ KB is given by the interpretation $\mathcal{I} = (\Delta^{\mathcal{I}}, \cdot^{\mathcal{I}})$ that consists of a non-empty set $\Delta^{\mathcal{I}}$ (the domain of $\mathcal{I}$) and the function $\cdot^{\mathcal{I}}$ as usual (e.g., see [15]).

The reasoning and decidability of $\mathcal{SHIQ}$ is also introduced in [15]. The interpretation $\mathcal{I}$ is the model of $\mathcal{R}$ and $\mathcal{T}$ if for each $R \sqsubseteq S \in \mathcal{R}$, $R^{\mathcal{I}} \subseteq S^{\mathcal{I}}$ and for each $C \sqsubseteq D \in \mathcal{T}$, $C^{\mathcal{I}} \subseteq D^{\mathcal{I}}$.

Let KB be a $\mathcal{SHIQ}$ knowledge base, N_P be a set of names such that all concepts and roles are in N_P. An *atom* $P(s_1, ..., s_n)$ has the form $P(s_1, \ldots, s_n)$, denoted as $P(\mathbf{s})$, where $P \in N_P$, and s_i are either variables or individuals from KB. An atom is called a *DL-atom* if P is a $\mathcal{SHIQ}$-concept or role; it is called *non-DL-atom* otherwise.

Definition 1 (Conjunctive Queries). *Let $x_1, \ldots, x_n$ and $y_1, \ldots, y_m$ be sets of distinguished and non-distinguished variables, denoted as $\mathbf{x}$ and $\mathbf{y}$, respectively. A conjunctive query $Q(\mathbf{x}, \mathbf{y})$ over a KB is a conjunction of atoms $\bigwedge P_i(\mathbf{s_i})$, where the variables in $\mathbf{s_i}$ are contained in either $\mathbf{x}$ or $\mathbf{y}$. We denote operator π [18] to translate $Q(\mathbf{x}, \mathbf{y})$ into a first-order formula with free variables $\mathbf{x}$: $\pi(Q(\mathbf{x}, \mathbf{y})) = \exists \mathbf{y} : \bigwedge (P_i(\mathbf{s_i}))$.* ◇

For $Q_1(\mathbf{x}, \mathbf{y_1})$ and $Q_2(\mathbf{x}, \mathbf{y_2})$ conjunctive queries, a *query containment* axiom $Q_2(\mathbf{x}, \mathbf{y_2}) \sqsubseteq Q_1(\mathbf{x}, \mathbf{y_1})$ has the following semantics:

$$\pi(Q_2(\mathbf{x}, \mathbf{y_2}) \sqsubseteq Q_1(\mathbf{x}, \mathbf{y_1})) = \forall \mathbf{x} : \pi(Q_1(\mathbf{x}, \mathbf{y_1})) \leftarrow \pi(Q_2(\mathbf{x}, \mathbf{y_2}))$$

Definition 2 (Conjunctive Query Answering). *An* answer *of a conjunctive query $Q(\mathbf{x}, \mathbf{y})$ w.r.t. KB is an assignment θ of individuals to distinguished variables, using $Ans(Q, KB)$ as a function, such that $\pi(KB) \models \pi(Q(\mathbf{x}\theta, \mathbf{y}))$.* ◇

We refer readers to [8,18,16] for further issues in conjunctive query answering for ontologies.

We follow the general framework of [17] to formalize the notion of a mapping system for DL ontologies, where mappings are expressed as correspondences between conjunctive queries[4] over ontologies.

Definition 3 (Ontology Mapping System). *An* ontology mapping system $\mathcal{MS}$ *is a triple $(\mathcal{O}_1, \mathcal{O}_2, \mathcal{M})$, where*

- *$\mathcal{O}_1$ is the* source *ontology, $\mathcal{O}_2$ is the* target *ontology,*
- *$\mathcal{M}$ is the mapping between $\mathcal{O}_1$ and $\mathcal{O}_2$, i.e. a set of assertions $q_S \rightsquigarrow q_T$, where q_S and q_T are conjunctive queries over $\mathcal{O}_1$ and $\mathcal{O}_2$, respectively, with the same set of distinguished variables $\mathbf{x}$, and $\rightsquigarrow \in \{\sqsubseteq, \sqsupseteq, \equiv\}$.*

An assertion $q_S \sqsubseteq q_T$ is called a sound *mapping, requiring that q_S is contained by q_T w.r.t. $\mathcal{O}_1 \cup \mathcal{O}_2$; an assertion $q_S \sqsupseteq q_T$ is called a* complete *mapping, requiring that q_T is contained by q_S w.r.t. $\mathcal{O}_1 \cup \mathcal{O}_2$; and an assertion $q_S \equiv q_T$ is called an* exact *mapping, requiring it to be sound and complete.* ◇

To have the same segmentation result for ontology integration system introduced later, we do not consider mapping transitivity here, i.e., if $\mathcal{O}_1$ and $\mathcal{O}_2$ have mapping $\mathcal{M}_{12}$; $\mathcal{O}_2$ and $\mathcal{O}_3$ have mapping $\mathcal{M}_{23}$, it does not imply the existence of mapping $\mathcal{M}_{13}$ between

---

[4] We denote a conjunctive query as $q(\mathbf{x}, \mathbf{y})$, with $\mathbf{x}$ and $\mathbf{y}$ sets of *distinguished* and *non-distinguished* variables, respectively.

$\mathcal{O}_1$ and $\mathcal{O}_3$. Furthermore, several mappings between two ontologies are considered as one single mapping. This form of mapping is decidable in inferencing while it is restricted to DL-safe mappings [11]. Mappings discussed in this paper are referred as DL-safe mapping by default. Further details about semantics and restrictions of ontology mapping system can be found in [11].

4 Segment-Based Conjunctive Query Answering over Distributed Ontologies

To discover the possible optimizations using approximate conjunctive query answering in the distributed environment, we need to analyze the distributed ontologies by considering their integration via mappings. We here define ontology integration system using mappings for distributed networking scenario. In the following, we denote $I = \{1, \dots, n\}, n \in \mathbb{N}$ and $i \neq j; i, j \in I$.

Definition 4 (Distributed Ontology Integration System (DOIS)). *A distributed ontology integration system (DOIS) is a triple* $(\{\mathcal{MS}_i\}, \mathcal{N}, \mathsf{Loc})$, *where*

1. $\{\mathcal{MS}_i\}$ *is a set of mapping systems. We denote* $\mathbf{O}$ *and* $\mathbf{M}$ *as the ontologies and mappings included in* $\{\mathcal{MS}_i\}$, *respectively.*
2. $\mathcal{N}$ *is a set of distributed nodes where the ontologies and mappings reside;*
3. $\mathsf{Loc} : \mathbf{O} \cup \mathbf{M} \to \mathcal{N}$ *is a location function such that* $N_i = \mathsf{Loc}(\mathcal{O}_i)$ *and* $N_{ij} = \mathsf{Loc}(\mathcal{M}_{ij})$, *where* $N_i, N_{ij} \in \mathcal{N}$ *and* $\mathcal{O}_i \in \mathbf{O}, \mathcal{M}_{ij} \in \mathbf{M}$. *This function aims to relate an ontology or mapping to a specific distributed node.* ◇

Given DOIS $(\{\mathcal{MS}_i\}, \mathcal{N}, \mathsf{Loc})$ *over* $\mathcal{O}$, *we use* $\mathsf{Ans}_e(Q, \{\mathcal{MS}_i\}, \mathcal{O})$ *to denote the complete set of answers for conjunctive query* $Q(\mathbf{x}, \mathbf{y})$ *over* $\mathbf{O} \cup \mathbf{M}$.

To simplify the presentation, in the following we introduce the notion of distributed system, which is inspired from [23] but not exactly the same.

Definition 5 (Distributed System). *Distributed system* $\mathfrak{D}$ *is a set of mapping systems* $\{\mathcal{MS}_i\}$. *Or equivalently,* $\mathfrak{D} = (\mathbf{O}, \mathbf{M})$ *where* $\mathbf{O} = \{\mathcal{O}_i\}$ *is a set of ontologies and* $\mathbf{M} = \{\mathcal{M}_{ij}\}$ *is a set of mappings between* $\mathcal{O}_i$ *and* $\mathcal{O}_j$ *in* $\mathbf{O}$. ◇

We introduce the notion of distributed system in addition to Definition 4 for better understanding our graph-based segmentation approach and algorithms, because by looking at Example 6, it is very easy to see $\mathfrak{D}$ is a graph with ontologies as vertex and mappings as edges.

Example 6. *The DOIS depicted in Figure 1 contains five ontologies distributed over five nodes, where* $\mathcal{O}_t$ *is the target ontology that has non-empty mappings with all other source ontologies.* $\mathcal{O}_1$, $\mathcal{O}_2$ *and* $\mathcal{O}_3$ *are connected by mappings, presented as dotted line.*

Given a DOIS, we know the following information: (1) A mapping $\mathcal{M}$ and its source and target ontologies; (2) the locations of those mappings or ontologies in the distributed network. Now the we are able to query over distributed ontologies that are integrated as a global ontology in mapping systems.

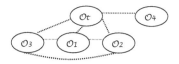

Fig. 1. Figure for Example 6. Given $\mathfrak{D} = (\mathbf{O}, \mathbf{M})$, source ontology $\mathcal{O}_i \in \mathbf{O}$, target ontology $\mathcal{O}_t \in \mathbf{O}$ and mappings (light-dot lines) between them. Circles are ontologies distributed on different nodes. $\triangle$

Because a DOIS consists of different ontologies, it is possible to improve the overall performance by dividing the distributed system of a DOIS into several segments (Example 9). Let's consider a simple example: The Law School and Faculty of Physics normally do not share projects or professors, or the shared information is not usually recognized by people. Therefore, ontologies describing these two departments are going to be grouped into different segments.

Definition 7 (Distributed Subsystem). *Given a distributed system $\mathfrak{D}$, distributed subsystem of $\mathfrak{D}$ is a pair $(\mathbf{O}', \mathbf{M}')$, denoted as $\mathfrak{D}' \sqsubseteq \mathfrak{D}$, iff $\mathbf{O}' \subseteq \mathbf{O}$ and $\mathbf{M}' = \{\mathcal{M}_{ij} \in \mathbf{M} : \mathcal{O}_i, \mathcal{O}_j \in \mathbf{O}'\}$. A distributed subsystem is also a distributed system.* $\diamond$

Definition 8 (Segment). *Given a distributed system $\mathfrak{D} = (\mathbf{O}, \mathbf{M})$, a segment of $\mathfrak{D}$ is distributed subsystem $\mathfrak{S} = (\mathbf{O}', \mathbf{M}')$ such that*

1. for all $\mathcal{M}_{ij} \in \mathbf{M}'$, we have $\mathcal{M}_{ij} \neq \varnothing$;
2. for any distributed subsystem $\mathfrak{S}'$ of $\mathfrak{D}$, if $\mathfrak{S} \sqsubset \mathfrak{S}'$, then $\mathfrak{S}'$ does not satisfy 1. $\diamond$

Different from [20], which aims to generate segments from a large domain ontology to facilitate ontology engineering, we can see a *segment* here is distributed subsystem $\mathfrak{S}$ of $\mathfrak{D}$ which satisfies the condition that all the ontologies in it are connected by non-empty mappings and any other distributed subsystem of $\mathfrak{D}$ which strictly includes $\mathfrak{S}$ does not satisfy this condition. This has two benefits:

1. If the ontologies of a distributed subsystem are not all connected by non-empty mappings, then this distributed subsystem can be divided into smaller distributed subsystems to improve performance.
2. We achieve completeness of answers as much as possible after segmentation by requiring the segment to be the inclusion maximal distributed subsystem which satisfies Condition 1 in Definition 8.

Therefore, our definition of segment *perfectly captures* the idea of balancing the trade-off between performance and completeness in querying distributed ontologies – the Example 9 presents how the segmentation is processed.

We are able to develop an algorithm for segmenting a distributed system $\mathfrak{D}$ directly based on existing algorithms (e.g., union-find algorithm [7]) to process graphs. Note that not all the complete subgraphs of G can be interpreted to segments, e.g., assuming a segment $(\{\mathcal{O}_1, \mathcal{O}_2, \mathcal{O}_3\}, \{\mathcal{M}_{12}, \mathcal{M}_{13}, \mathcal{M}_{23}\})$, apparently, $(\{\mathcal{O}_1, \mathcal{O}_2\}, \{\mathcal{M}_{12}\})$ forms a complete graph but it doesn't satisfy Definition 8. Let's see an example about how mappings affect segmenting result to form DOISs with different distributed nodes.

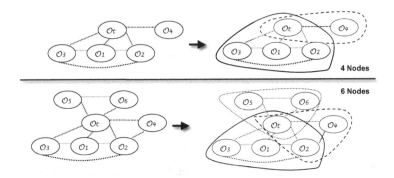

Fig. 2. Figure for Example 9. Given $\mathfrak{D} = (\mathbf{O}, \mathbf{M})$, source ontology $\mathcal{O}_i \in \mathbf{O}$, target ontology $\mathcal{O}_t \in \mathbf{O}$ and mappings (light-dot lines) between them.

Example 9. *In this example, we use 5 and 7 distributed ontologies for each DOIS, respectively. Next we will segment the DOIS to show how our segmenting approach works based on definitions above.*

1. *Let's first look at the case of four distributed nodes with one ontology on each node. Before segmenting the DOIS, we have mappings $\mathcal{M}_{12}$, $\mathcal{M}_{13}$ and $\mathcal{M}_{23}$ between ontologies $\mathcal{O}_1$ and $\mathcal{O}_2$, $\mathcal{O}_1$ and $\mathcal{O}_3$, $\mathcal{O}_2$ and $\mathcal{O}_3$, respectively. We also have mappings from target ontology to each source ontology $\mathcal{M}_{t1}$, $\mathcal{M}_{t2}$, $\mathcal{M}_{t3}$ and $\mathcal{M}_{t4}$. According to the definition of segments, we have the segmentation result depicted by following the arrow:*

   (a) $(\{\mathcal{O}_1, \mathcal{O}_2, \mathcal{O}_3, \mathcal{O}_t\}, \{\mathcal{M}_{t1}, \mathcal{M}_{t2}, \mathcal{M}_{t3}, \mathcal{M}_{13}, \mathcal{M}_{12}, \mathcal{M}_{23}\})$
   (b) $(\{\mathcal{O}_4, \mathcal{O}_t\}, \{\mathcal{M}_{t4}\})$.

   So there are two segments in this case.

2. *It is very often the case that the distributed nodes and extra mappings are added into the current system. Our seven nodes example indicates the segmentation status if we add $\mathcal{O}_5$ and $\mathcal{O}_6$ with mappings $\mathcal{M}_{24}$ and $\mathcal{M}_{56}$ to the five nodes distributed system. Mappings between target ontology and $\mathcal{O}_5$ and $\mathcal{O}_6$ are $\mathcal{M}_{t5}$ and $\mathcal{M}_{t6}$, respectively. Then we have the following segmentation result:*

   (c) $(\{\mathcal{O}_1, \mathcal{O}_2, \mathcal{O}_3, \mathcal{O}_t\}, \{\mathcal{M}_{t1}, \mathcal{M}_{t2}, \mathcal{M}_{t3}, \mathcal{M}_{13}, \mathcal{M}_{12}, \mathcal{M}_{23}\})$
   (d) $(\{\mathcal{O}_2, \mathcal{O}_4, \mathcal{O}_t\}, \{\mathcal{M}_{t2}, \mathcal{M}_{t4}, \mathcal{M}_{24}\})$
   (e) $(\{\mathcal{O}_5, \mathcal{O}_6, \mathcal{O}_t\}, \{\mathcal{M}_{t5}, \mathcal{M}_{t6}, \mathcal{M}_{56}\})$

   So there are three segments in this case.   △

As mentioned in the introduction, our aim is to find a balance between the completeness and performance for query answering over distributed ontologies by using segments of distributed systems. Thus, the union of the individual query answering result of each segment obviously *may not* be equal to the exact answers. We therefore introduce our approach to find out *segment-based answers* in querying distributed ontologies. In the mean time, for users who want to achieve complete answers, we also provide an alternative approach to compute them (Algorithm 2).

Definition 10 (Segment-based Query Answering). *Given a DOIS with* $\mathfrak{D} = (\mathbf{O}, \mathbf{M})$ *and a conjunctive query Q, let* $\mathbf{S} = \{\mathfrak{S}_i\}_{i=1,\ldots,n;\ n \in I}$ *and* $\text{Ans}_e(Q, \mathfrak{D}, \mathcal{O})$ *be complete answers to query over* $\mathcal{O}$, *where* $\mathcal{O}$ *is the target ontology and* $\mathcal{O} \in \mathbf{O}$. *Segment-based Query Answering in DOIS is defined as:*

$$\text{Ans}_a(Q, \mathfrak{D}, \mathcal{O}) = \bigcup_{i \in \{1,\ldots,n\}} \text{Ans}_e(Q, \mathfrak{S}_i, \mathcal{O})$$

where Ans_a *stands for segment-based query answering.*   ◇

Obviously, the set of segment-based answers are subset of the set of complete answers because the number of answers monotonically increases. Let's illustrate this by an example.

Example 11. *Assume ontologies from different departments* $\mathcal{O}_1$, $\mathcal{O}_2$ *and mapping* $\mathcal{M}$ *bridging them:*

- $\mathcal{O}_1 = \{\text{Professor} \sqsubseteq \text{Faculty} \sqcap \exists \text{teach.Course}\}$
- $\mathcal{O}_2 = \{\text{Professor} \sqsubseteq \text{Staff} \sqcap \exists \text{teach.Lecture}\}$
- $\mathcal{M} = \{1:\text{Professor} \sqsubseteq 2:\text{Professor}, 1:\text{Faculty} \sqsubseteq 2:\text{Staff}, 2:\text{Lecture} \sqsubseteq 1:\text{Course}\}$

Here, 1:Professor *means concept* Professor *in* $\mathcal{O}_1$. *If* $\mathcal{M}$ *holds and we ask for professors who teaches a certain course, we get complete answers because* $\mathcal{O}_1$ *and* $\mathcal{O}_2$ *are integrated with* $\mathcal{M}$ *as a global ontology. However, if* $\mathcal{M}$ *does not exist, then we can ask for segment-based answers (i.e.,* $\mathcal{O}_1$ *and* $\mathcal{O}_2$ *are divided into different segments). In this case the professors who give lectures with other departments are not included in the answers but can be computed in the end of our algorithm.*   △

5   Algorithms for Segment-Based Query Answering

There are three algorithms in this segment-based conjunctive query answering approach: (1) Segmentation, (2) query distribution and answering, and (3) termination and results collection. Different query distribution approaches are adopted for segments with different number of mapping elements for better allocating computing resources. The *cardinality* of $\mathfrak{S}$ is $\text{Card}(\mathfrak{S}) = |\mathbf{M}|$ that indicates the number of elements contained in $\mathbf{M}$. Then,

- a segment is called *single-element segment*, denoted as $\mathfrak{S}_s$, iff $\text{Card}(\mathfrak{S}_s) = 1$;
- a segment is called *multiple-element segment*, denoted as $\mathfrak{S}_m$, iff $\text{Card}(\mathfrak{S}_m) > 1$.

The *single-element segment* is a segment which contains only two ontologies (i.e., a target ontology $\mathcal{O}_t$ and a source ontology $\mathcal{O}_i$ with mapping $\mathcal{M}_{ti}$ between them, e.g., segment (b) in Example 6 is a single-element segment). On the other hand, the *multiple-element segment* consists more than one mapping ontologies connecting more than two ontologies (e.g., all segments in Example 6 except (b) are multiple-element segments). The exact conjunctive query answering are denoted as $\text{Ans}_e(Q, \mathfrak{S}_s, \mathcal{O})$ and $\text{Ans}_e(Q, \mathfrak{S}_m, \mathcal{O})$ for single/multiple-element segments, respectively, where Q is a conjunctive query over a common target ontology shared by all the segments of $\mathfrak{D}$.

Algorithm 1. Segmentation

Require: a DOIS $(\mathfrak{D}, \mathcal{N}, \text{Loc})$, a target ontology $\mathcal{O}_t$, empty ontology set **O**, empty mapping set **M** and empty graph set $\mathcal{G}$.

1: get all ontologies that have non-empty mappings with $\mathcal{O}_t$ and put them to **O**
2: get all mappings related $\mathcal{O}_t$ to **O** and put them to $\mathbf{M}_t$
3: get all mappings among ontologies in **O** and put them to **M**, add $\mathcal{O}_t$ to **O**, add $\mathbf{M}_t$ to **M**
4: establish graph G with n vertices **O** and edges **M**, where n is the number of ontologies in **O**
5: **for** $k = n; k \geqslant 2; k - -$ **do**
6:     get all complete subgraphs of G with number of vertices k and put them to $\mathcal{G}_k$
7:     remove all subgraphs that are already in $\mathcal{G}$ from $\mathcal{G}_k$
8:     add $\mathcal{G}_k$ to $\mathcal{G}$
9: **end for**
10: establish a set of segments **S** of $\mathfrak{D}$ based on $\mathcal{G}$
11: **for all** $\mathfrak{S}_l \in \mathbf{S}$ **do**
12:     put $\mathfrak{S}_l$ into $\mathbf{S}_s$ if $\mathfrak{S}_l$ is a single-element segment, else, put $\mathfrak{S}_l$ into $\mathbf{S}_m$
13: **end for**
14: output $\mathbf{S}_s$ and $\mathbf{S}_m$

Algorithm 1 starts with input of a DOIS with distributed system $\mathfrak{D}$ and a target ontology $\mathcal{O}_t$ over which the query is about to be executed. The system gets all ontologies connected to $\mathcal{O}_t$ and the corresponding mappings to establish a graph G (Step 1–4), and then it extracts the segments with k number of ontologies ($2 \leqslant k \leqslant n$) iteratively by computing the complete subgraphs of G with k vertices using classic *union-find algorithm* [7] (Step 5–12, this is exactly the segmentation procedure presented in Example 6). In the meanwhile, we need to eliminate those subgraphs of complete subgraphs of G with higher number vertices to make sure all the generated subgraphs are maximal complete subgraphs (Step 8). Then, we interpret the generated set of maximal complete subgraphs $\mathcal{G}$ back to a set of segments, written as **S** (Step 10). To facilitate query distribution, we classify the segments into single and multiple-element segments as input of the consequent Algorithm 2 (Step 11–14). We show all segments should contain $\mathcal{O}_t$.

Proposition 12. *Given a DOIS $(\mathfrak{D}, \mathcal{N}, \text{Loc})$, a target ontology $\mathcal{O}_t$ and a set of segments* **S** *generated in the Step 14 of Algorithm 1, the target ontology $\mathcal{O}_t$ is included in all the segments in* **S**.

Proof sketch. If G is complete graph without vertex $\mathcal{O}_t$, then G and $\mathcal{O}_t$ forms a complete graph. According to Definition 8, G can not be interpreted as a segment.   □

In Algorithm 2, for a single-element segment, the system sends the query to the distributed node where the source ontology in $\mathfrak{S}_{s_i}$ resides (Step 2–5, please note there is only one source ontology in each $\mathfrak{S}_{s_i}$ so the nodes are not occupied at this stage); for multiple-element segment, the system sends the query to an arbitrary unused node where an arbitrary source ontology resides (Step 6–11). To achieve exact query answering (Step 12) as supplement, the algorithm simply integrates all ontologies and mapping in $\mathfrak{D}$ on an unused remote node without segmentation and query over $\mathfrak{D}$. In Step 6–12, if all nodes are occupied, the algorithm waits until an arbitrary node finishes its

---

Algorithm 2. Query distribution and answering

Require: a DOIS $(\mathfrak{D}, \mathcal{N}, \mathsf{Loc})$, a conjunctive query Q, a target ontology $\mathcal{O}_t$, segments $\mathbf{S}_s$ and $\mathbf{S}_m$

1: create an empty list L_N for nodes that are idle (i.e., nodes that are not involved in executing query answering tasks).
2: **for all** $\mathfrak{S}_{s_i} \in \mathbf{S}_s$ **do**
3:     get ontology source ontologies $\mathcal{O}_i$ of $\mathcal{O}_t$ in $\mathfrak{S}_{s_i}$; get remote node $N_i = \mathsf{Loc}(\mathcal{O}_i)$, put N_i into L_N
4:     compute $\mathsf{Ans}_a(Q, \mathfrak{S}_{s_i}, \mathcal{O})$ on node N_i in parallel (i.e., parallel processing means this task is executed in parallel in different nodes over distributed network)
5: **end for**
6: **for all** $\mathfrak{S}_{m_j} \in \mathbf{S}_m$ **do**
7:     get source ontology set $\mathbf{O}_j$ of $\mathcal{O}_t$ in $\mathfrak{S}_{m_j}$
8:     find an arbitrary idle node $N_j = \mathsf{Loc}(\mathcal{O}_j)$, where $\mathcal{O}_j \in \mathbf{O}_j$
9:     put N_j into L_N
10:    compute $\mathsf{Ans}_a(Q, \mathfrak{S}_{m_j}, \mathcal{O})$ on node N_j in parallel
11: **end for**
12: get a random idle node N_i, compute $\mathsf{Ans}_e(Q, \mathfrak{D}, \mathcal{O})$ on N_i in parallel

---

Algorithm 3. Termination and results collection

Require: list of nodes L_N that is in-use (not idle), termination boolean signal U, timeout preset T

1: **while** $L_N \neq \varnothing$ **and** U =FALSE **and** $T \neq$ TIMEOUT **do**
2:     **if** N_i returns anwser **then**
3:         send out the answer from N_i, remove N_i from L_N
4:     **end if**
5: **end while**
6: final results collection

---

querying task and put the next query answering task to this node – this procedure managed independently by Algorithm 3 for termination and result collection.

Algorithm 3 returns the real time segment-based answers, manages the distributed nodes and monitors the terminating signal. Once all distributed nodes are not in use, or querying process are terminated. The system also terminates in a preset, maximum allowed execution time, considering one or several distributed tasks doesn't respond due to possible system or network failures.

In Algorithm 1, the optimized clique discovery problem to find complete subgraphs in graphs with size n, which is the number of ontologies, has complexity LOG-TIME [7]. The complexity of conjunctive query answering over $\mathcal{SHIQ}$ KB with size m, which is the number of concepts, is CO-NP-COMPLETE [8], which is the major computation in Algorithm 2. Our algorithms don't intend to reduce the computational complexity of query answering over distributed ontologies. Because the cope of our approach is to balance the trade-off between completeness and performance in distributed ontology query answering for actual web information systems, but not to optimize the query processing algorithms in local query processing.

6   Evaluation and Discussion

We have implemented our approach using KAON2 as the query answering engine for this evaluation[5]. Since our approach aims to improve the overall performance in querying distributed ontologies approximately, we need to evaluate the performance increases (presented by time saved) against the loss of completeness (presented by rates of the cardinality of the set of segment-based answers to that of the set of complete answers). The hypothesis for this evaluation is that the time saved outweighs the rates of approximation to complete answers. We show our evaluation by first introducing the setting for experiments, presenting and discussing the results afterwards.

6.1   Experiment Settings

We used 17 virtual distributed nodes to simulate a distributed network. The nodes are "virtual" because virtual machines were deployed on four actual computers, therefore the CPU power was shared. Each node held an ontology instance data set with fixed size. The instance data have different schemata that are either heterogeneous or homogenous. In our experiment, we are using four ontology schemata: (1) The Lehigh University Benchmark (LUBM) ontology[6]; (2) Proton ontology[7]; (3) SWRC ontology [22]; (4) FOAF ontology.[8] The instance data includes the following:

1. LUBM automatically generated ontologies that includes instance data of information about university life
2. Digital library data in Proton schema
3. Documents and publication data in Institute AIFB, University of Karlsruhe
4. Project and personal contacting information in FOAF schema

The ontology mappings were created manually for actual use in different projects (e.g., Proton–SWRC mapping was created for EU IST SEKT project[9], FOAF–LUBM mapping was created for Traditional Chinese Medicine project [6].)

Each node held a data set with either SWRC, Proton, LUBM or FOAF schema with corresponding ontology instance data with size approximately 1MB. Let's look at a 5 segments example of our experiment setting to see how the data and schemata are allocated. The 17 ontologies and the mappings between them constitute a distributed system $\mathfrak{D}$, whereas the segmentation process only applies to those ontologies that hold mappings with target ontology.

We used two conjunctive queries: One was to search the documents with their corresponding authors who were professors; the other was to find out all the abstracts of the documents written by Yimin Wang in 2006 with the associated projects. The data had many schemata created for different purpose, for example, data in SWRC schema were created for the local research group website portal, data in FOAF schema were created

---

[5] The experimental implementation is available upon request

[6] http://swat.cse.lehigh.edu/projects/lubm/

[7] http://proton.semanticweb.org/

[8] http://xmlns.com/foaf/0.1/

[9] http://www.sekt-project.com

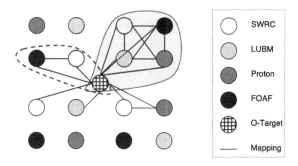

Fig. 3. Example of data allocation in 5 segments case. The circles stand for distributed ontologies with different schemata; O-Target means the target ontology.

for the Chinese Traditional Medicine project management, etc. Therefore, it was possible that Yimin Wang was working for a project information in FAOF schema and has publication information in SWRC schema – the mapping was used to infer complete information in this case.

```
1. SELECT ?x ?y
     WHERE { ?x rdf:type Professor . ?y rdf:type Document . ?y publicationAuthor ?x }

2. SELECT ?x ?y ?k
     WHERE { ?x year "2006". ?x author ?z . ?x abstract ?y .
             ?z name "Yimin Wang" . ?z worksFor ?k . ?k rdf:type Project}
```

An experiment unit was one execution of one query over a certain setting with different mappings covered. Based on Definition 8, it's easy to find when the number of mappings changes, the segmentation results are different. In this experiment, we set the DOISs to have four different settings with 16, 12, 9, 5 segments (Figure 3 depicts the 5 segments case), respectively. The two queries above were executed for 20 times for each experiment unit and the average execution time was computed, recorded and compared with the execution time held by exact query answering over same size of data. We compared times saved by using segment-based approach against the rates of approximation in different stages. We present both the time costs using global integration with and without segmentation approach applied.

6.2 Results and Discussions

In the experiments, the two queries above were executed in the distributed network and the results were collected and presented in Figure 4.

The X axis presents the time costs and the Y axis shows the corresponding rates of approximation (i.e. rate of approximation equals to $\frac{N_a}{N_e}$, where N_a and N_e are the numbers of segment-based and all answers, respectively).

The system computes the answers of segments on each distributed node and returns the answers one by one incrementally. Thus, in Figure 4, there are 16 points indicating the time saved against the rates of approximation. Similarly, $DOIS_2$, $DOIS_3$ and $DOIS_4$ have 12, 9 and 5 points in the figures, respectively. The black star and diamond symbols in two figures are time costs of querying global integration of all ontologies in $\mathfrak{D}$ in

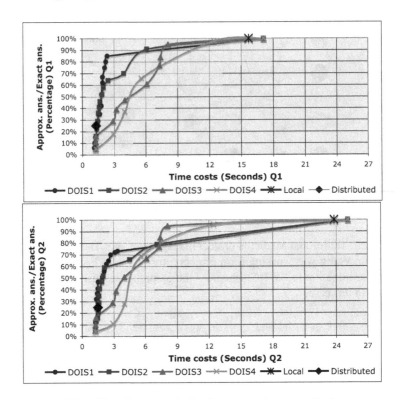

Fig. 4. Experiment results for the two queries. respectively.

local space, and querying complete distributed without considering mappings and integration. We can obviously see these two extreme situations are significantly short of either performance or completeness.

By aligning to the motivated scenario in Section 1 that is considered to be common on today's Web, there are three major messages delivered by this evaluation: First, in the motivated scenario, data in individual departments of a big organizations are usually reasonably (not heavily) interconnected. Moreover, the mapping among the departments are usually under control to avoid unintended information sharing. In this case, our approach have good rates of approximation with remarkable time saved. Taking $DOIS_1$ in Figure 4 for example, when the results have more than 80% rate of approximation to the set of complete answers by using fully global integration in local space, up to $85\%(\frac{15.7-2.3}{15.7})$ and 68% $(\frac{23.8-7.5}{23.8})$ execution time has been saved for the first and second queries respectively. Compare to the case without considering mappings and integration, we see the completeness is only about 25%. This is quite obvious: We have ontologies with four different schemata, so we can approximately get one of four answers if we do not integrate the distributed ontologies using mappings.

Second, however, we have also realized the rate of approximation relies on the queries. For instance, in Figure 4, the first query has better rate of approximation than the second one because the first query is less possible to access remote ontologies: The information about professors and their publications usually reside in individual departments, whereas in the second query the projects are very likely to be shared across departments or universities, resulting frequent usage of mappings. For certain queries, it is possible that the approximate process returns the complete answers, or doesn't give any result at all – that's why we also provide the exact query answering functionality.

Last but not least, personalization of query answering task is a key issue in our approach. It's configurable to make certain mappings integrated or not. People just need to tune the parameters to terminate the query answering procedure if they are satisfied with the segment-based answers, or wait for all the answers. Figure 4 indicates the performance is very promising if anticipated answers are achieved and process is terminated in the middle of runtime.

In a nutshell, our approach, which addresses segment-based conjunctive query answering over distributed ontologies, well meets our target, resulting in sound but *may* incomplete answers in a very efficient manner, especially when the answers are acceptable in practical scenarios.

7   Conclusions and Outlook

In this paper, we described a novel approach to address the problem of balancing the trade-off between completeness and performance in conjunctive query answering over distributed ontologies. The major contributions of this paper is following: (1) We introduced the notions of segment-based conjunctive query answering to achieve better performance with acceptable completeness by introducing distributed system segmentation for query distribution. (2) We designed and implemented three algorithms to support our approach. (3) We also evaluated the performance against the rate of approximation to the exact answers. The evaluation results indicated that our approach is very promising in the motivated scenario. We argued that this scenario is general on today's Web.

For future work, we plan to analyze the semantics of different ontology mappings to seek for other possible segmenting approaches to improve overall performance. There is also a potential opportunity for optimization if we employ a preprocessing procedure for identifying homogenous or redundant ontologies in DOIS.

References

1. Acharya, S., Gibbons, P.B., Poosala, V., Ramaswamy, S.: Join synopses for approximate query answering. In: Proc. of SIGMOD Conference, pp. 275–286 (1999)
2. Bao, J., Caragea, D., Honavar, V.G.: On the semantics of linking and importing in modular ontologies. In: Cruz, I., Decker, S., Allemang, D., Preist, C., Schwabe, D., Mika, P., Uschold, M., Aroyo, L.M. (eds.) ISWC 2006. LNCS, vol. 4273, pp. 72–86. Springer, Heidelberg (2006)
3. Cai, M., Frank, M.: Rdfpeers: a scalable distributed rdf repository based on a structured peer-to-peer network. In: Proceedings of the 13th international conference on World Wide Web, pp. 650–657 (2004)
4. Calvanese, D., Giacomo, G.D., Lenzerini, M.: A framework for ontology integration. In: Proceedings of the First Semantic Web Working Symposium, pp. 303–316 (2001)

5. Calvanese, D., Giacomo, G.D., Lenzerini, M., Rosati, R.: Logical foundations of peer-to-peer data integration. In: Proc. of PODS 2004, pp. 241–251 (2004)
6. Chen, H., Wang, Y., Wang, H., Mao, Y., Tang, J., Zhou, C., Yin, A., Wu, Z.: Towards a semantic web of relational databases: A practical semantic toolkit and an in-use case from traditional chinese medicine. In: Proc. of the 5th International Semantic Web Conference, pp. 750–763 (2006)
7. Galil, Z., Italiano, G.F.: Data structures and algorithms for disjoint set union problems. ACM Comput. Surv. 23, 319–344 (1991)
8. Glimm, B., Horrocks, I., Lutz, C., Sattler, U.: Conjunctive query answering for the description logic $\mathcal{SHIQ}$. In: Proc. of IJCAI 2007. AAAI Press, Menlo Park (2007)
9. Goasdoue, F., Rousset, M.C.: Querying distributed data through distributed ontologies: A simple but scalable approach. IEEE Intelligent Systems 18, 60–65 (2003)
10. Grau, B.C., Parsia, B., Sirin, E.: Combining OWL ontologies using epsilon-connections. J. Web Sem. 4(1), 40–59 (2006)
11. Haase, P., Motik, B.: A mapping system for the integration of OWL-DL ontologies. In: IHIS 2005, November 2005, pp. 9–16. ACM Press, New York (2005)
12. Haase, P., Wang, Y.: A decentralized infrastructure for query answering over distributed ontologies. In: Proc. of ACM Symp. on Appl. Comp., Seoul, Korea. ACM Press, New York (2007)
13. Halevy, A.Y., Ives, Z.G., Madhavan, J., Mork, P., Suciu, D., Tatarinov, I.: The Piazza peer data management system. IEEE Trans. Knowl. Data Eng (TKDE) 16, 787–798 (2004)
14. Hitzler, P., Vrandecic, D.: Resolution-based approximate reasoning for OWL DL. In: Cruz, I., Decker, S., Allemang, D., Preist, C., Schwabe, D., Mika, P., Uschold, M., Aroyo, L.M. (eds.) ISWC 2006. LNCS, vol. 4273, pp. 383–397. Springer, Heidelberg (2006)
15. Horrocks, I., Sattler, U.: Decidability of $\mathcal{SHIQ}$ with complex role inclusion axioms. Artificial Intelligence 160, 79–104 (2004)
16. Horrocks, I., Tessaris, S.: A conjunctive query language for description logic Aboxes. In: Proc. of AAAI/IAAI 2006, pp. 399–404. AAAI Press, Menlo Park (2000)
17. Lenzerini, M.: Data integration: a theoretical perspective. In: Proceedings of the twenty-first ACM SIGMOD-SIGACT-SIGART symposium on Principles of database systems, Madison, Wisconsin, pp. 233–246. ACM Press, New York (2002)
18. Motik, B., Sattler, U., Studer, R.: Query Answering for OWL-DL with rules. J. of Web Semantics 3(1), 41–60 (2005)
19. Pan, J.Z., Thomas, E.: Approximating owl-dl ontologies. In: Proc. of AAAI 2007, pp. 1434–1439 (2007)
20. Seidenberg, J., Rector, A.: Web ontology segmentation: Analysis, classification and use. In: Proc. of the 15th International World Wide Web Conference, Edinburgh (2006)
21. Serafini, L., Borgida, A., Tamilin, A.: Aspects of distributed and modular ontology reasoning. In: Proc. of IJCAI 2005, pp. 570–575 (2005)
22. Sure, Y., Bloehdorn, S., Haase, P., Hartmann, J., Oberle, D.: The SWRC ontology - semantic web for research communities. In: Jajodia, S., Mazumdar, C. (eds.) ICISS 2005. LNCS, vol. 3803, pp. 218–231. Springer, Heidelberg (2005)
23. Zimmermann, A., Euzenat, J.: Three semantics for distributed systems and their relations with alignment composition. In: Cruz, I., Decker, S., Allemang, D., Preist, C., Schwabe, D., Mika, P., Uschold, M., Aroyo, L.M. (eds.) ISWC 2006. LNCS, vol. 4273, pp. 16–29. Springer, Heidelberg (2006)

A Robust Ontology-Based Method for Translating Natural Language Queries to Conceptual Graphs

Tru H. Cao, Truong D. Cao, and Thang L. Tran

Faculty of Computer Science and Engineering
Ho Chi Minh City University of Technology
Vietnam
tru@cse.hcmut.edu.vn

Abstract. A natural language interface is always desirable for a search system. While performance of machine translation for general texts with acceptable computational costs seems to reach a limit, narrowing down the domain to one of queries reduces the complexity and enables better translation correctness. This paper proposes a query translation method that is robust to ill-formed questions and exploits knowledge of an ontology for semantic search. It uses conceptual graphs as the target language for the translation. As a logical inter-lingua with smooth mapping to and from natural language, conceptual graphs simplify translation rules and can be easily converted to other formal query languages. Experiment results of the method on the TREC 2002 and TREC 2007 data sets are also presented and discussed.

1 Introduction

A search engine has been an indispensable tool of computer users in the age of information technology, especially with the explosion of information on the World Wide Web. Although keyword-based search systems like Google and Yahoo have been really useful, a natural language query interface is still desirable and attracts much research effort. That is because it supports more natural and more precise query expressions by humans. To that end it resorts to the problem of automatic conversion of natural language queries into a logical form or to another formal query language sentences. It might be considered as a machine translation problem, whose solutions' performance for the general case appears to be saturated after many years of research. However, what can make the difference is limiting the domain of discourse. In particular, if only questions and querying phrases are considered, better performance can be achieved than for the domain of general texts.

Even for the limited input of query expressions, there are different approaches to the translation problem regarding the two following issues. First, it is about whether rigorous syntactic parsing is applied to a query expression before it is translated into a target language sentence. The disadvantages of the parsing approach are time consuming and requiring grammatically correct inputs, which is thus not robust to ill-formed queries. It is also not practical to require a user to always input a question without grammatical errors. Moreover, it may still face to the problem of syntactic

J. Domingue and C. Anutariya (Eds.): ASWC 2008, LNCS 5367, pp. 479–492, 2008.
© Springer-Verlag Berlin Heidelberg 2008

ambiguity, i.e., one sentence having more than one applicable syntax tree. Second, it is about whether a knowledge base (KB) is employed in the translation. For example, with the query *"What county is Modesto, California in?"*, given no knowledge base, *Modesto* and *California* can be tagged only as proper nouns and thus the implicit relation expressed by the comma between them cannot be interpreted. In contrast, with a knowledge base, they can be recognized as named entities (NE) of the types CITY and PROVINCE, respectively, whence the relation can be translated into one being a sub-region of the other.

Our proposed method in this paper does not rely on a strict grammar of querying sentences but does use an ontology and knowledge base for the translation. It provides knowledge not only for answering queries but also for their conceptual understanding, before they can be mapped to a formal language. There have been recent works on natural language interfaces for question answering systems. For instance, [11] implemented an ontology-based search system whose queries were lists of classes and instances and translated into expressions of SeRQL[1]. They were better than lists of normal keywords, but not as natural as human expressions. Meanwhile, accepting natural language queries, [4] followed the rigorous parsing approach using lambda calculus as an intermediate formal language for the translation. However, the focus of that work was on efficient porting interfaces between different domains rather on the translation itself.

The approach in [9] could be considered as closer to the syntax-free one. It used pattern matching of a natural language query to subject-property-object triples in a knowledge base, before converting the query to one of SPARQL[2]. For the example query therein *"What is a restaurant in San Francisco that serves good French food?"*, it first searched for those triples whose subjects, properties, and objects could match with *"restaurant"*, *"in"*, and *"San Francisco"*. That method thus could not produce a translation if the KB did not contain such a triple for the named entity *San Francisco*, although it existed in the KB. We argue that the translation step should not be mixed up with the answering step. That is, a query can have a translation in a target language although there is no matched answer to it in a knowledge base of discourse.

A closely related work to ours was [15], which followed the syntax-free approach to translate natural language queries into SeRQL expressions. It used the named entity recognition engine of GATE ([5]) and the PROTON ontology of KIM ([10]), but supplemented it with more entity types and relation types. The method was however just tested on the authors' manually collected 36 questions. In contrast, our method proposes conceptual graphs (CG [13]) as the target language and its robustness has been tested on the text retrieval data sets TREC 2002 and TREC 2007 with hundreds of diverse questions.

Conceptual graphs, based on semantic networks and Peirce's existential graphs, combine the visual advantage of graphical languages and the expressive power of logic. On the one hand, unlike many other graph-based languages, the formal order-sorted logic foundation of conceptual graphs offers it as a well-defined logical interlingua for translation to and from different formal languages (e.g. [2] mapping CGs to SeRQL). On the other hand, unlike many other formal languages, conceptual graphs

---

could be mapped smoothly to and from natural language ([14]). These are two salient points for which we propose conceptual graphs for translating and answering natural language queries.

Conceptual graphs have been used for solving problems in several areas such as, but not limited to, natural language processing, knowledge acquisition and management, database design and interface, and information systems. Tim Berners-Lee, the inventor of the World Wide Web, concluded in [1] that CGs could be easily integrated with the Semantic Web. It was also shown in [16] that there was a close mapping between CGs and RDF language. Regarding recent development of some CG systems, CGWorld ([7]) used CGs for queries, storage, and operations. In WebKB-2 ([12]), knowledge was stored in an extended model of CG, RDF, and terminological logics, and knowledge retrieval was performed as graph operations. In Corese ([3]), both knowledge and queries were represented in RDF extended with some CG features, which were then converted into CGs for matching using CG operations. There has been also research on automatic generation of CGs from general text in a specific domain, e.g. the rule-based method in [8] and the machine learning-based one in [17]. However, both of the works required syntactic parsing of input sentences and were evaluated mainly on semantic roles rather than whole sentences.

Next, for the paper being self-contained, Section 2 summarizes the basic notions of conceptual graphs. Section 3 presents in detail our proposed method for translating queries in English into conceptual graphs. Section 4 evaluates the performance of the method with experiment results. Finally, Section 5 sums up the paper and outlines our further work.

2 Basic Notions of Conceptual Graphs

A conceptual graph is a bipartite graph of *concept* vertices alternate with (conceptual) *relation* vertices, where edges connect relation vertices to concept vertices. Each concept vertex, drawn as a box and labelled by a pair of a *concept type* and a *concept referent*, represents an entity whose type and referent are respectively defined by the concept type and the concept referent in the pair. Each relation vertex, drawn as a circle and labelled by a *relation type*, represents a relation of the entities represented by the concept vertices connected to it. For brevity, we may call a concept or relation vertex a concept or relation, respectively. Concepts connected to a relation are called *neighbour concepts* of the relation. Each edge is labelled by a positive integer and, in practice, may be directed just for readability.

Fig. 1. An example conceptual graph

For example, the CG in Figure 1. says *"Cognac is a product. There is a province. France is a country. The province is a sub-region of France. Cognac is produced in the province."*, or briefly, *"Cognac is produced in a province in France"*. In a textual format, concepts and relations can be respectively written in square and round brackets as follows:

[PRODUCT: *Cognac*]→(PRODUCEDIN)→[PROVINCE: *]→(SUBREGIONOF)→[COUNTRY:
France]

Here, for simplicity, the labels of the edges are not shown.

In this example, [PRODUCT: *Cognac*], [PROVINCE: *], [COUNTRY: *France*] are concepts with PRODUCT, PROVINCE and COUNTRY being concept types, whereas (PRODUCEDIN) and (SUBREGIONOF) are relations with PRODUCEDIN and SUBREGIONOF being relation types. The referents *Cognac* and *France* of the concepts [PRODUCT: *Cognac*] and [COUNTRY: *France*] are *individual markers*. The referent "*" of the concept [PROVINCE: *] is the *generic marker* referring to an unspecified entity.

In the textual format, a CG can also be split into sub-graphs containing only one relation for each, using variable symbols to link identical concepts with the generic marker. For example, the above CG can be written as follows:

[PRODUCT: *Cognac*]→(PRODUCEDIN)→[PROVINCE: *x*], and

[PROVINCE: *x*]→(SUBREGIONOF)→[COUNTRY: *France*].

Corresponding to the notion of sorts in order-sorted predicate logic, concept types are partially ordered by the concept subtype order. For example, PROVINCE is a subtype of POLITICALREGION. Relation types can also be partially ordered. For example, SUBREGIONOF is a subtype of LOCATEDIN. Moreover, just as the notion of signature for a predicate in order-sorted predicate logic, which defines the sequence of its argument sorts (i.e., types), each relation type t has a signature denoted by $(t_1, t_2, ..., t_n)$ where n is its arity and t_i's are its argument types, which are concept types. Then a relation of type t in a CG is said to be *well-typed* only if, for every i from 1 to n, the type of the concept connected to the relation by the edge labelled i is more specific than t_i. A hierarchy of concept types and a hierarchy of relation types with signatures form a particular CG ontology. In this paper, we use interchangeably the terms class, entity type and concept type.

The semantics of CGs can be defined through the operator Φ that maps a CG to a first-order predicate logic formula. Basically, Φ maps each vertex of a CG to an atomic formula of first-order predicate logic, and maps the whole CG to the conjunction of those atomic formulas with all variables being existentially quantified. Each individual marker is mapped to a constant, each generic marker is mapped to a variable, and each concept or relation type is mapped to a predicate symbol.

Each concept of type t and referent m is mapped to:

$p(\Phi(m))$
where $p = \Phi(t)$.

Each relation of type t and neighbour concept referents $m_1, m_2, ..., m_n$ is mapped to:

$p(\Phi(m_1), \Phi(m_2), ..., \Phi(m_n))$
where $p = \Phi(t)$.

For example, let G be the CG in Figure 2.1, then $\Phi(G)$ is:

$\exists x \, (product(Cognac) \land province(x) \land country(France) \land$

$produced In(Cognac, x) \land subRegionOf(x, France))$

Partially ordered sets of concept and relation types are also mapped to formulas of first-order predicate logic. Each pair of a concept type t_1 and its super-type t_2 is mapped to:

$$\forall x\, (p_1(x) \Rightarrow p_2(x))$$

where $p_1 = \Phi(t_1)$ and $p_2 = \Phi(t_2)$.

Each pair of a relation type t_1 of arity n and its super-type t_2 is mapped to:

$$\forall x_1 \forall x_2 \ldots \forall x_n\, (t_1(x_1, x_2, \ldots, x_n) \Rightarrow t_2(x_1, x_2, \ldots, x_n))$$

where $p_1 = \Phi(t_1)$ and $p_2 = \Phi(t_2)$.

Using CGs for information retrieval, besides individual referents and the generic referent, we extend them with the *queried referent* denoted by "?", representing the named entities to be searched for. The generic referent in a query CG means that it does not care about a matched entity.

3 Construction of Query Conceptual Graphs

A query can be seen as expressing constraints in terms of relations between the queried entities and known ones. For example, consider the query *"What was the name of the movie that starred Sharon Stone and Arnold Schwarzenegger?"*. *Sharon Stone* and *Arnold Schwarzenegger* are NEs of the types WOMAN and MAN, respectively; *"the movie"* represents an unspecified entity of the type MOVIE; and the interrogative word *What* represents the queried NE of the type ALIAS. There are relations between the queried name and the movie, and between the movie and the two persons. As such, each NE can be mapped to a concept of the corresponding concept type and the whole query mapped to a query CG. Figure 2. show the query CG corresponding to this example query.

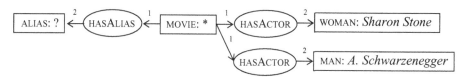

Fig. 2. An example query conceptual graph

As mentioned above, our proposed method to construct the CG corresponding to a query does not rely on a rigorous syntax of the query. The main task and focus is only to correctly recognize entities and their relations expressed by the query. The method composes of the following nine steps:

1. *Recognizing specified entities*: this step recognizes entities specified by names in a query. For instance, in the query *"What is the capital of Mongolia?"*, *Mongolia* is a specified entity.
2. *Recognizing unspecified entities*: this step recognizes entities represented by only words expressing entity types. For instance, in the example query *"How many counties are in Indiana?"*, *"counties"* represents unspecified entities of the type COUNTY.
3. *Extracting relational phrases*: this step finds out the phrases that represent relations between the entities in a query. For example, in the query *"What state is Niagara Falls located in?"*, *"located in"* is a phrase representing a relation between *Niagara Falls* and a state, which is the queried entity.

4. *Determining the type of queried entities*: this step determines the type of unknown entities represented by interrogative words such as *What* or *Which*. For example, in the query *"What is WWE short for?"*, the relation word *"short for"* corresponds to the relation type HASALIAS, which requires the range entity type ALIAS for the queried entity.

5. *Unifying identical entities*: this step groups the occurrences of the same entity into one. For example, in the query *"Who is the president of Bolivia?"* there are two identical entities represented by *Who* and *"the president"* to be grouped.

6. *Discovering implicit relations*: this step adds in relations that are not explicitly expressed by words in a query. For example, in the query *"What county is Modesto, California in?"*, there is an implicit relation between *Modesto* and *California*, meaning the former is a sub-region of the latter.

7. *Determining the types of relations*: this step maps the extracted relational phrases and discovered implicit relations to the appropriate relation types in the ontology of discourse. For instance, in the example query *"When was Microsoft established?"*, ESTABLISHMENTDATE is the suitable relation type for the relational word *"established"* in this query about time.

8. *Removing improper relations*: this step checks and removes improper relations constructed between entities in the previous steps. For example, in the query *"What city in Florida is Sea World in?"*, there are three entities represented by *"city"*, *"Florida"*, and *"Sea World"*. Without strict syntactic parsing, relations of the type LOCATEDIN may connect both the first two entities and the third one. However, that relation between *"Florida"* and *"Sea World"* is redundant in this case.

9. *Constructing the final CG*: this final step produces the CG corresponding to a query with respect to the ontology and KB of discourse.

Details of these steps are presented next.

3.1 Recognizing Specified Entities

There are various tools with respective ontologies and KBs that can be used for NE recognition, such as GATE, KIM, SemTag ([6]), ESPotter ([18]). Obviously, the performance of any system relying on named entities to solve a particular problem incurs errors of the NE recognition tool employed. However, in research for models or methods, the two problems should be separated. This work is not about NE recognition and we use GATE's semantic annotation tool OCAT and KIM's ontology PROTON and KB for experiments.

3.2 Recognizing Unspecified Entities

Unspecified entities are those that are not expressed by names in a query. However, they are represented by phrases implying entity types and thus can be recognized. We employ the ANNIE tool of GATE for this task by building a gazetteer of phrases and their corresponding entity types in the ontology of discourse.

3.3 Extracting Relational Phrases

Words or phrases expressing relations between entities are propositional and verbal ones like *"in"*, *"on"*, *"of"*, *"born"*, *"has"*, *"is"*, *"located in"*, etc. They can also be extracted by

ANNIE based on a gazetteer of phrases and their possible corresponding relation types in the ontology of discourse. For example, *"publish"* in a question can be mapped to the relation type DATEPUBLISH or HASPUBLISHER, and the suitable one depends on whether the question is about time (e.g. *"When was the first Wall Street Journal published?"*) or not (e.g. *"What company published Jasper Fforde's first book?"*).

3.4 Determining the Type of Queried Entities

The type of the entity represented by the interrogative word *What* (or *Which*) is determined by the following rules:

1. If *What* is followed by an entity type, then the type of the queried entity is that entity type. For example, in the query *"What province is Montreal in?"*, the word *"province"* specifies that the type of the queried entity is PROVINCE in the ontology of discourse.
2. Otherwise, the type is determined by the first NE after *What* and the relational phrase at the end of the query. For example, in the query *"What does Knight Ridder publish?"*, *Knight Ridder* is recognized as a company and the word *"publish"* entails that the queried entity is of the type PUBLISHEDMATERIAL.

The interrogative word *Who* may represent either a person or an organization. For example, in the query *"Who wrote the book Huckleberry Finn?"*, it represents a person. However, in the query *"Who provides telephone service in Orange County, California?"*, it means an organization. The appropriate entity type is determined on the basis of the involved relational phrases (e.g. *"wrote"* or *"provides"* in these examples) and the types of the entities after them (e.g. the book *"Huckleberry Finn"* or the service *"telephone"*).

Questions with the interrogative word *How* has two typical patterns. The first one is with an adjective to ask about a certain property of an entity. An example query of this pattern is *"How tall is the Sears Building?"*. Values of such properties are often represented by strings of the type STRING in an ontology like PROTON. In this example, the adjective is mapped to the corresponding property type HASHEIGHT. The second pattern is with *"many"* followed by an entity type to ask about the number of entities of that type involved in some relation. An example query of this pattern is *"How many counties are in Indiana?"*. Answering to such a query requires counting. One exception is queries asking about the population of a country, e.g. *"How many people live in Chile?"*, which is mapped to the property type POPULATIONCOUNT.

Time is also often represented by strings in data and knowledge bases. So, the interrogative word *When* in a query is mapped to a concept of the type STRING. For example, the signature of the relation type ESTABLISHMENTDATE is (ORGANIZATION, STRING).

3.5 Unifying Identical Entities

Two entities are considered as identical and unified under the following conditions:

1. One of them is an unspecified entity, and
2. The type of the unspecified entity is the same as, or a super-type of, the other entity, and

3. Between the two entities is the verb *"be"* in a particular form and tense such as *"is"*, *"are"*, *"was"*, *"were"*, etc.

For example, in the query *"Who is the president of Bolivia?"*, *Who* represents an unspecified entity of the type PERSON and *"president"* represents an entity of the type PRESIDENT, which is a subtype of PERSON. There is the relational word *"is"* between the two entities, so they are identical and can be unified.

3.6 Discovering Implicit Relations

If two entities are next to each other or separated by a comma, then there is an implicit relation between them. That relation is determined by the types of the entities and the relation types permitted for those two entity types in the ontology of discourse. For example, in the query *"In which US states has Barack Obama lived?"*, the type of *US* is COUNTRY and that of the unspecified entities represented by *"states"* is PROVINCE. Therefore, the appropriate type of the implicit relation between them is SUBREGIONOF.

3.7 Determining the Types of Relations

After the previous steps, the specified entities, unspecified entities, and relational phrases in a query are already recognized. The remaining task is to determine which relational phrase is between which two of the entities and what is the type of that relation. First, we present our approach to determine the appropriate relation type for a certain relational phrase in a query, with respect to the ontology of discourse. Let P_R be the relational phrase representing the relation between two entities of the types C_1 and C_2, and S_1 and S_2 be the original strings representing the two entities. We define the following sets of possible relation types:

1. R_1 is the set of possible relation types that correspond to P_R in the built-in gazetteer of relational phrases. For example, if P_R = *"publish"*, then R_1 includes DATEPUBLISH and HASPUBLISHER.
2. R_2 is the set of possible relation types between the entity types C_1 and C_2 as given in the ontology of discourse. For example, if C_1 = ORGANIZATION and C_2 = PERSON, then R_2 includes HASEMPLOYEE and HASFOUNDER.
3. R_3 is the set of possible relation types with respect to S_1 and P_R. For example, in the query *"Who is the founder of the Wal-Mart stores?"*, S_1 = *"founder"* and P_R = *"of"*, which derives HASFOUNDER as a possible relation type between *Wal-Mart stores* and the queried entity.
4. R_4 is the set of possible relation types with respect to P_R and S_2. For example, in the query *"Who was Charles Lindbergh's wife?"*, P_R = *"'s"* and S_2 = *"wife"*, which derives HASWIFE as a possible relation type between *Charles Lindbergh* and the queried entity.

The suitable relation types are then constrained within $R_1 \cap R_2 \cap R_3 \cap R_4$. For efficiency, we incorporate and encode all of these constraints into rules mapping relational phrases to suitable relation types in the ontology of discourse.

Second, we note that the phrase representing the relation between two entities can stand in different positions relative to those of the entities:

1. *In the middle*: for example, in the query "*Where is the location of the Orange Bowl?*", the relational word "*of*" is in the middle of the two entities represented by "*location*" and "*Orange Bowl*".
2. *After*: for example, in the query "*What state is the Filenes store located in?*", the relational word "*located in*" is after the second entity represented by "*Filenes store*".
3. *Before*: for example, in the query "*In what country is Angkor Wat?*", the relational word "*in*" is before the first entity represented by "*country*".

Therefore, for each pair of entities in a query, it is first checked if the relational phrase in the middle forms a proper relation between the two entities. If not, the relational phrases after and before the two entities are further checked.

3.8 Removing Improper Relations

Let E_1, E_2, ..., and E_N be the entities occurring in the left-to-right order in a query. We propose the following heuristic rules to remove improper relations extracted in the previous steps:

1. If E_i and E_{i+1} ($1 \leq i \leq N-1$) are next to each other, then E_i has only a relation with E_{i+1}, and all relations if assigned to E_i and other entities will be removed. For example, in the query "*In which US states has Barack Obama lived?*" ($E_1 =$ "*US*", $E_2 =$ "*states*", $E_3 =$ "*Barack Obama*"), there are three following possible relations extracted in the previous steps:
   [PROVINCE: ?*x*]→(SUBREGIONOF)→[COUNTRY: *US*], and
   [PERSON: *Barack Obama*]→(LIVEIN)→[PROVINCE: ?*x*], and
   [PERSON: *Barack Obama*]→(LIVEIN)→[COUNTRY: *US*],
   but the last one is to be removed.
2. If E_i and E_{i+1} ($1 \leq i \leq N-1$) are separated by a comma, then E_{i+1} has only a relation with E_i, and all relations if assigned to E_{i+1} and other entities will be removed. For example, in the query "*Who provides telephone service in Orange County, California?*" ($E_1 =$ "*Who*", $E_2 =$ "*telephone service*", $E_3 =$ "*Orange County*", $E_4 =$ "*California*"), there are four following possible relations extracted in the previous steps:
   [COUNTY: *Orange*]→(SUBREGIONOF)→[PROVINCE: *California*], and
   [TELEPHONESERVICE: *x]→(HASPROVIDER)→[COMPANY: ?], and
   [TELEPHONESERVICE: *x]→(LOCATEDIN)→[COUNTY: *Orange*], and
   [TELEPHONESERVICE: *x]→(LOCATEDIN)→[PROVINCE: *California*],
   but the last one is to be removed.
3. If there is the relational symbol "'*s*" between E_i and E_{i+1} ($1 \leq i \leq N-1$), then E_i has only a relation with E_{i+1}, and all relations if assigned to E_i and other entities will be removed. For example, in the query "*What is the name of Neil Armstrong's wife?*" ($E_1 =$ "*name*", $E_2 =$ "*Neil Armstrong*", $E_3 =$ "*wife*"), there are three following possible relations extracted in the previous steps:
   [MAN: *Armstrong*]→(HASWIFE)→[WOMAN: *x], and
   [WOMAN: *x]→(HASALIAS)→[ALIAS: ?*y*], and
   [MAN: *Armstrong*]→(HASALIAS)→[ALIAS: ?*y*],
   but the last one is to be removed.

4. If an entity is assigned relations to more than one entity standing before it, then only the relation with the nearest unspecified entity is retained. For example, in the query *"What city in Florida is Sea World in?"* (E_1 = *"city"*, E_2 = *"Florida"*, E_3 = *"Sea World"*), there are three following possible relations extracted in the previous steps:

[CITY: ?*x*]→(SUBREGIONOF)→[PROVINCE: *Florida*], and

[COMPANY: *Sea World*]→(LOCATEDIN)→[CITY: ?*x*], and

[COMPANY: *Sea World*]→(LOCATEDIN)→[PROVINCE: *Florida*].

However, since the entity *Florida* is already identified, the entity *Sea World* actually modifies the identity of the queried city, rather than *Florida*. Therefore, the last relation above is redundant and to be removed.

4   Evaluation Experiments

We have tested the proposed method on the data set TREC 2002 with 440 queries. The test uses the PROTON ontology with about 300 entity types, 100 relation and property types, and KIM World KB with over 77,000 named entities. The correctness of each constructed CG is manually justified with respect to the employed ontology and KB and the actual meaning of the corresponding query in natural language. Translation errors may occur due to one of the following causes:

1. The employed NE recognition engine like GATE's does not recognize all the named entities in a query precisely and completely. We call this an *R-error*.
2. The ontology and KB of discourse lack certain entity types, relation types, or named entities mentioned in a query. We call this an *O-error*.
3. The current CG query language is not expressive enough to represent certain queries. We call this an *Q-error*.
4. The proposed method itself does not construct of a correct CG. We call this an *M-error*.

Table 1. shows the number and percentage of each error type on the data set. The type *Other* is for those queries that do not have interrogative words, e.g. *"Name an art gallery in New York"*. With the original PROTON and KIM KB, there are 47 *R-errors* and 269 *O-errors*. In order to test the actual accuracy of the proposed translation method, we have then manually corrected the wrongly recognized NEs due to GATE, and supplemented PROTON and KIM KB with 27 entity types, 72 relation types, and 288 named entities. After that, there are only 22 *O-errors* left. The method itself causes 15 *M-errors*. The number of *Q-errors* is 57 in both cases.

Since we have used TREC 2002 queries as samples for analysis, there might be some bias in development of the translation rules. Therefore, we have had another test of the proposed method on TREC 2007. Table 2. presents the results on this data set. Before the wrongly recognized NEs are fixed and the ontology and KB are further enriched, there are 22 *R-errors* and 273 *O-errors*. After the *R-errors* are fixed and 18 more entity types, 67 more relation types, and 63 more named entities are added, the number of *O-errors* is reduced to 59. There are more *Q-errors* for TREC 2007 as compared with those for TREC 2002. However, the defined translation rules are still robust with only 5 *M-errors*. If not counting queries with *O-errors* and *Q-errors*, then

Table 1. Performance of the proposed method on TREC 2002, before and after wrongly recognized named entities fixed and ontology and knowledge base enriched

a. *Before*

Query Type	Number of Queries	Correct CGs	R-errors	O-errors	Q-errors	M-errors
What	201	38	32	113	18	0
Which	3	0	0	3	0	0
Where	62	21	7	33	1	0
Who	67	5	2	48	12	0
When	45	2	5	31	7	0
How	38	1	1	20	16	0
Other	24	0	0	21	3	0
Total	440	67 (15.23%)	47 (10.68%)	269 (61.14%)	57 (12.95%)	0 (0%)

b. *After*

Query Type	Number of Queries	Correct CGs	R-errors	O-errors	Q-errors	M-errors
What	201	162	0	11	18	10
Which	3	2	0	0	0	1
Where	62	61	0	0	1	0
Who	67	54	0	0	12	1
When	45	33	0	4	7	1
How	38	18	0	4	16	0
Other	24	16	0	3	3	2
Total	440	346 (78.64%)	0 (0%)	22 (5%)	57 (12.95%)	15 (3.41%)

Table 2. Performance of the proposed method on TREC 2007, before and after wrongly recognized named entities fixed and ontology and knowledge base enriched

a. *Before*

Query Type	Number of Queries	Correct CGs	R-errors	O-errors	Q-errors	M-errors
What	173	20	8	117	28	0
Which	15	1	2	8	4	0
Where	13	3	2	8	0	0
Who	57	8	1	37	11	0
When	13	1	0	11	1	0
How	56	0	1	5	50	0
Other	118	5	8	87	18	0
Total	445	38 (8.54%)	22 (4.94%)	273 (61.35%)	112 (25.17%)	0 (0%)

b. *After*

Query Type	Number of Queries	Correct CGs	R-errors	O-errors	Q-errors	M-errors
What	173	120	0	23	28	2
Which	15	9	0	2	4	0
Where	13	9	0	2	0	2
Who	57	36	0	9	11	1
When	13	10	0	2	1	0
How	56	4	0	2	50	0
Other	118	81	0	19	18	0
Total	445	269 (60.45%)	0 (0%)	59 (13.26%)	112 (25.17%)	5 (1.12%)

the translation accuracies are $346/(346+15) \approx 96\%$ and $269/(269+5) \approx 98\%$ for TREC 2002 and TREC 2007, respectively.

On the basis of the experiment results, we now analyse and discuss on the above mentioned four types of translation errors and how they can be overcome. Firstly, *R*-errors solely depend on the accuracy of an employed NE recognition engine, whose improvement is a separate problem. Whereas, the proposed method is robust to the test data sets, so the small number of *M*-errors is not of primary concern now. The others, *O*-errors and *Q*-errors, are addressed below.

Completeness of an ontology and knowledge base

Certain missing named entities and entity types can be easily supplemented to cover entities occurring in queries on a particular domain. Meanwhile, there are two following shortcoming cases of relation types in an ontology to be considered:

1. *Qualitative properties of entities*: Values of some properties of an entity are represented by adjectives in natural language. For example, answering the query *"What famous model was married to Billy Joel?"* requires the ontology of discourse to be able to represent the fame property of models. One way is to define the class FAMOUSMODEL, for instance, for those models who are famous. It would create various subclasses of models for different degrees on a fame scale. Another way is to define the relation type FAMEPROPERTY whose domain class is MODEL and range class is STRING, for instance. This way would raise the problem of matching string values later on. A choice depends on a consistent design of the whole ontology right at the beginning.

2. *Non-binary relation types*: In practice, there are relations with arities greater than two. An example is the query *"What year did the U.S. buy Alaska?"*, where *"buy"* actually is a 3-ary relation of *U.S.*, *Alaska*, and the queried year. However, in ontology and KB languages, such as RDF and OWL, only binary relations are directly supported. So, in order to represent an *n*-ary relation, one way is to define a reified relation type, which is an entity type that has *n* binary relation types with *n* entity types of that relation[3]. Then, for instance, this example query can be represented by the following query CG:

$$[\text{COUNTRY: } U.S.] \leftarrow (\text{SUBJECT}) \leftarrow [\text{BUY: *}] \rightarrow (\text{OBJECT}) \rightarrow [\text{PROVINCE: } Alaska]$$
$$\downarrow$$
$$(\text{TIME})$$
$$\downarrow$$
$$[\text{YEAR: ?}]$$

Correspondingly, our proposed method needs to be extended to recognize if a relation in a query is reified or not.

Expressiveness of the CG query language

In this paper, the introduced CG query language is still simple and can represent only basic relational constraints. It is not expressive enough to represent queries like *"How many counties are in Indiana?"* or *"What is the longest suspension bridge in the U.S.?"*. For the former query, the current language can represent the relation between a county and *Indiana* by the CG:

---

[3] http://www.w3.org/TR/swbp-n-aryRelations/

[COUNTY: *]→(SUBREGIONOF)→[PROVINCE: *Indiana*]

For the latter query, the current language can represent the relation between a suspension bridge and *U.S.* by the CG:

[STRING:*]←(HASLENGTH)←[SUSPENSIONBRIDGE:?]→(LOCATEDIN)→[COUNTRY:*US*]

To obtain complete query CGs, it needs aggregate functions such as COUNT and MAX applied to these basic CGs, respectively. Queries involving logical disjunction and negation connectives are also out of the scope of this paper.

5 Conclusion

We have presented our method to translate natural language queries into conceptual graphs with three following salient points. First, the method does not require grammatically correct querying sentences. In fact, it only concerns what are entities and relations in a query, whereas their relative positions are not too important. Second, it exploits an ontology to identify entities and their respective relations in a query. Since the ontology constraints valid relation types between certain entity types, it makes the method robust to ill-formed queries, not too dependent on relative positions of relations and entities. Third, we use conceptual graphs as the target formal language for the translation. With smooth mapping to and from natural language, conceptual graphs simplify the translation rules. As an interlingua, conceptual graphs can also be further translated to other formal query languages.

The experiment statistics show that the proposed method is robust to diverse structures and contents of questions in the test data sets, provided that the ontology and knowledge base of discourse cover well entities and relations in the domain. Still, as analysed above, to handle more query patterns, the ontology needs to be enriched to support qualitative properties and *n*-ary relations. The constructed translation rules then need to be revised to recognize relations that are reified in a query. Furthermore, the introduced CG query language needs to be extended with aggregate functions and logical connectives. These are the primary issues that we are currently investigating.

References

1. Berners-Lee, T.: Conceptual Graphs and the Semantic Web (Initially created: January 2001, Last change: April 2008), http://www.w3.org/DesignIssues/CG.html
2. Cao, T.H., Do, H.T., Pham, B.T.N., Huynh, T.N., Vu, D.Q.: Conceptual Graphs for Knowledge Querying in VN-KIM. In: Contributions of the 13th International Conference on Conceptual Structures, pp. 27–40. Kassel University Press (2005)
3. Corby, O., Dieng-Kuntz, R., Faron-Zucker, C.: Querying the Semantic Web with Corese Search Engine. In: Proceedings of the 3rd Prestigious Applications Intelligent Systems Conference (2004)
4. Cimiano, P., Haase, P., Heizmann, J.: Porting Natural Language Interfaces between Domains – An Experimental User Study with the ORAKEL System. In: Proceedings of the 12th ACM International Conference on Intelligent User Interfaces, pp. 180–189 (2007)
5. Cunningham, H., et al.: Developing Language Processing Components with GATE Version 3 (a User Guide). University of Sheffield (2006)

6. Dill, S., et al.: SemTag and Seeker: Bootstrapping the Semantic Web via Automated Semantic Annotation. In: Proceedings of the 12th International Conference on the World Wide Web, pp. 178–186 (2003)
7. Dobrev, P., Toutanova, K.: CGWorld - architecture and features. In: Priss, U., Corbett, D.R., Angelova, G. (eds.) ICCS 2002. LNCS, vol. 2393, pp. 261–270. Springer, Heidelberg (2002)
8. Hensman, S., Dunnion, J.: Using linguistic resources to construct conceptual graph representation of texts. In: Sojka, P., Kopeček, I., Pala, K. (eds.) TSD 2004. LNCS, vol. 3206, pp. 81–88. Springer, Heidelberg (2004)
9. Kaufmann, E., Bernstein, A., Fischer, L.: A Naïve but Domain-Independent Natural Language Interface for Querying Ontologies. In: The 4th European Semantic Web Conference, pp. 1–2 (2007)
10. Kiryakov, A., Popov, B., Terziev, I., Manov, D., Ognyanoff, D.: Semantic Annotation, Indexing, and Retrieval. Journal of Web Semantics 2 (2005)
11. Lei, Y., Uren, V.S., Motta, E.: SemSearch: A search engine for the semantic web. In: Staab, S., Svátek, V. (eds.) EKAW 2006. LNCS, vol. 4248, pp. 238–245. Springer, Heidelberg (2006)
12. Martin, P., Eklund, P.: Knowledge Representation, Sharing and Retrieval on the Web. In: Zhong, N., Liu, J., Yao, Y.Y. (eds.) Web Intelligence, pp. 243–276. Springer, Heidelberg (2003)
13. Sowa, J.F.: Conceptual Structures – Information Processing in Mind and Machine. Addison-Wesley, Reading (1984)
14. Sowa, J.F.: Matching Logical Structure to Linguistic Structure. In: Houser, N., Roberts, D.D., Van Evra, J. (eds.) Studies in the Logic of Charles Sanders Peirce, pp. 418–444. Indiana University Press (1997)
15. Tablan, V., Damljanovic, D., Bontcheva, K.: A natural language query interface to structured information. In: Bechhofer, S., Hauswirth, M., Hoffmann, J., Koubarakis, M. (eds.) ESWC 2008. LNCS, vol. 5021, pp. 361–375. Springer, Heidelberg (2008)
16. Yao, H., Etzkorn, L.: Conversion from the Conceptual Graph (CG) Model to the Resource Description Framework (RDF) Model. In: Contributions of the 12th International Conference on Conceptual Structures, pp. 98–114 (2004)
17. Zhang, L., Yu, Y.: Learning to generate cGs from domain specific sentences. In: Delugach, H.S., Stumme, G. (eds.) ICCS 2001. LNCS, vol. 2120, pp. 44–57. Springer, Heidelberg (2001)
18. Zhu, J., Uren, V.S., Motta, E.: ESpotter: Adaptive named entity recognition for web browsing. In: Althoff, K.-D., Dengel, A.R., Bergmann, R., Nick, M., Roth-Berghofer, T.R. (eds.) WM 2005. LNCS, vol. 3782, pp. 518–529. Springer, Heidelberg (2005)

Snippet Generation for Semantic Web Search Engines

Thomas Penin[1], Haofen Wang[1], Thanh Tran[2], and Yong Yu[1]

[1] Department of Computer Science & Engineering
Shanghai Jiao Tong University, Shanghai, 200240, China
{tpenin,whfcarter,yyu}@apex.sjtu.edu.cn
[2] Institute AIFB, Universität Karlsruhe, Germany
{dtr}@aifb.uni-karlsruhe.de

Abstract. With the development of the Semantic Web, more and more ontologies are available for exploitation by semantic search engines. However, while semantic search engines support the retrieval of candidate ontologies, the final selection of the most appropriate ontology is still difficult for the end users. In this paper, we extend existing work on ontology summarization to support the presentation of ontology snippets. The proposed solution leverages a new semantic similarity measure to generate snippets that are based on the given query. Experimental results have shown the potential of our solution in this problem domain that is largely unexplored so far.

Keywords: Snippet, Ontology summarization, Semantic measure.

1 Introduction

More and more ontologies are available on the Semantic Web. A significantly growing part is concerned with specific domains comprising ontologies designed to be useful to companies. To develop a semantic application, an engineer can draw from a large set of reusable ontologies. However, a main question remains: how to find and select the exact ontology matching the requirements? While the retrieval of potential candidate ontologies can be conveniently achieved through semantic search engines such as Sindice[1], Falcons[2], Swoogle[3], Watson[4], etc., the final selection still presents to be a difficult problem.

Our engineer can be considered as a *content curator* [1]. While he may be an expert in his domain, he does not necessarily have a deep understanding of Semantic Web technologies. When considering the result page of a search engine, his concern is to find out how these documents representing ontologies entail the query, which topics are covered, and if there are classes missing or that should not be included in the solution.

---

[1] http://www.sindice.com/
[2] http://iws.seu.edu.cn/services/falcons/objectsearch/index.jsp
[3] http://swoogle.umbc.edu/
[4] http://watson.kmi.open.ac.uk/WatsonWUI/

J. Domingue and C. Anutariya (Eds.): ASWC 2008, LNCS 5367, pp. 493–507, 2008.
© Springer-Verlag Berlin Heidelberg 2008

To avoid the burden of downloading all potentially interesting documents and to be confronted to the even more laborious process of opening them with an ontology visualization tool to examine their internal organization, most of the search engines offer certain facilities to get an idea of the ontology content from the result page. Sindice provides for example the main topic and metadata that can be explored by the user. Falcons associates labels with every document, while Swoogle displays an extract of the classes' names related to the searched terms. Watson offers a list of instances and classes matching the query.

Despite these features, these systems do not seem to exactly match the needs of the engineer. Sindice is limited to one topic per ontology and often only presents the document title. While the extracted classes' names in Swoogle can provide useful information, they can not serve as an adequate overview of the given ontology. Labels provided by Falcons are simply derived from the file name while Watson does not consider other resources than those related to the query.

To address the needs of our engineer, we propose a new snippet generation system for semantic web documents (ontologies), that brings the following contributions:

- A new measure for assessing the similarity of RDF sentences which is exploited for topic identification. Contrary to most of the measures we are aware of, this measure does not limit its scope to nouns from entities and class names but includes verbs and adjectives from triple's subjects, predicates, and objects. It does also not only consider sentences as bags of words, but use their internal structure to improve its accuracy.
- An extension of the work of [2] on ontology summarization and [3] on semantic similarity between words, showing that both can be successfully applied to the problem of snippet generation.
- A snippet generation system that can be tested through the web interface at http://snippet.apexlab.org.
- A user evaluation shows that our approach to snippet generation brings about promising results.

This paper will be organized as follows. In section 2, we will describe the elements and structures that will be used to define our semantic similarity measure in section 3. Section 4 will then describe the snippet generation process and present the structure of our system and its test interface. The quality and the efficiency of our solution will be explored in section 5. Finally, section 6 will discuss the related research and section 7 will conclude about the future work.

2   RDF Sentence Graph and Topic Graph

In this section, we present the definitions of RDF sentences and RDF sentence graphs, as discussed in [2]. Then, we extend this work to define RDF topics and RDF topic graphs.

Let O be an ontology, we call T the set of its RDF triples and B the set of its blank nodes. For $t \in T$, we note by $subj(t)$, $pred(t)$ and $obj(t)$ the subject, the predicate and the object of t respectively. We define the set of triples of O that contains blank nodes by $T_B = \{t \in T,\ subj(t) \in B$ or $obj(t) \in B\}$. It is clear that $T_B \subseteq T$. For $b \in B$, let $T_b \subseteq T_B$ be the subset of triples containing b.

We say that $t_i, t_j \in T$ are b-connected if they satisfy one of the following conditions:

- $\exists\, b \in B$ such that $t_i,\ t_j \in T_b$;
- $\exists\, t_k \in T$, such that t_i and t_k are b-connected and t_j and t_k are b-connected.

Definition 1. *An RDF sentence s is a set of triples such that:*

1. *$\forall i, j \in s$, i and j are b-connected;*
2. *$\forall i \in s, \forall j \notin s$, i and j are not b-connected.*

Intuitively, a RDF sentence is formed by a main RDF statement and all the other RDF statements b-connected to it. In particular, a RDF statement whose subject is not a blank node is called a main RDF statement. We call S the set of RDF sentences of O. For all $s \in S$, we define:

- $Subj(s) = \{subj(t)$ such that $t \in s$ and $subj(t) \notin B\}$;
- $Pred(s) = \{pred(t)$ such that $t \in s\}$;
- $Obj(s) = \{obj(t)$ such that $t \in s$ and $obj(t) \notin B\}$.

A reason to choose sentences rather than triples is to avoid the case of subjects and objects that are blank nodes. This would have made it more difficult to paraphrase them with natural language.

RDF sentences preserve the connections of the initial RDF graph. In order to rank RDF sentences, we can define links between them, based on their common nodes. [2] proposed to consider two kind of links (*sequential* – predicate or object of a sentences being the subject of another – and *coordinate* – several sentences with the same subject) and defined a parameter p to assign importance to each of them.

We reuse these notions of links to obtain what is called an *RDF sentence graph*. Like [2], we define the weight of the link from $s_1 \in S$ to $s_2 \in S$ by

$$w(s_1, s_2) = p * seq(s_1, s_2) + (1 - p) * cor(s_1, s_2), \tag{1}$$

where $seq(s_1, s_2)$ and $cor(s_1, s_2)$ are equal to 0 or 1, respectively depending on the existence or not of a sequential and a coordinate link between both sentences.

In an ontology, it is often possible to identify several topics. In a text document, a topic can be defined as a set of words or sentences sharing a certain semantic proximity. Similarly, we extend [2] and come with the following definition:

Definition 2. *Given a threshold θ, an RDF topic is defined as a set of RDF sentences such that the pairwise semantic similarity between these sentences is greater than or equal to θ.*

A definition of the similarity considered here is given in section 3.

As for sentences, we build an *RDF topic graph*. Let T_1 and T_2 be two topics and $|T_i|$ the number of sentences of T_i. Using equation 1, we define the weight of the link between T_1 and T_2 by

$$w(T_1, T_2) = \frac{1}{|T_1| + |T_2|} * \sum_{(s_i, s_j) \in T_1 \times T_2} w(s_i, s_j). \tag{2}$$

The main idea behind this definition is to be able to transfer the weight of the links between sentences to their topics. Note that we make use of an average weight.

3   Semantic Similarity Measure

Our main goal is to derive the aboutness of an ontology, i.e. the topics covered by an ontology. To do so, there is the need for a semantic similarity measure that can be used to decide whether two given RDF sentences belong to the same topic or not.

There already exist a lot of measures computing the similarity between onto-logical structures in order to achieve tasks such as ontology selection or alignment. Two commonly taken approaches are a comparison at the conceptual level (ontological structure) or at the lexical level (vocabulary). In some work, e.g. [4], both these levels are leveraged. Among the drawbacks of comparing ontologies at the conceptual level, there is the fact that structures depend on the way the ontologies are built. Patterns can differ and do not always reflect the complete meaning since concepts are usually not only expressed by logical structures but also by the choice of the vocabulary. Moreover, it is often required to consider structures complex enough to draw meaningful conclusions. To improve the relevance, works like [4] introduce lexical level comparison, using string [5] or semantic [6,3,7] similarity metrics. Such propositions rely however on taxonomies and often only compare class and instance names, and consider them as simple bags of words.

We propose another solution gathering both approaches, based on the definition of RDF sentence given in section 2. This choice allows a comparison between structures that have the same pattern whatever the ontology and whatever its size through the definition of subjects, predicates and objects for RDF sentences. Conceptualization is embedded within each sentence and concepts are not limited to class and instance names. Instead, the whole content of the ontology is available. Moreover, as explained in the current section, sentences are not limited to bags of words and their internal semantics is used to measure their similarity. Rather than relying on a basic string similarity, we use the semantic similarity defined between two words by [3]. Our choice was based on the performance of their solution compared with other propositions like [6] and [7]. However, we slightly extended its functionalities to be able not only to compare nouns, but also verbs, adjectives and adverbs, which is something commonly ignored by the state-of-the-art.

To design our measure, we try to apply some principles proposed by [8]: consider both commonalities and differences between sentences and ensure that the maximum is reached when they are identical. [5] gave other interesting features. A measure shall be fast (polynomial), stable, intelligent and discriminating. Section 5 shows how much we achieved these goals.

3.1   Similarity between Two Lists of Words

To compare sentences, we first need to be able to compute similarity between lists of words, considered as bags of words. Lists of words may or may not share common concepts and can have different length. Our idea, following [8], is to consider both commonalities and differences.

Let w be a given word and $w_1, \cdots, w_n$ be a series of words belonging to a list L. We define the semantic similarity between w and L by

$$sim(w, L) = \max\{sim(w, w_i) \text{ for } i = 1..n\}, \tag{3}$$

where $sim(w, w_i)$ is the result returned by the measure defined by [3].

The word w is related to concepts that may or may not be found in L. By computing the similarity with any word of L, we will get a high score if the list contains a concept close to w. If not, the score will be low, even with the max function. The maximum allows to prove that w and L share commonalities or differences, without actually considering their respective weight.

To improve the efficiency of the measure, we decided not only to consider nouns but also verbs, adjectives and adverbs. We used the WordNet taxonomy to find the noun and the verbal base corresponding to each given adjective, adverb or conjugated verb. This approach consisting to only compare nouns and verbal base can be considered as a certain loss of semantic meaning. However, we consider that it is far better than simply ignore these words and that a significant part of the meaning is still conveyed by the nouns.

Let L_1 and L_2 be two lists of words with respective length n_1 and n_2. Let w_{ij} be the ith element of the list L_j. The semantic similarity between L_1 and L_2 is defined by

$$sim(L_1, L_2) = \frac{1}{n_1 + n_2} \left(\sum_{i=1}^{n_1} sim(w_{i1}, L_2) + \sum_{j=1}^{n_2} sim(w_{j2}, L_1) \right), \tag{4}$$

where sim is defined by equation 3.

Now, if k pairs of words share the same semantic meaning, they will account for k times in the overall semantic similarity, which ensures that the weight of the different concepts is taken into consideration. We can see that if a concept is only present in one list and not in the other, it will tend to reduce the overall result. Finally, the average function keeps our measure between 0 and 1. We can clearly see that if the two lists have nothing in common, their similarity will be 0 and that it will be 1 if and only if they exactly share the same concepts.

3.2   RDF Sentence Semantic Similarity

According to section 2, an RDF sentence s can be seen as a virtual triple constituted of three elements: its subject $Subj(s)$, its predicate $Pred(s)$, and its object
$Obj(s)$. Each of these elements is a list of words. An immediate benefit is that
it is possible to consider RDF sentences for what they really are: sentences that
have a meaning, and that similarly to natural language sentences have a subject
and an object linked by a predicate.

In natural languages however, subject, predicate, and object do not share
the same importance. This seems to offer interesting perspectives if we consider
the opportunity to modify the different coefficients to give to different possible
matches: subject-subject, object-object, predicate-predicate, or subject-object.
Subject-object comparison is important to apprehend chiasms. Consider the
sentences *Alice hasMother Anne* and *Anne hasDaughter Alice*. Direct subject-
subject and object-object matches would lead to the conclusion that they are
not semantically close, which is a mistake. Comparing their subjects with their
objects shows that they are in fact very similar.

We define w_{ss}, w_{pp}, w_{oo}, and w_{so} as being the respective weights attributed
to subject-subject, predicate-predicate, object-object, and subject-object comparisons. To ensure that our measure will stay between 0 and 1, we assume
that

$$w_{ss} + w_{pp} + w_{oo} = w_{pp} + 2 * w_{so} = 1. \tag{5}$$

Let s_1, $s_2 \in S$. Let $sim_{ss}(s_1, s_2)$, $sim_{pp}(s_1, s_2)$, $sim_{oo}(s_1, s_2)$, $sim_{s_1 o_2}(s_1, s_2)$,
and $sim_{s_2 o_1}(s_1, s_2)$ be respectively the similarity between $Subj(s_1)$ and $Subj(s_2)$,
$Pred(s_1)$ and $Pred(s_2)$, $Obj(s_1)$ and $Obj(s_2)$, $Subj(s_1)$ and $Obj(s_2)$, and
$Subj(s_2)$ and $Obj(s_1)$. Since subjects, objects and predicates are lists of words,
this similarity is computed like explained in section 3.1.

Definition 3. *To be able to handle chiasms, we consider that if $sim_{ss}(s_1, s_2) +
sim_{oo}(s_1, s_2) \geq sim_{s_1 o_2}(s_1, s_2) + sim_{s_2 o_1}(s_1, s_2)$, the semantic similarity between
s_1 and s_2 is defined by*

$$sim(s_1, s_2) = w_{ss} * sim_{ss}(s_1, s_2) + w_{pp} * sim_{pp}(s_1, s_2) + w_{oo} * sim_{oo}(s_1, s_2).$$

Otherwise, it is defined by

$$sim(s_1, s_2) = w_{so} * (sim_{s_1 o_2}(s_1, s_2) + sim_{s_2 o_1}(s_1, s_2)) + w_{pp} * sim_{pp}(s_1, s_2).$$

In our system, very common words such as RDF keywords are excluded, since
they can give a high similarity to sentences that significantly differ in their
meaning.

3.3   Topic Similarity

Topics are lists of semantically close RDF sentences. As such, semantic similarity
between two topics is obtained like that between two lists of words:

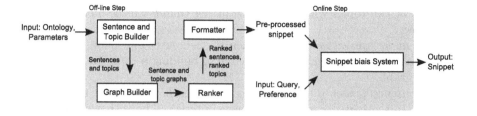

Fig. 1. Architecture of our snippet generation system

Definition 4. *Let T_1 and T_2 be two topics. Let L_1 and L_2 be the lists of their RDF sentences, with respective length n_1 and n_2. Let s_{ij} be the ith element of the list L_j. The semantic similarity between T_1 and T_2 is defined by*

$$sim(T_1, T_2) = \frac{1}{n_1 + n_2} \left(\sum_{i=1}^{n_1} sim(s_{i1}, L_2) + \sum_{j=1}^{n_2} sim(s_{j2}, L_1) \right)$$

where the similarity between a sentence and a list of sentences is the maximum similarity between this sentence and all the sentences of the list.

4   Snippet Generation Process

As explained in section 1, our system was designed to give the user ways to quickly find out whether an ontology suits his needs or not. It required the determination of the different topics and a ranking of both topics and RDF sentences in a snippet matching user preference. The determination of the different topics is achieved thanks to a hierarchical clustering algorithm. This choice was made since no knowledge of the number of topics is required in advance (the topic threshold is enough) while still being simple to implement. Some performance issues (see section 5) were however raised by experimental results, making it incompatible with the generation of a snippet that should almost instantly displayed once the results found by the search engine are known. Operations needed are more complex than in the case of traditional snippets for text documents [9].

Since most of the work of ontology summarization is unrelated to the query provided by the user, we increased the response time of our interface by splitting the process into two steps: one off-line and one online. The parsing of the ontology file is achieved by the Jena library[5].

The figure 1 shows the general organization of our system.

4.1   Off-Line Step

The *sentence builder* creates RDF sentences from the triples of an ontology. The *topic builder* gathers semantically similar sentences, given a threshold θ. Two

---

[5] http://jena.sourceforge.net/

sentences or two topics with a similarity higher than θ will be merged into the same topic.

The *sentence graph builder* takes then the sentences and build the sentence graph. The *topic graph builder* is in charge of building a topic graph using the sentence graph.

The *sentence ranker* ranks sentences within each topic, using the in-degree centrality in the sentence graph as salience criteria. [2] have shown that the in-degree centrality gave the best ranking results as soon as $p \geq 0.3$ in the sentence graph (see section 2). We take $p \geq 0.7$ and apply their algorithm. Even if a module of our demo system allows the user to change p, this is in practice not necessary. The *topic ranker* ranks the topics considering the topic graph and the in-degree centrality of its nodes.

The *formatter* takes the RDF sentences and apply natural language processing techniques in order to make them more easily readable by the final user. The result is output to the disk for further use. In our demonstration system, pre-processed snippet are stored under XML format and loaded for the online step of the snippet generation process.

4.2 Online Step

The *snippet bias* is applied when the snippet is generated. Sentences matching the user's query are selected. It is made sure that they will be visible in the final snippet, while respecting the user preference. As described in the section 4.4, several options are available to customize both the length of the snippet as well as the relevance of its content.

4.3 Natural Language Output

To improve the readability of the snippet, a system inspired from [10] and [11] was implemented to generate NL sentences. RDF sentences themselves are transformed into triples composed of a subject, a predicate and an object, obtained by considering all possible path in the sentence and by aggregating successive predicates.

RDF keywords are replaced by more natural formulations. X subClassOf Y becomes for instance X is a kind of Y. Sentences matching certain patterns are also transformed. The patterns are obtained by parsing the sentence using WordNet.

4.4 Snippets

In section 1, we considered some questions that the snippet shall answer.

For the user to identify the topics, they are clearly separated, their respective weight is given and their importance is shown by their rank. Within each topic, sentences are written using NL techniques and ranked according to their salience. The user has the possibility to choose the maximum number of topics

☐ Uploaded snippet   132 triples - 129 sentences

☐ Topics (3/4)			
Rank	Weight	Sentences	Sorted extract with sentence rank
1	2 %	2	1. A Category Travel *is a* concept.
			2. An Use it *is a subject of* category travel.
2	67 %	86	1. A Category Travel *is broader than* category service industries.
			2. A Category Travel *is broader than* category leisure.
			3. A Category Travel *is broader than* category transportation.
3	2 %	2	1. A Premium economy *is a subject of* category travel.
			2. An Economy class *is a subject of* category travel.

☐ Uncategorized sentences

Fig. 2. Sample snippet for `Travel.rdf` for the query *travel*

and sentences per topic to display. Different degrees of summarization can then be proposed. Finally, only query-related topics and sentences can be displayed or a rule can be applied, ensuring that the most salient topics and sentences will always be shown as soon as at least half of them are related to the query.

Figure 2 shows an example of snippet.

4.5   Test Interface

A demonstration system was implemented[6] as a Java web application. It simulates a search engine result page for predefined queries, allows the upload of an ontology, the personalization of all the parameters of the process to get a query-biased snippet and provides a test component for our semantic similarity measure.

5   Evaluation

To assess the quality of the system, we successively investigated the performance of our similarity measure, the quality of the snippets and to what extend the solution was promising from a user point of view.

The tests were carried out through our demonstration interface (see section 4.5) on our local gigabyte network. The server was a 2.4 GHz personal computer, with 1 GB memory, running Microsoft Windows XP.

Our test panel was composed of nine members from our laboratory. They have a certain familiarity with ontologies without being experts, and as such were more interested in technical details than the end-users targeted by our solution. This choice was made because the technical capacities required by the tests did not match those of casual end-users. Since our system is not yet integrated into a functional semantic search engine, we could not really test according to the use case as defined in section 1. As a result, the bias was not considered as too important.

---

[6] `http://snippet.apexlab.org`. Tested with Firefox.

5.1   Semantic Similarity Measure

This evaluation aimed both at assessing that our measure can match the appreciation of our test panel and at finding the right parameters to do so.

We selected six ontologies , as shown by table 1. To get close to the diversity of the real Semantic Web, they were chosen to exhibit different characteristics:

- *Topic number.* Since the overall semantic of the ontology is not considered to express the similarity between sentences, this aimed at investigating the stability of the measure and its coherence with the user opinion for different levels of homogeneity. We considered ontologies with more than 30 topics as having a high number of topics.
- *Vocabulary complexity.* It ranges from commonly used words to highly specific terms. This was thought to make sure that our measure stay coherent when considering sentences containing numerous unknown words. We considered vocabulary to be complex when more than 10% of the words were unknown.
- *The internal structure of the ontology.* While some contain complex relations between classes and entities, some offer a catalog-like organization (qualified here as simple). Ontologies from DBpedia for instance contain RDF sentences that often have very similar objects (category name). This should not harm the capacity of the measure to mimic human judgment.

Table 1. Composition and characteristics of the test set

Ontology	Topic Number	Vocabulary	Structure
AKTiveSAOntology.owl	High	Complex	Complex
animalsA.owl	Low	Simple	Complex
cv.rdfs	Low	Simple	Simple
History_of_China	Low	Complex	Simple
terrorism.owl	High	Complex	Complex
Travel-OilEdExportRDFS.rdfs	High	Simple	Simple

For each ontology, testers were asked to estimate the similarity between pairs of randomly-extracted sentences. They had the choice between *"nothing in common"*, *"somewhat related"*, *"rather similar"* and *"very close"*. Simultaneously, the system computed the similarity for different parameter configurations ♯1, ♯2, ♯3 and ♯4, defined with w_{ss}, w_{pp}, w_{oo} and w_{so} values respectively equals to 0.7, 0.1, 0.2 and 0.45 (strong subject, weak predicate), 0.3, 0.4, 0.3 and 0.3 (strong predicate), 0.45, 0.1, 0.45 and 0.45 (weak predicate) and 0.6, 0.2, 0.2 and 0.4 (strong subject).

Figure 3 (a) describes the opinion of the user and the system judgment for the different configurations. All results between 0.0 and 0.25 were considered as having *"nothing in common"*, between 0.25 and 0.5 as being *"somewhat related"*, between 0.5 and 0.75 as being *"rather similar"* and between 0.75 and 1.0 as being *"very close"*.

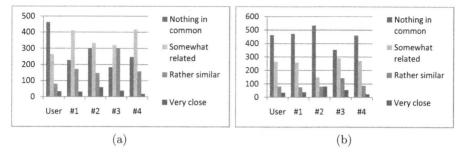

Fig. 3. Results (a) before and (b) after interval adjustment

The system appears more optimistic than users. This is apparently due to the max functions in our measure and to the fact that it does not know the context and may consider some words as semantically close even if the user disagrees. If we except the sentences having *"nothing in common"*, we can notice that configurations ♯1 and ♯4 have a behavior somewhat similar to that of the user.

We focused on these two candidates. To assess their quality, we refined our results – without changing user judgment – by redefining the intervals representing the similarity appreciation. Figure 3 (b) shows the results obtained for [0.0, 0.39],]0.39, 0.58],]0.58, 0.72] and]0.72, 1.0]. ♯1 and ♯4 did then match the appreciation of the users rather well. To check that they really were good parameters, we separately considered the results for the different ontologies. These are shown by figure 4.

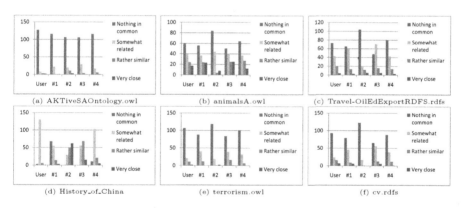

Fig. 4. Results after interval adjustment for each test ontology

Figure 4 (a) and (b) show that both configuration ♯1 and ♯4 behave well with different numbers of topics. The fact to not consider the overall semantics does not seem to play a too important role.

Results presented by (c) and (d) are more interesting, since they show a strong difference mainly due to vocabulary complexity. While ♯1 and ♯4 behave rather well with low term complexity, ♯1 appears to be more unstable than ♯4 when the system is confronted to unknown words. This indicates the good potential of ♯4.

Finally, (e) and (f) show that our measure is generally not affected by the internal complexity of the ontologies, and comfort the choice of ♮4, which mimics rather well human judgment. This comfort also our idea to consider all aspects of the triples and not only subjects, since ♮1 has more weight on subject-subject match than ♮4.

5.2 Snippet Quality

Using configuration ♮4, we considered three test activities. Users had to first find the best threshold for the topics. They were then asked to evaluate the quality of the clustering. Finally, we asked them to rank topics and sentences related to a query and compared their result with that of the system.

To find the best threshold, we choose three small ontologies and divided our users into groups of three. Each group was given an ontology. Each person worked alone and was provided a screen capture of the RDF graph obtained with a visualization plugin of Protégé[7]. After studying the ontology, they were asked to use the upload interface of our demo system repeatedly and to look for the threshold providing the best clustering. Their answers varied between 0.65 and 0.75, with an average of 0.71.

The quality of the clustering itself was assessed by a questionnaire. Testers should indicate on a four-level scale if the topics were conformed to the choice they would have made, if they estimated the topics to be coherent and what they thought about the overall quality of the clustering. A majority of users (5 out of 9) considered the result as conform to what they would have done. Others considered that they would probably have made some adjustments. The coherence of the topics gave approximately the same result, while everybody agreed on a good quality of the clustering, two users even considering it as very good.

It appeared that even with some slight clustering differences, the result of the biased snippets matched rather well the selection made by the users.

5.3 Performance

Performance for different ontologies is given by figure 5. We can notice the important time needed by the clustering phase w.r.t. the size. This justifies the off-line step but will also encourage further optimizations. The comparison between cv.rdfs and History_of_China points out the role of the vocabulary complexity. The more distinct words an ontology contains, the more access to the WordNet database are needed, which slows down both online and off-line steps. It appeared during our experiments that the time of the online module mostly depends on disk and network access speed. The fact to save the pre-processed snippet and to load it again represents up to a few seconds.

5.4 User Feedback

Our testers were asked some questions and were free to leave comments. While they all agreed on the readability and accuracy of the snippets, two users were

---

[7] http://protege.stanford.edu/

Ontology	Size	Triples	Sentences	Clustering	Total	Snippet
Cat_health.rdf	5 KB	25	22	3.484 s	4.171 s	0.141 s
animalsA.owl	8 KB	129	89	7.156 s	7.547 s	0.515 s
cv.rdfs	23 KB	419	248	17.03 s	18.343 s	1.110 s
History_of_China	52 KB	275	272	74.433 s	75.293 s	2.531 s
terrorism.owl	188 KB	2382	1438	737.472 s	754.156 s	0.187 s

Fig. 5. Clustering time, total pre-processing time and snippet generation time

not fully convinced, even if they did not deny the advantages brought by the system. One of them proposed to further investigate with a larger set of users and documents, which is in our opinion an interesting further step for our work along with its integration into a real search engine. Others were rather pleased with the potential of the proposed solution.

6   Related Work

To the best of our knowledge, query-biased snippet generation from ontologies is still a largely unexplored field. [12] recently proposed a solution to generate snippets based on term occurrence. Contrary to our approach, topic identification or similarity measure were not considered. Without precisely considering the case of snippets, [13] and [14] investigated ontology evaluation and selection, and illustrated the importance and the openness of this issue. Snippet generation is also not limited to ontologies and is still actively discussed. [9] proposed strategies to increase performance of snippet generation through document caching, which are interesting for further optimize our system, since disk storage and access have shown to cost up to a few seconds per snippet.

Document summarization and clustering are two fields closely related. Work on multiple text documents summarization, like [15] that use a centroid-based approach for topic determination, comes now along with ontology summarization like [2]. Our extension of the later with the addition of topics and topic graphs is inspired by techniques used in text summarization. Ranking ontological structure is also required to generate snippets of different length that still contain the essence of the ontology, as discussed by [2]. While we reuse their results, we also include the possibility to bias the ranking results according to the query provided by the user.

Similarity between concepts plays an important role in domains such as Information Retrieval, Natural Language Processing or even Genetics [16], and has been studied a lot in the literature. It also aims at facilitating ontology merging and aligning. [4] proposes an approach to compare ontology structures both from the conceptual and the lexical point of view. Even if we shared this idea, we did not separate both aspects like they did, since we considered structure and meaning to be closely related within the particular structure of RDF sentences. [8] has considered the similarity from a theoretical point of view and was an interesting methological help in the design process of our measure.

Different methods to measure semantic similarity between words are investigated by [7]. To outperform the traditional edge counting approach, [6] proposed a semantic similarity measure between terms using WordNet. Inspired by this work, [3] describes a new measure, used by our system[8]. However, its limitation to nouns and verbal bases brought us to add the possibility to consider adjectives, adverbs and conjugation as well. [17] designed a measure to compare different ontological structures, which differs from our approach since we do not only consider class names.

Finally, our system includes some characteristics of NL paraphrasing of ontologies, as investigated in [10] and proposed by [11]. The ideas described by this related work on NL go far beyond what we decided to implement but show what can be achieved in a near future.

7   Conclusion and Future Work

In this paper, we proposed a solution to the problem of ontology selection for non-specialists. Our system relies on a new semantic similarity measure, that exploits the semantics of structures called RDF sentences, does not limit its scope to class or entity names, and considers all words in the ontology resources without being limited to nouns and verbs. It also extends and gives an application to works like [2] and [3], and introduces a new and friendly way to consider results returned by semantic web search engines. A user evaluation assessed its potential and highlighted a few tracks for further development.

The next steps involve improvements in the clustering process, since our test data shows that it takes most of the running time. It is also planned to improve the natural language results, to be closer to the ideas expressed in [11]. The main future achievement will be the inclusion of the system into a real search engine, to further improve user support and to benefit from information such as relevance score computed by the engine.

References

1. Battle, L.: Preliminary inventory of users and tasks for the semantic web. In: 3rd Intl. Semantic Web User Interaction Workshop (2006)
2. Zhang, X., Cheng, G., Qu, Y.: Ontology summarization based on RDF sentence graph. In: Proceedings of the 16th international conference on World Wide Web (2007)
3. Seco, N., Veale, T., Hayes, J.: An intrinsic information content metric for semantic similarity in wordnet. In: Proceedings of 15th ECAI (2004)
4. Maedche, A., Staab, S.: Measuring similarity between ontologies. In: Proceedings of the European Conference on Knowledge Engineering and Knowledge Management (2002)
5. Stoilos, G., Stamou, G., Kollias, S.: A string metric for ontology alignment. In: International Semantic Web Conference (2005)

---

[8] http://eden.dei.uc.pt/~nseco/javasimlib.tar.gz

6. Resnik, P.: Semantic similarity in a taxonomy: An information-based measure and its application to problems of ambiguity in natural language. Journal of Artificial Intelligence Research 11 (1999)

7. Petrakis, E.G.M., Varelas, G., Hliaoutakis, A., Raftopoulou, P.: X-similarity: Computing semantic similarity between concepts from different ontologies. Journal of Digital Information Management (JDIM) (2006)

8. Lin, D.: An information-theoretic definition of similarity. In: Proc. 15th International Conf. on Machine Learning (1998)

9. Turpin, A., Tsegay, Y., Hawking, D., Williams, H.E.: Fast generation of result snippets in web search. In: Proceedings of the 30th annual international ACM SIGIR conference on Research and development in information retrieval (2007)

10. Hewlett, D., Kalyanpur, A., Kolovski, V., Halaschek-Wiener, C.: Effective nl paraphrasing of ontologies on the semantic web. In: Workshop on End-User Semantic Web Interaction, 4th Int. Semantic Web conference, Galway, Ireland (2005)

11. Wilcock, G.: Talking owls: Towards an ontology verbalizer. In: Proceedings of the ISWC Workshop on Human Language Technology for the Semantic Web and Web Services (2003)

12. Huang, Y., Liu, Z., Chen, Y.: Query biased snippet generation in XML search. In: Proceedings of the ACM SIGMOD International Conference (2008)

13. Sabou, M., Lopez, V., Motta, E., Uren, V.: Ontology selection: Ontology evaluation on the real semantic web. In: Proceedings of WWW (2006)

14. Alani, H., Brewster, C.: Ontology ranking based on the analysis of concept structures. In: Proceedings of the 3rd international conference on Knowledge capture (2005)

15. Radev, D.R., Jing, H., Styś, M., Tam, D.: Centroid-based summarization of multiple documents. In: Information Processing and Management, vol. 40 (2004)

16. Couto, F.M., Silva, M.J., Coutinho, P.M.: Semantic similarity over the gene ontology: Family correlation and selecting disjunctive ancestors. In: Proceedings of the CIKM Conference (2005)

17. Rodriguez, M.A., Egenhofer, M.J.: Determining semantic similarities among entity classes from different ontologies. IEEE Transactions on Knowledge and Data Engineering (2003)

Semantic Telecommunications Network Capability Services

Xiuquan Qiao, Xiaofeng Li, Tian You, and Lihao Sun

State Key Laboratory of Networking and Switching Technology,
Beijing University of Posts and Telecommunications, Beijing, 100876, China
{qiaoxq,xfli}@bupt.edu.cn, {youtian1985,leehow.sun}@gmail.com

Abstract. The providing of user-centric services in the B3G/4G network presents a great challenge in the service architecture. To narrow the semantic gap of Telecommunications Network and Internet in the service layer, we introduce the vision of User Centric Intelligent Service Environment (UCISE). The ubiquitous service ecosystem and the semantic service integration architecture of telecommunications network and Internet are proposed. To provide the semantic Telecommunications Network Capability Services (TNCS), we present a semantic description approach for TNCS by using the profile hierarchy of OWL-S. Then based on this approach, we demonstrate the cheapest click-to-call application case. In this way, the ontology-based accurate discovery, matching for telecommunication network capability services can be supported. This will facilitate the semantic convergence of telecommunications network and Internet in the service layer.

Keywords: User Centric Service, Semantic Parlay X, Semantic Parlay, Semantic Telecommunication Network Capability Service, Telecommunication Service Domain Ontology, Service-Oriented Architecture, Ontology, Semantic Web Service, OWL-S.

1 Introduction

The future ubiquitous convergent network is one user-centric harmony communication network. It will integrate various heterogeneous networks [1], [2]. The aim of ubiquitous convergent network is to provide user-centric pervasive service. As the goal of user-centric service is to facilitate the work and daily life of human, it will involve with various domain services, such as telecom domain services, financial services, and transport services. Service will become more intelligent. This requires that the intelligent service agent could accurately discover, automatically compose and invoke the services provided by telecommunication network or Internet based on the service context in the service-oriented architecture [3].

Therefore, the Telecommunications Network Capability Services (TNCS) like call control and message sending/receiving must be described in the semantic level to support the accurate service discovery. However, the open interface specifications (such as Parlay/OSA [4], Parlay X [5]) of telecommunication network are still in the syntactic description level. The WSDL-based open interface specifications lack a

J. Domingue and C. Anutariya (Eds.): ASWC 2008, LNCS 5367, pp. 508–523, 2008.
© Springer-Verlag Berlin Heidelberg 2008

semantic description for TNCS and contain no information about the capabilities about the described services. Therefore, the service matchmaking can only be done by string matching on the defined attributes such as name or address of service provider. It is very difficult for service agent to achieve the accurate discovery and automatic invocation of TNCS. So the current TNCS interface specifications based on WSDL are unable to support the provision of advanced intelligent service.

Today, semantic web service, as a new research paradigm, is generally defined as the augmentation of web service description through semantic annotation, to facilitate the higher automation of service discovery, composition, invocation and monitoring in an open environment. As a submitted draft to W3C, OWL-S [6] is OWL [7] -based ontology description framework for web service. The integration of Semantic Web Service (SWS) technology and telecommunications systems is currently the subject of intensive research. So far, a lot of research activities have been done in applying ontology and semantic web service technologies to mobile service domain. To solve the lack of QoS/QoE considerations of the existing Parlay X web service, Sungjune Hong [8], [9] presented the extended Parlay X with QoS/QoE for 4G network. In fact, the authors imported the context-awareness technology into the Parlay X gateway. Songtao Lin [10] proposed to provide VHE service by the Intelligent Mobile Service Platform based on semantic web technologies. Tomas Vitvar [11] discussed how semantic web services technology can facilitate dynamic and optimal integration of voice and data services with specific characteristics conforming users needs and preferences. Villalonga, C. [12] provided an overview of the Mobile Ontology based on the need for a standardized ontology that describes semantic models of the domains relevant for scalable NGN service delivery platforms. This work, as a part of IST SPICE project [13], is a meaningful attempt to establish a standardized ontology for mobile service delivery in NGN. However, this mobile ontology is used to address the semantic sharing among the distributed components of SPICE service platform. So the mobile ontology has its limitation. Additionally, Alistair Duke [14] explored the use of semantic web service within the Operational Support System (OSS). However, this approach only focuses on the telecommunications management.

From the above analysis, we can see that the existing work has not explored the semantic description problem of TNCS with the semantic web services and ontology technology. In addition, the semantic service integration architecture of telecommunications network and Internet in the service layer has not been discussed so far. In this article, we introduce the vision of User Centric Intelligent Service Environment (UCISE). To provide the semantic TNCS in UCISE, we present TelecomOWL-S for the semantic description of TNCS by using profile hierarchy technology of OWL-S. Moreover, telecommunication service domain ontology (TSDO) is proposed to support the nonfunctional properties description for TNCS. TSDO establishes the foundation of ontology-based reasoning and domain knowledge sharing. The proposed approach enhances the accuracy of telecommunications network capability service description and matching, and narrows the semantic gap between telecommunications network and Internet in the service layer.

The rest of this paper is structured as follows. In section 2 we introduce the vision of UCISE. Section 3 we present the complete semantic description approach for telecommunications network capability services. In section 4, we introduce the experimental environment and the demo service. Section 5 discusses the relations of the semantic Parlay X web service and the general Parlay X web service. Finally, conclusions are then drawn.

2 The Vision of User Centric Intelligent Service Environment (UCISE)

2.1 The Blueprint of UCISE

The goal of UCISE is to provide the user-centric personalized services in the ubiquitous convergent network environment. UCISE will provide a comfortable working and living space for users, and liberate human from trivial things to pursue more advanced activities. The loosely-coupled service resources and dynamic binding, seamless interoperability, and end-to-end network reconfiguration are the main characteristics of UCISE. The application layer can dynamically select the suitable services to compose based on the personalized user requirements and current context environment. According to the service functional and nonfunctional requirements, the lower network can dynamically perform the resource allocation and scheduling to complete the end-to-end network reconfiguration. The terminals can freely switch among the different access networks and achieve the seamless mobility. The intelligent terminals can also be real-time reconfiguration, as well as aggregation with the ambient devices.

2.2 Ubiquitous Convergent Service Ecosystem

To achieve the aim of UCISE, a ubiquitous convergent service ecosystem is presented. Service ecosystem (see Fig 1) is one harmony service providing and consuming mechanism, like as the natural food chain. Service providers publish their service in standard format, and service consumers can easily utilize these published service information to dynamically select and bind to the concrete service provider. In this way, the coarse granularity service assembly era comes into being.

As the goal of UCISE is to facilitate the work and daily life of human, it will involve with various domain services, such as telecom domain service, financial service, transportation service. But the heterogeneity is the main feature of various domain systems, service network middleware platform technology need to be used to screen the heterogeneity problem. Service network middleware platform is a kind of software that lies between domain system software and value-added application software. Unified domain services provided by service network middleware platform can be used to develop various domain-related or cross-domain value-added applications. For example, telecom service network middleware platform, like Parlay Gateway, can cover diversified foundation networks (such as, fixed telephone network, mobile telephone network, as well as various emerging networks, i.e. sensor network etc.) to provide cross-carriers, cross-networks, seamless-mobility telecom domain service.

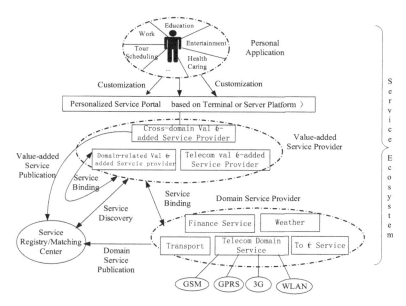

Fig. 1. Ubiquitous Convergent Service Ecosystem

After service providers have developed services, how can service consumers select and invoke these services according to their requirements? In order to bridge the gap between service providers and service consumers, one new service trading mechanism is needed. It's service-oriented computing architecture. Service providers (domain service providers and value-added service providers) publish their services to Service Registry, and Service Requestors find the needed services in Service Registry according to their requirements, and then bind to Service Provider according to the service descriptions information. In this way, value-added service providers can use all kinds of domain services or value-added services provided by other value-added service providers to compose user-oriented services, and then may publish them again. Thus, a virtual computing environment that supports dynamic resource allocation and heterogeneous network collaboration is formed. Nowadays, there are two problems needed to be resolved. Firstly, a trusty, controllable service management mechanism is absent; secondly, there is a great semantic gap in the description and management of service resource as well as service matching. This semantic gap has result in the inconsistent representations of service that lead to different understandings in both service provider and requestor sides. The absence of semantic info also causes the interoperation problem among different systems. Now, by combining the technologies of ontology and web service, semantic web services enable the software agents to discover, invoke, compose and monitor services automatically. This is one promising technology approach to address this problem, but it needs related domain ontology and common ontology to support.

Based on the above semantic service-oriented computing environment, a user-centric, personalization, intelligent service environment can be created. User can customize various services by the service portal to satisfy own personalization requirements, such as, education, work, entertainment, daily life. This smart service environment is characterized by ambience-awareness, personalization, intelligent, adaptive, consistency.

2.3 The Semantic Service Integration Architecture of Telecommunications Network and Internet

To complete the aim of UCISE, the intelligent agents need to accurately discover, automatically compose and invoke the services provided by telecommunications network and Internet based on the current service context. In this article, we import the semantic web service and ontology technologies into the telecommunications service domain. A semantic service integration architecture of telecommunications network and Internet is presented (see Fig 2). This architecture integrates the communication capabilities of telecom network with the rich information/content of Internet so as to provide the needed information to users in a suitable way. This will facilitate the semantic integration of telecommunications network and Internet in the service layer.

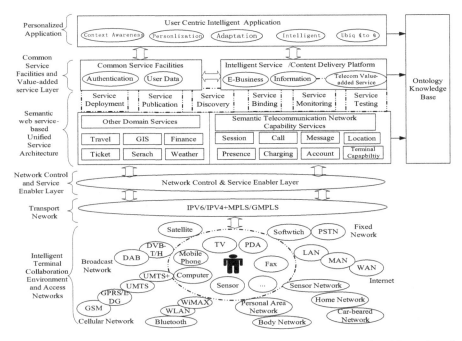

Fig. 2. The Semantic Service Integration Architecture of Telecommunication Network and Internet

The establishment of semantic service integration architecture will involved with the following aspects:

(1) **The provision of semantic telecommunications network capability services:** At present, semantic web has become the development trend of Internet, and semantic web service technology is applied to a variety of industry areas, such as travel services, search services, geographic information services, financial services,

meteorological services. Consequently, the telecommunication network capability services, such as call services, message services, positioning services, also need to evolve from WSDL-based specifications to semantic web service-based specifications. This will facilitate the semantic convergence of telecom network and Internet in the service layer.

(2) **The provision of value-added services/content:** Using the fundamental network services, the service platform can provide the user-oriented value-added services. The service platform needs to support the service development, deployment, management in the semantic web service environment.

(3) **The provision of service trading infrastructure:** In order to support this open, loosely coupled service computing environment, a public service trade mechanism is needed, which supports the service publication, discovery and invocation. In addition, some public service facilities are also indispensable, such as security authentication mechanism to ensure the security and the reliable service transactions, unified user data management mechanism to support the sharing of user data in heterogeneous domains.

(4) **The construction of ontology facilities:** To support the semantic interoperability in the service layer, a convergent service network related ontology knowledge base is needed. By defining the unified concepts and knowledge in the formal way, the heterogeneous systems can achieve the sharing of knowledge. This sharing ontology mainly consists of device ontology (such as the various terminals, telecommunications network equipment), network ontology (such as core network ontology, the access network ontology, service network ontology), service ontology (such as OWL-S, WSMO), network capacity service application ontology, as well as some application layer ontologies (such as the user profile, context ontology).

This innovative service integration architecture provides the new supporting environment and executing mechanism. And it laid the foundation for dynamic service adaption. Some old obstacles can be addressed, for example:

(1) **The loosely coupled problem of service logic and service resource:** In the traditional telecommunications networks, the service logic and its involved resources are tightly coupled. In this way, when users move to a new network domain, the service logic still need to use the specified resources in the procedure of the home network and can not use the resources which provide the same functionality of ambient network. This problem wastes the valuable network resources (such as network bandwidth) and increases the control complexity. In UCISE, the application logic can dynamically discover the satisfied service and invoke it. This architecture can satisfy the requirements of the context-awareness personalized service.

(2) **The sharing and unification of user data:** In the traditional telecommunications networks, user data and service specific data are the proprietary data of the concrete carrier. It brings the difficulty of application migration among different operators, and constraints the sharing of user data among different domain (such as telecom network, enterprise network). This service integration architecture completes the unification of user data and service personalized data. The user data is independent of the concrete domain and operators. For example, each user of UCISE will have a unique semantic identifier. When you call a friend with this

identifier, if he is in outdoors, the mobile phone number of this identifier will be used; if he is at home, the fixed phone number of this identifier will be used. In this way, the application system will have no need to concern the change of user data (such as the change of phone number or email address). This laid the foundation for the service mobility and the information consistency.

3 Semantic Description Approach for Telecommunications Network Capability Services (TNCS)

3.1 The Differences between TNCS and the Plain Web Services

Compared with the plain web services, the TNCS, especially that provided by mobile network, have some distinct domain-related characteristics.

From the terminals aspect, the TNCS provided by different networks maybe have different terminal requirements. The provision of telecommunication service is greatly dependent on the capabilities of terminals. As communication network technologies are evolving constantly, there are great capability differences among the different terminals, such as terminal browser, communication protocol. For example, the location service provided by CDMA network requires that the terminals support the GPSOne positioning technology. However, the location service provided by GSM network based on Cell-ID positioning technology almost has no special requirements for terminals. In order to deliver the seamless service on the user's different terminals, the service delivery platform must discover the appropriate TNCS based on the service context. So the semantic description of TNCS should consider the requirements on terminals.

From the network aspect, currently, there are a variety of communication networks, such as GSM, CDMA, fixed network, WLAN. The different networks have the different service quality. For example, the wireless networks are less reliable than line-based networks. To meet the different user's needs, service delivery platform should select the appropriate TNCS based on the Service Level Agreement (SLA). So the non-functional features like network type, service bandwidth, and communication mode are important for the semantic description of TNCS.

In addition, a significant difference between telecom service and Internet service is the charging pattern. Internet services are often free, however telecom services have various charging model, such as event-based, session-based, time, volume. In order to satisfy the different consuming levels of users, the intelligent agent or service delivery platform should query the suitable TNCS based on the user's cost preference. So, the charging model should be the important part of the semantic description of TNCS.

From the above analysis, we can see that the provision of TNCS is greatly constrained by the network condition, terminal capability and other non-functional features. The TNCS provided by a concrete network has its own using scope and conditions. Therefore, in order to provide the personalized services, the intelligent service platform should dynamically select the appropriate TNCS or other value-added services based on the service context info. To achieve this goal, the TNCS need to be accurately described in the semantic level so that the annotated semantic information can fully reflect its characteristics.

3.2   The Problem Statement of the Semantic Description of TNCS

In order to enable the service agent to accurately discover and automatically invoke the TNCS based on the service context, TNCS should be described in the semantic level, not the syntactic level. This work needs to solve two issues:

(1) **A Tailored Semantic Description Ontology for TNCS:** TNCS, especially that provided by mobile network, has some special non-functional features, such as network characteristics, billing policy, terminal capability requirements, and quality of service. These non-functional features are very important for context awareness mobile service. Currently, OWL-S provides a high-level ontology description framework for the general semantic web service. However, OWL-S is unable to fully support the important characteristics description of TNCS. In order to fully reflect the important features of TNCS, a tailored semantic description ontology for TNCS is needed. In this paper, we present TelecomOWL-S, which is a domain application of OWL-S, to solve this issue, see section 3.3.

(2) **The Ontology-based Formal Description of TNCS-Related Domain Concepts and Knowledge:** When semantically describing TNCS, its input/output parameters and some important service features like network type and terminal requirements involve with a lot of domain concepts and knowledge. In order to support the ontology-based reasoning and knowledge sharing, these concepts must be modeled in the formal way. However, telecommunications service domain includes a large number of concepts and terminology. How to organize these concepts and finally form a knowledge system is a great challenge. In this paper, we present Telecommunication Service Domain Ontology (TSDO) and TNCS-related Application Ontology to solve this issue, see section 3.4 and section 3.5.

3.3   Prescribing OWL-S for the TNCS Description

The semantic description of TNCS is the formal specification of the open interfaces provided by the telecommunications network. The concrete logic implementation of TNCS is not involved. So the ServiceModel and ServiceGrounding of OWL-S are sufficient for the semantic description of TNCS. Specifically, the atomic process ontology of ServiceModel can be directly used to model the network operations such as MakeACall, CancelCall. The ServiceGrounding ontology can be used to mapping the atomic process of ServiceModel to the operation of WSDL-based open interface specification like Parlay X. Therefore, the ServiceProfile of OWL-S needs to be extended to describe the related important features of TNCS.

In fact, the service profile is used to characterize a service for purposes such as advertisement, discovery, and selection. Service profiles may be published in various kinds of registries, discovered using various tools, and selected using various kinds of matchmaking techniques. OWL-S does not prescribe or limit the ways in which profiles may be used, but rather, seeks to provide a basis for their construction that is flexible enough to accommodate many different contexts and methods of use. In order to accurately describe the important features of TNCS, we present a tailored service ontology--TelecomOWL-S, which mainly extends the ServiceProfile ontology of OWL-S.

Similar to OWL-S, TelecomOWL-S also has three parts: TelecomServiceProfile, ServiceModel, and ServiceGrounding. As the semantic description of TNCS does not involve the complex service logic control, TelecomOWL-S adopts the same Ser-viceModel and ServiceGrounding ontology as those of OWL-S. The difference is that TelecomOWL-S redefines a new TelecomServiceProfile by extending ServiceProfile of OWL-S based on the characteristics of TNCS. Fig.3. depicts the high-level ontology overview of TelecomOWL-S.

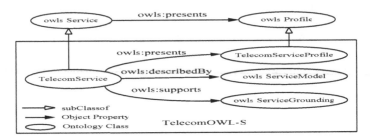

Fig. 3. The High-level Ontology of TelecomOWL-S

The TelecomServiceProfile is directly derived from the ServiceProfile of OWL-S. The extension ways mainly include two kinds, one is to limit the range of the existing property, and the other is to define the new special properties. Specifically, the classes and properties of TelecomServiceProfile ontology are shown in Fig.4.

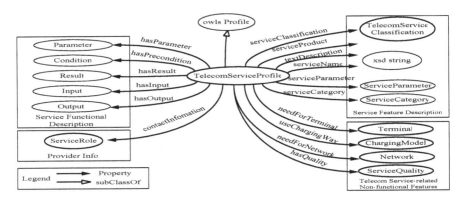

Fig. 4. TelecomServiceProfile Ontology

The TelecomServiceProfile mainly consists of 4 parts: Service Functional Descrip-tion, Service Provider Information, Service Feature Description and Telecom Service-related Non-functional Feature. The former three parts mainly inherits from OWL-S. Only the value ranges of some properties like serviceClassification, contactInforma-tion are constrained by the specific domain concepts defined in the TSDO. The last part is mainly used to describe some peculiar features of TNCS. Currently, four new object properties are defined.

(1) **needForTerminal:** This property can be used to depict the service requirements for user's terminal capabilities, such as terminal browser, terminal hardware, terminal software and WAP. The concepts and terminology about terminal capability are from the Terminal Capability Ontology of TSDO.

(2) **useChargingWay:** This property is defined to describe the service-related various billing policies and corresponding tariffs, such as time-based, volume-based, event-based, and flat fee. The concepts and terminology about charging are from the Charging Ontology of TSDO.

(3) **needForNetwork:** The characteristics of network providing service can be described by this property, such as the network type, network bandwidth. The concepts and terminology about charging are from the Network Ontology of TSDO.

(4) **hasQuality:** This property can describe the service quality, such as response time, connectivity, delay. The relevant concepts and terminology are from the Service Quality Ontology of TSDO.

3.4 Telecommunication Service Domain Ontology (TSDO)

The semantic descriptions of TNCS need the support of telecommunications service domain concepts and knowledge. Telecommunication Service Domain Ontology (TSDO) provides some shared domain concepts and knowledge about telecom service. This will facilitate the semantic sharing and interoperability of heterogeneous communication entities.

Considering the scalability and flexibility of TSDO, we construct the domain ontology in a hierarchical structure. TSDO only provides some core domain concepts. Based on these core domain ontologies, we can construct various concrete application-related ontologies, such as TNCS-related application ontology or service context ontology. Based on a modular design principle, it mainly comprises six sub-ontologies. Fig.5. shows the overview of TSDO. In addition, the construction of TSDO needs the support of some Common Ontology and other domain ontology, such as time ontology, location ontology.

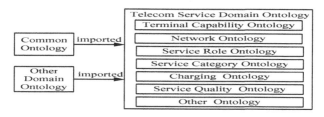

Fig. 5. Telecommunication Service Domain Ontology

(1)**Terminal Capability Ontology:** defines some main concepts about terminal software, terminal hardware, terminal browser and network characteristics supported by terminal. Currently, the UAProf [15] based on the W3C Composite Capabilities/Preferences Profile (CC/PP), which are defined by RDF (Resource Description Framework) language, are used to describe the mobile terminal capability. They cannot be directly used to describe the semantic TNCS. So we create the Terminal Capability Ontology by OWL language based on CC/PP and UAProf specifications.

(2)**Network Ontology:** mainly specifies the network concepts, network category, network features, as well as the relationships of various networks, such as mobile network, internet, and fixed network, GSM, CDMA, UMTS, WCDMA, WLAN.

(3) **Service Role Ontology:** describes the stakeholders' concepts of the service supply chain, for example, service provider, content provider, network operator, service user.

(4)**Service Category Ontology:** descriptions of telecommunication service classification. As a proposed service category standard by OWL-S, the UNSPSC (United Nations Standard Products and Services Code) [16] provides an open, global multisector standard for efficient, accurate classification of products and services. It often used in the E-commerce field. However, on the one hand, UNSPSC is not based on ontology, so it is only suitable for the serviceCategory property of ServiceProfile, not the serviceClassification property of ServiceProfile. On the other hand, UNSPSC has no concrete telecommunications service classification, only to Telecommunication Services (code: 81161700) level. Therefore, UNSPSC has no ability to enable the accurate telecommunications service query. So we construct the Service Category Ontology in TSDO to enable the accurate description of TNCS. It defines the relationship between various telecommunications services, like basic service, value-added service, voice service, data service, conference service, telecommunication network capability service, download service, browsing service, messaging service.

(5)**Charging Ontology:** defines the charging-related concepts about telecommunication service, includes payment way (such as prepaid and postpaid), charging type (such as time-based, volume-based, event-based, content-based), billing rate, as well as account balance.

(6) **Service Quality Ontology:** Telecommunication network must provide the services which have the end-to-end QoS guarantee. As the technical differences, the quality of service provided by different networks is different. Service Quality Ontology mainly defines the QoS-related concepts about telecommunication service, including access network QoS, core network QoS and user's QoE, such as call delay, message size, call through rate, positioning accuracy, network bandwidth.

As the length limit of this article, a part of Network Ontology is shown in Fig.6.

```
<owl Class rdf ID="Network"/>
<owl Class rdf about='#TelecomNetwork'>
   <rdfs subClassOf rdf resource="#Network"/>
</owl Class>
<owl Class rdf about='#MobileNetwork'>
  <rdfs subClassOf>
     <owl Class rdf about='#TelecomNetwork"/>
  </rdfs subClassOf>
   <owl disjointWith rdf resource="#FixedNetwork"/>
</owl Class>
<owl Class rdf ID="GSM'>
   <rdfs subClassOf rdf resource="#MobileNetwork"/>
   </owl Class>
<owl ObjectProperty rdf ID="operatedBy'>
   <rdfs range rdf resource="&ServiceRole;#NetworkOperator"/>
   <rdfs domain rdf resource='#TelecomNetwork"/>
</owl ObjectProperty>
```

Fig. 6. A part of Network Ontology

3.5 The Application Ontology of TNCS

The different TNCS has the different interface parameters, i.e. input/output. In order to describe the semantic TNCS, theses interface parameters must be described in the ontology format. Therefore, the related application ontology needs to be created, such as Parlay X Application Ontology or Parlay Application Ontology, see Fig.7.

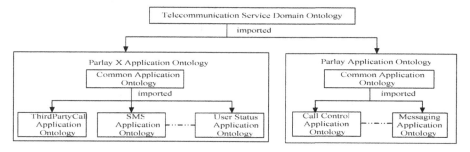

Fig. 7. The Application Ontology Overview of TNCS

4 The Experimental Environment and the Demo Service

4.1 The Prototype of Experimental Environment

We implemented an experimental prototype to validate the proposed approach. The framework of this environment is shown in Fig. 8.

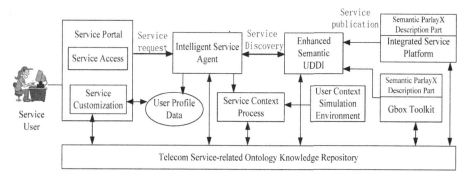

Fig. 8. The Experimental Environment Framework

In the experimental environment, we use two different ways to provide the semantic Parlay X services. ① The existing integrated service platform of our lab, which can support the voice and data value-added service, is extended. It can provide the semantic ThirdPartyCall, SMS, Conference service of Parlay X specification in this platform. ② We develop a semantic Parlay X gateway prototype using Parlay X SDK called

Gbox [17]. As Gbox can directly provide Parlay X web service, we only need to add the semantic description part.

To enable the publication of semantic Parlay X services, an enhanced semantic UDDI prototype is developed. With the comparison of the existing OWL-S/UDDI such as matchmaker [18], this semantic UDDI is a pure semantic UDDI. On the capability aspect, it can also support the registration of TelecomServiceProfile which is an extended OWL-S profile, besides OWL-S profile. In addition, this enhanced semantic UDDI can enable not only the matching of IOPE, but also the ontology-based matching of nonfunctional properties by extending OWL-S API toolkit.

The intelligent service agent is responsible for the execution of service logic. When received the service request, the intelligent service agent will acquire the user context information from service context process module. Then service agent will construct the service request based on the user preferences or profile, and send the request to the semantic UDDI. When received the returned service, the service agent will invoke semantic web services by service URI based on OWL-S API. Currently, this service agent cannot support the automatic service composition.

To facilitate the service access and service customization, we developed a service portal based on JSP/Servlet technology. This web portal provides a bridge between the service users and the intelligent service agent. The service users can login the service portal, initiate the service invocation and maintain own personalized preferences data.

Service context process module and user context simulation environment are used to provide the service context information. User context simulation environment provide a graphic interface and some service context information, such as network type, terminal type, location, activities, can be configured.

To enable the sharing of telecom service-related ontology knowledge among the service requester, semantic UDDI and service provider, we construct an ontology knowledge repository which consists of the telecommunication service domain ontology, Parlay X application ontology. All these ontologies are stored in the OWL file format and have the unified namespace. Currently, there are 320 telecom service-related ontology concepts and terminologies in this repository.

4.2 The Demo Service: The Context-Awareness Cheapest Click-to-Call Service

Now, the click-to-call service is becoming increasingly popular in Internet. The users can easily use a computer to initiate a telephone call by entering the caller and the callee number. However, the existing click-to-call services are always statically bundled with the call service provided by a concrete network operator. In technical aspect, the click-to-call services are generally based on the ThirdPartyCall service of Parlay X provided by the network carriers. As the different carriers have the different pricing policy, it is difficult for users to enjoy the cheapest click-to-call service. Generally speaking, the users always expect to enjoy the cheapest call service anywhere and any time. In addition, the existing click-to-call service needs to input the specific phone number. If this phone number (such as fixed phone) can not be connected, user must change the phone number (such as phone number) to try again. So the existing click-to-call service has no context-awareness feature.

In this experimental environment, the network operators publish their ThirdParty-Call service of the semantic Parlay X web service with detailed charging info on the semantic UDDI. Once the user initiates the service request, the service agent is responsible for submitting the request service profile to the UDDI behalf on the user. The semantic UDDI will match the service request with the service info stored in the repository and return the corresponding results. And then the service agent selects the appropriate service from the service list and dynamically invokes it. In addition, as we use the unique identifier to mark user. When using our context-awareness click-to-call service, user only needs to input this unique identifier and the service agent will select the corresponding phone number of this identifier to invoke the ThirdPartyCall service according to the user context information.

Compared with the existing Parlay X web services, the semantic Parlay X web services have the rich semantic description info, such as price policy, QoS, terminal capability requirement. All these added features unambiguously state the capabilities of TNCS and laid the foundation for the provision of the personalized intelligent services. In the experiment, we simulate four mobile operators to provide the semantic ThirdPartyCall services. The different charging policies are adopted during the different time periods. We publish these semantic ThirdPartyCall services on the semantic UDDI. The service agent queries the cheapest services on the UDDI, and then dynamically invokes the satisfied services provided by an operator. We play the role of users to initiate the Click-to-Call service on the service portal at different time. The results show that the operator which has the lowest charging rates is selected. Compared with the static binding and invoking way of the general Parlay X service, the user cost significantly reduces in the dynamically binding way.

5 Discussions

From the above analysis we can see that the semantic Parlay X web services have the rich semantic info. This is the foundation of the service precise matching and dynamic discovering. The semantic telecommunications network capability service is one of the key enabling technologies for the intelligent "user-centric" applications which have the context-awareness and self-adaptive features. The existing Parlay X web service is the cornerstone of the semantic Parlay X web service. Although the performance is reduced to some extent because of the introduction of the ontology process and the service matching, the influence will be reduce to a minimum degree by adopting the new computing technology or improving the equipment performance. Additionally, the general Parlay X web service and the semantic Parlay X web service are not conflict, but complementary. The telecommunication network capability service gateway can provide both the semantic Parlay X web service and the general Parlay X service.

In this article, we mainly researched the semantic description of TNCS based on OWL-S. Vitvar, T. [19] mentioned that WSMO-Lite also have begun to consider the description of domain-specific nonfunctional properties. In the following work, we will explore how to describe the TNCS by WSMO.

6   Conclusions

In this paper we introduced the vision of user centric intelligent service environment (UCISE) and discussed the service ecosystem as well as the semantic service integration of telecommunications network and Internet. To enable the provision of semantic telecommunications domain services, a semantic description approach for telecommunication network capability services is presented. In this way, the network operators can accurately describe the service capability based on the network and the operating conditions. This laid the foundation for the precise service discovery and the dynamic service invocation. Like the existing Parlay X or Parlay gateway, the new semantic Parlay X/Parlay gateway can provide the semantic TNCS. This will greatly eliminate the semantic gap between telecom network and Internet in the service layer. So far, we have implemented a prototype system, including semantic Parlay X gateway, telecom service domain ontology repository, Parlay X application ontology, semantic UDDI, intelligent agent. The cheapest third party call service is developed.

However, the research of semantic TNCS is still in an early phase. The TSDO need to be further enhanced. The standardization of semantic TNCS is still a long way. Currently, we are cooperating with Huawei Corp. to promote this work.

Acknowledgments. This work is performed in the Project "Service Intelligence for Convergent Network Environment" (No. 60672122) and "The Study of Service Context Prediction Theory and Key Technologies based on Human Cognitive Mechanism" (No. 60802034) supported by National Natural Science Foundation of China, as well as Specialized Research Fund for the Doctoral Program of Higher Education (No. 20070013026).

References

1. Zahariadis, T.: Trends in the Path to 4G. IEEE Communications Engineer. 1, 12–15 (2003)
2. Benali, O., et al.: A Framework for an Evolutionary Path toward 4G by Means of Cooperation of Networks. IEEE Communications Magazine 42, 82–89 (2004)
3. Qiao, X.Q., Li, X.F., Liang, S.Q.: Reference Model of Future Ubiquitous Convergent Network and Context-Aware Telecommunication Service Platform. The Journal of China Universities of Posts and Telecommunications 13, 50–56 (2006)
4. Parlay Group, Parlay API 4.0, http://www.parlay.org
5. Parlay x, http://www.parlay.org/en/specifications/pxws.asp
6. OWL-S.: http://www.w3.org/Submission/2004/07/
7. Web Ontology Language (OWL), http://www.w3.org/2004/OWL
8. Hong, S.J., Han, S.Y., Choi, B., Kim, Y.J., Ahn, C.H.: The semantic PARLAY for 4G network. In: 2nd International Conference on Mobile Technology, Applications and Systems, pp. 15–17 (2005)
9. Hong, S., Han, S., Song, K.-H.: The extended PARLAY X for an adaptive context-aware personalized service in a ubiquitous computing environment. In: Enokido, T., Yan, L., Xiao, B., Kim, D.Y., Dai, Y.-S., Yang, L.T. (eds.) EUC-WS 2005. LNCS, vol. 3823, pp. 288–297. Springer, Heidelberg (2005)

10. Lin, S.T., Chen, J.L.: Semantic Web Enabled VHE for 3rd Generation Telecommunications. In: 4th Annual ACIS International Conference on Computer and Information Science (2005)
11. Vitvar, T., Viskova, J.: Semantic-enabled Integration of Voice and Data Services: Telecommunication Use Case. In: Third IEEE European Conference on Web Services, ECOWS (2005)
12. Villalonga, C., Strohbach, M., Snoeck, N., Sutterer, M., Belaunde, M., Kovacs, E., Zhdanova, A.V., Goix, L.W., Droegehorn, O.: Mobile Ontology: Towards a Standardized Semantic Model for the Mobile Domain. In: Proceedings of the 1st International Workshop on Telecom Service Oriented Architectures (TSOA 2007) at the 5th International Conference on Service-Oriented Computing, Vienna, Austria, 17 September (2007)
13. IST SPICE project, http://www.ist-spice.org/
14. Duke, A., Richardson, M., Watkins, S., Roberts, M.: Towards B2B integration in telecommunications with semantic web services. In: Gómez-Pérez, A., Euzenat, J. (eds.) ESWC 2005. LNCS, vol. 3532, pp. 710–724. Springer, Heidelberg (2005)
15. WAP Forum UAProf (2001),
    http://www.openmobilealliance.org/tech/affiliates/
16. UNSPSC (United Nations Standard Products and Services Code),
    http://www.unspsc.org/
17. PARLAY X Service Creation Environment, http://www.appium.com
18. Matchmaker, http://www.daml.ri.cmu.edu/matchmaker/
19. Vitvar, T., Kopecký, J., Viskova, J., Fensel, D.: WSMO-lite annotations for web services. In: Bechhofer, S., Hauswirth, M., Hoffmann, J., Koubarakis, M. (eds.) ESWC 2008. LNCS, vol. 5021, pp. 674–689. Springer, Heidelberg (2008)

Understanding Semantic Web Applications

Kouji Kozaki, Yusuke Hayashi, Munehiko Sasajima, Shinya Tarumi,
and Riichiro Mizoguchi

The Institute of Scientific and Industrial Research, Osaka University
8-1, Mihogaoka, Ibaraki, Osaka, Japan
{kozaki,hayashi,msasa,tarumi,miz}@ei.sanken.osaka-u.ac.jp

Abstract. Ten years have passed since the concept of the semantic web was proposed by Tim Berners-Lee. For these years, basic technologies for them such as RDF(S) and OWL were published. As a result, many systems using semantic technologies have been developed. Some of them are not prototype systems for researches but real systems for practical use. The authors analyzed semantic web applications published in the semantic web conferences (ISWC, ESWC, ASWC) and classified them based on ontological engineering. This paper is a review of application papers published in Semantic Web conferences. We discuss a trend and the future view of them using the results.

Keywords: Ontology, Knowledge modeling, Knowledge Management.

1 Introduction

About 10 years after the birth of Semantic Web (SW), advocated by Tim Berners-Lee, fundamental technologies such as RDF(S) and OWL have been developed as well as their application systems. So many research and development projects on the basis of SW technologies have produced various applications from prototypes at the laboratory level to the practical full-scale systems. A survey report [1] says that about 300 companies provide SW technology related products. In spite of so many efforts on research and development of SW technologies, "Killer Application" of SW is still unknown [2]. Therefore, it would be beneficial for us to get an overview of the current state of SW applications to consider next direction of SW studies to realize semantic technologies which enhance utilization of knowledge on the web. Some researchers have already broadcasting information about SW studies (e.g., [3]), exhaustive information about each study makes it difficult to see and analyze the state of SW studies at a glance.

Fundamental features of SW include enabling computers to process semantics on various resources of WWW annotated by metadata, which is in turn defined by ontology. Since ontology is a fundamental and important technology for SW applications, this paper analyzes them from the view point of ontology. Especially we focus on "What type of ontologies is used" and "How ontologies are used." Following sections analyze the current state of SW applications to propose several directions for future research. Specifically, we classified 190 SW applications which utilize ontologies

J. Domingue and C. Anutariya (Eds.): ASWC 2008, LNCS 5367, pp. 524–539, 2008.
© Springer-Verlag Berlin Heidelberg 2008

Table 1. The number of SW applications which is analyzed in this paper

Conferences	Dates	Venues	Number of Apps
International Semantic Web Conference (ISWC)			
ISWC2002	Jun. 9-12, 2002	Sardinia, Italy	9
ISWC2003	Oct.20-23, 2003	Sanibel Island, FL, USA	19
ISWC2004	Nov. 7-11, 2004	Hiroshima, Japan	18
ISWC2005	Nov. 6-10, 2005	Galway, Ireland	25
ISWC2006	Nov.5-9, 2006	Athens, GA, USA	26
ISWC2007& ASWC2007	Nov.11- 15, 2007	Busan, Korea	18
European Semantic Web Conference (ESWC)			
ESWC2005	May29-Jun.1, 2005	Heraklion, Greece	24
ESWC2006	Jun.11-14, 2006	Budva, Montenegro	11
ESWC2007	Jun. 03 - 07, 2007	Innsbruck, Austria	17
Asian Semantic Web Conference (ASWC)			
ASWC2006	Sep.3- 7, 2006	Beijing, China	23

extracted from international conferences on SW (Table 1)[1]. In this paper, SW and ontology engineering tools such as ontology editors, ontology alignment tool, and so on, are not the target of the analysis because we focus on applications which utilize ontologies.

2   Related Work

Some researchers discuss trends and future prospects of Semantic Web and its applications. Alani et al. consider a killer application for SW from a viewpoint of business community [2]. They discusses that opportunities to make progress on cost, communities, creativity and personalization are important for appearance of a killer apps. And they argue SW researchers should focus on the four important areas. Léger et al. present 16 use cases from enterprises which are interested in SW technology and analyze 4 of them in detail [4]. Based on the analysis, they determine 12 knowledge processing tasks required in industry. These researches focus on specifying the requirement for SW applications in industry. We focus on analysis of the current state of SW applications in academic communities.

Motta and Sabou examine the current SW applications and introduce seven dimensions for analogizing them [5]. They compare some older and newer systems which they call the next generation of SW applications. As the result, they conclude that SW applications will have to deal with increasing heterogeneity of semantic sources and new web technologies such as social tagging and web services. We focus on usage of ontologies which are discussed in one of their dimensions because it is a key technology for all SW applications.

---

[1] ESWS2004 is excluded because it has been a symposium, not a regular conference.

Some researchers analyze and classify ontology-based applications. Uschold and Jasper classify ontology applications scenarios in three categories: 1)neutral authoring, 2)common access to information, and 3)indexing for search [6]. Mizoguchi enumerates the role of an ontology as follows: 1)a common vocabulary, 2)data structure, 3)explication of what is left implicit, 4)semantic interoperability, 5)explication of design rationale, 6)systematization of knowledge, 7)meta-model function, and 8)theory of content [7]. Kitamura discuss role of ontology of engineering artifacts mainly in knowledge modeling and categorize them into 1)shared vocabulary and taxonomy, 2)conceptual (standard) data schema, 3)metadata schema for documents, 4)semantic constraints for modeling, 5)generic knowledge and patterns, 6)interoperability and integration, 7)communication support and querying, 8)capturing implicit knowledge, and 9)basis of knowledge systematization [8]. We refer to these categories to consider types of usage of ontology and introduce 9 types discussed in next section.

3   The Method for Analyzing SW Applications

Fundamental feature of SW is to enable machines to process semantics on various resources of WWW by giving machine-readable meta-data, which is defined by ontology, to the resources. Since ontology is a fundamental and important technology for SW applications, we analyze them from the view point of ontology. Ontologies can be classified into several "types" according to several features: number of concepts, usage of models, target domain, depth of hierarchy, etc. An appropriate type of ontology for an application is specified by requirements of the application. Especially, quality of semantic information required for the application, i.e., intended semantic processing, constrains the type of the ontology.

This research analyzes SW applications from the view point of "Usage of ontology" and "Type of ontology used." In this section, the authors propose several types for both viewpoints with guidelines and classify ontology-based applications.

3.1   Types of Usage of Ontology for a SW Application

According to the application purpose, usage and requirements of the ontology differ. Here we introduce 9 types of usages. Basically, a SW application is categorized to one of the types. Some SW applications which use ontology for multiple ways are categorized to multiple categories.

(1) Usage as a Common Vocabulary

To enhance interoperability of knowledge content, this type of application uses ontology as a common vocabulary. Since this is the most fundamental and common usage for the following all categories from (2) to (9), applications in which ontologies play mainly a role to unify the vocabulary are categorized here. In life science, some large scale ontologies, such as GO[2] and SNOMED-CT[3] have been developed and used as common vocabularies for a long time. Recently, many SW applications use Linked Data[4] as represented by DBPedia[5] for sharing and connecting information on the SW.

---

[2] http://www.geneontology.org/
[3] http://www.ihtsdo.org/
[4] http://linkeddata.org/
[5] http://wiki.dbpedia.org/

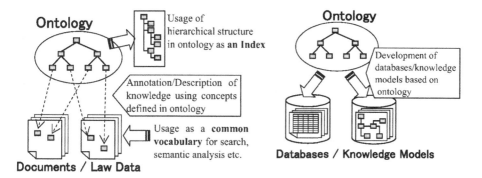

Fig. 1. Usage as a Common Vocabulary / an Index

Fig. 2. Usage as a Data Schema / a Rule Set for Knowledge Models

(2)Usage for Search

With appropriate metadata, we can realize search systems which use semantic information for searching. It depends on the metadata and the ontology to what extend the application can deal with the deep semantics. This is one of the most typical usages, thus an application whose main function is search will be categorized here. The function of semantic search is realized in some applications, such as a knowledge portal [9] and a knowledge management system [10], at early stage of SW. Recently, some semantic search services for web such as Powerset[6] and Hakia[7] are published.

(3) Usage as an Index

An application which uses ontologies as indexes for knowledge resources belongs to this category. An ontology is a system of vocabulary, thus the index becomes systematized one. Difference between this category and the categories (1) and (2) is that applications of this category utilize not only the index vocabulary but also its structural information (e.g., an index term's position in the hierarchical structure) explicitly when accessing the knowledge resources (Fig.1). For example, a semantic navigation system for information services [11] and a semantic view-based search engine [12] which navigates users utilizing such indexes with contextual structures defined by ontology are categorized here.

(4) Usage as a Data Schema

Applications of this category use ontologies as a data schema to specify data structures and values for target databases. Hierarchy of the ontologies specifies instances as data values, and each of the concept definitions specifies data structures (Fig.2). Typical applications of this category are a system for exchanging bibliography through network [13], various kinds of data management systems using semantics defined ontologies [14, 15, 16], and so on.

---

[6] http://www.powerset.com/

[7] http://www.hakia.com/

(5) Usage as a Media for Knowledge Sharing
Applications of this category aim at knowledge sharing among systems, between people and systems, or among people using ontologies and instance models about the target knowledge. Generally speaking, almost ontological systems aim at such knowledge sharing. To avoid confusion, this category includes such applications that stress enhancement of interoperability among different systems: applications for knowledge alignment, systems for knowledge mapping, communication systems among agents, support systems for communication, etc. From technical viewpoints, both of (i)ontology mapping systems using a reference ontology (Fig.3(a)), and (ii)systems which align ontologies behind the target knowledges and map target knowledges via the aligned ontologies (Fig.3(b)) are categorized here. For example, [17] takes the former approach to integrate distributed systems using a reference ontology which is called mapping ontology, and [18] takes the latter method to generate proper queries for web-based malls which have different product categories.

(6) Usage for a Semantic Analysis
Reasoning and semantic processing on the basis of ontological technologies enable us to analyze contents which are annotated by metadata. One of the most typical methods for such analysis is an automatic classifier for concept definitions using inference engines. For example, [19] discusses how SW technology can be used in biology to automate the classification of proteins through an experiment. Other examples are statistical analysis systems, validation system for scientific data (e.g. experimental result) [20]. Some applications have visualization tools for supporting the analysis [21]. Among the search systems in category (2), those systems which employ such ontological analysis belong to this category.

(7) Usage for Information Extraction
Applications which aim at extracting meaningful information from the search result are categorized here. Recommendation systems [22] which filter search results are example of the extraction. Other examples are a system extracting product features from web pages [23], a service which summarizes blogs and get useful information of products such as the total reputation and related products [24]. Comparing to other

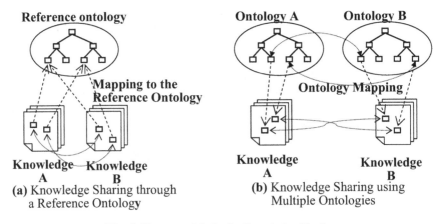

(a) Knowledge Sharing through a Reference Ontology

(b) Knowledge Sharing using Multiple Ontologies

Fig. 3. Usage as a Media for Knowledge Sharing

"search" categories, applications of the category (2) just output search results without modifications. Applications of the category (6) add some analysis to the output of (2), while those of this category (7) extract meaningful information before outputting for users.

(8) Usage as a Rule Set for Knowledge Models
We can use instance models, which are built upon definition of classes in ontologies, as knowledge models of the target world. In other words, we can use ontologies as meta-models which rule the knowledge (instance) models (Fig.2). Relations between the ontologies and the instance models correspond to that of the database and the database schema of category (4). Compared to the category (4), knowledge models need more flexible descriptions in terms of meaning of the contents.

From the viewpoint of ontology engineering, one of the most intrinsic roles of ontology is to rule knowledge models. For the purpose, a heavy-weight ontology which models the world appropriately with deep semantics helps the knowledge modeling and reasoning at a deeper level. For example, enterprise systems for healthcare delivery [25], scientific knowledge sharing [26], and e-government service [27] are developed based on ontologies. On the other hand, sometimes light-weight ontology helps them at a shallow level when the target world is large, shallow model is enough for reasoning and an efficient processing method is needed. Semantic MediaWiki[8] [28] is a typical example of the application.

(9) Usage for Systematizing Knowledge
Ontology provides semantic relationships among concepts. Putting them as the core conceptual structure, we can organize concepts of the target world which in turn becomes systematized knowledge. Referring to the concepts organized in ontology, we can build systems for managing knowledge. Typical applications of this category include integrated knowledge systems of category (1) to (8) such as knowledge management systems and contents management systems [29, 30].

3.2 Types of Ontology

This section categorizes ontologies without depending on target domains and their description languages. We introduce 5 categories from the viewpoint of semantic feature of ontologies. Although this categorization shares the way of thinking with an ontology spectrum by Lassila and McGuninness [31], our categorization does not have strict definitions because we focus on rough survey of ontologies used in applications. We are planning to refine the ontology types in the future.

(A) Simple Schema
This category includes simple schemas such as RSS and FOAF[9] for uniform description of data for SW applications, although they are not called ontology in a strict sense.

---

[8] http://semantic-mediawiki.org/
[9] http://www.foaf-project.org/

(B) Hierarchies of *is-a* Relationships among Concepts
One of the most fundamental elements of ontologies is a set of concepts identified in the target world with a hierarchical structure based on "is-a" relationship (rdfs:subClassOf of RDF(S) and OWL). Some portal sites navigate users by a menu with hierarchically organized topics, which is a kind of ontologies, is sometimes called light-weight ontology.

(C) Relationships other than *"is-a"* is Included
Ontologies can define concepts more explicit by using various relationships such as "part-of" (whole and part) and "attribute-of" (property). Ontologies of this category contains "is-a" relationships plus other various ones. Properties of RDF(S) or OWL are used for representing such relationships.

(D) Axioms on Semantics are Included
In addition to descriptions of the relationships of the category (C), ontologies can specify further constraints among the concepts or instance models by introducing axioms on such semantic constraints. Axioms on ontologies of this category include constraints about order such as transitivity and reflexivity, exclusivity of an instance, and so on. In case of OWL, constraints about relationships such as "transitiveProperty" and "inverseOf", constraints about sets such as "disjointWith" and "one of" are categorized to the axioms of ontologies in this category.

(E) Strong Axioms with Rule Descriptions are Included
Some ontologies require strong axioms which are necessary for further description of constraints on the category (D). Ontologies with rule descriptions, for example, by KIF or SWRL, are categorized here.

3.3 Steps for Analyzing SW Applications from Ontological Viewpoint

Three of the authors analyzed 190 papers introduced in section 1 according to the following steps:

(1) Giving short explanations about the application (One sentence for each).
(2) Identifying the usage of ontology (section 3.1)
(3) Identifying the target domain.
(4) Identifying types of ontology (section 3.2)
(5) Identifying the language for description
(6) Identifying the scale of ontology (number of concepts and/or instance models)

On the way of this analysis, the authors discussed about the criteria for classification of applications interactively. The rest of this paper describes the result.

4 Results of Analysis and Considerations

4.1 Distribution of Types of Usage of Ontology

Fig.4 shows the distribution of types of usage of ontologies in the systems presented in the papers surveyed. This graph shows that there is not so big difference among the ratios of each type of usage. However, comparing the amount of the applications from

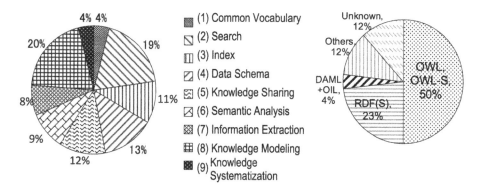

Fig. 4. The distribution of types of usage of ontology

Fig. 5. The proportion of ontology description language

types (1) to (7), which deal with "data" processing, and the those from types (8) to (9), which explicitly deal with "knowledge" processing, most of current studies in the Semantic Web application deal with "data" processing on structured data.

4.2   A Correlation between the Types of Usage and the Types of Ontology

Table.2 constitutes the relations between the types of usage and the types of ontology. We see from Table 2 and Fig. 4 that most of the Semantic Web applications use ontologies including a variety types of relations, that is, not only is-a relation but also the other relations. On the other hand, a few ontologies have complex axioms. However, only based on papers in the proceedings, it was very difficult for authors to properly assess whether an ontology contains axioms for semantic constraints or not. On this matter, we need further analysis of the axioms defined by each ontology. In addition, the more semantic processing the type of usage requires, that is, as the number of the type of usage raises, the more detailed definition the ontology requires.

Table 2. A Correlation between the Types of Usage and the Types of Ontology

	The Types of Ontology					
	(A) Simple Schema	(B) Is-a Hierarchies	(C) Other Relationship	(D)Axioms	(E) Rule Descriptions	Total
(1) Common Vocabulary	0	4	7	0	0	11
(2) Search	1	2	43	4	1	51
(3) Index	0	3	23	3	0	29
(4) Data Schema	0	0	32	5	0	37
(5) Knowledge Sharing	1	0	31	1	0	33
(6) Semantic Analysis	1	1	21	3	0	26
(7) Information Extraction	1	2	15	3	0	21
(8) Knowledge Modeling	0	1	36	9	8	54
(9) Knowledge Systematizatio	0	2	8	1	0	11
Total	4	15	216	29	9	273

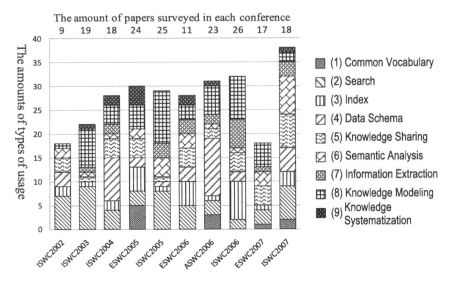

Fig. 6. The conference-by-conference transition of the types of usage

Fig. 5 shows the proportion of the usage of ontology description language[10]. Almost half of the systems use OWL or extended OWL. RDF(S) takes second place. This shows OWL is steadily lying as the foundation of the standard for the ontology description language. Yet it is unclear which sublanguage of OWL (i.e., Lite, DL or Full) is used in the ontology because most of papers do not specify the type of sublanguage in them.

4.3 The Conference-by-Conference Transition of the Types of Usage

Fig. 6 shows the conference-by-conference transition of the types of usage in the systems. In this chart, the conferences are sorted by the date therefore we can see the transition of semantic web application during the past five years. Although the amount of papers surveyed in each conference except ISWC2002 and ESWC2007 is about 20 and not so different from each other, the amount of types of usage are increasing year by year. This is because the development research of semantic web applications has matured for several years and still more features have been built in each system. Especially, while there is no significant change in the use of ontology as vocabulary or for retrieval ((1)-(3)), the numbers of the use for higher-level semantic processing ((4)-(9)) are increasing gradually.

As discussed in the section 4.1, there is only a moderate increase in the use of ontology for knowledge processing. We understand this indicates the difficulty of the

---

[10] An ontology used in an application fall into a single category according to ontology description language. Especially, distinction between the categories of RDF(S) and OWL/OWL-S is made by the usage of only RDF(S) or including OWL/OWL-S if only a little. If it uses both of RDFS and OWL, for example, it is classified into OWL/OWL-S.

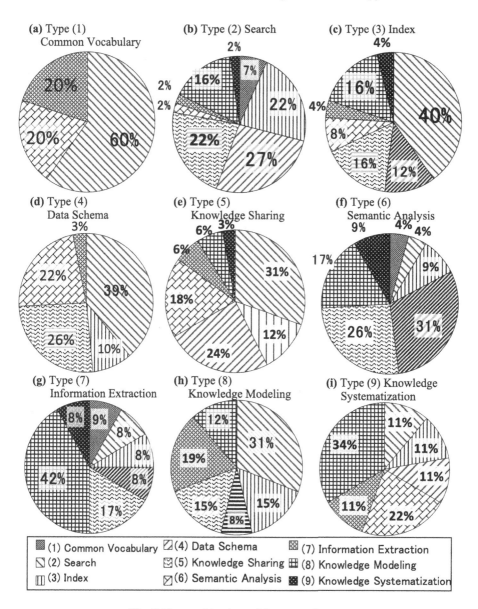

Fig. 7. The combinations of the types of usage

development of applications with knowledge processing. In other words, currently the mainstream of Semantic Web application development focuses on data processing, and overcoming the difficulty of knowledge processing might be a key to create killer applications.

4.4 The Combinations of the Types of Usage

As mentioned in section 3.1, an SW application might use ontologies for more than two purposes. However, among the nine types of usage of ontology, not all the combinations are found in applications and type-combination appearing in an application varies according to each type. Fig.7 shows such distributions of usage type combination for each usage type. For example, Fig.7(a) shows the distributions of the percentages of appearance of usage types (2), (6) and (7) together with type (1)[11]. Since the 3 types (i.e. (2), (6) and (7)) are usages mainly for semantic retrieval, common vocabularies tend to be used for search systems. However, the pie chart of usage type (2) for search in Fig.7(b) shows that the usages are combined with also others. For example, the combinations of usage type (2) for search and type (5) usage as a media for knowledge sharing imply integrated search across several information resources. Furthermore, Fig.7(g) shows that type (7) usages for information extraction are combined with type (8) usage as a rule set for knowledge models more frequently in compare with type (2) and (6). It shows that type (7) information extraction needs more detailed description of semantics than type (2) simple search and type (6) semantic analysis. The chart of type (9) usage for systematizing knowledge (Fig.7(i)) shows that the usages are combined with all other types systematically. This trend is consistent with description in section 2.1.

4.5 Application Domains

Fig.8 shows the distribution of the types of usage per a domain. Domains in which many applications have been developed are multimedia (image, movie, music, etc.), services (both of web services and services in the real world), software development, knowledge management, bioinformatics and medical care. In the domains in the business area, a variety of systems focusing on a particular subject, for example, product management, business process, and so on, have been developed. On the other hand, general-purpose systems, which are not focusing on a particular domain or subject, have been also developed. In Fig. 8, such systems are distinguished in the following ways; systems dealing with academic information such as papers and conference, or systems dealing with web resource and systems dealing with the other information. The distribution of the types of usage in each domain is not so different from each other. Therefore, semantic web technology is used in a variety of domains at all levels.

Here, we pick up several domains and discuss the distribution in detail. Fig. 9 shows the percentages of the type of usages in several domains which we can find In the software and service domains, the percentage of type (8) usage as a rule set for knowledge models is higher in comparison with scientific domains (scientific information, bio and medical). It implies that more heavy weight ontologies are developed in the software domain.

- In the business and bio domains, the percentage of usages for type (6) semantic analysis and type (7) information extraction is high. It implies that the needs of information analysis are higher in the domains than in others.

---

[11] The usage type (1) (usage as common vocaburay) does not cooccur with many of the other usages because we classified the applications into this type only if ontologies play mainly a role to unify the vocabulary in it.

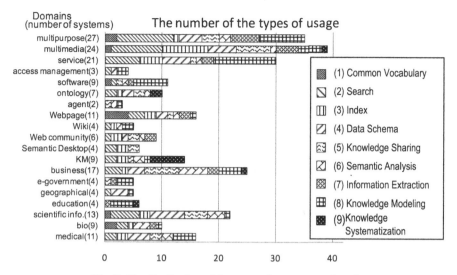

Fig. 8. The distribution of the types of usage per a domain

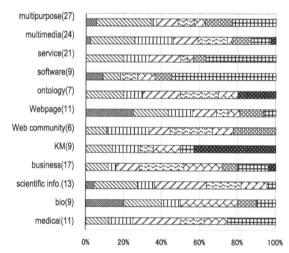

Fig. 9. The distribution of the types of usage per a domain (2)

Table 3. The transition of the type of usage in multimedia

Confrences	1)	2)	3)	4)	5)	6)	7)	8)	9)
ISWC2002	✓	✓							
ISWC2003	✓								
ISWC2003	✓								
ISWC2003	✓								
ISWC2004	✓	✓	✓	✓					
ISWC2004	✓		✓						
ESWC2005		✓							
ESWC2005			✓	✓					
ESWC2005	✓								
ISWC2005		✓					✓		
ISWC2005	✓								
ISWC2005	✓								
ESWC2006	✓		✓	✓					
ASWC2006		✓							
ASWC2006	✓								
ISWC2006			✓						
ISWC2006		✓							
ISWC2006	✓				✓				
ISWC2006	✓								
ESWC2007			✓						
ESWC2007		✓					✓		
ISWC2007	✓	✓							
ISWC2007	✓								
ISWC2007			✓					✓	
ISWC2007	✓			✓				✓	✓

Furthermore Table.3 shows the transition of the type of usage in multimedia domain. We can see a trend that the numbers of the use for higher-level semantic processing (types (4)-(9)) are increasing gradually.

4.6 Numbers of Concepts and Instances in Each Ontology

Generally, the size of an ontology and the numbers of instances used as metadata are expected to vary widely depending on the scale of an application. We cannot analyze the correlation of them enough because the number of concepts and instances are not written clearly in each paper surveyed. As far as what can be read out from the papers, many systems recently developed use the existing large-scale ontologies (or thesauri) such as DOLCE[12], WordNet[13] and DBPedia. Some systems use their own ontologies but it is difficult to analyze them because most of papers state only the outline of the ontologies. Some ontologies are available on the web so that we will be able to do further analysis with the ontologies downloaded from the sites.

5 Concluding Remarks and Future Work

For the purpose of providing basis for finding next direction of SW applications, this paper surveyed and analyzed the present state of SW applications from the viewpoint of usages and types of ontologies. Since the authors have depicted both strong points and weak points of current SW applications, we plan to continue additional survey and analysis of SW applications for future SW conferences and others related to SW, and report them at http://www.hozo.jp/OntoApps/. At the same time, we also plan to refine the viewpoints for the analysis proposed in this paper as follows.

1. Refinement of ontology types (cf. Section 3.2)
As described in section 3.2, we focus on semantic components of ontologies for their classification. To make relationships between type of the ontologies and usage of the ontologies in SW applications clearer, we need to classify ontologies in more detail. We plan to introduce a metadata set for such a classification proposed by [32], or formal types of ontologies in [31, 33], for example. Since applications which use existing ontologies or "Linked data" are increasing today (cf., section 4.5), survey on such existing resources is necessary.

2. Analysis on development process of ontologies and instance models
Since development process of ontologies and/or instance models is an important factor of development of SW applications, we plan to incorporate relevant viewpoints for further classification. For those applications which require deep and specialized knowledge (e.g. [19, 25]), domain experts commit ontology development. On the other hand, ontologies are (semi-)automatically built for those applications which require both scalability and wide range of knowledge on the web (e.g.[23, 24]).

To maintain SW applications, mechanisms for adding instance data are important and there are various approaches according to the applications' domains and/or goals. In some SW applications, an annotator creates metadata for additional instance data. While SemanticWiki [28] refers to collaborative knowledge, PiggyBank[34] creates instance data from existing databases efficiently. Furthermore, some applications aim at managing the lifecycle of ontologies and instances [35]. Viewpoints which differentiate these methods should give other perspectives for classification.

---

[12] http://www.loa-cnr.it/DOLCE.html
[13] http://wordnet.princeton.edu/

3. Classification of applied semantic technologies

SW applications require many other technologies: ontology development, semantic search by ontologies, inference mechanisms for DL, ontology matching, etc. Viewpoints about the applied technologies should be important issues. As a long term goal, the authors aim at supporting SW application development by providing an ontology of SW applications. For the development process, the ontology will help comparison among existing systems and a new system to be developed. Furthermore, guidelines for SW application development such as appropriate ontology type, usage and peripheral technologies, will be provided by the ontology.The result of our analysis is available at the URL, http://www.hozo.jp/OntoApps/.

References

1. Davis, M.: Semantic Wave 2008 Report: Industry Roadmap to Web 3.0 & Multibillion Dollar Market Opportunities (2008), http://www.project10x.com
2. Alani, H., Kalfoglou, Y., O'Hara, K., Shadbolt, N.R.: Towards a Killer App for the Semantic Web. In: Gil, Y., Motta, E., Benjamins, V.R., Musen, M.A. (eds.) ISWC 2005. LNCS, vol. 3729, pp. 829–843. Springer, Heidelberg (2005)
3. Möller, K., et al.: Recipes for Semantic Web Dog Food-The ESWC ans ISWC Metadata Projects. In: Aberer, K., Choi, K.-S., Noy, N., Allemang, D., Lee, K.-I., Nixon, L., Golbeck, J., Mika, P., Maynard, D., Mizoguchi, R., Schreiber, G., Cudré-Mauroux, P. (eds.) ASWC 2007 and ISWC 2007. LNCS, vol. 4825, pp. 802–815. Springer, Heidelberg (2007)
4. Léger, A., Nixon, L., Shvaiko, P.: On Identifying Knowledge Processing Requirements. In: Gil, Y., Motta, E., Benjamins, V.R., Musen, M.A. (eds.) ISWC 2005. LNCS, vol. 3729, pp. 928–943. Springer, Heidelberg (2005)
5. Motta, E., Sabou, M.: Next generation semantic web applications. In: Mizoguchi, R., Shi, Z.-Z., Giunchiglia, F. (eds.) ASWC 2006. LNCS, vol. 4185, pp. 24–29. Springer, Heidelberg (2006)
6. Uschold, M.: The Boeing Company: A Framework for Understanding and Classifying Ontology Applications. In: Proc. of the IJCAI 1999 Workshop on Ontologies and Problem-Solving Methods, KRR5 (1999)
7. Mizoguchi, R.: Tutorial on ontological engineering - Part 1. New Generation Computing 21(4), 365–384 (2003)
8. Kitamura, Y.: Roles of Ontologies of Engineering Artifacts For Design Knowledge Modeling. In: Proc. EDIProD 2006, pp. 59–69 (2006)
9. Corcho, O., Gómez-Pérez, A., López-Cima, A., López-García, V., del Carmen Suárez-Figueroa, M.: ODESeW. Automatic Generation of Knowledge Portals for Intranets and Extranets. In: Fensel, D., Sycara, K.P., Mylopoulos, J. (eds.) ISWC 2003. LNCS, vol. 2870, pp. 802–817. Springer, Heidelberg (2003)
10. Popov, B., Kiryakov, A., Kirilov, A., Manov, D., et al.: KIM-Semantic Annotation Platform. In: Proc. of ISWC 2003, pp. 834–849 (2003)
11. Dzbor, M., Domingue, J., Motta, E.: Magpie – Towards a Semantic Web Browser. In: Proc. of ISWC 2003, pp. 690–705 (2003)
12. Mäkelä, E., Hyvönen, E., Saarela, S.: Ontogator — A semantic view-based search engine service for web applications. In: Cruz, I., Decker, S., Allemang, D., Preist, C., Schwabe, D., Mika, P., Uschold, M., Aroyo, L.M. (eds.) ISWC 2006. LNCS, vol. 4273, pp. 847–860. Springer, Heidelberg (2006)

13. Haase, P., et al.: Bibster – A Semantics-Based Bibliographic Peer-to-Peer System. In: McIlraith, S.A., Plexousakis, D., van Harmelen, F. (eds.) ISWC 2004. LNCS, vol. 3298, pp. 122–136. Springer, Heidelberg (2004)

14. Sidhu, A.S., Dillon, T.S., Chang, E.: Protein Data Sources Management Using Semantics. In: Proc. of ASWC2006, pp. 595–601 (2006)

15. Chen, H., et al.: Towards a semantic web of relational databases: A practical semantic toolkit and an in-use case from traditional chinese medicine. In: Cruz, I., Decker, S., Allemang, D., Preist, C., Schwabe, D., Mika, P., Uschold, M., Aroyo, L.M. (eds.) ISWC 2006. LNCS, vol. 4273, pp. 750–763. Springer, Heidelberg (2006)

16. Fox, P., et al.: Semantically-Enabled Large-Scale Science Data Repositories. In: Proc. of ISWC2006, pp. 792–805 (2006)

17. Dimitrov, D.A., et al.: Information Integration via an End-to-End Distributed Semantic Web System. In: Cruz, I., Decker, S., Allemang, D., Preist, C., Schwabe, D., Mika, P., Uschold, M., Aroyo, L.M. (eds.) ISWC 2006. LNCS, vol. 4273, pp. 764–777. Springer, Heidelberg (2006)

18. Kim, W., et al.: Product information meta-search framework for electronic commerce through ontology mapping. In: Gómez-Pérez, A., Euzenat, J. (eds.) ESWC 2005. LNCS, vol. 3532, pp. 408–422. Springer, Heidelberg (2005)

19. Wolstencroft, K., et al.: A Little Semantic Web Goes a Long Way in Biology. In: Gil, Y., Motta, E., Benjamins, V.R., Musen, M.A. (eds.) ISWC 2005. LNCS, vol. 3729, pp. 786–800. Springer, Heidelberg (2005)

20. Wong, S.C., Miles, S., Fang, W., Groth, P.T., Moreau, L.: Provenance-based validation of E-science experiments. In: Gil, Y., Motta, E., Benjamins, V.R., Musen, M.A. (eds.) ISWC 2005. LNCS, vol. 3729, pp. 801–815. Springer, Heidelberg (2005)

21. Alani, H., et al.: Monitoring research collaborations using semantic web technologies. In: Gómez-Pérez, A., Euzenat, J. (eds.) ESWC 2005. LNCS, vol. 3532, pp. 664–678. Springer, Heidelberg (2005)

22. Ghita, S., Nejdl, W., Paiu, R.: Semantically Rich Recommendations in Social Networks for Sharing, Exchanging and Ranking Semantic Context. In: Gil, Y., Motta, E., Benjamins, V.R., Musen, M.A. (eds.) ISWC 2005. LNCS, vol. 3729, pp. 293–307. Springer, Heidelberg (2005)

23. Holzinger, W., Krüpl, B., Herzog, M.: Using Ontologies for Extracting Product Features from Web Pages. In: Proc. of ISWC 2006, pp. 286–299 (2006)

24. Kawamura, T., et al.: Ubiquitous Metadata Scouter – Ontology Brings Blogs Outside. In: Mizoguchi, R., Shi, Z.-Z., Giunchiglia, F. (eds.) ASWC 2006. LNCS, vol. 4185, pp. 752–761. Springer, Heidelberg (2006)

25. Kashyap, V., et al.: Definitions Management: A Semantics-Based Approach for Clinical Documentation in Healthcare Delivery. In: Gil, Y., Motta, E., Benjamins, V.R., Musen, M.A. (eds.) ISWC 2005. LNCS, vol. 3729, pp. 887–901. Springer, Heidelberg (2005)

26. Kraines, S., Guo, W., Kemper, B., Nakamura, Y.: EKOSS: A knowledge-user centered approach to knowledge sharing, discovery, and integration on the semantic web. In: Cruz, I., Decker, S., Allemang, D., Preist, C., Schwabe, D., Mika, P., Uschold, M., Aroyo, L.M. (eds.) ISWC 2006. LNCS, vol. 4273, pp. 833–846. Springer, Heidelberg (2006)

27. Della Valle, E., et al.: SEEMP: An semantic interoperability infrastructure for e-government services in the employment sector. In: Franconi, E., Kifer, M., May, W. (eds.) ESWC 2007. LNCS, vol. 4519, pp. 220–234. Springer, Heidelberg (2007)

28. Krötzsch, M., Vrandečić, D., Völkel, M.: Semantic mediaWiki. In: Cruz, I., Decker, S., Allemang, D., Preist, C., Schwabe, D., Mika, P., Uschold, M., Aroyo, L.M. (eds.) ISWC 2006. LNCS, vol. 4273, pp. 935–942. Springer, Heidelberg (2006)

29. Keller, R.M., et al.: SemanticOrganizer: A Customizable Semantic Repository for Distributed NASA Project Teams. In: McIlraith, S.A., Plexousakis, D., van Harmelen, F. (eds.) ISWC 2004. LNCS, vol. 3298, pp. 767–781. Springer, Heidelberg (2004)

30. Sevilmis, N., et al.: Knowledge Sharing by Information Retrieval in the Semantic Web. In: Gómez-Pérez, A., Euzenat, J. (eds.) ESWC 2005. LNCS, vol. 3532, pp. 471–485. Springer, Heidelberg (2005)

31. Lassila, O., McGuinness, D.: The Role of Frame-Based Representation on the Semantic Web, Technical Report KSL-01-02, Knowledge Systems Laboratory, Stanford University, Stanford, California (2001)

32. Hartmann, J., et al.: DEMO - design environment for metadata ontologies. In: Sure, Y., Domingue, J. (eds.) ESWC 2006. LNCS, vol. 4011, pp. 427–441. Springer, Heidelberg (2006)

33. Bechhofer, S., Volz, R.: Patching syntax in OWL ontologies. In: McIlraith, S.A., Plexousakis, D., van Harmelen, F. (eds.) ISWC 2004. LNCS, vol. 3298, pp. 668–682. Springer, Heidelberg (2004)

34. Huynh, D.F., Mazzocchi, S., Karger, D.R.: Piggy bank: Experience the semantic web inside your web browser. In: Gil, Y., Motta, E., Benjamins, V.R., Musen, M.A. (eds.) ISWC 2005. LNCS, vol. 3729, pp. 413–430. Springer, Heidelberg (2005)

35. Tran, T., Haase, P., Lewen, H., Muñoz-García, Ó., Gómez-Pérez, A., Studer, R.: Lifecycle-support in architectures for ontology-based information systems. In: Aberer, K., Choi, K.-S., Noy, N., Allemang, D., Lee, K.-I., Nixon, L., Golbeck, J., Mika, P., Maynard, D., Mizoguchi, R., Schreiber, G., Cudré-Mauroux, P. (eds.) ASWC 2007 and ISWC 2007. LNCS, vol. 4825, pp. 508–522. Springer, Heidelberg (2007)

A Formal Model for Classifying Trusted Semantic Web Services

Stefania Galizia, Alessio Gugliotta, and Carlos Pedrinaci

Knowledge Media Institute
The Open University, Walton Hall, Milton Keynes, MK7 6AA, UK
(S.Galizia,A.Gugliotta,C.Pedrinaci)@open.ac.uk

Abstract. Semantic Web Services (SWS) aim to alleviate Web service limitations, by combining Web service technologies with the potential of Semantic Web. Several open issues have to be tackled yet, in order to enable a safe and efficient Web services selection. One of them is represented by trust. In this paper, we introduce a trust definition and formalize a model for managing trust in SWS. The model approaches the selection of trusted Web services as a classification problem, and it is realized by an ontology, which extends WSMO. A prototype is deployed, in order to give a proof of concept of our approach.

Keywords: Semantic Web services, Selection, Trust, Classification.

1 Introduction

Semantic Web services (SWS) research aims at automating the development of Web service-based applications through semantic Web technology. By providing formal representation with well-defined semantics, SWS facilitate the machine interpretation of Web service (WS) descriptions. According to SWS vision, when a client expresses a goal that it wishes to achieve, the most appropriate Web service is automatically discovered and selected on the basis of the available semantic descriptions. Since the user does not know a priori the selected WS, the notion of trust should play an important role during the WS selection phase. However, the most common approaches for describing SWS, such as WSMO [0] or OWL-S [12], do not currently provide exhaustive means to model trust, and thus do not support trust-based selection of WS.

Notice that trust is a multifaceted concept. A trust understanding can indeed involve multiple – and not always the same - parameters, such as Quality of Service (QoS), reputation and security.

In our opinion, the main issue with representing trust in all its faces lies in its contextual nature – i.e. the same user may have different trust understandings in different contexts. For example, a user may trust a WS with a highly rated security certificate whenever she has to provide her credit card details. Conversely, the same user weights the opinions of past users about a specific WS in other situations – i.e. the evaluation of the WS reputation is a priority in the current trust understanding of the user. Moreover, distinct users may privilege different trust parameters in the same context; in this case, their priorities may depend on their different personal preferences.

J. Domingue and C. Anutariya (Eds.): ASWC 2008, LNCS 5367, pp. 540–554, 2008.
© Springer-Verlag Berlin Heidelberg 2008

In this paper, we introduce a definition and formalize an abstract model for trust in SWS that enables interacting participants – i.e. both WS users and providers - to represent and utilize their own trust understanding with a high level of flexibility, and thus take the possible multiple interacting contexts into account. The essential contribution of our approach is therefore a generally applicable yet completely automated mechanism for selecting trusted services according to context-specific criteria.

Specifically, in our model we characterize the trust-based WS selection as a classification problem. Firstly, all participants specify their own requirements and guarantees about a set of trust parameters. Then, at runtime, our goal is to identify the class of WS that matches the trust statements of involved participants, according to an established classification criterion. To accomplish this, we have based the proposed model on a general-purpose classification library.

In order to apply our approach to an existing SWS working environment – and thus verify its benefits - we represented our model within a specific ontology: Web Services Trust-management Ontology (WSTO). The latter makes use of WSMO as reference approach for SWS. WSMO is in fact the underlying model of IRS-III [2], the SWS execution environment developed within our research group. As a result, we enhanced IRS-III with the trust-based selection of WS.

It is worth highlighting that an earlier version of WSTO was described in a previous work [0]. Whereas in [5] we outlined the idea of characterizing trust as a classification process, in the present work we propose a more complex model that is able to accommodate multiple trust understandings and parameters. Moreover, while in [5] we proposed a general thesis on the different meanings of trust, we did not supply our definition of trust. We now provide our trust definition as well as its formal semantics.

The paper is organized as follows: Section 2 provides the background knowledge useful for placing our work; in Section 3, we outline our approach, while in Section 4 we provide the formal details of our methodology; in Section 5, we describe an implemented application; Section 6 compares our approach with related work and, finally, Section 7 concludes the paper and outlines our future work.

2 Background

In this section, we first outline WSMO, our basic vision of SWS, and its ontological specification. Then, we outline the different existing approaches on trust.

WSMO

The Web Service Modelling Ontology (WSMO) [3] is a formal ontology for describing the various aspects of services in order to enable the automation of Web service discovery, composition, mediation and invocation. The metamodel of WSMO defines four top level elements:

- *Ontologies* provide the foundation for describing domains semantically. They are used by the three other WSMO elements.
- *Goals* define the tasks that a service requester expects a Web service to fulfill. In this sense they express the requester's intent.
- *Web Service descriptions* represent the functional behavior of an existing deployed web service. The description also outlines how web services communicate (choreography) and how they are composed (orchestration).

- *Mediators* handle data and process interoperability issues that arise when handling heterogeneous systems.

One of the main characteristic features of WSMO is the linking of ontologies, goals
and web services by mediators which map between different ontological concepts
within specific WSMO entity descriptions. In order to facilitate appropriate mapping
mechanisms, four classes of mediators are considered within WSMO. For example, an
OO-mediator may specify an ontology mapping between two ontologies whereas a
GG-mediator may specify a process or data transformation between two goals.

Classification Library

The classification framework that we use and extend for our work is a library of generic, reusable components developed within the European project IBROW [9]. Its
purpose is to support the specification of classification problem solvers. The basic
structure is the UPML framework [4], on which WSMO is based. The library has
been specified in the OCML modelling language [8], and implemented in IRS-III [2].

Within the classification framework, we use the term 'observables' to refer to the
known facts we have about the object (or event, or phenomenon) that we want to classify. Each observable is characterized as a pair of the form (f, v), where f is a feature
of the unknown object and v is its value. Here, we take a very generic viewpoint on
the notion of feature. By feature, we mean anything which can be used to characterize
an object, such as a feature which can be directly observed, or derived by inference.
As is common when characterizing classification problems - see, e.g., [19], we assume that each feature of an observable can only have one value. This assumption is
only for convenience and does not restrict the scope of the model.

The solution space specifies a set of predefined classes (solutions) under which an
unknown object may fall. A solution itself can be described as a finite set of feature
specifications, which is a pair of the form (f, c), where f is a feature and c specifies a
condition on the values that the feature can take. Thus, we can say that an observable
(f, v) matches a feature specification (f, c) if v satisfies the condition c.

As we have seen, generally speaking, classification can be characterized as the
problem of explaining observables in terms of predefined solutions. To assess the explanation power of a solution with respect to a set of observables we need to match
the specification of the observables with that of a solution. Given a solution, *sol*:
$((f_{sol1}, c_1).....(f_{solm}, c_m))$, and a set of observables, *obs*: $((f_{ob1}, v_1).....(f_{obn}, v_n))$, four cases
are possible when trying to match them:

- A feature, say f_j, is inconsistent if $(f_j, v_j) \in obs$, $(f_j, c_j) \in sol$ and v_j does not
satisfy c_j;
- A feature, say f_j, is explained if $(f_j, v_j) \in obs$, $(f_j, c_j) \in sol$ and v_j satisfies c_j;
- A feature, say f_j, is unexplained if $(f_j, v_j) \in obs$ but f_j is not a feature of *sol*;
- A feature, say f_j, is missing if $(f_j, c_j) \in sol$ but f_j is not a feature of *obs*.

Given these four cases, it is possible to envisage different solution criteria. For instance, we may accept any solution, which explains some data and is not inconsistent
with any data. This criterion is called positive coverage [14]. Alternatively, we may
require a complete coverage - i.e., a solution is acceptable if and only if it explains all
data and is not inconsistent with any data. Thus, the specification of a particular classification task needs to include a solution (admissibility) criterion. This in turn relies

on a match criterion, i.e., a way of measuring the degree of matching between candidate solutions and a set of observables. By default, our library provides a match criterion based on the aforementioned model. That is, a match score between a solution candidate and a set of observables has the form (I, E, U, M), where I denotes the set of inconsistent features, E the set of explained features, U the set of unexplained features and M the set of missing features. Of course, users of the library are free to specify and make use of alternative criteria.

In many situations, specifying the conditions under which a candidate solution is indeed a satisfactory solution is not enough. In some cases, we may be looking for the best solution, rather than for any admissible one. In these cases we need to have a mechanism for comparing match scores and this comparison mechanism becomes then part of the specification of the match criterion. By default, our library includes the following score comparison criterion.

Given two scores, $S_1 = (i_1, e_1, u_1, m_1)$ and $S_2 = (i_2, e_2, u2, m_2)$, we say that S_2 is a better score than S_1 if and only if:

$$(i_2 < i_1) \lor$$
$$(i_2 = i_1 \land e_1 < e_2) \lor$$
$$(i_2 = i_1 \land e_2 = e_1 \land u_2 < u_1) \lor$$
$$(i_2 = i_1 \land e_2 = e_1 \land u_2 = u_1 \land m_1 < m_2)$$

In the notation above $x_i < x_j$ indicates that the set x_i contains less elements than the set x_j.

In conclusion, our analysis characterizes classification tasks in terms of the following concepts: observables, solutions, match criteria, and solution criteria.

Trust Approaches

Since trust can have different meaning in different contexts, several specifications can be found in literature. We can classify existing models into the following three main approaches:

- *Policy-based.* Policies are a set of rules that specify the conditions to disclose own resources;
- *Reputation based.* Reputation based approaches make use of rating coming from other agents or a central engine, by heuristic evaluations;
- *Trusted Third Party-based (TTP).* Trusted Third Party based models use an external, trusted, entity that evaluates trust.

These general approaches can be refined and/or combined in order to build a concrete trust establishment solution that can be deployed in a real system.

Many models [11, 15] formulate trust policies in semantic Web services by security statements, such as confidentiality, authorization, authentication. W3C Web service architecture [18 recommendations base trust policies on security consideration, even if the way to disclose their security policies is still not clear.

Some policy-based models rely on a TTP, which works as a repository of service description and policies [21] and meanwhile as an external matchmaker that evaluates service trustworthiness according to given algorithms.

Reputation-based models reuse concepts and approaches taken from Web-based social networks. In SWS as well as in social networks, trust is a central issue. In both the cases, interactions take place whenever there is trustworthiness. The idea is that involved participants express their opinion on other participants, by means of a shared vocabulary. Several algorithms for trust propagation and different metrics have been defined, most of them are more generically Quality of service (QoS) based [7; 17], since they consider the service ability the main trust statement.

3 Our Approach

We propose a formal approach for managing trust among semantic Web services, based on two ontologies: WSMO and the classification task ontology, both introduced in Section 2. We build an ontology - named Web Service Trust-management Ontology (WSTO) - that reuses the main concept of those ontologies, and extends them, for supporting SWS trust management. In our model, user preferences are the main elements on which Web service selection depends. Essentially, the user can decide which parameters should be considered in order to determine which class of Web services are trusted, in a given context. We embed trust-based SWS selection in a classification framework, whereas the classification task ontology provides the overall methodology that we adopt for managing trust. For our purposes, we classify Web services according to both the user and Web services trust requirements and guarantees.

In WSTO, the key concepts are *user*, *ws* and *goal*, where *user* denotes the service requester and *ws* is the service provider. Following the basic WSMO notions, a *goal* represents the service requester's desire or intention. The user usually expresses different trust requirements in achieving different goals. For example, she can be interested in data accuracy when retrieving timetable information, and security issues when disclosing her bank accounts. On the other hand, the *ws* aims to provide a set of trustworthy statements, in order to reassure the requester as well as to appear as attractive as possible.

The participants (*ws* and *user*) are associated with trust profiles, represented in WSTO by the class *trust-participant-profile*. A profile is composed of a set of trust requirements and guarantees. *Trust-guarantee* represents observables, pairs of feature and corresponding value *(f, v)*, while *trust-requirement* represents candidate solutions, pairs of feature and condition *(f, c)*.

We distinguish three logical elements in trust requirements: (i) a set of candidate solutions for expressing conditions on guarantees promised by the relevant parties; (ii) a candidate solution for requesting their reliability; and, (iii) a candidate solution for requesting their reputation evaluation. In a participant profile, the three elements are optional; choice depends strictly on the participant preferences in matter of trust.

In turn, the participant trust guarantees have three components: (i) a set of observables for representing the promised trust guarantees; (ii) an observable corresponding to the evaluation of the participant reliability; and, (iii) an observable for representing the reputation level of the participant. Whereas the promised trust guarantees are a set of promised values stated by the participant - such as *(execution-time, 0.9)* and *(data-freshness, 0.8)* - reliability and reputation guarantees are computed on-the-fly within dedicated execution environments (IRS-III in our use case, see Section 5). As mentioned earlier, a participant profile is composed of requirements as well as guarantees.

For example, a Web service may expose high *data-freshness* and strong *confidentiality* as guarantees. Moreover, the same Web service may define security requirements, such as conditions under which a service requester can access it.

Given observables and conditions, a classification criterion is now necessary to classify Web services and find the appropriate class that addresses both user and Web service requirements and guarantees. The classification match criterion we apply is the one described in Section 2, although other classification criteria can be easily represented in WSTO.

As solution admissibility criteria, we can apply *complete coverage* and *positive coverage*. The former demands that all requirements of the interaction have to be satisfied; the latter accepts that some requirements are fulfilled and no inconsistencies exist. Our classification library implements two different classification methods: *single-solution-classification,* and *optimal-classification*. The former implements a hill climbing algorithm with backtracking to find a suitable solution; the latter executes an exhaustive search for an optimal solution. We make use of the *optimal-classification-task* and redefine it as WSMO goal[1], *optimal-classification-goal*, whose *participant-profiles* and *trusted-ws* represent the pre-conditions and post-conditions of the goal, respectively.

Notice that the proposed model is comprehensive of all trust approaches listed in the previous section. In fact, it embeds a policy-based trust management, since the interacting participants express their trust policies in their – semantically described – profiles, while the adopted SWS broker will behave as a TTP by storing participant profiles and reasoning on them. Moreover, the reputation module enables a WS selection based also on reputation ontological statements.

4 The Formal Model

This section provides the formal definition of trust we adopt in our approach, as well as its semantics. Trust is a binary evaluation of trustworthiness: "trust" or "distrust". Whenever conditions for trustworthiness are established, the interaction between participants occurs; otherwise, it is not possible.

The trustworthiness $T_u^g(ws)$ that a user u perceives towards a Web service ws, when she invokes a goal g, is given by the expression:

$$T_u^g(ws) = \Psi\big(P_g(u), \Omega_g(ws)\big)$$

Ψ is a classification operator, that provides a class of Web services matching the user's trust requirements, according to the match criterion presented in Section 3. If trust requirements do not meet any Web service trust guarantees, Ψ returns a *null* value. This means that no trusted communication can occur.

Trust as perceived by the user u, can be either strong or weak. It is strong when the operator Ψ classifies Web services by adopting complete coverage as solution admissibility criterion. When the criterion selected is positive coverage, trust is regarded as weak. We did consider using two different operators - Ψ_s for strong trust, and

---

[1] WSMO goals can be seen as an evolution of UPML tasks.

Ψ_w for weak trust - however, we decided against this in order to increase the readability of our notations. Without losing generality, in the rest of this section, we assume only strong trust.

$P_g(u)$ is a function which selects a profile from the set of trust profiles associated (provided or accepted) with the user, according to the current goal. As mentioned earlier, a user can have different trust preferences in different contexts. The *current-selected-profile* is used to associate a trust-profile with a goal, according to the user's ontological statements.

The user trust profile suitable for a given goal is represented by a list of user requirements: $(f_1, c_1), .., (f_n, c_n), (f_{rL}, c_{rL}), (f_r, c_r)$. The user requirements $(f_1, c1), .., (f_n, c_n)$ are conditions on WS promised guarantees. They can involve QoS statements, or concern security issues. Moreover, the user could be interested to know more about the reputation and the reliability of the Web services she will interact with. The requirements (f_{rL}, c_{rL}) and (f_r, c_r) respectively express conditions on reliability and other user preferences concerning Web service behaviors.

Given a goal g, $\Omega_g(ws)$ is a complex operator that provides information about the WS profile, where ws satisfies g. The operator provides thus (i) the guarantees promised by the WS, (ii) a record of WS monitored behavior, and (iii) WS behavior as evaluated by other users. For conformity, we also refer to components (ii) and (iii) as guarantees, however they are automatically calculated by IRS-III, and are not strictly speaking guarantees. We should also note that, in principle, ws reputation and its historical evaluation may not always reassure the user.

More formally, Ω_g has three components:

$$\Omega_g = \left(\Pi_p^g, \Pi_h^g, \Pi_r^g \right)$$

Given a Web service ws, satisfying a goal g, $\Pi_p^g(ws)$ supplies the component of the ws profile published by ws itself. The ws guarantees, are pairs *(feature, value)*: $\{(f_{p1}, v_{p1}), ..., (f_{pm}, v_{pm})\}$. The published guarantees can involve QoS parameters, certificated security parameters issued by Certification Authorities, or any ontological statements certificated by TTP or simply provided by the WS for trust assurance purposes. The values $v_{p1}, ..., v_{pm}$ are normalized and homogenized. They are normalized to non-negative real numbers in the range [0,1]. Moreover, we assume that they are homogeneously scaled, where higher values correspond to higher performance. For example, higher performance for the parameter "execution time" would normally be indicated by a smaller value, but we normalize to a scale where a higher value indicates better performance. We are aware that this process can increase complexity, especially for those guarantees related to security issues, however, describing the normalization process is out of the scope of our current work. For alleviating the difficulty of representing numerical normalized values, we use a number of previously described heuristics.

$\Pi_h^g(ws)$ assigns the value v_{rL} to the ws reliability f_{rL}, by stating the observable (f_{rL}, v_{rL}). Reliability is a measure of how the Web service behaviour conforms with its related guarantees. Let $F_p = \{f_{p1}, ..., f_{pm}\}$ be the set of features with associated values of guarantees, and $F_h = \{f_{h1},...,f_{hk}\}$ the set of the monitored features for *ws*. We define F_{ph} = $F_p \cap F_h = \{f_1,...,f_j\}$, as the set of both promised and monitored features for *ws*. Whenever a feature is monitored more than once, we consider only the last observed, because we assume that Web service performance can alter, and the last value is closer to its predicted behavior.

We calculate the feature *conformance*, as defined by [16]. For every feature f_i belonging to F_{ph} with $1 \leq i \leq j$, we determine the conformance of f_i by value

$$\delta = \frac{v_i^h - v_i^p}{v_i^p}$$
, where v_i^h is the normalized monitored value associated to f_i, and

v_i^p is its normalized promised value, for the Web service *ws*.

The conformance value δ_i falls in the range [-1, 1]. It is a negative value when the promised value is better than the monitored one. It holds 0 when $v_i^p = v_i^h$, i.e., the promised value corresponds to the monitored one. Finally, when the Web service behaviour around the feature f_i is better than promised, δ_i is a positive value.

If $\delta_i \geq 0$ for every $1 \leq i \leq j$, then $\Pi_h^g(ws)$ will have the value 1, the maximum value for reliability, otherwise, reliability is represented by the normalized arithmetic average of each feature's conformance:

$$\Pi_h^g(ws) = v_{rL} = \left\| \frac{\sum_{i=1}^{j} \delta_i}{j} \right\|$$

Web service reliability, evaluated through the operator $\Pi_h^g(ws)$, provides a value for the feature f_{rL}, where the observable (f_{rL}, v_{rL}) is a component of the *ws* profile. v_{rL} is a guarantee that will be automatically generated by the adopted SWS broker by processing the *ws* published and monitored guarantees. For example, IRS - our reference SWS broker - automatically logs all interactions with Web services [13] and thus reliability is straightforward to compute.

The operator $\Pi_r^g(ws)$ provides a measure of Web service reputation. Users who have previously interacted with WS can supply ontological statements for describing perceived trustworthiness. These statements are observables - pairs *(feature, value)* - as annotated by users.

We introduce a reputation evaluation for making our model as context/user oriented as possible, because some users may be interested in the opinions that come from previous requesters. Nevertheless, we do not intend to emphasize this aspect of our trust evaluation because reputation statements may derive from malicious users interested in providing false evaluations for a variety of reasons. Therefore, we consider only reputation statements that have high conformance.

Let $F_r = \{f_1,...,f_{rr}\}$ be a set of features, and $\{(f_i, v_{i1})..., (f_i, v_{ij})\}$ the corresponding *ws* observables for the feature f_i, with $1 \leq i \leq rr$, respectively reputed by the users $\{u_{i1},.. u_{ij}\}$. We consider the reputation around the feature f_i can be estimated if and only if the standard deviation SDi from the average of the normalized values $\{v_{i1}.., v_{ij}\}$ is lower than a given threshold D.

We can now define $\Pi_r^g(ws)$ as the average trustworthiness perceived by the users towards the Web service ws:

$$\Pi_r^g(ws) = v_r = \frac{\sum_{i=1}^{rr} w_i \bar{v}_i}{\sum_{i=1}^{rr} w_i}$$

Where w_i is a weight that excludes the reputation statements that cannot be estimated. It can hold $\{0,1\}$: $w_i = 1$ when the $SDi \leq D$, otherwise its value is 0. The value v_r is assigned to the feature f_r, where the observable (f_r, v_r) is a component of the ws profile, computed within IRS-III.

Having extracted the participant profiles, the operator Ψ classifies Web services, i.e., it solves the problem of finding a class that best explains a set of known Web service guarantees, according to user trust requirements. The output is binary: the WS class exists or not, which corresponds to the trust or distrust value for the function $T_u^g(ws)$.

5   Case Study: A Trusted Virtual Travel Agent

The proposed formal model has been implemented within an existing SWS execution environment: IRS-III [2]. The reasons for adopting IRS-III are the following: firstly, this framework has been designed and built within our institution; secondly, WSMO (Section 2) has been incorporated and extended as the core IRS-III epistemological framework; finally, the classification library we use and extend (Section 3) is represented in OCML [8], the ontological representation language used by IRS-III.

IRS-III is a platform and a broker for developing and executing SWS. By definition, a broker is an entity which mediates between two parties and IRS-III mediates between a service requester and one or more service providers. A core design principle for IRS-III is to support capability-based invocation. A client sends a request which captures a desired outcome or goal and, using a set of semantic Web service descriptions, IRS-III will: a) discover potentially relevant Web services; b) select the set of Web services which best fit the incoming request; c) mediate any mismatches at the data, ontological or business process level; and d) invoke the selected Web services whilst adhering to any data, control flow and Web service invocation constraints. Additionally, IRS-III supports the SWS developer at design time by providing a set of tools for defining, editing and managing a library of semantic descriptions and also for grounding the descriptions to either a standard Web service with a WSDL description, a Web application available through an HTTP GET request, or code written in a standard programming language (currently Java and Common Lisp).

In our work, we implemented a new IRS-III module that exploits WSTO and thus enhances the current functional-based (i.e. based on pre and post conditions, assumption and effect descriptions) selection mechanism of IRS-III with a trust-based selection mechanism. Given several Web services, semantically annotated in IRS-III and all with the same functional capability, but different trust guarantees, the class of Web services selected will be the one that matches closest with the user trust requirements, according to the classification mechanism introduced in the previous section.

As a test-bed for our module, we deployed a prototype application (the Virtual Travel Agency) and compared the existing version of IRS-III (non trusted) with the improved one (trusted). In the proposed scenario, IRS-III acts as SWS execution environment as well as TTP, by storing participant profiles and reasoning on them. The current prototype considers participant observables and needs, but it does not include the reputation module and the historical monitoring. The prototype is implemented in OCML and Lisp. The goal is to find the train timetable, at any date, between two European cities. Origin and destination cities have to belong to the same country (European countries involved in our prototype are: Germany, Austria, France and England). The client that uses this application in IRS-III publishes her trust-profile, with trust requirements and/or trust guarantees. In our prototype, we provide three different user profiles and three different Web services, able to satisfy the user goal. User profiles are expressed through trust requirements, without trust guarantees. All user requirements are performed in terms of security parameters: *encryption-algorithm*, *certification-authority* and *certification-authority-country*. Every user expresses a qualitative level of preference for every parameter.

```
USER4
(def-class trust-profile-USER4 (trust-profile)
      ((has-trust-guarantee :type guarantee-USER4)
       (has-trust-requirement :type requirement-USER4)))

(def-class requirement-USER4 (security-requirement)
      ((encryption-algorithm :value high)
       (certification-authority :value medium)
       (certification-authority-country :value medium)))

USER5
........
(def-class requirement-USER5 (security-requirement)
      ((encryption-algorithm :value medium)
       (certification-authority :value low)
       (certification-authority-country :value low)))

USER6
........
(def-class requirement-USER6 (security-requirement)
      ((encryption-algorithm :value low)
       (certification-authority :value high)
       (certification-authority-country :value high)))
```

Listing 1. User Profiles

For instance, the *user4* would like to interact with a Web service that provides a high security level in terms of encryption algorithm, but she accepts medium value for Certification Authority (CA) and CA country. Representing user requirements in a

```
ENCRYPTION-ALGORITHM HEURISTIC
(def-instanceencryption-algorithm-abstractor abstractor
((has-body '(lambda (?obs)
            (in-environment
              ((?v . (observables-feature-value ?obs
                                  'encryption-algorithm)))
              (cond ((== ?v DES)
                     (list-of 'encryption-algorithm 'high
                               (list-of (list-of
                                 'encryption-algorithm ?v))))
                    ((== ?v AES)
                     (list-of 'encryption-algorithm 'medium
                               (list-of (list-of
                                 'encryption-algorithm ?v))))

                    ((== ?v RSA)
                     (list-of 'encryption-algorithm 'low
                               (list-of (list-of
                                 'encryption-algorithm ?v)))))))))
```

Listing 2. Encryption Algorithm Heuristic

qualitative way seems to be more user-friendly. Heuristics are necessary for expressing quantitative representations in qualitative form. The listing below is an example of heuristic.

The heuristic *encryption-algorithm-abstractor* establishes that whenever the encryption algorithm adopted by a Web service provider is like *DES*, then its security level is considered high. Whenever both User and Web service describe their profiles, they implicitly agree with the qualitative evaluation expressed by the heuristic. In turn, whenever the Web service provider makes use of an algorithm like *AES*, according to the heuristic in Listing 2, its encryption ability is deemed medium, otherwise, if the adopted algorithm is like *RSA*, the security level is low. Other heuristics provide qualitative evaluations of *CAs*, and *CA countries*. For instance, security level of *globalsign-austria* is retained high, conversely German CAs are considered medium-secure.

The user can apply these heuristics, or define her own, sharing her expertise and knowledge with other users. Alternatively, the user can even express her requirements in a precise/quantitative way, by specifying the exact values expected from Web service guarantees, for example, the CA issuing security token has to be *VeriSign*. Given several Web services, semantically described in IRS-III, all with the same capability, but different trust profiles, the class of Web service selected will be the one that matches closest with the user trust profile.

We developed a user-friendly Web application to test our implementation, which is available at http://lhdl.open.ac.uk:8080/trusted-travel/trusted-query.

The snapshot in Figure 1 shows the Web application interface. The user who would like to know train timetable between two European cities enters the desired city names and date. The user owns a trust profile associated to her name: *dinar* is instance of *user4* trust profile, *vanessa* of *user5*, *stefania* of *user6*.

Whenever the application starts, IRS-III recognizes from the user name, the trust user profile. In the prototype, the requirements expressed by the user are treated as candidate solutions within the classification goal.

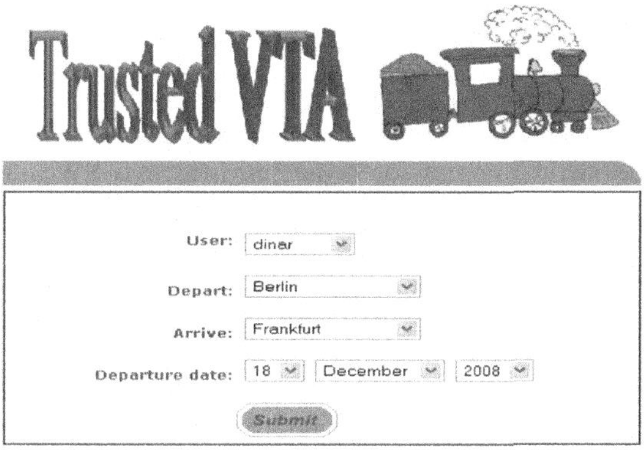

Fig. 1. Web Application

Fig. 2. Web Application Output of the user "dinar" Invocation

The class of Web services whose trust guarantees best match with user requirements is selected. As we applied the complete coverage criterion, the match is strict, that means every user requirement is explained (matches with a Web service trust guarantee) and none is inconsistent.

Figure 2 is a snapshot of the resulted trusted VTA booking. The application returns the list of Web services able to satisfy the user goal, and that one invoked, which matches with *dinar* trust requirements. It follows the Web service output, the requested timetable. The application can easily be tested with the other user instances implemented, *vanessa* and *stefania*. It can be noticed that *vanessa* trust profile matches with Web service class *get-train-timetable-service-T3*, while *stefania* with *get-train-timetable-service-T2*.

The "non-trusted" based version of the application is available at http:// lhdl. open. ac.uk:8080/trusted-travel/untrusted-query. This application implements a virtual travel agent based on the standard IRS-III goal invocation method. The output returns only the train timetable requested, without any trust-based selection.

6 Related Work

A number of current approaches model social aspects of trust [6], while some recent efforts in the last few years concern service-oriented views of trust [1]. However, few approaches provide methodologies for managing trust in a SWS, and none comprehensively incorporate all possible approaches of trust (policy, reputation TTP), as we do in WSTO.

The work proposed by Vu and his research group [16,17], who use WSMX [20] as an execution environment, is closely related to the work reported here. Vu et al. [16, 17] propose a methodology for enabling a QoS-based SWS discovery and selection, with the application of a trust and reputation management method. Their approach yields high-quality results, even under behaviour that involves cheating. With respect to their work, the methodology we propose is less accurate in terms of service behaviour prediction. However, their algorithm is wholly founded on reputation mechanisms, and is therefore not suitable for managing policy-based trust assumptions. Currently, policy-based trust mainly considers access control decisions via digital credentials. Our framework, by enabling participants to declare general ontological statements for guarantees and requirements, is also able to accommodate a policy-based trust framework.

Olmedilla et al. [11] propose a methodology for trust negotiation in SWS. They employ PeerTrust [10], a policy and trust negotiation language, for establishing if trust exists between a service requester and provider. The main aspect, which distinguishes their methodology from ours, is that they assume that trust is solely based on policy. They do not propose any mechanism for managing reputation or monitoring past service behaviour, as we do. Similar to our approach, they use WSMO as the underlying epistemology. Moreover, they assume delegation to a centralized trust matchmaker, where the participants disclose policies. Similarly, in our approach, we assume that IRS-III plays the role of trust matchmaker. Furthermore, they also address negotiation, which is an important issue in SWS interaction. We do not propose a formal negotiation mechanism here, but, as both requester and provider disclose their guarantees as credentials within IRS-III, we are able to automatically enable an implicit negotiation process.

There are other approaches for managing trust in SWS which are less closely related to ours such as KAoS [15]. Within KAoS a set of platform-independent services that enable the definition of policies ensuring the adequate predictability and controllability of both agents and traditional distributed systems is proposed. Even though they present a dynamic framework, and recognize trust management as a challenge for policy management, the framework is not specifically tailored to trust management in SWS.

7 Conclusions and Future Work

In this paper, we have presented a formal model for managing trust in SWS and have envisaged Web service selection and invocation as a classification problem, where the solution takes the form of a class of Web services matching participating trust profiles. Embodied within the trust profiles are the participant priorities with respect to trust, which can be related to reputation, credentials, or actual monitored behavior. Our definition of trust is described through a binary measure: whenever participant trust profiles match, a trusted interaction can occur, otherwise trusted interaction is deemed to not being feasible. We have adopted WSMO as underlying epistemology for WS description, and used IRS-III as an execution environment.

Trust has different meanings within different contexts: trust can be based on service ability or on reliability. In other contexts, trust can be related to reputation or delegated to TTP evaluations. The main contribution of our approach is to provide a framework that enables a comprehensive range of trust models to be captured. In fact, the framework can easily capture the multiple trust parameters that characterize a specific scenario, by simply specializing the WSTO reference ontology. Future work will extend our implementation to incorporate a comprehensive management suite for WS reputation and reliability. Additionally, we also plan to import a range of sophisticated reputation algorithms, and to improve the monitoring component.

References

1. Anderson, S., et al.: Web Services Trust Language (WS-Trust), May 2004, vol. 1.1 (2004), http://msdn.microsoft.com/ws/2004/04/ws-trust/
2. Domingue, J., Cabral, L., Galizia, S., Tanasescu, V., Gugliotta, A., Norton, B., Pedrinaci, C.: IRS-III: A Broker-based Approach to Semantic Web Services. Journal of Web Semantics 6(2), 109–132 (2008)
3. Fensel, D., Lausen, H., Polleres, A., De Bruijn, J., Stollberg, M., Roman, D., Domingue, J.: Enabling Semantic Web Services: Web Service Modeling Ontology. Springer, Heidelberg (2006)
4. Fensel, D., Motta, E., Benjamins, V.R., Decker, S., Gaspari, M., Groenboom, R., Grosso, W., Musen, M., Plaza, E., Schreiber, G., Studer, R., Wielinga, B.: The Unified Problem-solving Method Development Language UPML. In: IBROW3 Project (IST-1999-19005), vol. 1.1 (1999)
5. Galizia, S.: WSTO: A classification-based ontology for managing trust in semantic web services. In: Sure, Y., Domingue, J. (eds.) ESWC 2006. LNCS, vol. 4011. Springer, Heidelberg (2006)
6. Golbeck, J., Hendler, J.: Inferring trust relationships in web-based social networks. ACM Transactions on Internet Technology (2006)
7. Maximilien, E.M., Singh, M.P.: Toward Autonomic Web Services Trust and Selection. In: Proceedings of 2nd International Conference on Service Oriented Computing (ICSOC 2004), New York (November 2004)
8. Motta, E.: Reusable Components for Knowledge Models: Principles and Case Studies in Parametric Design Problem Solving. IOS Press, Amsterdam (1999)
9. Motta, E., Lu, W.: A Library of Components for Classification Problem Solving. In: Proceedings of PKAW 2000 - The 2000 Pacific Rim Knowledge Acquisition, Workshop, Sydney, Australia, December 11-13 (2000)
10. Nejdl, W., Olmedilla, D., Winslett, M.: PeerTrust: Automated trust negotiation for peers on the semantic web. In: Jonker, W., Petković, M. (eds.) SDM 2004. LNCS, vol. 3178, pp. 118–132. Springer, Heidelberg (2004)

11. Olmedilla, D., Lara, R., Polleres, A., Lausen, H.: Trust Negotiation for Semantic Web Services. In: 1st International Workshop on Semantic Web Services and Web Process Composition in conjunction with the 2004 IEEE International Conference on Web Services, San Diego, California, USA (2004)

12. OWL-S working group, OWL-S: Semantic Markup for Web Services. OWL-S 1.2 Pre-Release (2006), http://www.ai.sri.com/daml/services/owl-s/1.2/

13. Pedrinaci, C., Lambert, D., Wetzstein, B., Lessen, T., Cekov, L., Dimitrov, M.: SENTI-NEL: A Semantic Business Process Monitoring Tool. In: Workshop Ontology-supported Business Intelligence (OBI 2008) at 7th International Semantic Web Conference (ISWC 2008), Karlsruhe, Germany (2008)

14. Stefik, M.: Introduction to Knowledge Systems. Morgan Kaufmann, San Francisco (1995)

15. Uszok, A., Bradshaw, J.M., Johnson, M., Jeffers, R., Tate, A., Dalton, J., Aitken, J.S.: KAoS Policy Management for Semantic Web Services. IEEE Intelligent Systems 19(4), 32–41 (2004)

16. Vu, L.H., Hauswirth, M., Aberer, K.: QoS-based Service Selection and Ranking with Trust and Reputation Management. In: 13th International Conference on Cooperative Information Systems (CoopIS 2005), Agia Napa, Cyprus, Oct. 31 - Nov. 4 (2005)

17. Vu, L., Hauswirth, H., Porto, M., Aberer, F., K.: A Search Engine for QoS-enabled Discovery of Semantic Web Services. International Journal of Business Process Integration and Management (IJBPIM) (2006)

18. W3C (2004). Web Services Architecture. W3C Working Draft (February 11, 2004), http://www.w3.org/TR/ws-arch/

19. Wielinga, B.J., Akkermans, J.K., Schreiber, G.: A Competence Theory Approach to Problem Solving Method Construction. International Journal of Human-Computer Studies 49, 315–338 (1998)

20. WSMX working group, Overview and Scope of WSMX (2005), http://www.wsmo.org/TR/d13/d13.0/v0.2/

21. Zhengping, W., Weaver, A.C.: Using Web Service Enhancements to Bridge Business Trust Relationships. In: Fourth International Conference on Privacy, Security, and Trust (PST 2006), University of Toronto, Institute of Technology, Markham, Ontario, Canada, October 30-November 1 (2006)

Author Index